The
BIOLOGY
of
PARAMECIUM

ANTONI VAN LEEUWENHOEK,
LID VAN DE KONINGHLYKE SOCIETEIT IN LONDON.

The
BIOLOGY
of
PARAMECIUM

RALPH WICHTERMAN, Ph.D.

Professor of Biology,
Temple University,
Philadelphia

New York   THE BLAKISTON COMPANY, INC.   Toronto

PRINTED IN THE UNITED STATES OF AMERICA
AT THE COUNTRY LIFE PRESS, GARDEN CITY, L.I.

This Book Is
Dedicated to the Memory
of
Antony van Leeuwenhoek
of Delft, Holland
(1632–1723)

"Language and land and lineage are no bars to mutual and native under-standing. An honest man in any country is linked to all other honest men in all other countries. When a true man like Antony van Leeuwenhoek is born, the heavens are opened. Even when he dies he is not dead: his spirit glows with the divine light forever, and will forever be seen and understood—somewhere, sometime, by somebody."

CLIFFORD DOBELL

Preface

For more than 20 years, the *Protozoa*—both free-living and parasitic—have proved to be as fascinating "little animals" to this writer as they must have been to their discoverer, Antony van Leeuwenhoek, nearly three centuries ago. This book is written about one of them, namely *Paramecium,* of which many races of species have been my constant companions in the laboratory in unbroken existence for over 17 years.

Because of their universal occurrence in nature, the ease with which they may be cultivated, and the manner in which they lend themselves to biological study and experimentation, paramecia have long been favorite organisms for use in class instruction and investigation in the laboratory. It is to be remembered that *Paramecium,* like other representatives of the phylum *Protozoa,* is more than a single cell. Indeed, each is an entire organism performing, in an amazingly efficient manner, the basic physiological functions so characteristic of all animals. No other group of organisms may offer more than *Paramecium* and related forms toward the solution of many of our most fundamental problems in the field of biology. Now, more than ever before, these organisms are being employed to provide a better understanding of certain life processes.

The main purpose in writing this book is to present not only my own research but also the essential discoveries, results, and conclusions that are to be found in approximately 2,000 references from 1674 to the present wherein *Paramecium* has been used for the solution of biological and other problems. Much of the original literature appears in several languages and is widely scattered in many foreign journals of science. Many of the publications, especially the older ones, are not always easy of access. An examination of the Table of Contents will reveal that paramecium-research not only is concerned with every field of biology but also impinges greatly on biochemistry and biophysics, medicine, and pharmacology—indeed, even the fields of psychology and sociology. Noteworthy in this regard are the more recent investigations upon *Paramecium* dealing with genetics, sexuality, cytology, and serology which have broadened and extended our horizons of knowledge.

My labors upon the book cover a period of eight years which includes full teaching duties and research upon these organisms. From the beginning, it was soon found that limits had to be imposed in order to keep the book within reasonable bounds. A certain amount of repetition was unavoidable since a given process may have different aspects. Those who seek additional information or wish to consult original sources may refer to the Bibliography.

It is impossible to acknowledge adequately here, my indebtedness to many friends and colleagues who have aided directly and indirectly in the preparation of this book. However, I should like to publicly express my gratitude to those who have read and criticized portions of the manuscript lying mainly within their fields of specialization: Professors ASA A. SCHAEFFER, JAMES A. HARRISON, and Dean WILLIAM T. CALDWELL, Temple University; Professors DAVID H. WENRICH, LEWIS V. HEILBRUNN, WILLIAM F. DILLER, and Dr. ELEANOR W. FLICK, University of Pennsylvania; Professor TRACY M. SONNEBORN and Dr. RUTH V. DIPPELL, Indiana University; Professor CHARLES B. METZ, Yale University; Professor LAUREN C. GILMAN, University of Miami; Professor ALFRED M. ELLIOTT, University of Michigan; Dr. REBECCA C. LANCEFIELD, Rockefeller Institute; Dr. RUTH S. COOPER, Princeton University; Dr. HENRY HIRSHFIELD, University of Missouri, and Professor WILLIAM D. BURBANCK, Emory University. I should like to make it clear that no one but myself is to blame for any errors which may be found in the book.

The style of writing has been kept as simple as is consistent with the subject. Help in this regard has been given by Professor ELEANOR M. TILTON, Barnard College, to whom I am sincerely grateful. Thanks are also given to Professor ARTHUR C. GIESE, Stanford University, for permitting me to examine an unpublished manuscript, to MARGARET LOCHHEAD for the translation of certain Hungarian papers, and especially to URSULA GOLDSTEIN WHITT for aid in the careful translation of certain lengthy German works. Gratitude is expressed not only to Dr. JAMES B. LACKEY, then Science Editor of The Blakiston Company (presently Research Professor, School of Engineering, University of Florida), for his long-continued encouragement and counsel in the task, but also to IRENE CLAIRE MOORE, then Assistant Manuscript Editor, The Blakiston Company (presently Book Editor, United Lutheran Publication House, Philadelphia), for her invaluable aid in shepherding the book through its many editorial phases with a thoroughness seldom encountered.

Thanks are given to the OFFICE OF NAVAL RESEARCH, Department of the Navy, and to the COMMITTEE ON RESEARCH, Temple University, for continued support of research upon *Paramecium* at the University and at the Marine Biological Laboratory, Woods Hole, Massachusetts. Acknowledgment is also made to the BIOLOGICAL INSTITUTE of Temple University.

Finally, I wish to express my deep appreciation to my wife, BESSIE SWIFT WICHTERMAN, who, in countless ways, has helped in the production of this book from beginning to end.

Grateful acknowledgment is also made to the following investigators and publishers for the illustrations and tables from the sources cited; it is made in this manner because it provides more ready reference to individual illustrations or tables.

Anderson, T. F.: Previously unpublished, Figs. 26, 29.

Back, A., and L. Halberstaedter: *Am. J. Roentgenol. Radium Therapy,* Table 8; Fig. 50.

Ball, G. H.: *Biol. Bull.,* Table 37.

Bernheimer, A. W.: *Tr. Am. Microscop. Soc.,* Fig. 37.

Bernheimer, A. W., and J. A. Harrison: *J. Immunol.,* Tables 33, 34, 35.

Boell, E. J., and L. L. Woodruff: *J. Exp. Zoöl.,* Fig. 69.

Bullington, W. E.: *J. Exp. Zoöl.,* Figs. 7, 9.

Calkins, G. N.: *Biol. Bull.,* Fig. 124.

Calkins, G. N.: "Biology of the Protozoa," 1933, Philadelphia, Lea & Febiger, Fig. 88 (modified); After Michelson, Fig. 131.

Calkins, G. N., and S. W. Cull: *Arch. Protistenk.,* Figs. 89 (modified), 90.

Calkins, G. N., and F. M. Summers: "Protozoa in Biological Research," 1941, New York, Columbia University Press; After Diller, Fig. 102; After Hall, Figs. 61, 62; After Jennings, Table 29, Fig. 114; After Ludloff, Fig. 77 C; After Peebles, Fig. 125; After Richards, Figs. 57, 58, 59; After Wetzel, Fig. 136; After Woodruff and Erdmann, Figs. 82, 83; After Glaser and Coria, Figs. 39, 40.

Cattell, Jaques (Editor) (After Gause, et al.): Biological Symposia, IV, Copyright, 1941, The Ronald Press Company, Fig. 64; After Allee, Fig. 65.

Causey, D.: Univ. of California Pub. in Zool., Fig. 35.

Chatton, E., and S. Brachon: *Compt. rend. soc. biol.,* Fig. 13.

Chen, T. T.: Figs. 79, 92, 110, 118.

Chen, T. T., *J. Heredity,* Figs. 81, 98; *J. Morphol.,* Fig. 100.

Claff, C. L.: Previously unpublished, Fig. 70.

Claff, C. L.: *Physiol. Zoöl.,* Fig. 41.

Claparède, E., and J. Lachmann: "Etudes sur les Infusoires et les Rhizopodes," Fig. 11 A, B.

Cunningham, B., and P. L. Kirk: *J. Cellular Comp. Physiol.* Fig. 68.

Diller, W. F.: *J. Morphol.,* Figs. 84, 87, 94, 95, 99, 101, 102.

Dobell, C.: "Antony van Leeuwenhoek and His 'Little Animals,'" 1932, London, Staples Press Limited, Frontispiece.

Dragesco, J.: Previously unpublished, Fig. 27 D.

Emery, F. E.: *J. Morphol.,* Table 12.

Frisch, J. A.: *Arch. Protistenk.,* Tables 18, 19.

Gause, G. F: "The Struggle for Existence," 1934, Baltimore, The Williams & Wilkins Company, Figs. 66, 67.

Gause, G. F.: *Quart. Rev. Biol.,* Tables 20, 21.

Gaw, H. Z.: *Arch. Protistenk.,* Figs. 71, 72.

Gelei, J.: Allattani Kozlemenyek, Fig. 11 C.

Gelei, J.: *Arch. Protistenk.,* Figs. 17, 18, 21.

Giese, A. C.: Personal communication, Table 31.

Giese, A. C.: *J. Cellular Comp. Physiol.,* Fig. 48.

Giese, A. C., and P. A. Leighton: *J. Gen. Physiol.,* Fig. 47.

Gilman, L. C.: Previously unpublished, Table 30.

Hafkine, M. W.: *Ann. inst. Pasteur,* Fig. 137.

Halberstaedter, L., and A. Back: *Nature,* Table 11; Fig. 49.

Hargitt, G. T., and W. W. Fray: *J. Exptl. Zoöl.,* Tables 5, 7.

Harrison, J. A., and E. H. Fowler: *J. Exptl. Zoöl.,* Fig. 132.

Herfs, A.: *Arch. Protistenk.,* Table 24.

Jacobs, M.: *J. Exptl. Zoöl.,* Table 17 (modified).

Jakus, M. A.: *J. Exptl. Zoöl.,* Figs. 27 C, 28.

Jakus, M. A. (After Krüger): *J. Exptl. Zoöl.,* Fig. 27 B.

Jakus, M. A., and C. E. Hall: *Biol. Bull.,* Fig. 25.

Jennings, H. S.: "Behavior of the Lower Organisms," 1931, New York, Columbia University Press, Figs. 27 A, 74, 75, 76, 77 A, B, D, E; After Statkewitsch, Fig. 46.

Jennings, H. S.: Bibliographia Genetics, Fig. 126; After Stocking, Fig. 127.

Jennings, H. S.: *Genetics,* Figs. 112, 113.

Jennings, H. S.: *Proc. Am. Phil. Soc.,* Tables 13, 14, 15 (modified); Fig. 56.

Jones, E. P.: *Univ. of Pittsburgh Bull.,* Fig. 55.

Kahl, A.: "Die Tierwelt Deutschlands: Urtiere oder Protozoa," 1930, Germany, Gustav Fischer, Fig. 12.

Kalmus, H.: "Paramecium," 1931, Germany, Gustav Fischer; After Bütschli, Fig. 133,

Fig. 89 (modified); After Fortner, Figs. 53, 134; After Mast, Fig. 135.

King, R. L.: *Biol. Bull.,* Fig. 34.

King, R. L.: *J. Morphol.,* Figs. 30, 32, 33.

King, R. L., and H. W. Beams: *J. Morphol.,* Figs. 44, 45.

Kudo, R. R.: "Protozoology," Illinois, Charles C Thomas, Publisher, Fig. 73.

Landis, E. M.: *J. Morphol.,* Figs. 96, 97.

Liebermann, P. R.: *Tr. Am. Microscop. Soc.,* Table 2.

Lloyd, L.: *Proc. Leeds Phil. Lit. Soc.,* Fig. 128.

Loefer, J. B.: *Arch. Protistenk.,* Table 1; Figs. 14, 63.

Loefer, J. B.: *J. Exptl. Zoöl.,* Fig. 60.

Lund, E. E.: *J. Morphol.,* Fig. 19.

Lund, E. E.: *Univ. of California Pub. in Zool.,* Figs. 22, 23.

MacLennan, R. F., and H. K. Murer: *J. Morphol.,* Fig. 43.

Mast, S. O.: *Biol. Bull.,* Figs. 16, 51, 54.

Metz, C. B.: *Am. Naturalist,* Figs. 122, 123.

Pace, D. M., and K. K. Kimura: *J. Cellular Comp. Physiol.,* Table 23.

Parpart, A. K.: *Biol. Bull.,* Table 6.

Rees, C. W.: *Univ. of California Pub. in Zool.,* Fig. 20.

Richards, O. W.: American Optical Company, Fig. 8.

Sawano, E.: *Science Repts. Tokyo Bunrika Daigaku,* Fig. 42 (redrawn).

Schewiakoff, W.: *Z. wiss. Zoöl.,* Fig. 36.

Sonneborn, T. M.: "Advances in Genetics," 1947, New York, Academic Press, Tables 26, 27; Figs. 80, 85, 86, 107, 108, 109, 121.

Sonneborn, T. M.: *Am. Scientist,* Figs. 116, 117, 119.

Sonneborn, T. M.: Cold Spring Harbor Symposia on Quantitative Biology, Figs. 115, 120.

Sonneborn, T. M.: The Harvey Lectures, Springfield, Ill., Charles C Thomas, Publisher. Table 36.

Sonneborn, T. M.: *Heredity,* Fig. 114 A.

Staples Press Limited, Gabb Collection, London, Frontispiece.

A. H. Thomas Company, Fig. 141.

Unger, W. B.: *J. Exptl. Zoöl.,* Table 25.

Wenrich, D. H.: *Tr. Am. Microscop. Soc.,* Figs. 5, 6.

Woodruff, L. L.: *Biochem. Bull.,* Fig. 130.

Woodruff, L. L.: *Quart. J. Microscop. Sci.,* Fig. 38.

Woodruff, L. L.: *Tr. Conn. Acad. Arts Sci.,* Figs. 2, 3.

Worley, L. G.: *Proc. Nat. Acad. Sci. U.S.,* Fig. 24.

RALPH WICHTERMAN

Autumn, 1952
Philadelphia, Pa.

Contents

Chapter 1

History of the Genus

The development of our knowledge of *Paramecium* as indeed of all of the Protozoa and other microscopic organisms was dependent upon the development of the microscope. The important fact that lenses may be ground to magnify objects seems to have been established in classical antiquity. Later the field of optics was stimulated largely by a desire to improve spectacle lenses, the value of which was perhaps first appreciated by the illuminators of manuscripts. At first single lenses were used for magnification and it was not until the end of the sixteenth century that Hans and Zacharias Janssen, Dutch spectacle-makers, placed two convex lenses in proper positions in a tube to form the first true compound microscope. It was not until the seventeenth century, however, that the construction of microscopes was considerably improved. A number of clever craftsmen fashioned good instruments, the most outstanding person being Antony van Leeuwenhoek. Some of his microscopes could magnify up to 270 diameters.

Our knowledge of Protozoa in general and *Paramecium* in particular parallels this early development of the microscope. It is conceivable that these early microscopists, who brought nearly everything possible within the scope of vision of their simple lens systems, also saw species of *Paramecium*. These pioneers in the early discovery of microscopic animal life opened new, unexplored fields of science where literature, terminology and taxonomy as we now know them were non-existent and journals in which to report their fascinating discoveries were few. One of the oldest journals, *The Philosophical Transactions of the Royal Society,* which first appeared in 1665, was one of the first to publish descriptions of their findings. Ciliates were unquestionably seen and described in letters addressed to the Royal Society and observations were published in privately printed books. As might be expected, some of these descriptions have proved inadequate for purposes of identification. But in letters addressed to the Royal Society and published in its Transactions, in letters exchanged between the microscopists, and in privately printed books we find some definite evidence pertaining to the discovery and description of *Paramecium*.

It seems highly probable that Antony van Leeuwenhoek, the discoverer of the Protozoa and bacteria also discovered members of the genus *Paramecium* as early as 1674 and 1677. In an account of the early history of *Paramecium*, Woodruff (1945) reported that, according to newly published manuscripts Leeuwenhoek corresponded with Constantijn Huygens (senior) and informed him of some of his observations. Huygens passed on the information to his son, Christiaan, who repeated certain of the observations. In 1678 Christiaan wrote a letter accompanied by sketches,

1

to his brother Constantijn (Huygens, 1888), and described several Protozoa which he found in infusions, including one that can be identified as a species of *Paramecium* (Fig. 2 [1]). The letter was forwarded to Leeuwenhoek because he wrote to Constantijn Huygens (senior) and stated that he recognized all the animals observed by Christiaan as having previously been seen by himself. In a careful study of the writings of Leeuwenhoek, Dobell (1932) maintains that this discoverer of the Protozoa also discovered *Paramecium*.

A great deal of interest has been centered on the first individual to present a recognizable drawing of *Paramecium*. Buonanni (1691) published the first figure of a free-living ciliate, apparently a *Colpidium*. In 1693 King figured some animals which are so equivocal that they may or may not have been drawn to represent *Paramecium*. However, in 1703 an anonymous writer described and drew four figures of which two (a and b) have been generally regarded as the first published illustrations of a *Paramecium* (Fig. 2 [2]). It would be of considerable historical interest to know the identity of this early microscopist whose drawings were taken from an extract of some letters of September 29, 1702 and sent to "Sir C. H." relating to some microscopic observations. The description with drawings appeared in *The Philosophical Transactions of the Royal Society* in 1703.

In 1718, Joblot published an account with a figure generally accepted as representing a *Paramecium* and called by him "Chausson" (slipper) (Fig. 2 [3]). Joblot's figure is therefore the second to represent a drawing of *Paramecium* and he is the first to refer to *Paramecium* as the "slipper-shaped animalcule"—a term in use even today (Fig. 1).

Hill (1752) first attempted to apply scientific names to microscopic animals and was the first to use the genus name, *Paramecium*. Hill probably used the Greek adjective, παραμηχης=*parameces,* which means oblong or longish (to distinguish it from the rounded forms). He then added *ium* to the stem *paramec* to make a noun. His characterization of the members in the group *Paramecium* and descriptions of four species is vague. For *Paramecium,* he writes, "Animals which have no visible limbs or tails and are of an irregularly oblong figure." He describes each of his four species with a short single sentence and uses the word *oblong* in each definition as follows:

"The *Paramecium,* with an oblong, voluble body, obtuse at each end.

"The *Paramecium,* with an oblong body, smallest toward the head.

"The *Paramecium,* with an oblong body, narrowest toward the middle.

"The *Paramecium,* with a slender, oblong body."

His descriptions are of little importance. However, his drawings of "*Paramecium* Sp. 3" (Fig. 2 [4]) are of value but they are similar to those reported by the anonymous writer in 1703. The view is held generally that Hill's figures are copies of the 1703 figures (see figure 2 [2] and [4] for comparison).

It is possible that Ledermüller in 1763 might have seen species of *Paramecium,* judging from his descriptions. He speaks of these organisms thus, "die Classe der Infussions Thierlein," which represents the first usage of the term which was latinized to *Infusoria* and first employed as such by Wrisberg in 1765 (Fig. 2 [5]). Ellis in 1769 described *Paramecium* showing figures of the ciliate (Fig. 2 [6]). He refers to it however as "*Volvox terebrella* or the gimlet." He says, "It is visible to the naked eye, moves swiftly, turning itself around

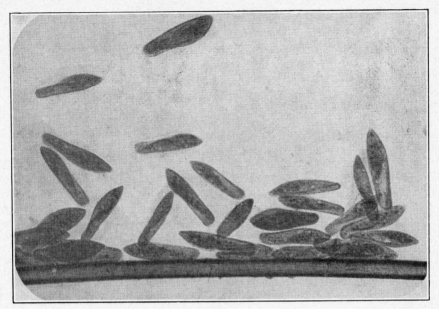

FIG. 1. Living specimens of *Paramecium caudatum* photographed next to a hair from author's head. Note how closely the paramecium to the right resembles the shape of a lady's slipper. (Wichterman)

as it swims, just as if boring its way." We have described here the first account of its method of swimming in a spiral manner. One of his figures is the first illustration of trichocysts—said by Ellis to be sketched from an animal "torpid with all its fins extended." Spallanzani (1776) described the contractile vacuoles of a species of *Paramecium* and noted the radiating canals.

Our present system of classification begins with the work of the Swedish biologist Linnaeus (Karl von Linné, 1707–78). His historic work, "Systema naturae," was first published in 1735 in Latin, the accepted scholarly language of the times. However, it was not until the tenth edition published in 1758 that the fundamental plan was presented upon which is based the system now used for the naming of animals. In his twelfth edition Linnaeus (1768) placed all "Animalcules" under three genera, namely *Volvox, Furia* and *Chaos*. The

genus *Chaos* contained all kinds of *Infusoria* under one species, *Chaos infusorium*.

The next great student of microscopic organisms was O. F. Müller, the first microscopist to apply the Linnean system of nomenclature to these forms. There are three parts to his important monographic studies. The first and second parts were published in 1773 and 1774 respectively wherein a section is devoted to the *Infusoria*. His last volume was published posthumously in 1786 and represents his most important contribution in the study of these lower organisms.

Müller presented a thorough treatment of this group and regarded members of the *Infusoria* as the simplest animals but included bacteria, diatoms, worms and rotifers in addition to Protozoa. In his last work, there are excellent figures of *Paramecium* (Fig. 2 [8]). He also portrayed conjugation in *Para-*

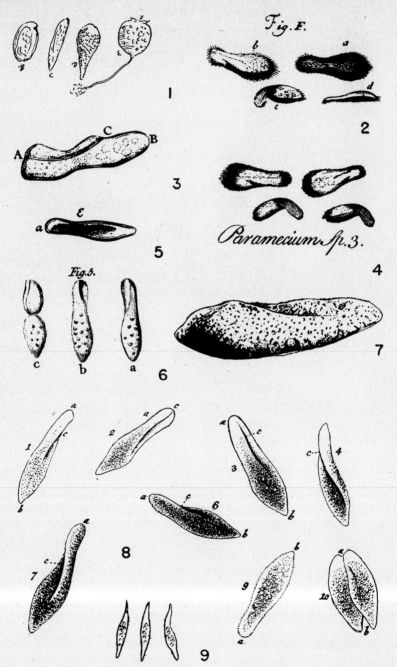

FIG. 2. The reproductions approximate the size of the figures in the original publications. (1) From Huygens, 1678. Fig. C. (2) From anonymous author, 1703. (3) Chausson. Joblot, 1718. (4) Paramecium, Sp. 3. Hill, 1752. (5) Animalculum pisciforme ex infusione apii. Wrisberg, 1765. (6) *Volvox terebrella*. Ellis, 1769. (7) Pantoffelthierchen. Gleichen, 1778. (8) *Paramecium aurelia*. Müller, 1786. (9) *Paramecium caudatum*. Hermann, 1784. (Woodruff)

4

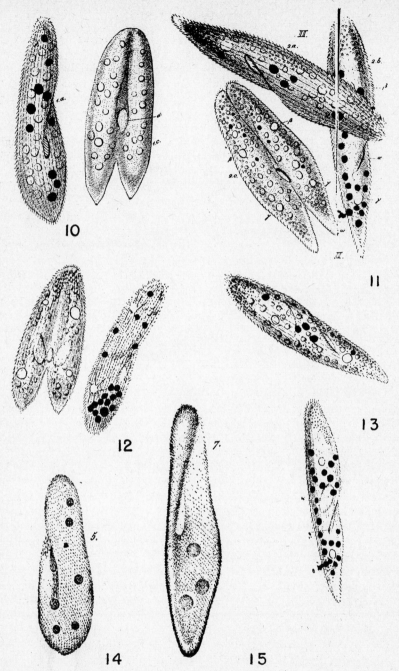

FIG. 3. The reproductions approximate the size of the figures in the original publications. (10) *Paramecium aurelia.* Ehrenberg, 1833. (11) *Paramecium caudatum.* Ehrenberg, 1833. (12) *Paramecium aurelia.* Ehrenberg, 1838. (13) *Paramecium caudatum.* Ehrenberg, 1838. (14) *Paramecium aurelia.* Dujardin, 1841. (15) *Paramecium caudatum.* Dujardin, 1841. (Woodruff)

mecium which however appears to have been first observed by Leeuwenhoek. In Müller's first treatise (1773) his earliest reference to *Paramecium* is characterized and spelled as follows:

"*Paramaecium*. Vermis inconspicuus, simplex, pellucidus, complanatus, oblongus." He lists two species, *Paramaecium aurelia,* which is described as "Paramaecium oblongum, antice in longitudinem plicatum," and *Paramaecium histrio* which he later removed from the genus in his final volume of 1786, not believing it to be a species of *Paramecium.*

This first description of *Paramecium* by Müller (1773), is not accompanied by figures but it identifies his *Paramaecium aurelia* with Hill's "*Paramecium* Sp. 3" which Hill apparently copied from the anonymous author of 1703. Here occurs the first change in spelling of the name *Paramecium,* as given by Hill (1752), to *Paramaecium.* This has led to considerable confusion in the literature even up to the present. The insertion of the letter *a* is etymologically incorrect since it is an error of transcription (Ludwig, 1930 a; Kalmus, 1931; Woodruff, 1945). Since Müller made direct reference to Hill's *Paramecium* but misspelled it *Paramaecium,* the latter is unacceptable according to Article 19 of the International Rules of Zoological Nomenclature which states, "The original orthography of a name is to be preserved unless an error of transcription, a lapsus calami, or a typographical error is evident." This error in spelling was recognized by Hermann (1784), Ehrenberg (1833, 1838) and Dujardin (1841). Ludwig (1930 a) makes the point that O. F. Müller's addition of the letter *a* is understandable because the Danes use the joined *ae* and do not greatly differentiate it from the letter *a.* Therefore the etymologically correct spelling of the genus

name is *Paramecium.* Of course there is no excuse whatever for the insertion of the letter *o* in the name, written *Paramoecium* by so many American authors. This appears to have originated as a typographic error since the joined *oe* so closely resembles the joined *ae* of Müller's original spelling although the letter *o* erroneously occurred in a listing of Müller's genera.

In his comprehensive work of 1786, Müller not only described but presented excellent figures of *Paramecium* which he chose to name *Paramaecium aurelia.* He described additional species which since have been placed in other genera. In this, his last work, is the following description:

PARAMAECIUM—Vermis inconspicuus, simplex, pellucidus, membranaceus, oblongus.
Paramaecium aurelia—Paramaecium compressum, versus antica plicatum, postice acutum.

The species name of *Paramecium aurelia* appears to have been taken from the resemblance of the ciliate to a golden-colored pupa or chrysalis of a butterfly: *aureus* (Latin) = gold or golden color; *chrysos* (Greek) = golden. It is the first species of *Paramecium* named under the Linnean system of nomenclature that has come down to the present time as *Paramecium aurelia* O. F. M. (1786).

A question must now be raised about the validity of *P. aurelia.* Upon reading Müller's description of *Paramecium aurelia* and examining his accurately drawn figures (reproduced here), we find the description partly includes the species now known as *Paramecium caudatum.* Maupas and a number of other investigators also recognized this.

About this time Gleichen (1778) observed specimens (Fig. 2 [7]) and reported the first observations of feeding in *Paramecium*. Using carmine, he was the first to note food-vacuole formation in this and other ciliates. With the close of the eighteenth century and the beginning of the nineteenth, interest in these microscopic animals increased with vigor. Hermann (1784) described and figured a species which he called *Paramecium caudatum* (Fig. 2 [9]) and which was accepted by Schrank (1803). Hermann's description more closely resembles that of the ciliate, *Amphileptus*, but it cannot be considered as being of *Paramecium*. Gruithuisen (1812) made the striking observation of cyclosis—the circulation of the endoplasm—in *Paramecium*, although a similar phenomenon had been seen in the plant cell by Corti in 1774.

During this period Goldfuss (1817) first used the term *Protozoa* which also included the Polyps and Medusae. However, it was not until 1845 that von Siebold employed the term in its present sense.

The next outstanding investigator of the Protozoa was C. G. Ehrenberg (1830) who in an early treatise on microscopic organisms acknowledged two species already named by Müller. These were *Paramaecium* (*Paramecium*) *Chrysalis* Müller and *Paramaecium Aurelia* Müller. Ehrenberg's Plate IV, Fig. 2 shows six colored illustrations of the animals he called *Paramaecium Chrysalis*. These figures are well drawn and resemble what is now recognized as *P. aurelia*. In this work he also listed the following as new species: *Paramaecium sinaiticum, P. compressum, P. ovatum* and "*Trichoda? Paramaecium n. sp.*" all of which have been removed from the genus. It is evident that he followed Müller's original incorrect spelling of the genus but he recognized the error and corrected it to *Paramecium* in his later works.

Then on June 11, 1832, Ehrenberg discovered a new species which he called *Paramecium caudatum* and described in 1833 as follows:

Paramecium caudatum N. sp.! (exclamation point his). Geschwänztes Längethierchen, Pantoffelthierchen. P. corpore subcylindrico, fusiformi, antico fine crassiore, rotundato, postico sensim attenuato, subcaudato.

He reported, "Die Form und Grösse dieses Thierchens ist der des *Param. Aurelia* ganz ähnlich, aber immer hinten schwanzartig zugespitzt, weniger cylindrisch, mehr spindelförmig und dabei mit gelblichem Farbetone." He further characterized and compared the new species with *P. aurelia*. It is, of course, the valid species which has come down to us as *Paramecium caudatum* Ehrbg. In his great monograph, "Die Infusionsthierchen als vollkommene Organismen" (1838), Ehrenberg gave the sizes, shapes, and figures of the two common species of *Paramecium* (Fig. 3 [10–13]) which he characterized as follows:

Paramecium Aurelia, Pantoffelthierchen. P. corpore cylindrico, subclavato, antica parte paullo tenuiore, plica longitudinali obliqua in os multum recedens exeunte, utrinque obtuso.

Under *Paramecium aurelia*, he lists the following homonyms and references to animals (p. 350) which he believed should be included under this species as follows:

Paramecium Aurelia, Pantoffelthierchen. Little animals longer than an oval, Leeuwenhoek (1675)
Animalcula in Pepper Water, Anonymous? Phil. Trans. (1703)
Chausson, Joblot (1718)
Animalcules in Pepper Water first sort, Baker (1742)
Paramecium species 3. and 1.? Hill (1751)
Würmer in Heuwasser, Ledermüller (1760)

Animalculum pisciforme, Wrisberg (1765)

Volvox Terebella, Ellis (1769)

Paramaecium Aurelia, Puppe-Aflangeren, Müller (1773)

Pantoffelartige Thiere in Heuinfusionen, Göze (1774)

Karkassenpolyp, Pelisson? (1775)

Animali elittici massimi a due stelluzze, Spallanzani (1776)

Pandeloquenthierchen, Gleichen (1777)

Pantoffel- und Pandeloquenthierchen, Gleichen (1778)

Paramecium Aurelia, Hermann (1784)

Paramaecium Aurelia, Müller (1786)

Paramaecium Aurelia, Schrank (1803)

Paramaecium, Treviranus (1803)

Grosse Pendeloquen, Gruithuisen (1812)

Paramaecium Aurelia, 1824 ⎫
Peritricha Pleuronectes, 1824 ⎪ Bory de St. Vincent (1824–31)
Bursaria Calceolus, 1826 ⎬
Polytricha Pleuronectes, 1831 ⎭

Paramaecium plures species, Losana (1832)

Paramaecium Aurelia, Abhandl. Akad. Wissensch. Berlin (1830)

Paramaecium Aurelia, Poggendorff's Ann. d. Physik (1832)

Paramaecium Chrysalis, Wagner (1832)

Paramecium Aurelia et pisciforme, Gravenhorst (1833)

The list with complete citations is followed with a large number of locations where these forms have been reported in Europe and Asia.

Going next to the new species, he reports as follows:

Paramecium caudatum, geschwänztes Pantoffelthierchen. P. corpore fusiformi, antica parte obtusiore, postica magis attenuata.

He described the larger one, pointed posteriorly as *Paramecium caudatum* while the smaller one with the posterior end less drawn to a point was placed in Müller's *P. aurelia.*

In the same manner, we find him listing under this species the following:

Paramecium caudatum, geschwänztes Pantoffelthierchen.

Paramecium caudatum, Hermann (1784). Amphileptus?

Paramaecium caudatum, Schrank (1803)

We see that the name *P. caudatum* was already taken before Ehrenberg's use of it and Woodruff (1945) makes the interesting point that according to modern rules of nomenclature the name "caudatum" was technically unavailable when used by Ehrenberg. On the other hand, the animals described by these men do not appear to have been species of *Paramecium.*

In his monograph, Ehrenberg (1838) gives a comprehensive review of the literature on these organisms up to the time of publication. The list is brought up to date in the present book. In addition to his descriptions, there are also shown good figures of *P. aurelia* and *P. caudatum* (including conjugation of both). These figures show clearly the cilia, food vacuoles, contractile vacuoles with canals, macronuclei, oral groove, mouth and cytopyge. It is interesting to note that one of his figures of *P. aurelia* shows three contractile vacuoles with canals.

Since some of the species of the genus *Paramecium* are so widely distributed over the earth and hence commonly encountered in random collections in bodies of water, it is likely that they must have been seen and reported by many of the old microscopists. One is not likely to be aware of this until some of the very old descriptions are examined. Most of these early reports first appearing in 1674 and continuing through and beyond the next century are not readily accessible to students in the field.

Altogether, 56 species of *Paramecium* are listed by Ehrenberg as occurring in the literature. Of this number, he believed 48 to be homonyms. Ehrenberg listed, described, and figured the remaining eight species which he believed to be valid, exactly as follows:

1. *Paramecium Aurelia*
2. *Paramecium caudatum*
3. *Paramecium Chrysalis*
4. *Paramecium Colpoda*
5. *Paramecium ? sinaiticum*
6. *Paramecium ? ovatum*
7. *Paramecium compressum*
8. *Paramecium Milium*

Of these eight, *P. aurelia* and *P. caudatum* are the only species accepted today.

It is conceivable that Müller earlier may have combined the characteristics of both *P. aurelia* and *P. caudatum* into his so-called *P. aurelia*. The very early taxonomists and students of *Paramecium* were concerned mainly with general body shape. Internal structure was generally ignored. They did not see micronuclei for instance, which are of prime diagnostic value. Attempts to make changes now on the basis of strictly inadequate descriptions would only lead to greater confusion in the taxonomy of *Paramecium*. One must resist the temptation, however great, of reading into these early sketches and descriptions, features that may be implied but that do not actually exist. Therefore the present author is inclined to follow the scheme of Ludwig (1930 a) in that, according to Article 24 of the International Rules of Zoological Nomenclature, the classification of these two species be accepted as follows:

Paramecium aurelia, O. F. M. partim Ehrbg.
Paramecium caudatum, Ehrbg.

Focke in 1836 gave the name *Paramecium bursaria* to a valid species which may have been described by Ehrenberg under the genus *Loxodes* (*Loxodes bursaria* Ehrbg.).

Dujardin (1841) recognized and gave excellent figures of *P. aurelia* and *P. caudatum* (Fig. 3 [14, 15]).

Stein (1859–78) in his great treatise, "Der Organismus der Infusionsthiere" gave a remarkably clear characterization of the genus *Paramecium* which has value at the present time. However, it is bewildering to note that such competent observers as Stein, Claparède and Lachmann, Bütschli, Balbiani, Kent, Engelmann and others combine the two species of *P. aurelia* and *P. caudatum* into one and apply Müller's original name of *P. aurelia* for both.

In their monograph of the Protozoa, Claparède and Lachmann (1858–60) eliminated *P. caudatum* but included *P. aurelia* and *P. bursaria*. However, they also added the following as new species:

1. *P. putrinum*
2. *P. ovale*
3. *P. inversum*
4. *P. microstomum*
5. *P. glaucum*

In addition, they included the *P. versutum* of Müller. Excepting *P. aurelia* and *P. bursaria*, all of the remaining species of *Paramecium* listed by these two investigators are not now acceptable as valid species.

Engelmann (1862) described *P. ambiguum* n. sp. and its rediscovery is reported by Kahl (1928). Their accounts remind one of a colorless form of *P. bursaria* and on the basis of inadequate descriptions cannot be regarded as a valid species.

In his monograph on the *Infusoria*, Fromentel (1874) lists eight species of *Paramecium* as follows:

1. *P. aurelia*
2. *P. ovatum* (Ehrenberg)
3. *P. regulare* n. sp.

4. *P. flavum* n. sp.
5. *P. roseum* n. sp.
6. *P. milium* (Ehrenberg: Cyclidium milium, Müller)
7. *P. subovatum* n. sp.
8. *P. colpoda* (Ehrenberg: Kolpoda ren, Müller)

From Fromentel's description and figures, his *P. aurelia* may be that species or *P. caudatum*. The remaining seven species as designated by Fromentel are obviously unacceptable today. His *P. flavum* and *P. roseum* resemble species of *Blepharisma* and *P. subovatum,* a recently divided anterior daughter. *P. regulare* n. sp. appears to coincide with the description as given for *"P. putrinum"* by Claparède and Lachmann.

Kent (1882) in his "Manual of the Infusoria," recognized in addition to *P. aurelia; P. bursaria,* *"P. putrinum"* and *P. glaucum*—all of which were listed earlier by Claparède and Lachmann. Kent described another new species, *P. marina,* which has no standing at present.

Gourret and Roeser (1886) described *P. pyriformi* n. sp. but it is not a valid species of *Paramecium.*

During the nineteenth century—especially the latter part—a number of outstanding investigators presented more accurate descriptions than ever before, with emphasis being placed on the internal structure of *Paramecium.* Thus it became inevitable that nuclear components, which are important taxonomic characters, be clearly defined.

We are therefore greatly indebted to Maupas for his classical studies on *Paramecium,* especially his detailed investigations on the nuclear structure. Maupas (1883) noted that the so-called species *P. aurelia* (in reality *P. aurelia* and *P. caudatum*) contained individuals with two different sets of nuclear structures and wrote (1888) that earlier observers (mentioned above) as well as

himself confused the two species in question. He then proceeded to define in a lucid manner the characteristics of the two species. Maupas (1889) and Hertwig (1889), working independently, presented their classical studies of conjugation in *Paramecium* which further helped to set the two species on a firm foundation.

Stokes (1885) described *Paramecium trichium,* the fourth and last universally accepted species to be named before the end of the nineteenth century. We are indebted to Wenrich (1926) for a clearer definition of this species.

Bürger (1908) described a ciliate from Chile which he called *P. nigrum* but it, in reality, is a species of *Frontonia.*

At the beginning of the twentieth century, it is surprising to find that Calkins (1906) should again raise the question as to whether *P. caudatum* should be regarded as a mere variant of *P. aurelia.* He used the specific name *P. aurelia* to include both *P. aurelia* and *P. caudatum* and his account of conjugation of what he called *P. aurelia,* is in fact that of *P. caudatum* (Calkins and Cull, 1907). It is disconcerting to have Calkins (1906) report that *P. caudatum* may become *P. aurelia* and *P. aurelia* may in time lapse into *P. caudatum.*

Shortly after this, Jennings (1908) in his studies on heredity in *Paramecium* showed the existence of two main groups—an "aurelia" group and a "caudatum" group—based statistically on mean length measurements of a considerable number of pure-line cultures.

Making use of cytological observations and pedigreed individuals in isolation cultures over long periods of time, Woodruff (1911) recognized the species differences. Protozoölogists are familiar with Woodruff's famous culture of pedigreed *Paramecium aurelia* which had been kept under fairly continuous observation for over 25 years (Woodruff, 1932, 1943). The species has retained its morphologic characters throughout this long period of close observation.

In this regard, the author has maintained in pure-line culture for periods up to 15 years, all four valid species described before 1900 as follows: *P. aurelia, P. caudatum, P. bursaria,* and *P. trichium.* In repeated observations up to the present time, the species characters have been constant.

Powers and Mitchell (1910) described the next valid species of *Paramecium* which they choose to call *P. multimicronucleata (multimicronucleatum)* and which was more clearly defined by Landis (1925) in his account of conjugation of this species.

In 1919, Woodruff discovered the sixth valid species which he named *P. calkinsi* (1921).

Shortly after, Woodruff and Spencer (1923) described another valid member of the genus, the seventh, which they called *P. polycaryum.*

Gelei (1925, a) added a new species to the list which he called *"P. nephridiatum"* but this now appears to be of questionable status.

In 1928, Wenrich described the eighth valid species called by him *P. woodruffi.*

In a study of species of the genus *Paramecium,* Wenrich (1928 a) considered eight of the species mentioned in this history as valid ones. Ludwig (1930 a) and Kalmus (1931) each list 10 while Kahl (1930) lists as many as 17, five of which he mentions as having uncertain status. The new species added by Kahl are as follows: *P. traunsteineri, P. chilodonides, P. ficarium, P. pseudoputrinum* and *P. chlorelligerum.* Kudo (1947) considers only nine in his textbook on protozoölogy.

CHRONOLOGICAL ORDER OF DESCRIBED SPECIES OF PARAMECIUM

(The reader is referred to the monograph of Ehrenberg (1838) for a comprehensive bibliography of the 57 species or references to species made prior to 1838. Of that number, Ehrenberg considered most of them as homonyms and listed as valid only eight as listed below*.)

SPECIES	AUTHOR	DATE
*P. aurelia**	Müller	1773
P. histrio	Müller	1773
P. versutum	Müller	1786
*P. chrysalis**	Müller	1786
P. oceanicum	Eysenhardt and Chamisso	1820
*P. caudatum**	Ehrenberg	1833
P. bursaria	Focke	1836
*P. kolpoda**	Ehrenberg	1838
*P. sinaiticum**	Ehrenberg	1838
*P. ovatum**	Ehrenberg	1838
*P. compressum**	Ehrenberg	1838
*P. milium**	Ehrenberg	1838
"P. putrinum"	Claparède and Lachmann	1858–60
P. ovale	Claparède and Lachmann	1858–60
P. inversum	Claparède and Lachmann	1858–60
P. microstomum	Claparède and Lachmann	1858–60
P. glaucum	Claparède and Lachmann	1858–60
P. ambiguum	Engelmann	1861
P. regulare	Fromentel	1874

SPECIES	AUTHOR	DATE
P. flavum	Fromentel	1874
P. roseum	Fromentel	1874
P. subovatum	Fromentel	1874
P. marina	Kent	1880–82
P. trichium	Stokes	1885
P. pyriformi	Gourret and Roeser	1886
P. nigrum	Bürger	1908
P. multimicronucleata= (multimicronucleatum)	Powers and Mitchell	1910
P. calkinsi	Woodruff	1921
P. polycaryum	Woodruff and Spencer	1923
"P. nephridiatum"	Gelei	1925
P. woodruffi	Wenrich	1928
"P. Pseudoputrinum"	Kahl[1]	1930
"P. ficarium"	Kahl	1930
"P. traunsteineri"	Kahl	1930
"P. chilodonides"	Kahl	1930
"P. chlorelligerum"	Kahl	1930
"P. duboscqui"	Chatton and Brachon	1933; 1936

[1] In addition to the five new species of Kahl (1930), he lists 12 others whose names appear in the present listing.

The reader is referred to the papers of Cole (1926), Ludwig (1930 a), and Woodruff (1911, 1945) for a discussion of certain interesting and controversial points of nomenclature concerning the species of the genus. The validity of the described species of *Paramecium* is considered in the next chapter.

Chapter 2

Classification and Species of Paramecium

A. Taxonomy
B. Genus Characteristics Emend
C. "Aurelia" and "Bursaria" Groups
D. Key to the Species of Paramecium
E. Well-defined Species of Paramecium
    "AURELIA" GROUP
      1. P. caudatum
      2. P. aurelia
      3. P. multimicronucleatum
    "BURSARIA" GROUP
      4. P. bursaria
      5. P. trichium
      6. P. calkinsi
      7. P. polycaryum
      8. P. woodruffi
F. Uncertain Species of Paramecium
      1. P. putrinum
      2. P. nephridiatum
      3. P. traunsteineri
      4. P. chilodonides
      5. P. ficarium
      6. P. pseudoputrinum
      7. P. chlorelligerum
      8. P. duboscqui

A. Taxonomy

Members of the genus *Paramecium* are unicellular organisms and hence are placed in the phylum Protozoa. The early name Animalculae was given to all forms of microscopic life in which *Paramecium* was of course included. Ledermüller (1763) utilized the name *Infusoria* to include all members of the entire phylum of Protozoa. As the name implies, *Infusoria* was at first used to embrace all those forms commonly found in infusions of decaying animal and vegetable matter. Dujardin (1841) divided the "Infusoires" into rhizopods, flagellates, and ciliates—a classification adopted in the main by Bütschli (1887) who limited the use of the term *Infusoria* to Protozoa bearing cilia at some period of their life history. Although Perty (1852) introduced the term *Ciliata,* Claparède and Lachmann (1858–60) clearly defined this class and added still another, the *Suctoria.* At the present time the following two classes of ciliates are generally recognized: class *Ciliata,* possessing cilia throughout trophic life and class *Suctoria* with cilia during the early developmental stages only and tentacles in the adult stage. Following is the classification of *Paramecium* based on a number of taxonomic works.

Kingdom: *Animalia* (Linnaeus, 1758)
Phylum: *Protozoa* (Goldfuss, 1817; emend von Siebold, 1845)
Subphylum: *Ciliophora* (Doflein, 1902)
Class: *Ciliata* (Perty, 1852) = *Infusoria* (Ledermüller, 1763) (emend Bütschli, 1889)
Subclass: *Euciliata* (Metcalf, 1923)
Order: *Holotricha* (Stein, 1859)

Suborder: *Trichostomata*
(Bütschli, 1889; emend
Kahl, 1930)
Family: *Parameciidae* (Kent,
1881)
Genus: *Paramecium* (Hill,
1752; emend Stein 1860;
emend Wichterman)

Although the term Protozoa was first used by Goldfuss in 1817, he included in the group the Polyps and Medusae. Siebold (1845) employed the term in its modern sense by establishing the unicellular concept of the Protozoa.[1]

Doflein (1902) introduced the subphylum, *Ciliophora,* to include Protozoa with cilia (or čirri) which serve as structures of locomotion and food-capture. Metcalf (1923) divided the class *Ciliata* into two subclasses based primarily on nuclear structure or organization. In the first subclass, namely *Protociliata,* are placed ciliates with two or more nuclei of one kind such as are found in *Opalina.* The second subclass, *Euciliata,* contains ciliates possessing macronuclei and micronuclei as found in the species of *Paramecium.* The order *Holotricha,* to which *Paramecium* belongs, was originated by Stein (1859) for the reception of all those ciliates that possess uniform ciliation in which the cilia differ but slightly from one another over the body surface. Most holotrichous ciliates possess a cytostome but an adoral zone of membranelles does not occur; others are astomatous. Asexual and sexual reproduction, and encystment are common for members of this order. Nutrition is holozoic,

[1] In this regard, it is of interest to report the non-cellular conception of the Protozoa as proposed by Dobell (1911). He considered the protozoön—i.e. paramecium—as a non-cellular but complete organism differently organized in both structure and function when compared with cellular organisms of the Metazoa. The vast majority of protozoölogists, however, consider the Protozoa as unicellular animals.

saprozoic and holophytic and its members include parasites as well as free-living forms inhabiting fresh, brackish and salt waters. Bütschli (1889) created the suborder *Trichostomata* which was modified by Kahl (1930) to include those holotrichous ciliates possessing a peristome lined with rows of cilia. The family *Parameciidae* (*Paramaeciidae,* Kent, 1881) originally contained free-swimming, asymmetric ciliates, roughly cylindrical or flattened and included besides *Paramecium,* a number of genera since removed. Dorsal and ventral surfaces are distinct with an oral groove or depression running from anterior left to middle right of body on the ventral surface. The body and oral groove are covered with cilia. At present, the family *Parameciidae* contains a number of well-defined species belonging to the genus *Paramecium* but others of doubtful validity have sometimes been included. Kahl (1930) added one other genus to the family, namely *Physalophrya.* According to Kahl, the genus *Physalophrya* is without a peristome but the mouth is said to be located near the anterior half of the body. An undulating membrane is absent from the cytopharynx although a ciliary row occurs on its left dorsal wall. The taxonomic position of this genus is not clear. Kahl lists only one species, *Physalophrya spumosa,* found in fresh water.

B. Genus Characteristics Emend

Free-swimming, ovoid, elongate or cigar-shaped ciliates of medium length and visible to the naked eye; natation generally vigorous and rotatory—characteristically spiralling to the left except *P. calkinsi* which spirals predominantly to the right; rounded or obliquely truncated anteriorly with posterior end rounded or conic; in shape, sometimes resembling imprint of a foot; asymmetric due to the presence of an oblique depression, the oral groove on the ventral (oral) surface; a thin but definite membrane, the pellicle, covers the body and consists of small, roughly rectangular or hexagonal fields; usually one— sometimes two—cilia emanate from the middle of each field. Rows of cilia cover the body and oral groove and are of fairly uniform length except at the posterior end where they are longer and less active than elsewhere, frequently forming a tuft. The oral groove generally extends about one-half the length of the body beginning anteriorly where it is widest, then tapering toward the mouth or cytostome which is also ventral. Leading from the mouth is a short S-shaped cylindrical cavity referred to in its entirety as the pharynx or cytopharynx. This structure is widest near the mouth and becomes narrow posteriorly. This wide, funnel-shaped and normally ciliated region near the mouth is sometimes called the vestibulum which decreases in diameter to form the esophagus. Thus it is seen that the posterior end of the cytopharynx and the structure called the esophagus are the same. An undulating membrane, frequently mentioned as being present, is absent.

There is a thin outer cortical layer known as the ectoplasm (ectosarc) which lies directly beneath the pellicle and encloses a larger inner fluid portion, the endoplasm (endosarc). In the ectoplasm and covered by the thin pellicle are the thickly set spindle-shaped bodies, the trichocysts which are placed at right angles to the body surface. Upon extrusion, the trichocysts become very long and thin. The fluid endoplasm contains granules, food vacuoles, and crystals of different size. The darkly appearing crystals are optically active and generally more numerous in the anterior or posterior end of the body. Cyclosis is usually conspicuous. Contractile vacuoles are generally two in number and located peripherally on the aboral surface at a distance approximately one quarter of the entire body length from the anterior and posterior ends. (*P. multimicronucleatum* possesses frequently three or more.) These vacuoles collapse alternately and empty the liquid contents to the outside of the body through pores. The contractile vacuoles are of two types. In one, called canal-fed vacuoles, a number of small canals (radial canals) radiate from the vacuole and empty into it as in *P. caudatum*. The other is a vesicle-fed type in which a number of smaller vesicles or vacuoles lie close to the larger vacuole and empty into it, as in *P. trichium*. An anus (cell-anus, potential-anus, anal spot, cytopyge, cytoproct) is ventral, and posterior to the mouth. A single, large, granular macronucleus which is roughly ellipsoidal or kidney-shaped is present slightly anterior to the middle of the body. One or more small, compact, or vesicular micronuclei are found generally close to or impressed against the macronucleus and are of important taxonomic value.

Asexual and sexual reproduction involving binary fission, endomixis, hemixis, autogamy, cytogamy, opposite mating types leading to conjugation, and cyst formation have been described.

Distribution universal with species inhabiting fresh, brackish, and salt water.

C. "Aurelia" and "Bursaria" Groups

Woodruff (1921) called attention to the fact that the species of *Paramecium* fall naturally into two clearly defined groups when based on body shape. He listed them as the "aurelia" group and the "bursaria" group respectively as follows (Fig. 4a):

1. *"Aurelia" group:* Individuals characterized by a relatively long spindle- or cigar-shaped body, round or cir-

cular in cross section with a somewhat pointed posterior end. Included in this group are the well-defined species *P. aurelia, P. caudatum,* and *P. multi-micronucleatum.*

2. *"Bursaria"* group: Individuals characterized by the shorter and broader form of the body, dorsoventral flattening and more rounded posterior end with somewhat obliquely truncated anterior end. Included in this group are *P. bursaria, P. calkinsi, P. woodruffi, P. polycaryum,* and *P. trichium.*

Within each group one finds two characteristic types of micronuclei of important taxonomic value as follows (Fig. 4b):

a. *"Caudatum" type:* Individuals in which the micronucleus is relatively large and composed of a rather compact mass of chromatin bounded by a nuclear membrane. This is the type of micronucleus found in *P. caudatum, P. bursaria,* and *P. trichium.*

b. *"Aurelia" type:* Individuals in which the micronucleus is relatively small and vesicular. It consists of an extremely small concentrated mass of chromatin called the endosome centrally located with a distinct space between it and the nuclear membrane. This distinction is based on fixed and stained specimens because in living ones it is difficult to identify micronuclei of the "aurelia" type, although micronuclei of the "caudatum" type may be seen easily

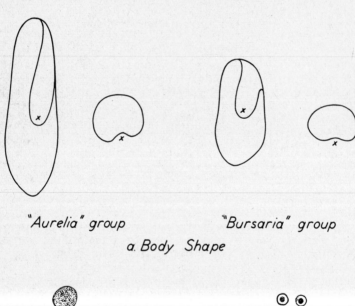

"Aurelia" group   "Bursaria" group

a. Body Shape

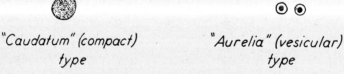

"Caudatum" (compact)   "Aurelia" (vesicular)
type   type

b. Form of Micronucleus

FIG. 4. (a) General body shape of "aurelia" and "bursaria" groups. The letter "x" represents the oral groove in the ventral view of animal and its respective cross section. (b) Characteristic appearance of micronuclei in "caudatum" (compact) and "aurelia" (vesicular) types of Paramecium. (Wichterman)

in the living condition. This is the type of micronucleus found in *P. aurelia, P. multimicronucleatum, P. woodruffi, P. calkinsi,* and *P. polycaryum.*

Wenrich (1928 a) made the important observation that the position of the anus or cytopyge is different in the two groups (Figs. 5 and 6). In members of the "bursaria" group, the anus is terminal and slightly to one side of the posterior end. In the "aurelia" group, it is subterminal, situated on the ventral side between the posterior end of the oral groove and the posterior end of the body. In *P. caudatum* and *P. multimicronucleatum,* the cytopyge lies between these two points or slightly nearer the end of the oral groove. In *P. aurelia,* the cytopyge is nearer the posterior end of the body.

D. Key to the Species of Paramecium (Figs. 5 and 6)

1. Slender, cylindrical, or cigar-shaped animals ("aurelia" group); bluntly rounded, anteriorly; somewhat pointed or conic, posteriorly; widest region about two-thirds body length behind anterior end. 2

1A. Shorter and wider animals ("bursaria" group) somewhat dorsoventrally flattened with obliquely truncated anterior end and broadly rounded posterior end. 4

2. One large, compact micronucleus ("caudatum" type); body pointed or conic, posteriad; two canal-fed contractile vacuoles; length 170–290 microns. . . . . *P. caudatum*

2A. More than one micronucleus and of vesicular type. 3

3. Usually less than 175 microns in length; two small vesicular ("aurelia" type) micronuclei; posterior end of body less pointed than *P. caudatum;* two canal-fed contractile vacuoles; smallest of the "aurelia" group; length 120–170 microns. . . . . *P. aurelia*

3A. Usually longer than 175 microns; three or four (occasionally up to seven), small, vesicular micronuclei; commonly more than two (frequently three) canal-fed contractile vacuoles; posterior end of body pointed or conic as in *P. caudatum;* largest of "aurelia" group; length 180–310 microns. . . . . . . *P. multimicro-nucleatum*

4. Animals nearly always colored green because of zoöchlorellae within; cyclosis characteristically rapid; single large compact micronucleus; two canal-fed contractile vacuoles; length 85–150 microns. . . . . *P. bursaria*

4A. Animals not green in color. 5

5. Two conspicuous vesicle-fed contractile vacuoles deeply set in endoplasm, each leading to exterior by means of convoluted outlet canal; radial canals absent; single compact micronucleus; smallest of "bursaria" group; length 70–90 microns. . . . . *P. trichium*

5A. Animals without conspicuous vesicle-fed contractile vacuoles and convoluted outlet canals but with radial canals. 6

6. Rotation of body on long axis mainly in direction of right

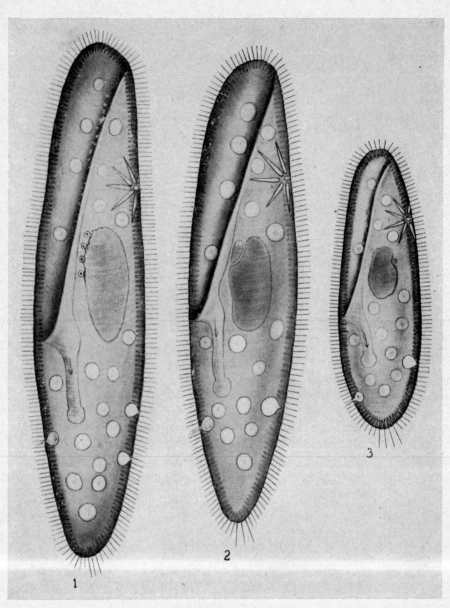

Fig. 5. This and the next figure are drawn to scale with the largest representing *P. multimicronucleatum* (1) and the smallest, *P. trichium* (8). (1) *P. multimicronucleatum.* Note large size, extra posterior contractile vacuole and the small vesicular micronuclei. (2) *P. caudatum.* Note character of micronucleus. (3) *P. aurelia.* Note smaller size and vesicular micronuclei, which are regularly two in number. (× 500.) (Wenrich)

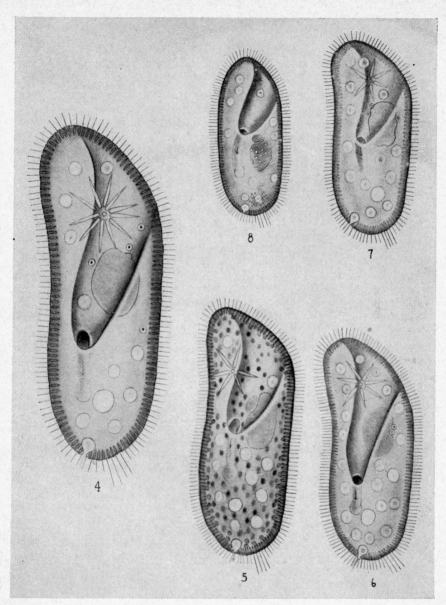

Fɪɢ. 6. (4) *P. woodruffi*. This is the largest of the bursaria form group. Note posterior position of mouth and the scattered position of the several vesicular micronuclei. (5) *P. bursaria*. In this figure the zoöchlorellae are represented by dark bodies which contrast sharply with the lighter and larger food vacuoles. (6) *P. calkinsi*. In shape, this is very similar to *P. woodruffi*, but is much smaller and has usually but two vesicular micronuclei. (7) *P. polycaryum*. In comparison with *P. calkinsi*, this species is not only smaller and a little wider for its length, but has the mouth farther forward and has the additional micronuclei (four appears to be the characteristic number). (8) *P. trichium*. This smallest of the species has a contractile vacuole apparatus near each end that is different from all the others. (× 500.) (Wenrich)

spiral; usually two (occasionally one to five) vesicular micronuclei; two canal-fed contractile vacuoles; fresh and brackish water; length 110–140 microns. .... *P. calkinsi*

6A. Rotation of body on long axis in direction of left spiral. 7

  7. Length generally less than 115 microns; usually four (occasionally three to eight) small vesicular micronuclei; mouth near center of body; two canal-fed contractile vacuoles; length 70–110 microns. .... *P. polycaryum*

7A. Length generally greater than 115 microns; mouth posterior to center of body; three or four (occasionally up to eight) scattered vesicular micronuclei; two canal-fed contractile vacuoles; brackish water; largest of "bursaria" group; length 120–210 microns. .... *P. woodruffi*

E. Well-defined Species of Paramecium

In the literature we find many descriptions and references to species of *Paramecium* (see list of described species in the preceding chapter). Most of these are so ill defined that we are not warranted in placing them in the genus. It is possible, in some cases, to determine that the author of a so-called new species was erroneously studying a ciliate from another genus. As an example, this appears to have been the case with the species described by Bürger (1908) as *P. nigrum*—in all probability a species of *Frontonia*. Other described species appear to have been based on observations of degenerate or abnormal individuals. This apparently is the case in the species described as *P. glaucum* (Claparède and Lachmann, 1858–60) although in the original illustration, an ellipsoidal organism is shown which possesses two contractile vacuoles with canals. Still other so-called species lack an accurate cytological study based upon too few individuals.

It is well known that for a given species of *Paramecium*, considerable form and size differences may exist. Statistical measurements of *P. aurelia, P. caudatum,* and *P. multimicronucleatum* show a mean length distinctive for each species. However, there may be considerable overlapping in size measurements when we deal with large or small races of each species. The food supply and temperature are still other factors which may govern size as well as shape. Well-fed paramecia generally are plump; starved ones, thin. Also it is the writer's experience that preconjugants —organisms ready to join in conjugation—are smaller than other, vegetative, forms. This is the case in a large number of preconjugant stages in ciliates of other genera, e.g. in the conjugation cycle of *Nyctotherus cordiformis* (Wichterman, 1937 b). It should be remembered that the early accounts of paramecium were based mainly on size and form, which in part explains the confusion existing in taxonomy. Little or no attention was given to internal structure, especially in regard to nuclei. It is not surprising therefore that at the close of the nineteenth century and at the beginning of the twentieth, *Paramecium aurelia* was applied indiscriminately to the species we know now as *P. aurelia* and *P. caudatum*. We are indebted to the careful work of Maupas (1889), Hertwig (1889), Jennings and Hargitt (1910), and Woodruff (1911) for a precise determination of these well-known species and the placing of each of them on a firm foundation. Detailed cytological studies of the conjugation process of *P. bursaria* by Hamburger (1904) proved beyond a doubt the validity of this species.

We may raise the question—what are the distinguishing characters which should be considered for the inclusion or rejection of the present species and those which are to be described in the future? An adequate description should include in addition to size and shape of the protozoön, a detailed account of structure both nuclear and cytosomal. In the study of speciation, we find that the micronuclei, particularly their size, shape, structure, and position are extremely important taxonomic characters (Table 3). The account should be followed by carefully made drawings and photographs. Attempts should be made to isolate the ciliate and grow it in culture. The newer researches on serology have been shown to have value in species identification. In proposing a new method of distinguishing species of *Paramecium,* Bragg (1935 a) identifies *P. caudatum, P. aurelia,* and *P. trichium* by the distinctly characteristic manner in which recently formed food vacuoles are passed from the pharynx to the cytoplasm.

Direction of spiraling in paramecium was claimed by Bullington (1925, 1930) to be a specific characteristic of sufficient taxonomic value to demand consideration in future classifications; this feature is considered in the present classification.

"Aurelia" Group

1. **Paramecium caudatum** (Ehrenberg) (Figs. 5 [2] and 15).

This is the "slipper-shaped animalcule" of the early microscopists that is so widely distributed and extensively studied. It was clearly recognized by Ehrenberg (1833) as distinct from *P. aurelia* on the basis of the pointed posterior end. It measures commonly 170–290 microns and is cigar- or spindle-shaped with a bluntly rounded anterior end and somewhat pointed or conic posterior end, roughly forming an angle of 45°–60°. In cross section it is circular. Its general shape and size may show considerable variation depending upon the race, food supply, and temperature. Some races show the posterior end more pointed or sharply conic than others. Some individuals, instead of being gracefully slender and streamlined, are plump. Generally the animals show the greatest diameter of the body in the region about two-thirds their length from the anterior end.

The ventrally located oral groove extends slightly more than half the length of the body in the direction of a right spiral. Near the posterior region of the oral groove is found the mouth (cytostome) which leads into the S-shaped cytopharynx. Two contractile vacuoles with approximately seven radial canals are present. These vacuoles are found on the aboral surface and lie in about the first and last quarters of the body in a line parallel with the ciliary rows. The two vacuoles contract alternately and empty their contents to the outside of the body by means of pores. The cytopyge is ventral and found medially between the posterior end of the oral groove and the posterior end of the body.

The large, compact, smoothly ellipsoidal or kidney-shaped macronucleus contains chromatin granules which stain rather intensely with nuclear dyes. It is generally found slightly anterior to the middle of the body. The single micronucleus is spherical to ovoid in shape, compact, but with chromatin granules seemingly finer than those of the macronucleus. It measures approximately 8 microns in diameter and stains readily with nuclear dyes so that it is easily visible with the low powers of the microscope. The micronucleus is located close to, or frequently lodged in, a shallow cavity of the macronucleus.

Rows of cilia cover the body and are of fairly uniform length except those at the extreme posterior tip of the body. Here they are considerably longer, forming a conic bundle the cilia of which move in a vibratory manner. This species spirals characteristically to the left (Fig. 7 A).

Trichocysts are present.

This species is world-wide in distribution and commonly found in ponds and bodies of stagnant and fresh water.

2. Paramecium aurelia (Müller partim Ehrenberg) (Fig. 5[3]).

This is the first valid species to be named that has come down to us as *P. aurelia* O. F. Müller (1773). As mentioned in the history of the genus, Müller's description and figures (1786) equally fit the present day *P. caudatum* and it is possible he may have been dealing with both species. Ehrenberg (1833) recognized these differences hence credit is due him for the species distinction.

Members of this species commonly measure 120–170 microns and while in some rare cases the sizes may overlap with those of *P. caudatum,* the former is usually much smaller. It is cigar- or spindle-shaped with the posterior end rounded as well as the anterior end. However, the posterior region is considerably broader than the anterior end. Therefore, because of its smaller size and body shape, it is generally possible to recognize it in the living condition.

The ventral, oral groove extends slightly posterior to the middle of the body where one finds the mouth and cytopharynx. Two contractile vacuoles with radial canals are found on the aboral surface, one behind the other in the direction of the ciliary lines. These vacuoles are located in about the first and last quarters of the body and empty to the outside by means of excretory pores. The cytopyge is ventral and subterminal and rather close to the posterior tip of the body.

The macronucleus is of a compact type which stains readily with nuclear dyes. It is smoothly ellipsoidal or kidney-shaped in or slightly above the center of the body. Of important diagnostic character are the two very small micronuclei of the vesicular type and measuring approximately 3–5 microns. These are usually found lying close to the macronucleus. One should stain the micronuclei to make certain of species identification. The fact that there are two micronuclei of comparatively small size along with the knowledge these organisms are the smallest of the "aurelia" group should guide one to exact species recognition.

As in *P. caudatum* the cilia exist in rows and are of moderate, uniform length except at the posterior tip where they are longest. This species spirals characteristically to the left (Fig. 7B). Trichocysts are present.

This species appears to be world-wide in distribution and specimens are found in ponds and bodies of stagnant and fresh water.

3. Paramecium multimicronucleatum[2] (Powers and Mitchell) (Fig. 5 [1]).

This species was first described by Powers and Mitchell (1910) and is probably the same one referred to by Hance (1917) and Thapar and Choud-

---

[2] The name of this well-defined species has been referred to as *"Paramecium nucleatum"* by Mast (1947) and Seaman (1947) in violation of Article 19 of the International Rules of Zoological Nomenclature (Wichterman, 1947 a).

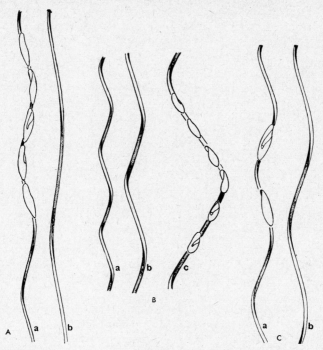

FIG. 7. Spiraling in Paramecium. (A) Spiral paths of *Paramecium caudatum;* (a) short narrow spiral of greatest speed; (b) average spiral. (B) Left spirals of *Paramecium aurelia;* (a) short narrow spirals of greatest speed; (b) average spirals; (c) spirals of greatest length and width. (C) Spiral paths of *Paramecium multimicronucleatum;* (a) short narrow spiral of greatest speed; (b) average spiral. (Bullington)

hury (1923). In his account of conjugation, Landis (1925) placed the species on a firm foundation. In general form the species resembles *P. caudatum* but *P. multimicronucleatum* is generally larger, measuring commonly between 180–310 microns, although longer ones measuring up to 400 microns have been reported. In a study of *P. multimicronucleatum,* Köster (1933) reported that this species is on the average 100–150 microns larger than *P. caudatum.* However, some races have mean lengths similar to *P. caudatum* making it impossible to distinguish the two species on the basis of size. The study of Landis (1925) revealed that the anterior end is a little more blunt, the posterior end a little more pointed and the oral groove a little

shallower than the corresponding structures of *P. caudatum.* The cytoplasm generally appears dense and opaque in this species.

As in the previously described species, the oral groove extends slightly beyond the middle of the body and at its posterior end one finds the mouth and cytopharynx. This species spirals characteristically to the left (Fig. 7C).

Unique for members of this species is their tendency to form extra contractile vacuoles similar in structure to those of *P. caudatum.* Most individuals appear to have two or three, but some show up to six. The vacuoles are aboral and arranged in a straight line parallel to the long axis of the body. Commonly one is found in the anterior region and the

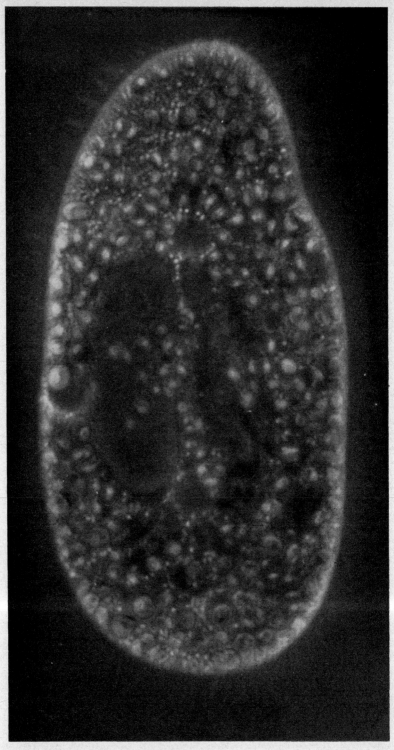

Fig. 8. (*See facing page for legend.*)

extra ones rather close together in the posterior region. It has been shown that members of a given race possess a varied number of vacuoles and there does not appear to be any correlation of number of vacuoles with size of the animals. As with size, one cannot depend upon actual numbers of contractile vacuoles as a species character since some show only two vacuoles and are therefore similar to *P. caudatum* in this respect.

In addition, races of *P. caudatum* have been reported in which extra contractile vacuoles occur (Bhatia, 1923; DeGaris, 1927; Wenrich, 1928 a, and Wichterman, 1946 c). The radial canals of the contractile vacuole are generally longer and more conspicuous than in *P. caudatum*. The cytopyge is ventral and about midway between the posterior end of the oral groove and the posterior end of the body.

The macronucleus of *P. multimicronucleatum* is generally larger than that of *P. caudatum*. It is somewhat ellipsoidal or kidney-shaped but occasionally irregular and at times very long with one irregular edge. Of a compact type, it is composed of chromatin granules and the entire structure stains more lightly than the other species, even with the Feulgen nucleal reaction. The most important diagnostic character of the species is the size and variable number of micronuclei. They are extremely small, of the vesicular type, and measure 0.7–2.5 microns. Powers and Mitchell (1910) describe the number of micronuclei from two to seven, Köster (1933) from two to eight—occasionally without micronuclei—Woodruff (1921) from six

to nine while Landis (1925) gives four as the typical number. In two races maintained by the present author, the typical number of micronuclei is three and the size larger than originally stated, namely 2.5 microns. In these races they were large enough to be detected with the 16 mm. objective of the microscope in Feulgen preparations. It seems, therefore, that the number of micronuclei is variable for the species but that it is likely to be constant in a given race. They are frequently found together— sometimes in a row—some distance away from the macronucleus. Trichocysts are present. The species is worldwide in distribution and commonly found in ponds and bodies of stagnant and fresh water.

"BURSARIA" GROUP

4. **Paramecium bursaria** (Focke) (Figs. 6 [5] and 8).

Individuals of this species may have been first listed under the genus *Loxodes* (*L. bursaria*) by Ehrenberg (1838) but Kahl (1930) is of the opinion that the *Loxodes bursaria* of Ehrenberg more closely resembles a species of *Climacostomum* than a species of *Paramecium*. One cannot be certain of the organism from the early descriptions. However, we are indebted to Focke (1836) for the adequate description of a green *Paramecium* which he called *Paramecium bursaria* (bursa = pouch). Claparède and Lachmann (1858–60), Kent (1880–82), and Schewiakoff (1896) recognized this form and redescribed it in their monographs. In her

---

Fig. 8. Photograph of living specimen of *Paramecium bursaria* (× 1300) as taken through Spencer phase microscope (bright contrast). Observe cilia and large number of zoöchlorellae in body, trichocyst layer beneath pellicle and extruded trichocysts to right of animal. Also cytopharynx (*right center*), large macronucleus (*left center*), smaller micronucleus (*right of macronucleus and above cytopharynx*), anterior and posterior contractile vacuoles. (Richards)

detailed account of conjugation, Hamburger (1904) set the species on a solid foundation. It is the basic type of the "bursaria" group in which others of similar shape are placed.

Members of the species commonly measure 85–150 microns. Like other members of this group, it is relatively wide and short being obliquely truncated or rounded anteriorly, broadly rounded posteriorly and dorsoventrally compressed. It is sometimes referred to as being foot-shaped, i.e. similar to the imprint of a foot. Bragg (1936) described, with the aid of figures, the great variations in shape, which are likely to depart widely from the characteristic form of the species. The most readily distinguishable species character is its green color due to the presence of many small green unicellular algae, the so-called zoöchlorellae. These zoöchlorellae which live symbiotically with the paramecia are species of *Chlorella* belonging to the *Chlorococcales*. Study of a large number of races by the author shows the zoöchlorellae to be comparatively few in some while most show great numbers of tightly packed algae. The intensity of the green color of the chlorophyll varies with the different races. Oehler (1922b), Wichterman (1941, 1943, 1946 a, 1947 b, 1948 b), and others have reported races of *P. bursaria* which were green but became colorless when all the zoöchlorellae disappeared. Such colorless forms appear perfectly healthy and normal in all respects and grow well in culture.[3]

The ventral oral groove extends about one-half the length of the body and near its posterior end is the mouth and cytopharynx. Two contractile vac-

uoles with radial canals are located as follows: one anteriorly often near the middle of the body and the other posterior. The contractile vacuoles empty to the outside by means of pores.

The large, smoothly ellipsoidal or kidney-shaped macronucleus is of a compact type with large chromatin granules which stain intensely. It is located in or slightly above the center of the body. The single, lens-shaped micronucleus measures approximately 7 microns in length and is of a compact ("caudatum") type staining less intensely than the macronucleus. The micronucleus is nearly always found close to the anterior end of the macronucleus and frequently impressed into it.

Cyclosis is characteristically rapid in this species. The rows of cilia are more closely set than in the other species; those at the extreme posterior tip being longer than elsewhere. Like the cilia, trichocysts appear to be more concentrated per unit of area than in members of other species.

P. bursaria is characteristically left spiraling but can spiral in either direction (Fig. 9A).

This species is world-wide in distribution and commonly found associated with vegetation in ponds, streams, and large bodies of fresh water.

5. **Paramecium trichium** (Stokes; emend Wenrich) (Fig. 6 [8]).

P. trichium is the smallest of the species, the majority of specimens measuring 70–90 microns; Stokes' measurements average 83 microns. It was originally described by Stokes (1885) who was so impressed with the number and prominence of the trichocysts that this feature is incorporated in the name. We are indebted to Wenrich (1926) for a detailed restudy of this species with the result that it now rests upon a firm

---

[3] Races of colorless *P. bursaria* devoid of all zoöchlorellae and which were once green have been maintained in continuous culture by the writer for over five years. They are grown in the same lettuce infusion as green ones and are exposed to light like the others.

foundation. The general shape is like that of others of the "bursaria" group except that in many specimens, the anterior end is rounded and less obliquely truncated. However, there appears to be considerable variation in shape. Some races have sides parallel or nearly so, yet others may show the anterior or the posterior regions widest. The ventral, oral groove is wide anteriorly and tapers gradually in an oblique manner to the mouth which is found usually slightly anterior to the middle of the body on its right side. Leading from the mouth is the comparatively long cytopharynx which extends well into the posterior region of the body.

The structure of the contractile vacuoles is of important taxonomic value and is considered in greater detail on page 70. There are two contractile vacuoles in *P. trichium,* one located anteriorly and one posteriorly. In the members of this species examined by the author, the vacuoles appear nearer the anterior and posterior extremities of the body than in other species. Characteristic of this species and differing from all others is the presence of conspicuous vesicular-fed vacuoles instead of canal-fed ones. Stokes reported that the two contractile vacuoles, instead of being placed one in each body-half as in *P. bursaria,* are here anterior and close

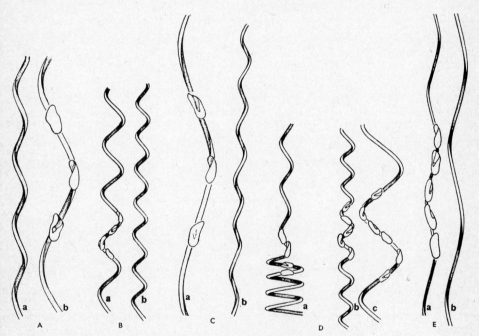

Fig. 9. Spiraling in Paramecium. (A) Normal left spirals of *Paramecium bursaria;* (a) extremely narrow spiral of greatest speed; (b) extremely long spiral of about average speed. (B) Spiral paths of *Paramecium trichium;* (a) spiral of maximum size; (b) short narrow spirals of greatest speed. (C) Spiral paths of *Paramecium calkinsi;* (a) maximum size of characteristic right spiral; (b) short narrow right spirals of greatest speed. (D) Spiral paths of *Paramecium polycaryum;* (a) right spiral changing to left; (b) average left spiral; (c) left spiral of maximum size. (E) Spiral paths of *Paramecium woodruffi;* (a) left spiral of greatest speed; (b) left spiral, average size. (Bullington)

together and contracting quickly with the one beginning to form again almost before the completion of the other's systole. Wenrich (1926) reported the discovery of the second posterior contractile vacuole. Stokes probably saw two vesicles of the anterior contractile vacuolar apparatus and overlooked the posterior one. This is possible because the vacuole is small and the posterior region of *P. trichium* is thickest and contains numerous dark granules and opaque crystals. A detailed account of this unique contractile vacuolar apparatus is given by King (1928). Briefly, each anterior and posterior apparatus consists of feeding vesicles, contractile vacuole, excretory tube, and pore. The contractile vacuole is surrounded by a series of smaller vacuoles or vesicles which coalesce and join the contracting vacuole (Fig. 34). The contractile vacuoles are found in the endoplasm and move about slightly. Each communicates to the exterior by means of a long convoluted tube which ends in a small pore on the dorsal surface of the body between ciliary rows (Fig. 33). The cytopyge is subterminal, very close to the extreme posterior end of the body, egestion being in consequence easily observed in this species.

The macronucleus is ellipsoidal or kidney-shaped and found generally in the center of the body. Instead of being composed of chromatin granules of uniform size as in the other species, the surface shows a pocketed condition. Each pocket contains a globule of nuclear material differing in stainability from the nuclear matrix surrounding it. The compact micronucleus, which stains intensely, measures from 4–7 microns, is broadly oval or spherical, and of the "caudatum" type. It is usually found close to the macronucleus but is seen occasionally a considerable distance from it.

Wenrich noted the marked flexibility of the body which is confirmed here. Stokes reported that trichocysts are very abundant, and are so arranged that they elevate the cuticular surface of the body into minute hemispherical bosses. While other species show this crenated profile to a greater or lesser degree, it is especially conspicuous in *P. trichium* and the trichocysts are prominent and thickly set. Cilia are of moderate length except for those at the posterior extremity which are longest. This species spirals characteristically to the left (Fig. 9B).

P. trichium is found in ponds and bodies of stagnant water and streams of fresh water.

6. Paramecium calkinsi (Woodruff) (Figs. 6 [6] and 10).

This species was discovered in 1919 by Woodruff (1921) and named in honor of Prof. G. N. Calkins. The general shape is characteristic of the "bursaria" group with the broadest part anterior. Members average 120 microns in length and 50 microns in breadth and are flattened dorsoventrally. However, it is probable that racial size differences will be found to occur when new races are discovered. The cell surface in profile shows a slightly crenulated outline resembling *P. trichium* in this respect. Each of the small elevations represents the position of a trichocyst and the depressions between represent the points of origin of cilia. Trichocysts are evenly distributed in the cortical layer of protoplasm except at the anterior end where they are more dense. Cilia are of moderate length except at the extreme posterior tip where they are longest. In contrast to all other species which spiral characteristically to the left when swimming, *P. calkinsi* spirals to the right on its long axis (Bullington 1925, 1930)

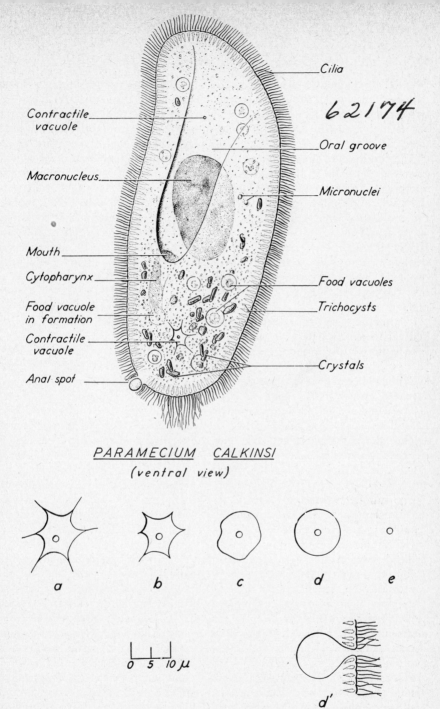

PARAMECIUM   CALKINSI
(ventral view)

FIG. 10. (_Top_) (× 750.) (_Bottom_) Shape changes of contractile vacuole when specimens of _Paramecium calkinsi_ are examined in original environmental water of great salt concentration. Stellate condition in diastole (_a_) and contraction (_b, c, d_) through _pore_ (_e_). Soon after assuming spherical shape in _d_, vacuole almost immediately contracts in systole leaving only persistent pore shown in all stages. Side view of stage _d_ is shown in _d'_ with contractile vacuole opening to outside of body through the trichocyst layer and pore. (Wichterman)

29

(Fig. 9C). A detailed account of the ecology, cultivation, structural characteristics and mating types of this species has been given by the author (Wichterman, 1951).

The unusually long, ventral, oral groove is very wide at the truncated anterior end. It tapers diagonally to the right side and ends slightly below the center of the body. The mouth is located at the posterior end of the oral groove and leads into the cytopharynx which extends well into the posterior end of the body. The cytopyge is subterminal near the posterior extremity of the animal.

Contractile vacuoles are two in number, each with radial canals and found in the anterior and posterior parts of the body respectively. Wenrich (1928 a) and Wichterman (1951) observed that the radial canals of the contractile vacuoles are not as well defined as in other species.

The large, compact-type macronucleus is ellipsoidal or kidney-shaped and is found in or slightly above the center of the body.

The number, structure and type of micronuclei are important diagnostic characters in this species. Two micronuclei commonly are found in or near a slight depression of the macronucleus. Woodruff reports that frequently there are two tiny depressions each with a micronucleus; sometimes the micronuclei lie at a considerable distance from the macronucleus. The relative size and structure of the micronuclei are similar to those of *P. aurelia,* measuring approximately 3–5 microns and of the vesicular ("aurelia") type. Races with a single micronucleus have been observed by Metz (1947) and the author.

Wenrich (1928 a) reported that specimens found in fresh water when gradually transferred to increasing concentrations of sea water lived and behaved normally when pure sea water was finally reached. If transferred directly from fresh water to sea water, they died. He found that when specimens were transferred directly from brackish pond water to fresh water, no harmful results could be noticed. Bullington (1930) however could transfer his race directly from fresh water to sea water. See also Spencer (1924).

This species is found in fresh-water ponds, brackish water, and sea water of high salinity content (Wichterman, 1951).

7. Paramecium polycaryum (Woodruff and Spencer) (Fig. 6 [7]).

This species was discovered in 1922 in some material collected in Louisiana and reported by Woodruff and Spencer (1923). They described the organism as similar in form and structure to *P. calkinsi* but that the former has more than two micronuclei. Wenrich (1928 a) reported that the length ranges from 70–110 microns. In comparison with *P. calkinsi* which it more closely resembles, *P. polycaryum* is smaller and a little wider. It is slightly broader at the truncated anterior end and rounded at the posterior end. Like other members of the "bursaria" group, it is dorsoventrally flattened.

The ventral oral groove is wide at the truncated anterior end and extends to the approximate center of the right side of the body. Coming from the posterior region of the oral groove is the centrally located mouth which leads into the cytopharynx. Opton (1942) reported the presence of 15 ciliary rows which comprise the penniculus of *P. polycaryum* in contrast to the eight ciliary rows commonly found in members of the "aurelia" group. The cytopyge is subterminal and close to the posterior extremity of the body.

Contractile vacuoles are two in number, each with radial canals and situated in the anterior and posterior ends respectively.

The macronucleus is of a compact type and similar to others in the same group—somewhat ellipsoidal to kidney-shaped. It is located in or slightly above the center of the body. Of diagnostic value and embodied in the specific name are the micronuclei which measure approximately 3 microns. These are of the vesicular ("aurelia") type and vary from three to eight although four appears to be the typical number. They are found close to the macronucleus. This variation in number of micronuclei but similarity in structure reminds one of *P. multimicronucleatum*. In this respect, Woodruff (1921, 1921 a) makes the point that this animal holds the same position in the "bursaria" group as *P. multimicronucleatum* holds in the "aurelia" group.

Trichocysts are present and cilia are of moderate length except at the posterior extremity where they are longest. This species spirals characteristically to the left (Fig. 9D).

Careful cytological studies upon this species are needed.

Found in fresh water.

8. **Paramecium woodruffi** (Wenrich) (Fig. 6 [4]).

P. woodruffi is the largest of the "bursaria" group of paramecia. The length ranges from 120–210 microns with most individuals measuring between 160–180 microns. It was discovered in 1927 in brackish water at the Marine Biological Laboratory, Woods Hole, Massachusetts by Wenrich (1928 b) and named in honor of Prof. L. L. Woodruff.

In general morphology, it belongs to the "bursaria" group being obliquely truncated anteriorly, rounded posteriorly and though dorsoventrally flattened is more cylindrical than other members of this group. Wenrich (1928 b) noted that the ciliate was slightly yellow but that the color disappeared when the organism was placed in fresh-water cultures. The yellow color was partly restored when a small amount of sea water was again added. It appears, therefore, that in their native habitat, the slightly yellowish color is characteristic.

The ventral oral groove is unusually long, as it is in *P. calkinsi*, extending posteriorly about two thirds of the body length. At the posterior end of the oral groove is found the oval mouth which leads into the cytopharynx. The cytopyge is found near the extreme posterior tip of the body.

Two contractile vacuoles with radial canals are present, one in the anterior and the other in the posterior third of the body. According to Wenrich, each contractile vacuole possesses 10–12 radial canals—a number larger than is found in most other species. There is a single outlet pore for each contractile vacuole.

The large, compact-type macronucleus is ellipsoidal and found usually anterior to the middle of the body. Members of the species are reported as having a variable number of micronuclei—up to eight—but most specimens possess three or four. They are of the vesicular ("aurelia") type and measure approximately 4 microns. The positions of the micronuclei are characteristically varied as noted by Wenrich. They may be found anywhere in the body instead of the usual place close to the macronucleus as in the other species. However, the majority of micronuclei in a given animal are to be found in the vicinity of the macronucleus.

Trichocysts are present and the cilia are of moderate length except for the

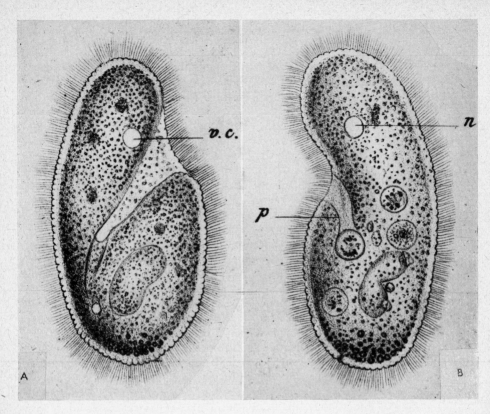

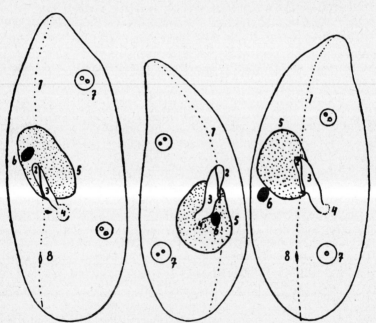

Fig. 11. (*See facing page for legend.*)

tuft of longer cilia at the posterior tip of the body.

Specimens were cultured in fresh water by Wenrich who noted a greater variation in size than when cultured in brackish water. He believed that fresh-water cultures to which a small amount of sea water has been added appear to be more favorable than cultures of pure fresh water. Members of this species were discovered in the same body of water which contained *P. calkinsi*. Bullington (1930) cultured *P. woodruffi* in both sea water and fresh water. This species spirals characteristically to the left (Fig. 9E).

Found in brackish water but can live in fresh water.

F. Uncertain Species of Paramecium

1. Paramecium putrinum (Claparède and Lachmann) (Fig. 11A, B).

In their large monograph, Claparède and Lachmann (1858–60) described a new species of *Paramecium* from decaying or putrid vegetable matter. This feature of the organism's habitat was used for the choice of the specific name —*Paramecium putrinum*. These investigators stated that *P. putrinum* measured 87–122 microns in length, that it corresponded closely in form with *P. bursaria* but differed from that species by the presence of only one contractile vacuole, located anteriorly, and the absence of trichocysts. The cytopyge is described as being located on the ventral surface but not as far forward as in *P. aurelia*. Their figures (Fig. 11A, B) and description show the macronucleus to be kidney-shaped and located in the

posterior part of the body below the oral opening. A micronucleus is located near the macronucleus. Claparède and Lachmann described with figures the pocketed condition of the macronucleus as Wenrich (1926) had done for *P. trichium*. However, their interpretation of it was very ingenious but erroneous. They believed that the small droplets of macronuclear substance were early stages of embryo formation and they contrasted these so-called stages in *P. putrinum* with those of *P. bursaria*. In reality, these "embryos" appear to have been parasitic stages of *Suctoria*—some may even have been small ciliates that were ingested. It is noteworthy to add that these investigators believed there were not enough structural differences to warrant separation of *P. putrinum* from *P. bursaria* but that the outstanding specific distinction was shown in the different so-called "embryos." It appears, therefore, that had they been aware of their fallacious interpretation of "embryonic development," they would not have created the species *P. putrinum* on a strictly morphologic basis.

A number of European workers have reported on ciliates called by them *P. putrinum*. Bütschli (1876) stated that the *P. putrinum* he encountered and found in conjugation possessed trichocysts as well as an anterior and a posterior vesicle-fed contractile vacuole. This species may well have been *P. trichium*. Schewiakoff (1893) described *P. putrinum* as measuring 120–140 microns, with an anteriorly located vesicle-fed contractile vacuole. The size appears to be too great for it to have been *P.*

FIG. 11. Uncertain species of Paramecium. Photographed from original sources. (A, B) *Paramecium putrinum* (Claparède and Lachmann); (p) esophagus; (v.c. and n) contractile vacuole. (C) *Paramecium nephridiatum* (Gelei); (1) ventral suture line; (2) mouth; (3) cytopharynx; (4) food vacuole in process of formation; (5) macronucleus; (6) micronucleus; (7) contractile vacuoles showing excretion pores; (8) anus.

trichium. Roux (1901) reported that his *P. putrinum* possessed an anterior and posterior contractile vacuole each with radiating canals but without trichocysts. Since this species did not have vesicle-fed contractile vacuoles, it could not have been *P. trichium* but might have been a colorless *P. bursaria*. Roux may have overlooked the trichocysts. Doflein in his "Lehrbuch der Protozoenkunde" (4th ed., 1916) states, "*P. putrinum* Cl. u. L. ist kürzer und plumper als die vorigen, (*P. aurelia*) sonst sehr ähnlich, Länge 80–140 μ." With this indefinite description, a figure is shown of it which is nearly identical to the original except small vesicles have been drawn around the contractile vacuole. None is shown around the contractile vacuole in the original illustration. In another section Doflein shows in great detail and accuracy many conjugation stages in a species called *P. putrinum* but the stages are strikingly similar to those of conjugation reported by Diller (1934) and Wichterman (1937 a) for *P. trichium*. Alverdes (1923) described *P. putrinum* as having one contractile vacuole with radial canals near the middle of the body. Wenrich (1926) is of the opinion that the species described by Schewiakoff, Roux, and Alverdes are all different species but that the one of Bütschli is probably *P. trichium*.

The *P. putrinum* of Claparède and Lachmann resembles *P. trichium* in some respects such as the conspicuous crenulated body surface and pocketed macronucleus. However, specimens of other genera also show these characters. Although Claparède and Lachmann describe an oral groove for *P. putrinum*, their figures do not show one, so that they may not have been dealing with a species of *Paramecium*. On the other hand, it is possible that these two workers overlooked the posterior contractile vacuole as Stokes (1885) appears to have done for *P. trichium*. In that event, one cannot be certain of the single contractile vacuole as a species character. Perhaps the *P. putrinum* of Claparède and Lachmann may have been a race of *P. bursaria*, which it is claimed to resemble so closely, but was devoid of all zoöchlorellae. The author's experience with certain races of *P. bursaria* which have lost their zoöchlorellae is that frequently large numbers of opaque crystals accumulate in the posterior end of the paramecia (Wichterman, 1941, 1948 c). Crystals may have prevented Claparède and Lachmann from finding the posterior contractile vacuole. Their report of the absence of trichocysts for this species may be in error. They speak of observations on other species of *Paramecium* which at certain times did not show trichocysts. To check this point, the author applied commonly used dyes containing acid to living specimens of *P. bursaria* and *P. trichium*—species which *P. putrinum* is supposed to resemble. In most cases, the trichocysts of each were expelled but in a number of cases, the animals were killed before trichocysts could be discharged.

Wenrich (1928 a) stated that Europeans apparently apply the name of *P. putrinum* to any form that is like *P. bursaria* in shape but without zoöchlorellae. *P. putrinum* is listed as a valid species in the textbooks of protozoölogy by Calkins (1933) and Kudo (1947). It is listed as an uncertain species by Kahl (1930) in his taxonomy of the ciliates. It is not included in the survey of the genus by Wenrich (1928 a) and is rejected by Ludwig (1930 a) who believed it to be a degenerated, univacuolar form in an abnormal environment.

After reading the inadequate description of *P. putrinum* by Claparède and Lachmann wherein great emphasis was placed on the erroneous concept of "embryo-formation" as a species character and the contradictory accounts of subsequent authors it is the writer's opinion that the species *P. putrinum* be rejected on the basis of incomplete and inconclusive observation.

2. Paramecium nephridiatum (Gelei, 1925) (Fig. 11C).

Gelei (1925 a) described a new species which he called *P. nephridiatum* after the Greek word *nephros* meaning kidney. It is reported as being similar to *P. caudatum* in a number of respects such as being quite narrow in the anterior region but widely rounded posteriorly and more yellow in color.

According to Gelei, it is as long as the longest "caudatum" forms but always thicker than the thickest of them. Kalmus (1931) who places it in the "bursaria" group, gives the size as being over 150 microns in length. Actually in body shape, it appears to be a combination of the "aurelia" and "bursaria" types. The macronucleus is described as being rounded and the single micronucleus of medium size and oval in shape. Trichocysts are reported as few in number.

Gelei was impressed with a number of features which he believed his new species possessed that were not found in any others. The first, which he apparently considered most important, concerned the contractile vacuole. In *P. nephridiatum* (Gelei, 1925 a), the contractile vacuole with its radial canals is said to possess two excretory pores instead of one. In this regard, Kalmus (1931) reports that he has at times been able to observe two pores at one contractile vacuole in *P. caudatum* and King (1935) has seen two in *P. aurelia*. Kahl (1930) is of the opinion that *P. bursaria* also possesses two excretion

pores for each vacuole. This feature of double pores scarcely can be used for valid species identification if it occurs in *P. bursaria* and *P. caudatum*. Wenrich (1928 a), on the other hand, studied eight species of paramecia including *P. bursaria* but did not find two outlet pores per vacuole in any species.

P. nephridiatum was first described by Gelei as lacking the characteristic tuft of longer cilia at the posterior end of the body but he recognized their presence in later research (Gelei, 1926 a, b). Another feature stated for this species is the presence of two cilia emanating from each ciliary field. Later, Gelei recognized examples of double cilia in single ciliary fields of *P. caudatum* so that this species character may be challenged. His original paper in Hungarian presents several conflicting accounts in regard to morphological characters.

It appears that *P. nephridiatum* is in need of careful cytological study especially in regard to nuclear structure and general body shape. Although the species is listed in the works of Ludwig (1930 a), Kahl (1930) and Kalmus (1931) it is not included in the textbooks of protozoölogy by Calkins (1933) or Kudo (1947). Kahl reports that it is rare in Germany; it has not been reported from the United States. It is reported as being found in water of low-salt concentration (brackish).

On the basis of structure, as given in the literature, it is the conviction of the author that *P. nephridiatum* cannot be accepted as a valid species of *Paramecium*.

3. Paramecium traunsteineri (Kahl, 1930) (Fig. 12A).

This species is listed by Kahl (1930) after having been observed by Baumeister with the statement that the description has not yet been published. Kahl's figure (Fig. 12A) is of a slender organism approximately 120 microns in length and with a somewhat obliquely truncated anterior end and rounded posterior end. The oral groove extends well beyond the center of the body. The macronucleus is roughly spherical and anterior to the middle of the ciliate. A spherical micronucleus measuring 3 microns is close to the macronucleus.

In this species, emphasis is placed upon the complete absence of radial canals or vesicles around the anterior and posterior contractile vacuoles.

Further cytological studies are needed before this species can be accepted as a valid one.

4. Paramecium chilodonides (Kahl, 1930) (Fig. 12B).

This species, like the preceding one, was first observed by Baumeister but a description of it was not published. According to Kahl, it measures about 120 microns in length and is comparatively wide with parallel sides (Fig. 12B). It is flattened dorsoventrally and truncated anteriorly to the left at an angle of 45 degrees. Trichocysts are said to be numerous.

Kahl's only figure shows an exceedingly long oral groove which extends well into the posterior region of the ciliate. There are two contractile vacuoles each with radial canals. The macronucleus is roughly spherical and located above the anterior half of the body. Near it is the single micronucleus.

Since the organism is about the same size and shape of *P. bursaria* it is possible that we have here a colorless race of this species. More information is needed before this organism is acceptable as a new species.

5. Paramecium ficarium (Kahl, 1930) (Fig. 12C).

Under the heading of uncertain species, Kahl (1930) adds a few lines wherein he states his uncertainty of this so-called species which he discovered in 1928. No description or size is given but Kahl reports that the organism is found in salt water and that conjugation occurs. His inadequate figure (Fig. 12C) shows a club-shaped form widest at the anterior end. Also shown are two contractile vacuoles, oral groove, macronucleus, micronucleus, and caudal cilia.

It appears to be a deformed *Paramecium* of the "caudatum" type but in the absence of an adequate description, it most certainly cannot be listed as a valid species and should be rejected.

6. Paramecium pseudoputrinum (Kahl, 1930) (Fig. 12D).

Kahl (1930) reports that this species was named by Baumeister but that a description has not yet been published. It is compared with *P. trichium* which it is supposed to resemble except *P. pseudoputrinum* is considerably wider and contains few spindle-shaped trichocysts.

Kahl's figure of *P. pseudoputrinum* (Fig. 12D) shows the ciliate to be extremely wide of the "bursaria" group and with a long oral groove extending beyond the middle of the body. Two vesicle-fed contractile vacuoles are represented, as well as the long caudal cilia. The ellipsoidal macronucleus is centrally placed and the large oval micronucleus is shown near it.

No other information is given and on the basis of this inadequate description, it seems likely that the species in question is simply a race of plump individuals belonging to *P. trichium*. For these

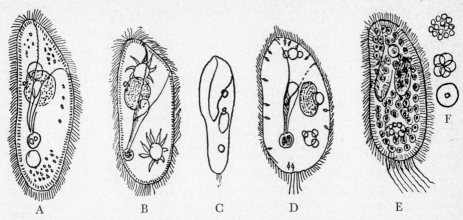

FIG. 12. Uncertain species of Paramecium. Photographed from original sources. (A) *P. traunsteineri*. (B) *P. chilodonides*. (C) *P. ficarium*. (D) *P. pseudoputrinum*. (E) *P. chlorelligerum*. (F) Three phases in the formation of the contractile vacuole in *P. chlorelligerum*. (Kahl)

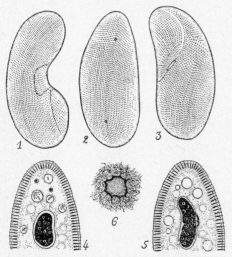

FIG. 13. *Paramecium duboscqui* n. sp. (Chatton and Brachon). (1–3) Shape and ciliature in different views of ciliate with two contractile vacuole openings in 2. (4) Anterior end of *P. duboscqui sphaerocaryum* showing the two vesicular micronuclei anterior to the macronucleus. (5) *P. duboscqui bactrocaryum* showing the two bacilliform micronuclei to the right of the macronucleus. (6) Unique contractile vacuole and its "spongiome."

reasons *P. pseudoputrinum* of Kahl is unacceptable as a valid species of *Paramecium*.

7. Paramecium chlorelligerum (Kahl, 1930) (Fig. 12E).

Kahl described this species as measuring 80–100 microns in length, of a green color, and similar to *P. bursaria* but differing from it in being slender. Zoöchlorellae are not mentioned as being the algae within the paramecium which give it the green color. The two vesicle-fed contractile vacuoles said to be present in this species (Fig. 12F) are supposed to resemble those of *P. trichium*. The macronucleus is described as being oval with a spherical micronucleus near it, measuring 3–4 microns in diameter. It is reported as being found in the detritus of a marsh.

The description as given above by Kahl is inadequate. Since it falls within the size limits of *P. trichium*, it may be a race of this species in which the members had recently ingested green organisms. If Kahl's observations on the vesicle-fed contractile vacuoles are correct, the organism could hardly be considered *P. bursaria* which possesses radial canals. Because of insufficient information pertaining to the structure of this ciliate, it is the author's opinion that *P. chlorelligerum* is unacceptable as a valid species.

8. Paramecium duboscqui (Chatton and Brachon, 1933) (Fig. 13).

In 1933, Chatton and Brachon briefly described a new species which they named *P. duboscqui*. Its length was given as 100–150 microns and the organism was said to differ from all other species in possessing a distinctly kidney-shaped body when viewed laterally.

The macronucleus was described as being ellipsoidal, sometimes notched or unequally divided into two parts with some specimens showing small balls of chromatin as a result of macronuclear fragmentation.

Specimens possess two micronuclei which differ in two types of the species. In one type, called *Paramecium duboscqui sphaerocaryum,* two smaller vesicular micronuclei are present which resemble those of *P. aurelia.* However, the micronuclei are found in the anterior end of the animal and at a considerable distance from the macronucleus. In the remaining type called *Paramecium duboscqui bactrocaryum,* the two micronuclei are curved, slender, bacilliform structures which exist close to the macronucleus.

Two contractile vacuoles are described in which one is located anteriorly and the other posteriorly with both opening on the dorsal surface of the body. They are of a unique type and differ in structure from those found in other species of paramecium (Fig. 13 [6]).

After three years in culture, Chatton and Brachon (1936) reported the passage of type *"sphaerocaryum"* into type *"bactrocaryum."* They considered *P. duboscqui bactrocaryum* as the specific stable type and *P. duboscqui sphaerocaryum* to be merely a purely somatic variation due to external factors.

On the basis of their description, the ciliate *P. duboscqui* cannot be considered as a valid species of the genus *Paramecium.*

Chapter 3

Morphology and Cytology of Paramecium

A. General Morphology
  1. Size, Shape, and Color
  2. Mouth, Vestibulum, Pharynx, and Esophagus
  3. Anus
B. The Ectoplasm, Endoplasm, Pellicle, Silverline, Fibrillar, and Neuromotor Systems
C. The Cilia
D. The Trichocysts
E. The Contractile Vacuoles
F. Cytoplasmic Inclusions
  1. Mitochondria (Chondriosomes, Cytomicrosomes, etc.)
  2. Golgi Bodies (Apparatus) and the Vacuome
  3. Crystals
  4. Food Vacuoles and Digestive Granules
  5. Other Cytoplasmic Inclusions (Granules, Chromidia, Chromatophores, etc.)
G. The Nuclei
  1. The Macronucleus
  2. The Micronucleus
  3. Nuclear Variation

A. General Morphology[1]

1. Size, Shape, and Color.

Paramecia are microscopic organisms, each consisting of a single cell and visible to the naked eye as a minute, elongate body. Members of the species range in length from approximately 80

microns, as in *P. trichium,* the smallest, to approximately 350 microns, as in *P. multimicronucleatum,* the largest. When a large number of measurements are taken of the individuals of a given species of paramecium and the data treated statistically, it is found that an average size of the species can be determined. However, racial differences exist in respect to size and it is possible for the mean lengths of races of different species to overlap such as in *P. caudatum* and *P. multimicronucleatum* although the latter is generally larger.

It is well known that certain environmental factors markedly influence size and shape. Food is one of these factors. Wallengren (1902) reported that the sizes of paramecia which were undergoing starvation changed. This observation was corroborated by Lipska (1910) and others. Zingher, Narbutt, and Zingher (1932) reported an average fluctuation of 62 microns between starved and normal strains of a large race of paramecium. Jennings (1908) observed that when the food in a culture decreased rapidly, the paramecia became smaller. He reported that when animals were placed in fresh hay infusion they became at first conspicuously shorter and thicker. With one exception, Jennings reported that his races of paramecia did not undergo fission until they had become thinner and more elongated again (Jennings and Hargitt, 1910). According to Jennings, increased

[1] For surface, volume, and weight determinations of paramecium, see page 120.

nutrition increases the body length but the result is not always the same. Increased nutrition produces two main effects—to increase directly the size of the adults and to accelerate muliplication through fission. In all cases, increase of nutrition increases the breadth (thickness of the body); decrease of nutrition decreases breadth and the change is immediate and very marked. He found that by changes in nutrition, the mean length of a given race of paramecia in culture could be changed in a week from 146 microns to 191 microns; the breadth from 31 microns to 54 microns. These differences due to environmental action are not inherited. Preconjugation stages of paramecia and other ciliates are conspicuously smaller than typical vegetative ones. In fact one method of inducing conjugation is by the immediate termination of the food supply in a culture of well-fed animals, which in turn results in smaller organisms. Very young or recently divided animals are of course much smaller than their parents. Jones (1933) reported that average length decreased as division rate increased, and vice versa, prior to the inanition period. In a study of population and size changes within a pure line of P. multimicronucleatum, he found that food supply was the variable which chiefly controlled both population concentration and size within the culture.

We are indebted to Jennings for his statistical treatment of size and shape of his "caudatum" (P. caudatum) and "aurelia" (P. aurelia) races. In his study of these two groups (species) Jennings (1908) made the remarkable prediction which follows, ". . . If my impression is correct, most lines will have a mean length either below 145 microns or above 170 microns; rarely will lines be found whose mean falls between these values. Such at least has been my experience in a large amount of work. Furthermore, I am inclined to believe that those belonging to the smaller group (mean length below 145 microns) will be found to have as a rule two micronuclei; those belonging to the large group but one micronucleus. This matter is worthy of special examination" (Jennings, p. 500, 1908). This view was made in contradiction to that of Stein, Claparède, Balbiani, Koelliker, Engelmann, Bütschli, Gruber, and Kent —outstanding protozoölogists of the late nineteenth century who reunited the two species under the name of P. aurelia and eliminated P. caudatum. Maupas (1888) however, recognized the species differences based upon micronuclear characters.

In carefully controlled experiments, Loefer (1938) showed a close correlation of size to pH of the culture medium in bacteria-free cultures of Paramecium bursaria. He reported that organisms grown at pH 6.0–6.3 were the longest in the series and averaged 129.3 microns while those in more acid or alkaline media were considerably shorter. The shortest specimens averaged 86–87 microns in length from cultures at pH 7.6–8.2 (Table 1; Fig. 14).

Another factor having a direct bearing on size is the temperature of the environment. Jollos (1913 a, b, 1914) found in numerous experiments that most of the races under investigation produced small animals if the temperature of their environment was raised.

In spite of environmental conditions, which of course must be taken into consideration, size is a fairly constant species character when measurements of large numbers of animals are treated statistically.

Since P. caudatum is perhaps the best-known representative of the genus, we shall discuss its morphology in considerable detail and speak of important

Table 1

<small>AVERAGE SIZE OF *Paramecium bursaria* IN RELATION TO pH (Loefer, 1938)</small>

Initial pH	Final pH	Length in Microns	Coefficient of Variation-length	Width	Remarks
4.6	4.6	97.5 ± 3.32	31.5	51.4 ± 1.03	no growth
5.1	5.3	110.6 ± 1.04	10.7	48.4 ± 0.462	growth
5.5	5.7	126.6 ± 0.687	6.2	43.2 ± 0.280	”
6.0	6.3	129.3 ± 0.829	7.7	44.4 ± 0.350	”
6.5	6.8	115.8 ± 0.729	7.3	44.4 ± 0.448	”
6.9	7.2	102.9 ± 0.632	7.0	44.2 ± 0.428	”
7.4	7.4	100.2 ± 0.786	9.0	44.7 ± 0.611	”
7.7	7.6	86.0 ± 0.784	9.8	46.0 ± 0.561	”
8.2	8.0	87.0 ± 0.780	9.4	54.7 ± 0.781	”
—	—	Avg. = 106.2	—	Avg. = 46.8	—

differences in other species by comparison. It is easily seen with the naked eye when observed in a drop of fluid over a black background. It appears light gray or white, measuring commonly between 170–290 microns, being slightly smaller than the dot of the period at the end of this sentence. *P. caudatum* is commonly referred to as being slipper- or cigar-shaped. Under comparatively low powers of the microscope, one can observe it to be about four times as long as broad and somewhat cylindrical with distinctly different ends. The forward moving anterior part is slender with a blunt or rounded end while the posterior end is more noticeably pointed or cone-shaped (Fig. 15). The widest part of the organism is in a region about two-thirds body length behind the anterior end. The body of the animal is asymmetric in form, showing a well-defined oral or ventral surface and an aboral or dorsal one. On the oral surface is found a broad, shallow, obliquely directed groove, the *oral groove* which extends from the extreme anterior end of the body where it is widest, to the region two thirds of the body length from the anterior end. The entire body, including the oral groove, is covered with fine protoplasmic extensions, the *cilia*, which are set in fairly uniform rows. When the animal is observed laterally or ventrally, the groove can be seen twisting in the direction of a right spiral. Thus the asymmetric ciliate is seen to have well-defined an-

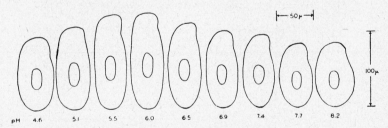

FIG. 14. Relation of pH to size of *Paramecium bursaria*. Diagrammatic reconstructions from camera lucida drawings and size measurements shown in Table 1. Organisms were grown in a Bacto-tryptone medium. The numbers designate initial pH of the cultures from which the ciliates were taken. (Loefer)

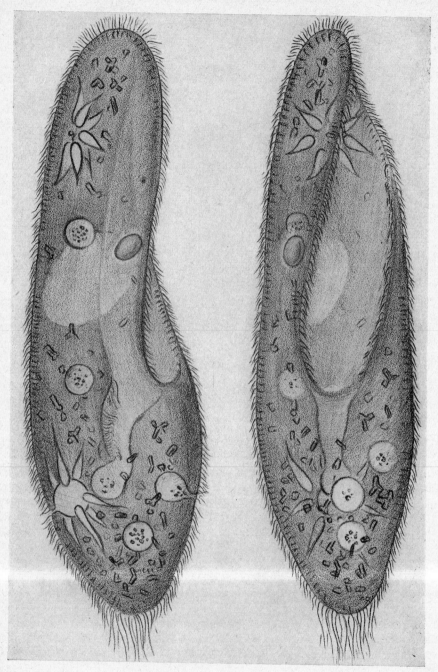

FIG. 15. Lateral view (*left*) and ventral view (*right*) of *Paramecium caudatum*. (× 650.) (Wichterman)

terio-posterior differentiation, dorsal and ventral surfaces and a right (dextral) and left (sinistral) side.

The body form of the remaining species of the "aurelia" group namely *P. aurelia* and *P. multimicronucleatum* are somewhat similar in shape to that described here for *P. caudatum* except that in *P. aurelia* the size is smaller and the posterior end is less pointed. As mentioned earlier all members of the "bursaria" group are somewhat dorsoventrally compressed or flattened. While the oral groove in members of the "bursaria" group varies in relative length depending on the species, the structure in every case is twisted in the direction of a right spiral.

In respect to color, it is well known that *P. bursaria* is green because of the many zoöchlorellae existing in the endoplasm. However, exceptions have been noted by the author and by a number of investigators who have reported zoöchlorellae-free (colorless) clones of this species.

P. woodruffi, as found in its native habitat—brackish water—is of a pale yellow color but when transferred to fresh water, the yellow color disappears (Wenrich, 1928 a, b). The yellow color is partly restored when a small amount of sea water is added to fresh-water cultures.

Occasionally *P. multimicronucleatum,* the largest of the species, appears densely opaque under the low powers of the microscope, while other species appear translucent and somewhat colorless to light gray.

2. Mouth, Vestibulum, Pharynx, and Esophagus.

In all species of *Paramecium,* the mouth or *cytostome,*[2] like the oral

groove, is found on the ventral surface. Leading from the mouth is a tubular structure through which the food passes. This food-passageway has been referred to in the literature by different names which, briefly enumerated, are: (1) the gullet, (2) the pharynx, (3) the cytopharynx, (4) the esophagus; or that it is composed of the following parts: (5) cytopharynx and gullet, (6) cytopharynx and esophagus, (7) vestibulum and esophagus, (8) vestibulum, cytopharynx (pharynx), and esophagus. It is apparent that confusion is certain to exist when the different names are used interchangeably for identical structures. Due to the investigations of Bozler (1924 a, b), Gelei (1925, 1934), Lund (1933, 1941) and Mast (1947), our knowledge of the morphologic details of the pharyngo-esophageal complex in paramecium has been extended in no small measure.

Bozler studied *P. caudatum;* Gelei, *"P. nephridiatum"* and *P. caudatum;* Lund, *P. multimicronucleatum,* and Mast investigated *P. caudatum, P. multimicronucleatum, P. aurelia,* and *P. trichium.* In general, there is agreement in descriptions of the various parts of the food canal and minor differences are probably due to species characters or interpretation. In the light of these newer and more detailed researches, an attempt is made here to give some degree of uniformity to the descriptive terminology of this admittedly complex structure.

The *vestibulum* leads directly into the fixed, oval-shaped opening called the *mouth* (cytostome). Extending directly from the mouth toward the center of

---

[2] A term proposed by Haeckel for the distinction of the oral aperture of unicellular

animals (Gr. *kutos,* cell; *stoma,* a mouth). Bozler prefers to think of the so-called cytostome as a "secondary mouth" opening. He is of the opinion that the true or "primary mouth" is located at the extreme end of the pharyngo-esophageal tube where the developing food vacuole is formed.

the body is the wide *pharynx* (cyto-pharynx). The pharynx then turns sharply posteriad to become the slender, tapering *esophagus*. Thus the esophagus is roughly parallel to the body surface of paramecium except at its posterior extremity. Here the esophagus turns again toward the center of the animal to lead into the forming food vacuole (Fig. 15). If one thinks of the pharynx as resembling the bowl of a pipe, the esophagus would have as its counterpart the stem of the pipe. In shape, the pharynx with the esophagus more closely resembles a sigmoid curve directed posteriorly and located to the right of the midline of the body of paramecium.

Fig. 17. Ciliary fibrillar pattern and the relation of the mouth region to the vestibulum in *P. caudatum*. In the center opening of the vestibulum are the four rows of fibers of the "Vierermembran" showing to the right fibers of the penniculus. (Gelei)

Gelei (1925) gave the name vestibulum[3] to the cytostomal region where the pellicle of the body curves gradually inward. It is seen that Gelei's funnel-shaped vestibulum, in the way it was first used, includes the widest part of the food canal that is frequently referred to by others as the cytopharynx. Gelei's present description of the vestibulum (Figs. 16, 17) consists of the invaginated body-pellicle leading up to the mouth and what is here considered to be the anterior part of the pharynx.

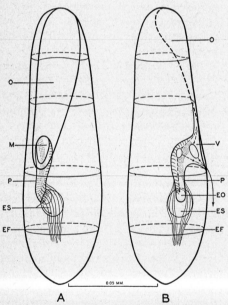

A   B

Fig. 16. Camera outline of *Paramecium aurelia* showing the feeding apparatus and its relation to the rest of the body. (*A*) Oral groove and mouth facing upward. (*B*) Oral groove and mouth facing to the right. (*O*) Oral groove. (*M*) Mouth. (*V*) Vestibulum. (*EO*) Esophageal opening. (*P*) Pharynx. (*ES*) Esophageal sac. (*EF*) Esophageal fibers. (Mast)

[3] When first introduced, Gelei (1925) considered the food-passageway to be composed of two parts, the outer vestibulum and the inner esophagus without reference to a cytopharynx. Later (1934) he used the term vestibulum to include the wider portion of the ciliated funnel which leads into the cytopharynx (pharynx) and which is followed by the esophagus.

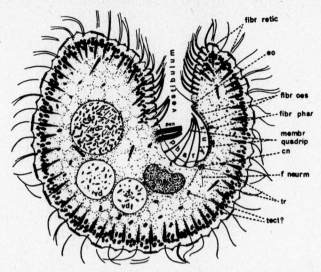

FIG. 18. Diagrammatic transverse section of *P. caudatum* through region of cytopharynx showing relationship of vestibulum to penniculus and "Vierermembran." (Gelei)

The region bears shorter cilia than on the body proper and these with the subpellicular fibrils are seen to be arranged in a crescentic manner around the mouth. This wide vestibulum, which according to Gelei is free of trichocysts, then merges into the pharynx and curves sharply posteriad to join the esophagus. Actually the vestibulum is not sharply delimited from the anterior end of the cytopharynx; indeed, most investigators speak only of the cytopharynx and esophagus so that the term cytopharynx might include the vestibulum of Gelei. Long, powerful cilia, found in the dorsal surface of the pharynx, have been observed by a number of investigators. Gelei (1934) reported the presence of four rows which he calls, "Vierermembran: membrana quadripartita" and Lund (1941) observed at least four rows (Figs. 17 and 18).

A structure called the *penniculus* is found on the left wall of the cytopharynx and spirals through approximately 90 degrees so that its posterior extremity is on the ventral (oral) surface of the esophagus. According to Lund, the penniculus consists of eight rows of cilia arranged in two closely set blocks of four each. Gelei (1925) gave as its function, the forcing of food elements into the body (Fig. 19).

At the posterior end of the pharyngo-esophageal cavity, there is found a blind pouch or pocket first observed by Bozler (1924 a, b). This pocket, called the *esophageal process* by Lund, is described as being heavily ciliated on its ventral surface with the effective stroke of the cilia directed anteriorly. On the other hand, the existence of such a blind pocket in this region is doubted by Mast (1947) who has made a detailed study of this region in living organisms. Bozler is the first to have called attention to the presence of long fibers found in the posterior region of the esophagus. These have been seen since by Gelei, Lund, and Mast. A detailed study of these fibers, which are attached to the dorsal side and posterior region of the esophagus, was made by Lund who called them paraesophageal and postesophag-

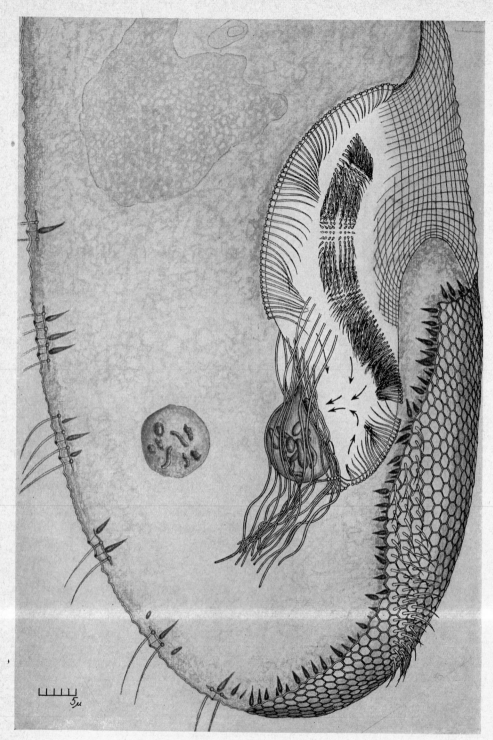

FIG. 19. (*See facing page for legend.*)

eal fibers. The distal extremities of these fibers are found to end freely in the endoplasm and appear to aid in the passage of food particles (principally bacteria) into the developing food vacuole. The detailed fibrillar structures of the pharyngo-esophageal complex is found on p. 53. According to Lund (1941) the food vacuole first appears as a growing bulge on the dorsal side of the esophagus and is retained by eight or ten long postesophageal fibers.

Parts of the food passageway are heavily ciliated and the ciliature has been referred to in the literature as an "undulating or oral membrane." Maupas (1883) first described, in paramecium, an undulating membrane which possessed a basal plate composed of many rows of basal granules resting upon a homogeneous basal lamella. Maier (1903) described two zones of membranelles in the cytopharynx and called the anterior one an undulating membrane. Wetzel (1925) also mentions such a membrane in paramecium. Rees (1922 a) showed the zone to possess a series of flat paintbrush-like cilia unlike a typical undulating membrane. In their studies of the cytopharynx, Bozler (1924 a, b), Gelei (1925), Kahl (1930), and Kalmus (1931) deny correctly the existence of an undulating membrane but describe instead a great concentration of short cilia. Gelei's (1925) detailed study of the ciliature of this region revealed what certain of the others had observed, namely a heavy brush of cilia which he called a "peniculus motorikus"—now known as the peniculus. Lund (1933, 1941) has given an excellent account of this structure, including the ciliation of the cytopharynx and esophagus (Fig. 19). Investigators differ regarding the exact location of the peniculus but its precise position may be a species character. Since the existence of a heavily ciliated region has been demonstrated conclusively and the cilia do not function in the manner of an undulating membrane, that term cannot be applied to the cilia in the cytopharynx or esophagus of paramecium.

3. **Anus (Potential-anus, Cell-anus, Anal Spot, Cytoproct, Cytopyge[4]).**

In all members of the "aurelia" group of paramecium, the anus is to be found on the ventral (oral) surface of the body, almost vertically behind the mouth. In the living condition, one can observe it only when the animal is in the process of egestion. However, the general region where defecation of food occurs is fairly constant for a given species. Using silverline impregnation methods, its location can be observed clearly (Fig. 21B). Wenrich (1928 a) made the important observation that the position of the anus in members of the "bursaria" group is near the posterior extremity—actually a little to one

[4] Cytopyge is a term in wide usage which was first introduced by Haeckel for the distinction of the anal aperture of unicellular animals. (Gr. *kytos,* cell; *pyge,* the rump.)

FIG. 19. Longitudinal view of the cytopharynx of Paramecium with a portion of the right side of the organism cut away. The peniculus, which consists always of eight rows of cilia arranged in two closely set blocks of four, is on left wall of the cytopharynx and spirals through approximately 90° so that its posterior extremity is on the ventral (oral) surface of the esophagus. Only one row of the long cilia of the "Vierermembran" is shown along the dorsal surface of the cytopharynx. Cilia of the esophageal pouch are shown at posterior end of peniculus. Note heavy paraesophageal fibers enmeshed in forming food vacuole. (Lund)

side of the terminal pole. On the other hand, members of the "aurelia" group show the anus between the posterior end of the oral groove and the posterior end of the body (Figs. 5 and 6). Thus in *P. caudatum* and *P. multimicronucleatum*, the anus is approximately centrally located between these two points, or perhaps closer to the end of the oral groove. In *P. aurelia*, it is nearer the posterior end of the body but still not as far posterior as in members of the "bursaria" group. Wenrich (1928 a), noted that

P. aurelia, with its somewhat more rounded posterior end, as compared to the other species in this group, and its cytopyge farther toward the posterior end than in those species (of the "aurelia" group), occupies a position intermediate between *P. caudatum* and *P. multimicronucleatum* on the one hand, and the entire "bursaria" group on the other.

B. The Ectoplasm, Endoplasm, Pellicle, Silverline, Fibrillar, and Neuromotor Systems

The cytoplasm of paramecium is differentiated into a narrow, external, or cortical zone called the *ectoplasm* (ectosarc) and a larger, internal, or medullary region called the *endoplasm* (endosarc). The ectoplasm[5] is a distinctly permanent part of the body, strikingly delimited from the endoplasm. It is the part of the ciliate which comes in direct contact with the environment and therefore the region that is first to receive external stimuli. It contains the trichocysts, cilia, and fibrillar structures and

is bounded externally by a body covering or envelope called the *pellicle*. Because of these distinctive features, it is sometimes called the *cortex*. The endoplasm is the more fluid, voluminous part of the cytoplasm which contains many cytoplasmic granules as well as other inclusions and structures of a specialized nature to be dealt with later.

If paramecium is observed in the living condition under low powers of the microscope, the pellicle appears homogeneous. However, if carefully stained or impregnated animals are examined under high magnification, the pellicle and the ectoplasmic region are found to consist of a very complex structural organization. For this reason, Gelei (1925) preferred to eliminate the term pellicle and substitute cuticula or cuticle instead. The pellicle of paramecium has been thoroughly and ingeniously studied by a large number of investigators. As a result, extensive literature has appeared on the elaborate techniques and complicated structures concerned with the pellicle, deeply staining fibrils, and other ectoplasmic structures. There has not been complete agreement on descriptions of structures and perhaps less in the interpretations placed upon these descriptions.

Maupas (1883) described the pellicle of paramecium as a pattern of rhombohedral or rectangular fields with the cilia placed in their centers. On the other hand, Bütschli (1887–89), Kölsch (1902), and Maier (1903) believed the fields to be hexagonal with the cilia correctly placed in the centers of the hexagons. Schuberg (1905) described a pellicular pattern of hexagons on the dorsal side of paramecium and rhomboids on the ventral side. He located the trichocysts in their correct positions, namely in the centers of the anterior and posterior sides of the polygons in respect to polarity of the animal. In

---

[5] Brown (1940), in his study of protoplasmic viscosity in paramecium, has stated that the microscope centrifuge gives evidence on the condition of the ectoplasmic material just under the pellicle. He points out that in centrifuging at low speeds (forces) a vacuole stops when it reaches the ectoplasmic layer, and requires a higher speed to start it moving again, thus indicating an elastic or plastic nature of the cytoplasmic region.

addition, he found the longitudinal fibrils connecting the basal granules. Khainsky (1910) confirmed the work of Schuberg and expressed the view that the quadrilaterals were formed from the hexagons due to lateral stress of unequal strengths. He showed the cilia correctly placed in the centers of the polygons, but placed the trichocysts at the corners of them. Rees (1922 a) described a so-called neuromotor center with fibrils in *P. caudatum* and showed the cilia of the pellicle to be arranged in longitudinal rows; those on the aboral surface being almost parallel and those on the oral side slightly oblique. He demonstrated that the ciliary lines from opposite sides join in a series of V's, the apices of which are placed in a line, called the *ciliary suture* which runs the

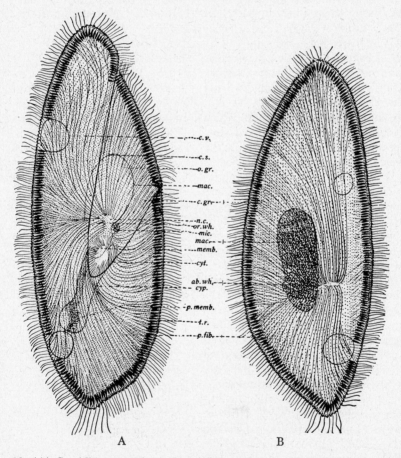

A   B

FIG. 20. (A) Semidiagrammatic sketch of *Paramecium caudatum*, oral view, showing oral whorl of peripheral fibers and the ciliary lines, ciliary suture, and trichocyst ridges. (B) Aboral view of *Paramecium caudatum*, showing aboral whorl, ciliary lines, and trichocyst ridges. Abbreviations: (*c. v.*) contractile vacuole; (*c. s.*) ciliary suture; (*o. gr.*) oral groove; (*mac.*) macronucleus; (*c. gr.*) ciliary grooves; (*n. c.*) neuromotor center; (*or. wh.*) oral whorl; (*mic.*) micronucleus; (*memb.*) cytopharynx membranelle; (*cyt.*) cytostome; (*cyp.*) cytopharynx; (*t. r.*) trichocyst ridges; (*p. fib.*) peripheral fibers. (Rees)

length of the animal and passes through the cytostome (Fig. 20).

The pronouncement by Sharp (1914) of a "neuromotor apparatus" in the complex ciliate, *Epidinium* (*Diplodinium*) *ecaudatum,* marked the beginning of and served as the stimulus for a great deal of investigation on neuromotor or fibrillar systems in ciliates generally. The reader is referred to the reviews of silverline and fibrillar systems in ciliates by Klein (1927 a, b), Gelei (1929), Lund (1933), and Taylor (1941). There has come into wide usage a silver impregnation method for demonstrating the pellicular surface and fibrils in species of paramecium and other related forms. Basically, it is a silver impregnation method in which the organisms are first fixed by drying. Silver nitrate is then applied and, in the presence of sunlight, colloidal silver is deposited on certain surfaces in the reduction. As far back as 1875, Golgi and then Ramon y Cajal (1903) used a silver impregnation method. Many modifications have been made since but the "dry method" of Klein as reported in his many papers is used rather extensively. This method has been altered further by Gelei and Horváth (1931) who show additional advantages with their "wet silver method" (for silver impregnation techniques see Chapter 14). The structures thus impregnated with silver are referred to in their entirety by Klein and others as the *silverline system*. Besides Klein, other investigators, namely Gelei (1929–39), Jirovec (1926), Lynch (1930), Jacobson (1931), Lund (1933), Kidder (1933, 1934) and many others have utilized the technique for the study of paramecium and many other ciliates.

When impregnated or stained, the pellicular surface of paramecium is found to be composed of a pattern of polygons of which most are hexagons. On the ventral surface, however, this hexagonal nature is lost, especially in the region of the cytostome, cytopyge, and ciliary suture, where frequently quadrilateral figures may be seen (Fig. 21). All the polygons are slightly concave to form the so-called dimples or "Grübschen" of the German workers. In these depressed centers of the pellicular polygons are found the cilia, each possessing at the proximal end a basal granule or kinetosome which is embedded in the ectoplasm. Such a polygon with its cilium is known as a *ciliary field*. It has been shown by Schuberg (1905) and many later investigators that the basal granules are connected by longitudinal fibrils resembling the beads on a string. Brown (1930), Gelei (1925), and Lund (1933) have described transverse fibers that join the longitudinal ones. The transverse fibrils pass from one basal granule to another in adjacent rows. They are seen commonly in the region of the cytostome.[6]

THE SILVERLINE AND FIBRILLAR SYSTEMS OF PARAMECIUM

When specially treated, one is able to identify two sets of elements in paramecium as follows: (1) the silverline system which depends upon the reduction and deposition of colloidal silver and (2) the fibrillar system which depends upon the marked affinity of certain structures for soluble stains.

The Silverline System

The silverline system may be divided into two parts. One part called by Klein (1928, 1931) the "indirectly connected"

---

[6] Gabor von Gelei (1937) described an additional system of fibrils in *P. caudatum, P. multimicronucleatum,* and *P. trichium* at approximately the same level of the basal granules or slightly below them. He reported that the fibrils extend throughout the entire body surface.

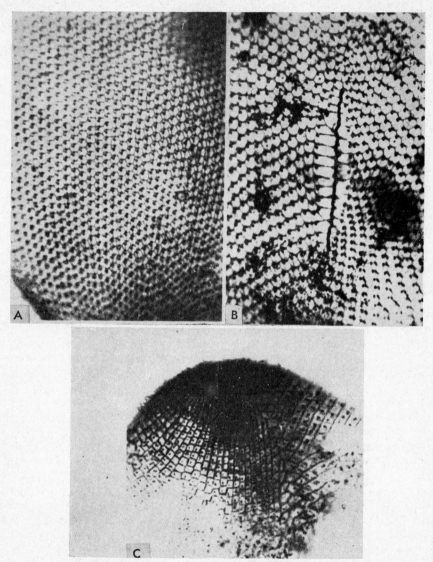

FIG. 21. Photographs of pellicular surface in Paramecium. (A) Hexagons in posterior, dorsal region of body. Note excretion pore to right of center. (B) Pellicular surface around cytopyge. (C) Pellicular pattern at anterior end of body. (Techniques of Gelei and Horváth.) (Gelei)

conductile system, consists of a series of closely set polygons, usually hexagons but flattened into rhomboids or other quadrilaterals in the regions of the cytostome, cytopyge, and sutures. The hexagons occur in longitudinal rows with one side of the hexagon toward the anterior part and another side toward the posterior part of the body. The walls of the hexagons are ridges surrounding depressions or dimples approximately 1 micron in depth. According to Lund

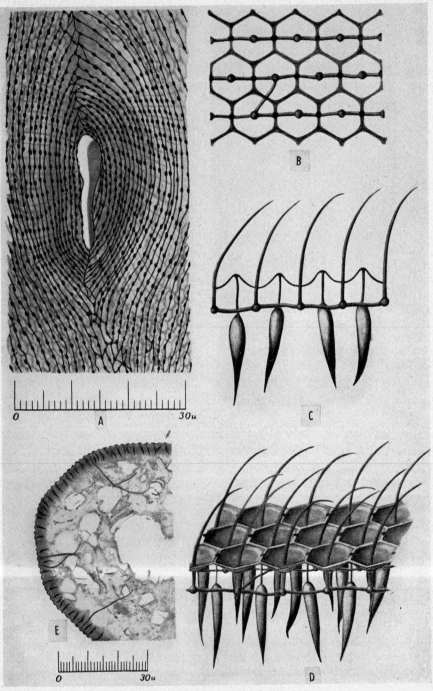

Fig. 22. (*See facing page for legend.*)

(1933) the distance between the sides of the hexagon from ridge to ridge varies from 1.8 microns to 6 or more microns. It is in the centers of the anterior and posterior sides of the hexagons or rectangles that the deeply stained granule is found which marks the outer end of the trichocyst. The granule therefore represents the attachment of the trichocyst to the pellicle (Fig. 22). The second part of the silverline system called by Klein the "directly connected" conductile system, consists of longitudinal lines linking into a chain all the basal granules in one longitudinal row of hexagons (or quadrilaterals) and of the basal granules themselves.

The Fibrillar System[7]

A number of nuclear stains, especially Heidenhain's, also show the surface of paramecium to consist of the series of polygons. The basal granules are always at or near the centers of the polygons forming longitudinal chains of the body fibrils. These run nearly parallel to the long axis of the body except in the regions of the sutures and the cytostome where they curve sharply inward.

[7] Sharp (1914) used the term "neuromotor apparatus" to designate a complex system of fibrils which he considered conductile in the ciliate *Epidinium* (*Diplodinium*) *ecaudatum*. Ten Kate (1927) suggested the term *sensomotor apparatus* instead of neuromotor apparatus. See also Brown (1930) in regard to the neuromotor apparatus of *Paramecium*.

The Pharyngeal Complex of Fibrils and the "Neuromotorium"

A large number of fibrils has been described in the region of the cytopharynx and esophagus. According to Bozler (1924 a, b), Gelei (1925), and Lund (1933, 1941) there are found, leading posteriorly from one side of the end of the esophagus, approximately 10 long parallel fibers which are believed to form a chute-like structure down which the food vacuoles slide (Fig. 19). Some believe the wall of the cytostome to be smooth; others that it possesses a pellicular pattern. Gelei reports that the vestibulum and esophagus (pharynx) of *"P. nephridiatum"* contains a network of subpellicular fibrils which are connected with the body surface. It is in the ectoplasm and anterior to the cytostome that Rees (1922 a) reported the position of a double body which he called the "neuromotor center." Gelei reported two "nerve centers" in approximately the same region. Both investigators described fibrils emanating from these centers and passing into the endoplasm.

We are indebted to Lund (1933) for the detailed study of the "neuromotorium" in *P. multimicronucleatum* and the fibrillar system associated with the cytopharynx and esophagus, which he called the pharyngeal complex (Fig.

FIG. 22. All figures are of *Paramecium multimicronucleatum*. (A) The cytostomal region from a specimen prepared by Klein's method but not allowed to dry completely. Showing the longitudinal body fibrils, the circular cytostomal fibrils, the basal granules, and, more faintly, the pellicular ridges. A portion of both the preoral and the postoral suture is shown, and the outline of the cilia of the left cytostomal wall. Below is a scale of 30 microns. ($\times$ 1700.) (B) Semidiagrammatic drawing of the pellicular structure and the basal granules and longitudinal body fibrils. Based on drawings and measurements taken with camera lucida from a tangential section of the anterior dorsal surface ($\times$ 7300.) (C) Same as above, but viewed in side elevation. Based on camera lucida drawings of longitudinal sections of anterior dorsal region. ($\times$ 7300.) (D) Composite drawing based on figs. 12 and 13. ($\times$ 7300.) (E) Showing internally discharged trichocysts in a specimen where they can be positively identified as such and not mistaken for ciliary rootlets or internal fibrils. Below is a scale of 30 microns. ($\times$ 915.) (Lund)

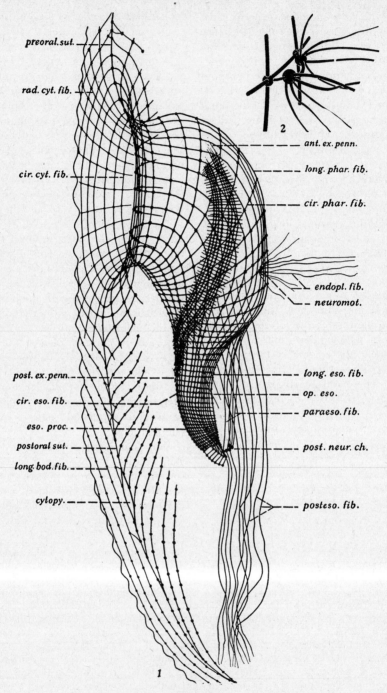

FIG. 23. *(See facing page for legend.)*

23). He lists the parts of the complex as follows: (a) the pharyngo-esophageal network, (b) the neuromotorium and associated fibrils, (c) the penniculus, (d) the esophageal process, (e) the paraesophageal fibrils, (f) the posterior neuromotor chain, and (g) the postesophageal fibrils. Based upon Lund, their descriptions follow and the reader should refer directly to the figures.

(a) *The pharyngo-esophageal network.* The pharyngo-esophageal network is a trumpet-shaped structure lying in the walls of the cytopharynx and esophagus. It is composed of the basal granules of the cytopharynx and esophagus, the longitudinal pharyngeal and esophageal fibrils which are the inner portions of the radial cytostomal fibrils and the circular pharyngeal and esophageal fibrils. These circular fibrils of the cytopharynx and esophagus correspond to the circular cytostomal fibrils which have their origin in the converging longitudinal body fibrils. Thus, the longitudinal fibrils of the general body surface become the circular fibrils of the pharyngeal region, while the longitudinal fibrils of the pharyngeal region become the radial fibrils of the body surface. The intersections of the fibrils of each set with those of the other are marked by basal granules and from most or all of these granules, the cilia protrude into the cavity of the cytopharynx and esophagus.

Cilia vary in length. Some cilia are prominent on the left wall of the cytopharynx but they are difficult to separate from the ciliation of the penniculus which is discussed later. The right wall also has a narrow band of long cilia. With the exception of the dorsal wall, most of the cytopharynx and esophagus have either exceedingly short, fine cilia or none at all. The dorsal cilia are long and heavy forming a continuous band throughout the posterior two thirds of the cytopharynx and as far down the esophagus as its dorsal wall is intact. According to Lund, the esophagus does not expel the food from its posterior extremity, but from an opening on its dorsal side, leaving beyond that a blind pocket which is the esophageal process.

(b) *The neuromotorium and associated fibrils.* On the left dorsal wall of the cytopharynx at about the level of the posterior margin of the cytostome is a very small, bilobed mass, *the neuromotorium* (Lund). Each lobe is slightly larger than the basal granules of the pharyngeal region and is connected by at least one, possibly two short, fine fibrils to one of these near-by basal granules. In some specimens a considerable space between the two lobes permits one to see a connecting fibril, in

FIG. 23. (1) Diagrammatic representation of the pharyngo-esophageal network, with its associated fibrils of the body surface and the endoplasm, in *P. multimicronucleatum*. The cilia have been omitted for the sake of clearness. (× 2250.) (*ant. ex. penn.*) Anterior extremity of penniculus; (*cir. cyt. fib.*) circular cytostomal fibril; (*cir. eso. fib.*) circular esophageal fibril; (*cir. phar. fib.*) circular pharyngeal fibril; (*cytopy.*) cytopyge; (*endopl. fib.*) endoplasmic fibril; (*eso. proc.*) esophageal process; (*long. bod. fib.*) longitudinal body fibril; (*long. eso. fib.*) longitudinal esophageal fibril; (*long. phar. fib.*) longitudinal pharyngeal fibril; (*neuromot.*) neuromotorium; (*op. eso.*) opening of esophagus; (*paraeso. fib.*) paraesophageal fibril; (*posteso. fib.*) postesophageal fibril; (*post. ex. penn.*) posterior extremity of penniculus; (*post. neur. ch.*) posterior neuromotor chain; (*postoral sut.*) postoral suture; (*preoral sut.*) preoral suture; (*rad. cyt. fib.*) radial cytostomal fibril. (2) Diagrammatic representation of the neuromotorium, in approximately the position shown in large figure. (× 9000.) (Lund)

others the two granules lie adjacent, and many specimens either reveal no such bodies or they are too well concealed by other structures to be studied. From the neuromotorium, fibrils radiate into the endoplasm. Of these, four or more usually pass almost to the dorsal body wall, but the rest, which seem to be variable in number, are shorter and not definite in position. All are termed endoplasmic fibrils. Approximately half of the endoplasmic fibrils come from each granule. Their function is unknown, but they may coördinate the feeding movements of the oral cilia with the lack or abundance of food in the endoplasm, thus serving a nutrition-regulating function according to Lund. Their distal ends are unattached.

As reported by Lund, it is difficult to identify the neuromotorium and the endoplasmic fibrils described here as being similar to those found and figured by Rees (1922 a) but both the bilobed body and its accompanying fibrils are without doubt identical with the structures described and figured by Gelei (1925).

(c) *The penniculus.* In members of the "aurelia" group the penniculus is commonly composed of eight rows of basal granules, four equally spaced rows being separated by a slightly greater gap from four additional equally spaced rows. Opton (1942) reports the presence of 15 ciliary rows making up the penniculus in *P. polycaryum.* The basal granules, usually numbering over 60 to each longitudinal row, are connected by both longitudinal and transverse fibrils. The ends of these fibrils of both sets extend a short distance beyond the last basal granules and end apparently without specific attachment. The entire structure lies slightly deeper than the pharyngeal network and proceeds from the anterior, left dorsal wall of the cytopharynx, around the left wall in a

slightly oblique direction to terminate on the anterior, ventral wall of the esophagus. Since each basal granule bears a cilium, the result is a heavy band of cilia. The connection of the penniculus to the pharyngo-esophageal net is not known but Lund believes that the cilia of the penniculus beat independently of the rest of the pharyngeal cilia, aiding indirectly the passage of food into the esophagus, while those of the lower end help in forming food vacuoles. Gelei (1925) is also of the opinion that the cilia of the penniculus are independent.

(d) *The esophageal process.* At the posterior extremity of the gullet the circular fibrils once more completely enclose the cavity forming a tube. Also the longitudinal fibrils converge at the posterior end and unite in a double row of granules bound by transverse fibrils closing the end of the tube to form a blind pouch, the *esophageal process.* On the ventral wall of the esophageal process is a heavy tuft of long cilia. Lund believes that they probably aid the cilia at the posterior part of the penniculus in directing the course of food and in forming food vacuoles.

(e) *The paraesophageal fibrils.* Proceeding posteriorly from the dorsal wall of the cytopharynx at its junction with the esophagus are five or more heavy fibrils. These terminate near the posterior end of the esophageal process in five large granules which are connected in a chain. These fibrils are said by Lund to be neither tense nor very lax as shown in stained specimens. They pass over the dorsal opening of the esophagus.

(f) *The posterior neuromotor chain.* The chain of five granules in which the paraesophageal fibrils end, posteriorly, is usually connected only on the left side to the ladder-like terminal portion of the esophageal process. Since both ends sometimes appear connected and in a few cases neither end is attached, their exact relation in the living animal is uncertain. From it pass three, occasionally four, heavy fibrils toward the posterior end of the animal which indicates some function other than merely that of anchoring the paraesophageal fibrils according to Lund.

(g) *The postesophageal fibrils.* In addition to the three fibrils from the posterior neuromotor chain, from eight to 12 fibrils pass posteriorly from the right wall of the cytopharynx toward the caudal end of the body. Usually six to eight of these can be traced beyond the esophageal process so that in all, about 10 fibrils pass back and are lost as they approach the posterior end of the animal. It is claimed that the distal ends are unattached in stained preparations and that the fibrils are quite lax. Lund expresses the view that these fibrils comprise the "Schlundfadenapparat" of Bozler (1924 a, b) and are the "Stomodaealfasern" of Gelei (1925). Bozler and Lund are of the opinion that these fibrils assist in the formation of food vacuoles and also direct the course of the vacuoles after they are formed. In the living animal, the fibrils can be seen only a short distance away from the gullet. Their distal ends appear to be free in the endoplasm.

Discussion and Interpretations of the Silverline and Fibrillar Systems

The first part of Klein's silverline system ("indirectly connected" conductile part) appears to be the same as the typical pellicular pattern described by many workers. The point is made by Lund that the silverlines in the hexagonal pattern shown by Klein mark only the ridges of the pellicular polygons. Since the pellicle is slightly thicker at the top of these elevations, more silver is likely to collect there; this is more strikingly shown in well-dried specimens. Optically it may appear that fibrils pass immediately beneath the pellicle—sometimes through the polygons but fibrils are actually non-existent. The "wet silver" methods rarely show the pellicular pattern. Klein's "directly connected" conductile system is actually the subpellicular fibrillar system so ably demonstrated by Lund. Both Gelei and Klein, in later papers, agree that the fibrillar system of paramecium is entirely subpellicular. There may be some justification for the disharmony in descriptions of purely morphologic parts as set forth by the various investigators. It should be remembered that Rees studied *P. caudatum;* Gelei, *P. caudatum* and *"P. nephridiatum";* Lund, *P. multimicronucleatum,* and Klein investigated *P. aurelia.* It is entirely possible that definite species differences exist in respect to pellicular and fibrillar patterns, i.e. two basal granules per ciliary field in *"P. nephridiatum"* while only one is generally found in the other species.

It is an established fact that fibrils and a fibrillar apparatus exist in paramecium. Parts of it can be seen in the living condition although greater detail is dependent upon fixed and stained preparations. Worley (1933) studied intracellular fiber systems by microdissection in which the ectosarc of paramecium was removed and studied in the living, unstained state. He demonstrated that the direct and indirect connecting fibers in the vicinity of the cytostome are not artifacts (Fig. 24).

What is the significance of the complicated fibrillar systems found in paramecium or, for that matter, in other ciliates? Investigators have not been lax in suggesting functions ascribed to these fibrils. In a number of cases, interpretations have been based solely upon conclusions drawn from an examination of

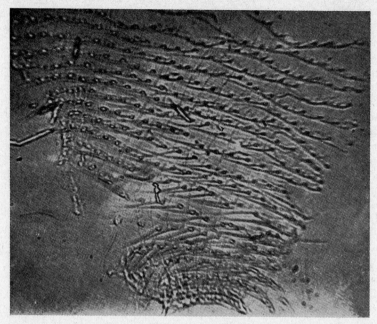

FIG. 24. Longitudinal or direct connecting fibers (horizontal) and cross or indirect connecting fibers (vertical) near the cytostome (lower margin) of Paramecium. Photograph from a fresh preparation. (Worley)

purely morphologic elements with very little or no experimentation. The interpretations are so unsound scientifically, that it reminds one of Ehrenberg's (1838) fallacious theories in which he claimed identity of organs in the *Infusoria* to those common in higher animals.

Fibrils in the Protozoa have had the following functions given to them: (1) mechanical support, (2) elasticity, (3) contractility, (4) conductivity, and (5) "metabolic influence." The last function to be added to the list, namely metabolic influence, was proposed by Parker (1929) for the fibrillar complex in paramecium. Schuberg (1905) suggested that the fibrils connecting the basal granules in paramecium function in the metachronism of ciliary movement. Sharp (1914) discussed the fibers and their possible functions in the com-

plex ciliate, *Epidinium* (*Diplodinium*) *ecaudatum*. He considered the possible functional roles fibers might conceivably play, such as supportive, contractile, and conductile. He eliminated two of them and considered that the fibers are conductile in *Epidinium*. Rees, (1922 a) came to a similar conclusion in that the morphology and staining characteristics suggest that the fibers in *P. caudatum* are conductile.

It should be mentioned that the fibrillar system of Rees was later investigated by Jacobson (1931). She came to the conclusion that certain of the elements called fibrils by Rees were in fact internally discharged trichocysts, the extrusions of which were caused by the killing agent (Fig. 22E). Although Rees performed some experimental studies they were limited in scope and inconclusive. He severed the fibrils joining the cytopharyngeal cilia with the neuromotor center and noticed a marked difference in rate and amplitude between the cilia anterior and posterior to the cut. When the neuromotor center was destroyed, coördinated movement of body cilia was interrupted. The experimental work should be repeated in greater detail. In his study of the neuromotor (fibrillar) system of *P. multimicronucleatum,* Lund (1933) expressed the view that the system has primarily a conductile function based on a purely morphologic analysis and without any experimental foundation.

According to Gelei's interpretation, the outer pellicular system (Klein's indirect system) functions as a network for maintenance of body form. He believed the internal fibrillar complex is a system for integrating basal granules of cilia, the trichocysts, and even the pore of the contractile vacuole. Klein's (1926 a–31) interpretation of the inner, fibrillar complex is essentially similar to that of Gelei. Klein's view of reversal in the effective stroke of cilia is ingenious. He believes this is accomplished by means of a kind of primitive reflex arc in which the axial filament of the cilium is the receptor, the basal granule (and its two secondary granules) a kind of "relator" and the protoplasmic sheath of the cilium, the effector.

In the last analysis, the author of this book emphasizes the fact that since very few interpretations have been made upon experimental evidence in paramecium, it must frankly be admitted that the specific function or functions of many of the fibrils are not known. An attack in the direction of experimental investigation and functional observation was undertaken by Worley (1934) and Lund (1941). If continued it is likely to yield very profitable results on the nature and function of fibrils. Worley dissected paramecia after they were first

quieted with a 0.5 per cent solution of Novocain. Realizing the difficulty of cutting single, longitudinal, subpellicular fibers, he was able to sever several rows simultaneously at different points on the body and observe the effects. After an operation of this type, he found that continuity of the propagated ciliary waves was broken. Metachronal waves were found to pass forward from the posterior end of such an operated paramecium and were dissipated at the cut surface. Coördinated ciliary activity did not reappear for the remainder of the length of the animal. Since the cilia of this area beat synchronously after the operation, it was concluded that the ectoplasm and not the endoplasm is important in the conduction of metachronal impulses. The region of synchronism was visible only directly anterior to the point at which a cut was made. It was found that on adjacent surfaces, metachronal ciliary activity continued, and propagated ciliary waves, which originated at the posterior end of the animal, succeeded in reaching the anterior tip. In all cases, the cilia anterior and in line with the cut, beat synchronously while none of the other cilia was affected. On the basis of this and other operations, it was concluded by Worley that coördinated ciliary activity in paramecium follows straight conduction paths over the surface of the ciliate following the longitudinal arrangement of the "silverline fibers" (subpellicular fibers) of the body. While it is clear that the fibrils or silverline system is of prime importance in the conduction of metachronal impulses, it appears possible that the reversal impulses are transmitted by the undifferentiated ectoplasm according to Worley. Lund (1941) has described how the cilia in the penniculus function as well as the fibrils associated with the posterior part of the cytopharynx. He has shown how these elements play an important role in the movement of food and food vacuoles. The suggestion is made that better success may be the reward of those investigators who wish to cultivate species of paramecium or other ciliates in bacteria-free, nutrient media by making a close study of the feeding mechanism and associated fibrils. Within limits, one is able to predict the nature of the food from an examination of the detailed structure of the ingestatory organelles.

Finally, to ascribe functions to the various fibrils without experimentation is incompatible with sound scientific principles. Perhaps with the highly specialized instruments now available for microdissection used in the study of single cells, some of the problems associated with fibrillar functions will be answered.

C. The Cilia (Lat. *cilium;* an eyelash)
(See also Chapter 8)

Perhaps the most remarkable structures of paramecium, or, for that matter, in the whole field of protozoölogy are the cilia. They are short, fine protoplasmic processes that completely cover the body of paramecium and serve as organs of locomotion and food-capture. Movement of the body is possible because of their large number and manner of beating (Fig. 73). When paramecium is observed under the microscope, the cilia and the organism appear to be moving with great rapidity. In reality, such is not the case. Bidder (1923) makes the interesting comment that we forget, as we look through a microscope, that though distance is magnified, time is not magnified. Because of the fact that cilia move with low velocity, it is not surprising to realize that the speed of movement of paramecium is given as 1.3 mm. per second, according to Tabulae Biologicae.

It appears that Leeuwenhoek, in 1677, was the first to observe cilia and recognize their use in locomotion. He mentions them frequently in his writings. A number of Leeuwenhoek's contemporaries also saw ciliates and cilia but it was not until 1703 that they were shown on paramecium by the anonymous author who first sketched the organism. He referred to the cilia as "very minute feet" as Leeuwenhoek did earlier.

When examined closely under high magnification, a single specimen shows the cilia to be arranged in parallel rows. Considerable variation is found in the length of cilia on one organism. On the body proper they are of moderate and fairly uniform length except for those at the extreme posterior tip which are longest. There is also wide range in length of cilia found in the cytopharynx and esophagus. Cilia are found to vary from 0.1 to 0.3 micron in diameter and may reach a length of approximately 16 microns. Bütschli (1889) estimated that a single paramecium possesses as many as 2500 cilia. Accounts, however, vary for according to Kalmus (1931) Maupas gives 350 cilia in ciliary fields for small paramecia; Jensen claims 3500 and Schumann up to 14,000. Gelei (1925) states *"P. nephridiatum"* possesses approximately 18,000 cilia. This variation in number is understandable in part, for it would depend not only upon the species of paramecium investigated but also upon its size.

Liebermann (1929) made a comparative study of the ciliary arrangement in the eight well-defined species of *Paramecium*. A count was taken of the number of rows of cilia in a measured unit of width as well as the number of cilia in a measured unit of length (Table 2). He noted that all the species are similar in their arrangement of ciliary rows on the dorsal surfaces of

Table 2

TABLE SHOWING COUNTS OF ROWS OF CILIA IN TRANSVERSE DISTANCE OF 12 MICRONS AND NUMBER OF CILIA IN THE LONGITUDINAL DISTANCE OF 12 MICRONS AND THE TOTAL NUMBER OF CILIA IN AN AREA OF 144 SQUARE MICRONS. THE NUMBERS ARE AVERAGES OF 15 COUNTS MADE ON SEVERAL INDIVIDUALS. OIL IMMERSION OBJECTIVE AND 10× OCULAR WERE USED IN MAKING THE COUNTS. (Liebermann, 1929)

DORSAL VIEW

Animal	Width of 12 Microns No. of Rows	Length of 12 Microns No. of Cilia	No. of Cilia 144 Square Microns
1. P. bursaria	9.75	6.6	64.45
2. P. polycaryum	9.2	6.8	62.56
3. P. caudatum	9.1	6.0	54.63
4. P. aurelia	7.9	6.6	52.14
5. P. trichium	7.8	6.0	46.80
6. P. calkinsi	7.7	6.0	46.20
7. P. woodruffi	9.0	5.0	45.00
8. P. multimicronucleatum	7.1	6.1	43.31

paramecium. On the dorsal surface the rows are parallel to the long axis of the body except toward the ends where the rows converge. On their ventral surfaces, there is a preoral and a postoral line or suture against which rows of cilia terminate. It is possible to differentiate between the members of the "aurelia" and "bursaria" groups by ciliary arrangement. In the former the rows of cilia on the right side of the preoral suture run parallel to this line (left side in a ventral view). In the latter group none of the ventral rows of cilia is parallel to the preoral suture in the region anterior to the cytostome but they bend toward and terminate against this line.

Liebermann reported that when the total number of cilia within an area of 144 square microns is counted, the relative numbers may be expressed as follows: P. bursaria > P. polycaryum > P. caudatum > P. aurelia > P. trichium > P. calkinsi > P. woodruffi > P. multimicronucleatum.

Cilia have been studied extensively in the living condition and in fixed and stained or impregnated preparations. It is to be realized that because of their minuteness, accounts differ concerning their structure. They are found to be optically homogeneous when observed by transmitted and polarized light in the living condition. It is claimed by some that a single cilium consists of a central axis filament or fiber surrounded by an elastic sheath of protoplasm. Movement is said to be due to the active contraction in one plane of the axial filament and recovery due to the elasticity of the enveloping sheath. Local thickenings occur at intervals along the axial filaments when silverline impregnation methods are used. Gelei (1925) reported the presence of lipoid substance in cilia of paramecium. But one wonders how far the structure of a cilium as revealed by fixation and subsequent staining is a reliable counterpart to the structure of the living element.

Jakus and Hall (1946) have made electron microscope observations of cilia using a shadow-casting technique with chromium. They show a cilium of paramecium to consist of a bundle of about 11 fibrils extending its full length (Fig. 25). The diameter of the dried fibrils is between 300 and 500 Å. They report

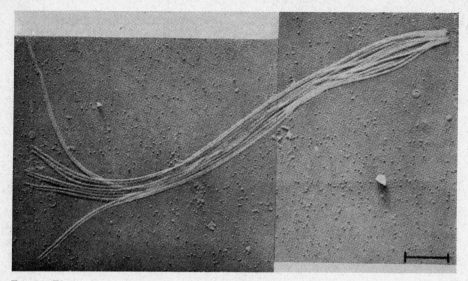

FIG. 25. Electron micrograph of a cilium from Paramecium shadow-cast with chromium. (× 11,000.) (Jakus and Hall)

that in fixed preparations, the component fibrils usually adhere to form a compact bundle but in unfixed cilia, they separate. While no conclusive evidence is presented for the existence of a sheath surrounding the bundle of fibrils, they report that poorly defined cross-striations have been observed in two or more adjacent fibrils. These striations may represent the remnants of some enveloping structure or membrane. A membrane is clearly visible in specimens prepared by the "critical point" drying method (Anderson, 1951) in which the disruptive effects of surface tension are minimized. Electron micrographs, like that of Fig. 26, are then obtained showing cilia as long tapering structures bounded by an external membrane which surrounds a number of internal fibers, one of which extends to the tip.

Each cilium is shown to have at its proximal end, a *basal granule* or small, spherical corpuscle called a *kinetosome* embedded in the ectoplasm. These basal granules or kinetosomes are cytoplasmic

"self-reproducing" systems which always reproduce themselves by division. Lwoff (1950) has shown that the kinetosomes of ciliates are also endowed with genetic continuity. Not only are they able to divide and to produce cilia but according to Lwoff they are able to secrete fibers, and give rise to other granules producing trichocysts or trichites.

The region occupied by the granule can be observed in the living condition by a point of high refractile index; basal granules stain conspicuously with a number of dyes. There is commonly only one cilium for each pellicular hexagon (ciliary field) (Fig. 22C and D). Gelei (1925) and others have found instances of two cilia in polygons in *P. caudatum*. In biciliated polygons, it is found that each cilium possesses a basal granule. Klein (1926 b, 1931) and Gelei (1932) report that in some individuals, cilia are present with groups of two or three granules at their bases instead of the characteristic single basal granule. Gelei observed that a ring and not a basal granule is frequently encountered. However, Lund (1933) is of the opinion that Gelei's basal ring is an effect produced by viewing the outer deposit on one large granule or of a cluster of granules perhaps with connecting fibrils.

In a study of the morphogenesis of cilia in *P. aurelia*, Downing (1951) reported that neither new rows of cilia are produced during growth nor are new cilia added in the rows during growth of non-dividing specimens. Instead, new cilia are produced only immediately prior to active division first by doubling and second by the moving apart of the doubled cilia. This separation of the daughter cilia is confined to the same ciliary meridian as the parent cilium and this doubling of cilia on the ventral surface is the first indication that divi-

Fig. 26. Electron micrograph of cilia (C) and trichocysts (T) of *Paramecium caudatum* after "critical point" method of Anderson. (× 18,600.)

sion will follow. Beginning in the region of the mouth, the zone of ciliary doubling spreads around the center of paramecium in a narrow band, then broadens anteriorly and posteriorly until both ends of the animal are reached.

It is outside the sphere of this book to set forth the numerous theories of ciliary movement. For them, the reader is referred to the reviews and monographs of Pütter (1904), Erhard (1910), Prenant (1913), and Gray (1928, 1929, 1930).

Cilia of paramecium are shed readily if the organism is injured. Peter (1899) demonstrated the important fact that when a ciliate is crushed or dissected into small fragments, the cilia continue to beat if they are in connection with a piece of the cytoplasm. He showed that the beating of the cilium is independent of the cell-body as a whole. Since the basal granules are present in the ectoplasm, this has led to the theory that the kinetic seat of ciliary activity resides in these granules (Henneguy-Lenhossek theory). Generally, cilia are not capable of prolonged movement when isolated from the cell-body. It is of interest to report that Erhard (1910) claims the basal granules can be destroyed by heat without affecting the motility of the cilia.

Schäfer (1904, 1905) presented a hypothesis on the organization of a cilium and utilized working models to illustrate his views. He presented the idea that a cilium is a hollow curved extension of the cell which contains hyaloplasm bounded by a very delicate elastic membrane. The hypothesis further contends that a rhythmic flowing of hyaloplasm from the body of the cell into and out of the cilium would result in an alternate extension and flexion of it. In this regard, Alverdes (1922 a–e) by pressure, forced some material out of paramecium which he claimed promoted move-

ments of cilia. This material which he called "hyaline drops" appeared on the surface of the flattened animal. It was noted that cilia whose basal granules rested in this hyaline material beat not only longer and more vigorously than others in the culture medium, but were correlated in their movements. The experiments of Harrison and Fowler (1946) also are worthy of note in this connection. In serologic studies of the antigenic characters of conjugating pairs of *P. bursaria*, they reported the reaction of individual paramecia in the presence of effective antiserum. They observed that first the distal ends of a few cilia became enlarged. This was followed by two or more cilia which became attached at the enlargements and which moved erratically. Following this, a gelatinous precipitate was collected in the interciliary spaces, resulting in ciliary tangling over the whole body surface. In these observations, one wonders whether the enlargements of amorphous material have come from the body of the organism and through the cilia or whether it is simply a surface reaction.

Ciliary movement and locomotion as applied to paramecium are considered on page 233.

D. The Trichocysts (Gr. *thrix, trichos,* hair; *kystis,* a bladder)

Trichocysts, when at rest, are refractile, fusiform, or carrot-shaped elements embedded in the ectoplasm of paramecium and located at right angles to the body surface (Figs. 20 and 22). They are located in the centers of the anterior and posterior walls of the pellicular polygons. Resting trichocysts measure about 4 microns in length and are found approximately 1–2 microns below the surface of the pellicle to which each is attached by a very delicate connecting element. Trichocyst extrusion may be evoked by a number of different kinds of stimuli such as chemicals, electric shocks, mechanical injury as in cutting or with pressure and desiccation. Upon stimulation, the trichocysts are extruded very rapidly—within several milliseconds. The degree of extrusion varies. Some trichocysts are only partly extruded with a given stimulus, some are completely extruded but attached to the body, while still others are completely extruded and free of the protozoön (Figs. 27 and 28). Fully extruded trichocysts are needle-like, six to 10 times the length of resting ones. They may reach a length of 40 microns and a width of less than a micron. In experiments upon trichocyst stimulation, the writer has observed some individuals which failed to extrude trichocysts when the majority of specimens responded. Such specimens appeared to be killed before the trichocysts were capable of being extruded. Paramecia undergoing fission and conjugation are capable of trichocyst extrusion as well as typical vegetative forms (Wichterman, 1946 a).

Ellis (1769) appears to have been the first to demonstrate the extrusion of trichocysts but incorrectly interpreted their function. In his figures, he shows unmistakably one paramecium after "geranium infusion" had been added and comments, "this animal discovers its *fins* (trichocysts) and becomes torpid." Schmidt (1849) compared the similarity of these ectoplasmic structures with the rhabdites of the *Turbellaria* and suggested that the unicellular animals be classified with the worms. He believed that the extruded forms represented a fibrous slime which might be poisonous. We are indebted to Allman (1855) who named these bodies *trichocysts*. He believed as did Maupas (1883) that the trichocysts consisted of spirally wound threads—like the nematocysts of coelenterates—which un-

coiled during the extrusion. Kölsch (1902), Maier (1903), Verworn (1915), Schuberg (1905), Mitrophanow (1904, 1905), Khainsky (1910), and Tönniges (1914) have added to our information on the structure of trichocysts. More recently and with improved methods, Krüger (1929, 1930, 1931, 1936), Schmidt (1939), Jakus (1945), and finally Jakus and Hall (1946) have greatly increased our knowledge on the structure and biochemical properties of trichocysts. Using the dark-field microscope, Krüger has shown detailed structure of the trichocysts at rest and extruded (Fig. 27B). A study of these figures shows that it is the swelling of the "Quellkörper" that appears to account for the elongation during extrusion. In explaining extrusion, he maintains that the first stage is the lifting of the cap from the tip. This is immediately followed by a taking up of enough water to cause a large volume increase. The view was expressed that platelet micelles contain a swelling substance in resting trichocysts and the extruded shaft represents the swelling. Schmidt (1939) observed birefringence in the body of the resting and tip of the extruded trichocyst. He expressed the opinion that the birefringent component is the swelling substance which becomes fibrous upon extrusion. His analogy with extruded threads of myosin is worthy of consideration.

Using the dark-field and electron microscope, Jakus (1945) made an exhaustive and valuable study of the detailed structure and biochemical properties of the trichocysts of paramecium. She reported that under dark-field (Fig. 27C) the trichocyst is found to consist of two parts in which the larger part is roughly oval and bounded by a wide bright line. From the broader end there extends a smaller elongated structure delimited by a thin bright line which

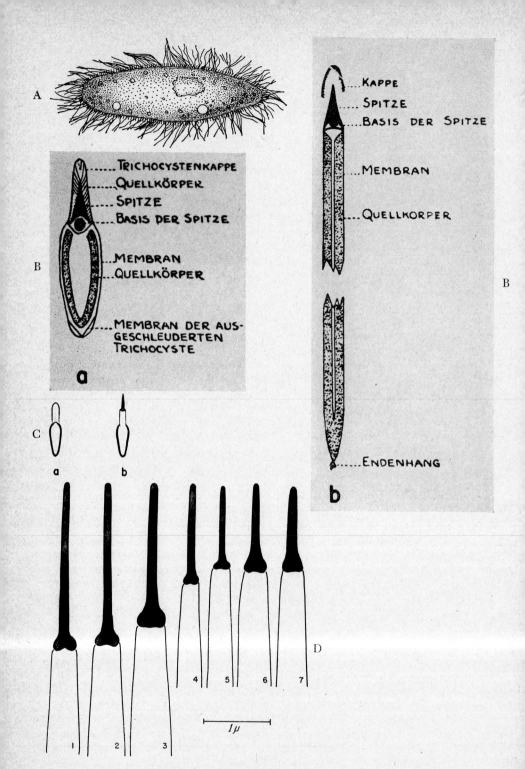

A

B

TRICHOCYSTENKAPPE
QUELLKÖRPER
SPITZE
BASIS DER SPITZE

MEMBRAN
QUELLKÖRPER

MEMBRAN DER AUS-
GESCHLEUDERTEN
TRICHOCYSTE

a

KAPPE
SPITZE
BASIS DER SPITZE

MEMBRAN

QUELLKÖRPER

B

ENDENHANG

b

C

a   b

D

1   2   3   4   5   6   7

1μ

FIG. 27. (*See facing page for legend.*)

66

almost fades out proximally. It is suggested that this is the trichocyst cap described by Krüger (Fig. 27B) as lying over the tip although the tip itself is not visible. According to Jakus, the brighter body of the resting trichocyst forms the shaft when extrusion occurs and the tip is bright and well defined, apparently having lost the cap which previously covered it. She reports that a narrow, pale shaft can barely be distinguished, extending from the base of the tip to the body; at the junction of the two, the bright periphery of the body is interrupted. Probably in the resting stage this opening was occupied by the base of the tip. In the electron micrographs, Jakus shows the tip to consist of two parts as follows: (1) a spike which tapers to a sharp point, and (2) a broader, more opaque base. Jakus and Hall (1946) report that in some specimens, the tips appear regularly cross-striated with striations about 300 Å apart. Jakus, Hall, and Schmitt (1942) and Jakus (1945) describe the shaft below the tip as being cross-striated (Fig. 28). Pease (1947) also studied the structure of trichocysts as revealed by the electron microscope.

The structure and composition of the extruded shafts are worthy of note in this study. Electron micrographs show the tip generally very dense and straight and the shaft spindle-shaped with cross-striations of alternating darker and lighter bands resembling individual striated muscle fibers of metazoa. The spacing between bands of like density is relatively constant being on the average 554

Å apart. Jean Dragesco of the Collège de France has made electron micrographs (unpublished) of the trichocyst tips in seven species of *Paramecium* and has shown the existence of species differences in the structures (Fig. 27D). Fig. 29A represents a stereoscopic pair of electron micrographs of the trichocysts of *P. caudatum* showing what appears to be a structure composed of a stack of disks arranged on a four-fold helical axis. A stereoscopic pair of trichocysts of *P. aurelia* is shown in Fig. 29B revealing a coarser spacing than that of *P. caudatum* (Anderson, T. F., unpublished). Additional information upon trichocysts is given by Knoch and König (1951) and Wohlfarth-Bottermann (1950).

The reader is referred to Jakus (1945) for detailed information on the effects of enzymes, acids, alkalies, electrolytes, and other chemicals upon the trichocysts. She concluded that the tip as well as the shaft is protein in nature. Although the trichocyst membrane protein is not identical with any known protein, it is fibrous and appears to resemble collagen more closely than any other fibrous protein. X-ray studies of the fibrous proteins indicate that they fall almost exclusively into two main configuration groups as follows: (1) the keratin-myosin group and (2) the collagen group. From the studies made thus far, it appears that the chemical composition of the trichocysts falls into the latter group although the writer is unaware of any x-ray defraction studies made upon these elements. Jakus makes the suggestion that because of the high

FIG. 27. (A) Paramecium with trichocysts extruded, as a result of the application of picric acid. (B) Hypothetical longitudinal section of (a) the resting trichocyst, and (b) the extruded trichocyst of Paramecium. (C) Diagram of (a) a resting trichocyst, and (b) a partially extruded trichocyst as seen in the dark-field microscope. (D) Drawings of electron micrographs of trichocyst tips to show the existence of species differences in these structures. (1) *P. caudatum*. (2) *P. multimicronucleatum*. (3) *P. bursaria*. (4) *P. aurelia*. (5) *P. polycaryum*. (6) *P. "porculus"* (Fauré-Fremiet). (7) *P. trichium*.

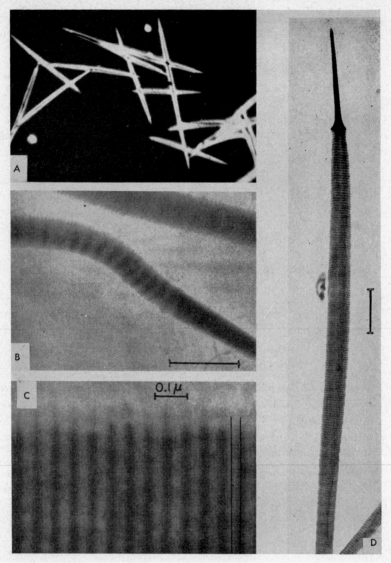

Fig. 28. (A) Dark-field photomicrograph of trichocysts stained with methylene blue. (× 1300.) (B) Electron micrograph of trichocysts extruded with a trace of phosphotungstic acid. (× 24,000.) (C) Electron micrograph of portion of trichocyst shaft, from 3-day-old suspension of trichocysts extruded by electrical shock, dried and stained with phosphotungstic acid at pH 2. (× 113,000.) (D) Electron micrograph of trichocyst extruded by drying, and stained with phosphotungstic acid at pH 2. (× 15,000.) (Jakus)

velocity of trichocyst extrusion, the membrane is preformed in the resting trichocyst and is extended by the sudden uptake of water during the process.

According to Brodsky (1924) the fundamental force is not the mechanical pressure in trichocyst extrusion but the expansion of the colloidal substances upon

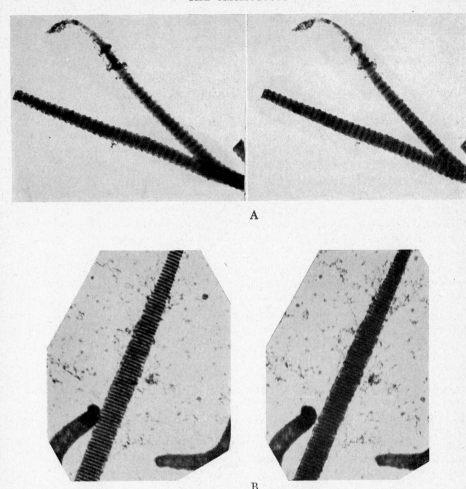

A

B

FIG. 29. Stereoscopic pairs of electron micrographs of the trichocyst shafts in *P. caudatum* (A) and *P. aurelia* (B). (× 20,412.) (Anderson)

stimulation in the extrusion through the pellicle.

Although we now have available a fairly good body of knowledge of its detailed structure and biochemical nature, we know comparatively little about the origin, development, and function of trichocysts.[8] Mitrophanow (1905) is of the opinion that the trichocysts in paramecium are formed in the endoplasm near the macronucleus. They are then said to migrate toward the periphery of the body and assume a radial position in the ectoplasm. This view is also held

[8] In the ciliate, *Frontonia leucas,* Tönniges (1914) reported the presence of chromatic material in the endoplasm. Because of its structure he called this material "trichochromidia" and interpreted it as developing stages of trichocysts migrating to the ectoplasm. This actually appears to be extruded macronuclear chromatin. In this regard, Chatton, Lwoff, and Lwoff (1931) review briefly other theories of trichocyst origin and report that the trichocysts originate from basal granules in the *Foettingeriidae,*

by Brodsky (1908, 1924) who believed in addition that they are composed of colloidal excretory substances. Lwoff (1950) is of the opinion that kinetosomes in ciliates are able to give rise to granules which produce trichocysts. In one of his figures of paramecium, Gelei (1925) shows trichocysts in the endoplasm which he believed to be in various stages of development. The explanation of trichocyst origin in paramecium is as yet unsatisfactory. It should be taken into account that in the microtome-sectioning of paramecium it is possible for trichocysts to become dislodged from the ectoplasm and fall into the endoplasmic regions.

The function of trichocysts is still uncertain. There does not appear to be any toxic action associated with trichocyst extrusion in paramecium but such toxic action has been described for other ciliates (Visscher, 1923). Although trichocysts in paramecium are generally considered to be organelles of defense, little security is given them from their chief enemy, Didinium. Mast (1909) has demonstrated that when great masses of trichocysts are extruded, the protective screen given paramecium results in some measure of safety (Fig. 135). Observations by the author support the view that extrusion of trichocysts represents a reaction to injury.

Saunders (1925) and Jones and Cloyd (1950) report that on coming into contact with a solid surface paramecium frequently attaches itself by extensions of the trichocysts. Saunders observed that adhesion was more frequent around pH 7.9 and expressed the view that alkali acts on the paramecium thereby increasing its permeability and osmotic pressure. This in turn causes the extrusion of small quantities of trichocyst fluid which solidifies on contact and permits anchorage. Jennings (1931, p. 60) reports a similar phenomenon of attachment except that the cilia are used instead of the trichocysts.

Brodsky (1908) described the peculiar phenomenon of internally discharged trichocysts in Frontonia and Khainsky (1910), Jacobson (1931), and Lund (1933) described them for P. caudatum (Fig. 22E). The elements described and figured by Rees (1922 a) as neuromotor fibers enervating the trichocysts may in reality be internally discharged trichocysts.

E. The Contractile Vacuoles

In the first separate treatise devoted to the Protozoa, Joblot (1718) reported the discovery of the contractile vacuoles. Spallanzani (1776) gave a good description of them in a ciliate that appears to have been paramecium. He reported their rhythmic changes, radiating canals and even went so far as to suggest they have a respiratory function. Contractile vacuoles have been studied since by a large number of investigators who have added much to our knowledge of their structure, operation, and function.

Bütschli (1887–89) classified contractile vacuoles into two general classes. He called the first class vesicle-fed vacuoles which are formed by the coalescence of a number of smaller or accessory vacuoles. The second class was called canal-fed vacuoles. These consist of from one to 10, commonly five to seven, slender radiating canals which are found in one plane. These radial canals form a characteristic rosette about each larger, roughly spherical vacuole and empty into it. However, King (1935) has shown for P. aurelia and P. multimicronucleatum that the radial canals form small vesicles which increase in size, coalesce and form the contracting vacuole. If this method of vacuolar formation is the same in all other species of Paramecium which possess radial canals,

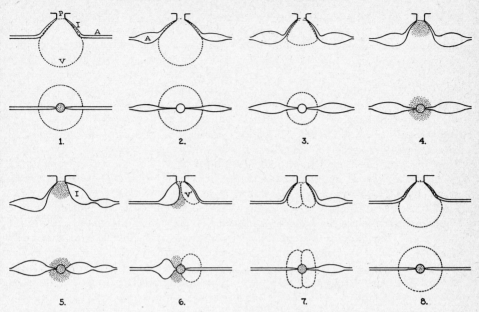

FIG. 30. Diagrams of successive stages in the cycle of the contractile vacuole in *Parame-cium multimicronucleatum*. (*Upper figs.*) Side views. (*Lower figs.*) Face views of the vacu-ole. Temporary structures represented by dotted lines; permanent structures by entire lines. (V) Vacuole. (V¹) Vesicle. (P) Pore. (I) Injection canal. (A) Ampulla. (1) Full diastole, with membrane closing pore. (2) Beginning of diastole, membrane closing pore ruptured. (3) Middle systole, with ampullae filled. (4) End of systole, membrane closing pore. (5) Contents of ampulla passing into injection canal. (6) Formation of vesicles from injection canals. (7) Fusion of vesicles to form contracting vacuole. (8) Contracting vacuole in full diastole. (King)

the distinction made by Bütschli would not hold. Nevertheless, a distinction can be made on the presence or absence of radial canals which surround the vac-uole.

Species of *Paramecium* normally pos-sess two contractile vacuoles, except for *P. multimicronucleatum,* which com-monly has two to three but may show up to seven. Contractile vacuoles are located directly underneath the ecto-plasm on the dorsal surface with the exception of *P. trichium,* where they are more deeply situated in the endo-plasm. The vacuole which is found nearer the anterior end of the body and in the vicinity of the macronucleus is sometimes called the nuclear vacuole.

The other, located posteriorly and nearer the peristome, is occasionally re-ferred to as the peristomial vacuole. In the case of *P. multimicronucleatum* where there are three or more vacuoles, one is found in the anterior region and the remaining ones in the posterior re-gion of the body, all in nearly perfect linear alignment.

The functional cycle of these struc-tures is as follows: A contractile vac-uole becomes enlarged with fluid—mainly water—until it reaches its max-imal size; the phase being called *dias-tole*. This phase is quickly followed by another called *systole* which is charac-terized by the sudden collapse of the vacuole. Upon contraction, the vacuolar contents are discharged to the outside of the body through a pore which is located directly above each vacuole and is fixed in position (Fig. 30). Contradictory reports have appeared, especially within the past 100 years, on the nature, operation and permanence of the contractile vacuoles.[9] One finds in the literature a large number of investigators including Carter, Degen, Ehrenberg, Claparède, Lachmann, Siebold, Stempell, and more recently Nassonov, Gelei, Lloyd, and Young who favor the idea of a permanent membrane surrounding the contractile vacuole. Opposing this view in favor of a contractile vacuole without a permanent membrane are Dujardin, Meyen, Stein, Perty, Maupas, Rhumbler, Schmidt, Zenker, Wrześniowski, Bütschli, Lankester, Kent, Khainsky, and more recently Taylor, Day, King, and Dimitrowa.

While this difference in opinion may be due in part to the inadequacy of microscopes of the time, observations upon different species, or upon living or killed and stained specimens, technique,[10] and the fallibility of the observer, it is now possible to reconcile some differences of opinion concerning membranes in the light of newer knowledge in physiologic chemistry. Also, it now appears that certain parts of a contractile vacuolar system are permanent and other parts only temporary to be formed anew after each systole. Probably no one today would deny the existence of contractile vacuolar membranes

and their selective permeability—a property of all living membranes. But opinions vary on the visibility, thickness, and permanence of these vacuolar membranes. Certain membranes, conspicuous in living as well as fixed and stained animals, have been designated in the literature as morphologic membranes. On the other hand, most living or physiologic membranes are only a single molecule to a few millimicra thick. Such a thin membrane is actually a phase boundary or interface between two different or immiscible fluids and beyond the limit of visibility.

It occurs to the author that since membranes—whether they be referred to as physiologic or morphologic—function similarly, the distinction between them is a poor one. Microscopes and newer methods of the future are likely to demonstrate membranes not now visible. Their only difference then would be from a monomolecular layer to a multimolecular one—a difference primarily in degree rather than kind.

CONTRACTILE VACUOLES WITH RADIAL CANALS

A contractile vacuolar apparatus consists of from five to 12 radial canals (feeding canals, feeders) which empty into and ultimately form the contractile vacuole. A short tubule connects the vacuole to the pellicle where the tubule emerges as a pore. Vacuolar contents pass out this tubule first through the internal pore which is adjacent to the vacuole, then to the exterior by means of the external pore on the dorsal surface. According to Khainsky (1910) a thin, elastic region of cytoplasm called the *papilla pulsatoria*, bursts at the time of vacuolar collapse to form the pore in *P. caudatum*. That fluid actually leaves the vacuole was demonstrated by Jennings (1904).

The structure and operation of the

---

[9] For a detailed account of the structure and function of the contractile vacuole in Protozoa, the reader is referred to Weatherby (1941).

[10] Hirschler (1927) and MacLennan (1933) have shown that the warm method of impregnation as advocated by Nassonov (1925) tends to produce overimpregnation. This results in the formation of a heavy black zone resembling a membrane but it is in reality a granular osmiophilic layer.

canal-fed contractile vacuoles have been studied in detail by Nassonov (1924, 1925), Gelei (1925, 1928), King (1935, 1950, 1951), Fortner (1924, 1925, 1926 b, c), and others. The knowledge obtained has been based upon living specimens and upon stained, impregnated, and sectional material. It is generally held that the radial canals consist of three parts as follows: (1) the terminal (distal, posterior) part, (2) the ampulla, and (3) the injector (injecting canal) (Figs. 31 and 32). In the region of the terminal part, a highly specialized osmiophilic material is found from which, according to Nassonov (1924), hypertonic fluid is secreted into the lumen of the radial canals. He believes that this osmiophilic material is homologous to the Golgi material in the metazoan cell although this idea is rejected by King (1935). Gelei (1925, 1928) refers to this region and material as the "nephridial plasma" and compares the nephridial system of paramecium with that of higher forms of animal life.

The ampulla is a bulb- or flask-like structure when fully distended (diastole of ampulla), but at systole, it is of the same diameter as the terminal portion and cannot be sharply delimited from it. The fluid, which appears to be mainly water, is removed from the protoplasm surrounding the radial canal.

Fluid is then passed to, and temporarily stored in, the ampulla which, in diastole, now becomes flask-shaped. The work of Fortner and King (1935) suggests that since the ampulla is not always flask-shaped but at times composed of a series of bulb-like swellings, peristaltic action may occur by which the ampulla is finally filled with liquid (Fig. 32E).

When the ampulla is fully distended in diastole, it collapses and the fluid is then passed through the injector to form the contractile vacuole. Thus it is seen that the systole of the ampullae results in diastole of the vacuole. King (1935) has shown that the injectors lead up to the pore not only in systole but even in diastole. This would indicate that the ampullae do not fuse to form the contractile vacuole nor that ampullar fluid is forced into the vacuole. Instead, as the ampullae empty themselves, the injectors swell up and pass their fluid contents into the cytoplasm to form a vesicle immediately beneath the pore. The membrane of this vesicle is derived from a membrane which closed the end of the injecting canal. According to King, separate fluid vesicles exist for an instant, then coalesce to form the extended contractile vacuole now in full diastole. The fluid is then discharged through the delicate canal and finally to the exterior by means of the external

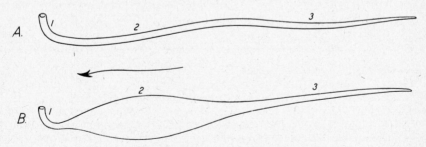

FIG. 31. Diagram of radial canal. (A) Empty. (B) Full; (1) injector, (2) ampulla, (3) terminal part. Arrow direction of flow. (Wichterman)

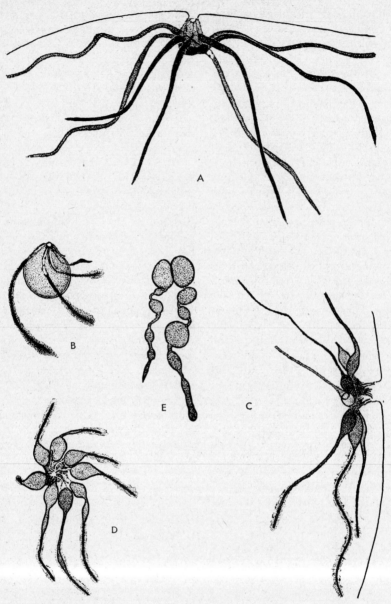

FIG. 32. All figures of *P. multimicronucleatum* (osmic impreg.). (× 1200.) (A) Lateral view of contractile vacuole in diastole or early systole. (B) Side view of vacuole in diastole with the injection tubules of feeding canals leading up to the pore. (C) Lateral view of vacuole in complete systole with injection canals leading up to pore, ampulla filled. (D) Face view of contractile vacuole in complete systole showing pore and feeding canals with injection tubules, ampullae and distal portion with specialized protoplasmic layer. (E) Feeding canal showing many ampulla-like swellings. (King)

pore. The pore is then closed by the remains of the membrane of the contractile vacuole.

Like King (1935) and Day (1930), Fortner (1924, 1926 b, c) believes the contractile vacuolar membrane is a temporary structure—that it disappears at systole and must be formed again during diastole. On the other hand Nassonov (1924) demonstrated osmiophilic membranes around the vacuoles in *P. caudatum* which he reported as permanent structures merely collapsing at systole, only to be refilled from the radial canals at diastole. With Gelei (1925, 1928), he is of the opinion that the apparent flowing together of the proximal ends of these canals represents only the openings of the permanent vacuolar walls. In this regard, Haye (1930) reports the presence of thin but definite contractile vacuolar walls in paramecium. Young (1924) came to the conclusion that the vacuolar system in *P. caudatum* is a permanent one, a view which is also held by Lloyd (1928). Howland (1924) reported that the walls of the contractile vacuoles in paramecium are composed of droplets which are formed by the dissolving of small granules. Fortner observed such granules surrounding the radial canals before they became empty. Fortner, with others, maintains that the proximal ends of the radial canals unite to form the contractile vacuole. According to King and Fortner, the radial canals are permanent structures, but the vacuole itself is a temporary structure, always created anew after each contraction.

In their experiments upon ultracentrifuging of paramecium, King and Beams (1937) report that the contractile vacuole is sometimes forced away from its outlet pore and into the endoplasm and that a new vacuole arises from the coalescence of the vesicles formed at the ends of the feeding canals. They present evidence that the main vacuole and its membrane are purely temporary, formed anew before each diastole by the fusion of feeding vesicles formed at the vacuolar ends of the feeding canals.

It is held by Young (1924) and others that the discharging tubule and its outlet pore, like the radial canals, are permanent structures. The pore, which measures approximately 0.8–1.7 microns, can be demonstrated in stained and living animals during systole and diastole (Fig. 33). Fortner claims the pores to be temporary structures in that the *papilla pulsatoria* are seen only during collapse of the vacuoles. His view that the pore is always found in a space devoid of cilia in from four to six ciliary fields is denied by King. Gelei (1928) observed a series of tangential lines surrounding the pore which gave the appearance of an iris diaphragm. He suggested that the lines may represent the closing mechanism of the pore. King (1935) however, has shown that these lines are actually the feeders leading from the ampulla.

Usually one pore is found for each contractile vacuole. Gelei (1925) reported that two pores are present at each contractile vacuole in "*P. nephridiatum*" and accordingly is a species character. However, Kalmus (1931) and King (1935) have reported the occasional presence of two pores at one contractile vacuole in *P. caudatum* and *P. aurelia* respectively.

King (1950) studied the origin and fate of vacuolar pores in nigrosin-stained dividing and non-dividing paramecia and noted a tendency for new pores to be multiple during the fission process as follows: *P. aurelia*, 8 per cent, *P. caudatum*, 57.4 per cent, and *P. multimicronucleatum*, 65.6 per cent. He found that in vegetative *P. aurelia*,

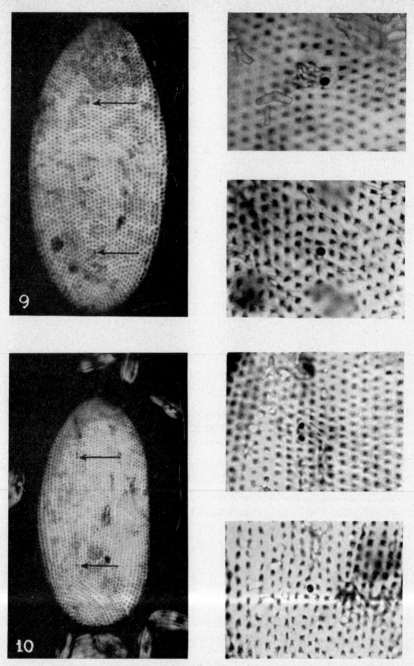

Fig. 33. (9) Dorsal view of *P. aurelia* showing pore of anterior and posterior contractile vacuoles marked by arrows. (× 500). To the right, enlarged views of the same. (× 1600.) Nigrosin preparation. (10) Dorsal view of *P. aurelia* showing two pores of the anterior vacuole and one of the posterior. (× 500.) To the right, enlarged views of the same. (× 1600.) Nigrosin preparation. (King)

the anterior pore was multiple in 1.4 per cent, the posterior in none; in *P. caudatum* the anterior pore was multiple in 2.8 per cent, the posterior in 1.8 per cent; in *P. multimicronucleatum,* the anterior pore was multiple in 34.7 per cent, the posterior in 24.5 per cent. In the two latter species, they often function in association with an extra vacuole which is passed on during the fission process. King concluded that extra contractile vacuoles so often seen in species of *Paramecium* result from this tendency to produce multiple pores and contractile vacuoles at division.

Contractile Vacuoles without Radial Canals: *P. trichium*

The conspicuous vesicle-fed contractile vacuolar system of *P. trichium* lacks radial canals, hence in this respect it is entirely different from all other species of paramecium.

In his original description of the species, Stokes (1885) reported two rapid and alternately contracting vacuoles in the anterior region of the ciliate. None was reported from the posterior region. In a study of this species Wenrich (1926, 1928 a) reported vacuoles in the posterior end as well, which were evidently overlooked by the discoverer. This omission by Stokes is understandable because the contractile vacuoles in *P. trichium* are situated deeply in the endoplasm, and the posterior end of this organism frequently possesses many dark granules and crystals which tend to obscure the posterior vacuole in the living condition. Stained preparations, of course, show both contractile vacuoles readily. Wenrich (1926) was uncertain whether the two main vacuoles emptied alternately into the same outlet tube or whether there existed an auriculo-ventricular relationship between them.

We are indebted to King (1928) for a detailed account of the structure and operation of the contractile vacuoles in this species. Using the relief method of staining as described by Bresslau (1921) and Coles (1927), King was able to demonstrate that each contractile vacuole is connected with the exterior by a long convoluted tubule (Fig. 34). The tube was found to terminate in a small pore located between longitudinal ciliary rows on the dorsal body surface. Each pore of the anterior and posterior contractile vacuole was found to open to the exterior between the same or adjacent ciliary rows. Unlike the contractile vacuoles in the other species, they do not appear to be fixed in position but move about slightly in the endoplasm. King found that the inner end of the convoluted tubule terminates in a cup-like valve with which the vacuole is in contact when undergoing systole. He observed that the vacuole takes about two seconds to empty (systole) and the time between completion of two systoles 3.1 seconds at 25° C. Upon contraction, the feeding vesicles grow (Fig. 34 [12a, b]). According to King, by the time the vacuole has half completed systole (Fig. 34 [12c]) the feeding vacuoles coalesce; upon completion of systole (Fig. 34 [12d]) the new contractile vacuole has reached its maximum size. Thus when it comes in contact with the cup-like valve, a membrane appears to pass over its surface, then it begins to contract (Fig. 34 [12e]). King has shown that the membrane of the contracting vacuole closes the inner end of the convoluted discharging tubule at the end of systole and therefore acts as a valve. The new vacuole, with enclosing membranes, is formed by the coalescence of the feeding vesicles. It is before systole of this new vacuole that the remnant of the old vacuolar wall, in closing the internal pore, joins with the adjacent wall

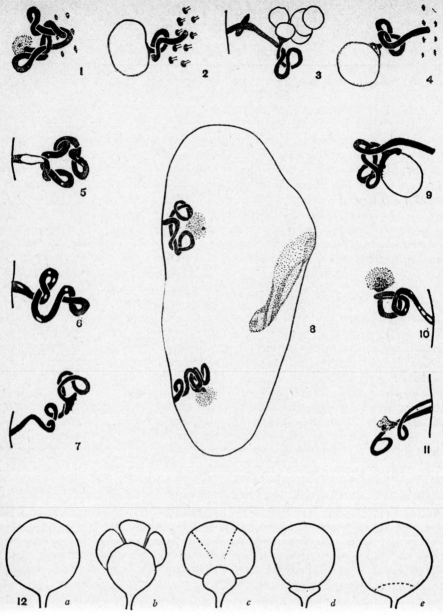

Fig. 34. Contractile vacuolar apparatus of *Paramecium trichium*, dried in 10 per cent China Blue unless otherwise stated. ($\times$ 1000.) (1) Posterior apparatus of animal. (2) Anterior apparatus of animal. (3) Posterior apparatus. (4) Posterior apparatus. (5) Anterior apparatus. (6) Posterior apparatus of same animal as Fig. 5. (7) Posterior apparatus of animal. (8) Entire animal showing cytopharynx, anterior and posterior vacuolar apparatus. (9) Anterior apparatus. (10) Anterior apparatus (10 per cent nigrosin). (11) Two tubules and pores from anterior end of same animal. (12) Diagrams of various stages of contracting vacuole: (*a*) beginning of systole; (*b*) systole half-completed, feeding vesicles grow larger; (*c*) fusion of feeding vesicles; (*d*) systole complete, cup-like inner end of tubule; (*e*) fusion of new contractile vacuole with cup-like inner end of tubule preparatory to systole. (King)

of the newly formed vacuole and ruptures inwardly. Fluid in this manner is allowed to escape through the tubule and external pore.

ORIGIN AND INHERITANCE OF CONTRACTILE VACUOLES AND PORES

In binary fission, the contractile vacuoles separate, one going to each daughter. The anterior daughter retains the anterior parental vacuolar apparatus as its posterior vacuole and a new one is formed anteriorly. The posterior daughter retains the posterior parental vacuolar apparatus as its posterior vacuole and a new one is formed anteriorly. Nassonov (1924) and Haye (1930) present some evidence that the new vacuoles originate from the old ones. However, the view is more generally held that the new contractile vacuoles arise de novo as reported by Dimitrowa (1928). King (1951) noted that in P. aurelia, one of the first signs of beginning division is the simultaneous appearance of two new smaller pores, each anterior to but not connected with each of the old pores. The new pores become the anterior pores of the new daughters. The old anterior pore becomes the posterior pore of the anterior daughter while the old posterior pore becomes the posterior pore of the posterior daughter. See also King (1952).

During conjugation, the author has observed that the contractile vacuoles undergo no change. They can always be seen to function normally throughout the entire process.

Dimitrowa (1928) induced the formation of extra contractile vacuoles in compressed specimens of P. caudatum by simple mechanical pressure on the cover glass. The extra vacuoles so produced appeared normal although in a few cases, no radial canals were found. It is interesting to report that in such pressurized paramecia, the usual number of vacuoles was restored at fission. On the other hand, if one had been induced by pressure at fission, only one new vacuole was formed. In the case where three extra vacuoles were induced, one daughter received three and produced a new one when the organism divided. According to Dimitrowa, a marked accumulation of metabolites occurs during fission which results in a disturbance of normal functioning of the animal. The result is that new contractile vacuoles are formed. After fission, the daughter cells now with two contractile vacuoles each are without an excess of metabolites, therefore the stimulus for the creation of extra vacuoles is absent until the next fission process. The author (Wichterman, 1946 b) has observed the tendency for the formation of extra contractile vacuoles in paramecia parasitized by microörganisms. It is conceivable that the metabolic waste products of the parasites so disturbs the host that supernumerary contractile vacuoles are formed.

Information on the origin and formation of extra contractile vacuoles has come from an observation made by Hance (1915, 1917). His work was reported on a species called P. caudatum but which appears to have been P. multimicronucleatum. This species should serve as a valuable tool because of its tendency to produce extra contractile vacuoles. Hance reported that very small vacuoles occur which have apparently just formed and which are usually at some distance from the others. These increase fairly rapidly in size until they reach the maximum size. During the growth of a new vacuole, the one nearest to it temporarily loses its regular contraction and when the new vacuole has reached full size it beats spasmodically a few times before it settles down to its regular rhythm. Soon the old and the new vacuoles become accustomed to the new conditions and the usual rhythmic beat begins. This is not always the case, since vacuoles have been observed to form without affecting the rhythmic beat of the older vacuoles near it.

The work of Hance (1915) on the inheritance of extra contractile vacuoles in this species of paramecium is worthy of note. He isolated single specimens each with three contractile vacuoles. In one pure-line culture several weeks after isolation, he found 8.6 per cent of the individuals with two vacuoles; 65.7 per cent with three vacuoles and 25.7 per cent with four contractile vacuoles. He reported that in cultures with very rapidly dividing paramecia, 59.1 per cent have only two contractile vacuoles but develop a third or even a fourth vacuole later. In cultures started with two-vacuoled specimens, progeny were found showing three and four vacuoles. The most common distribution at division in the three-vacuoled forms was three vacuoles to the posterior daughter and two to the anterior one—a new vacuole, of course, arising in each daughter.

It is suggested that *P. multimicronucleatum*, with its tendency to form extra contractile vacuoles, be studied in the precision microcompression chamber (Wichterman, 1940, Fig. 106s). With this instrument fine pressure adjustments to within a few microns can be made on a single living specimen over a long period of time. In this manner, valuable information on the origin of contractile vacuoles might be obtained.

In an examination of the literature on contractile vacuoles in species of paramecium, one observes the contradictory claims of a number of investigators. More exacting work is required, especially of an experimental nature, upon living organisms. One should exercise extreme care in the interpretation of killed specimens and be on guard against overstained or overimpregnated material. Whether they be invisible physiologic membranes or visible so-called morphologic membranes, the fact remains that there is bounding the contractile vacuole a membrane which demonstrates selective permeability. A number of questions remain unanswered. For instance, what prevents the fluid from passing back into the radial canals upon systole of the contractile vacuoles? Lloyd and Beattie (1928) report that it may occur although Kalmus (1931) is inclined to disagree.

The functions ascribed to contractile vacuoles are considered in Chapter 7.

F. Cytoplasmic Inclusions

1. **Mitochondria (Chondriosomes; Cytomicrosomes, etc.).**

Mitochondria are found in the cytoplasm of paramecium as well as in nearly all kinds of plant and animal cells. Usually they exist as single spherules or chains of spherules, short rods or filaments, or appear dumbbell-shaped and measure from 0.5 microns to 1.5 microns. In the first monographic treatment of mitochondria in Protozoa, Fauré-Fremiet (1910) showed that they exist as a separate group of cytoplasmic inclusions and are found practically universally in unicellular animals.

In life, mitochondria are refractile but in osmic acid techniques, become gray-brown or black, and are found to bleach faster than Golgi bodies. They stain with basic dyes after fixation in lipoid preservatives, and frequently stain lightly or not at all after fixatives containing acetic acid. Mitochondria become green in color when stained vitally with Janus green B.

Ordinarily mitochondria appear to be evenly distributed throughout the cytoplasm but in certain Protozoa, these bodies may be localized around storage granules, beneath the pellicle and around the contractile vacuole. Volkonsky (1934) observed a small concentration of mitochondria around the food vacuole of paramecium during the alkaline phase of digestion and Horning (1926) reports that they regularly occur between the basal granules.

Causey (1926) described the mitochondria of paramecium as rod-shaped and less than a micron in length (Fig. 35). Fauré-Fremiet (1910) reported

Fig. 35. Cross section of *P. caudatum* showing many rod-like mitochondria. (× 1300.) (Causey)

that they are spherical and larger than a micron. Horning (1926) in one of the few observations on living mitochondria reported the presence of dumbbell-shaped forms in division which have so frequently been observed in fixed and stained preparations.

It has been shown by a number of investigators, namely Fauré-Fremiet (1910), and Hirschler (1924, 1927) that mitochondria are composed of both lipoids and proteins as shown by their solubilities and reactions to stains.[11] The

[11] As is the case with lipoids in general, they are soluble in acetic acid, ether, and absolute alcohol, stain with Nile blue sulfate and the sudans as well as other fat-soluble dyes. It should be remembered that fatty acids, neutral fat, and phosphatides also respond to such tests.

statements made upon the exact chemical constitution of mitochondria vary according to the investigator. Some claim they consist of a phosphatide, unsaturated lipoids, fatty acids, neutral fat, albumin, and nucleic acid. Others claim they possess a large amount of lipoid to a small amount or none and that they possess various kinds of protein. Thus it appears that mitochondria of different organisms vary considerably in their solubilities due to a difference in ratio of lipoids to proteins. Giroud (1929) observed that the mitochondria of paramecium are resistant to lipoid solvents, therefore it is apparent that they should be visible after Formalin-alcohol fixation. MacLennan and Murer (1934) report that the typical mitochondrial rods of *P. caudatum* leave a distinct ash while the mitochondria of *Opalina* do not.

Many functions have been ascribed to mitochondria. They have been reported as being reserve food and possessing enzymes, thus playing a role in digestion. They have also been described as playing a part in cellular respiration and excretion. Zweibaum (1922) reported an increase in the amount of fatty acid in *P. caudatum* immediately before conjugation and suggested that mitochondria possess a reproductive function.

That mitochondria are other than cytoplasmic in origin is considered by Joyet-Lavergne (1926) who reported that they are bodies attached to protein granules which originate from nuclei. This interpretation, however, is questioned by Daniels (1938). For *Uroleptus,* Calkins (1930) reported that granules which he believed to be mitochondria were traceable to macronuclear fragments. Miller (1937) in studying amebas, raised the question whether not only mitochondria but Golgi bodies and certain other cytoplasmic inclusions may not be bacterial spores, fungi, yeasts, and indigestible material. While it is likely that some cytoplasmic inclusions reported by a few observers may have been foreign elements and not mitochondria, the fact remains that mitochondria are specialized, commonly encountered cytoplasmic constituents of paramecium and most protozoa.

2. Golgi Bodies (Apparatus) and the Vacuome.

In 1899, Golgi discovered and reported extremely small cytoplasmic inclusions which now bear his name. Structures referred to as Golgi bodies have been questioned by some who criticize the rigorous treatment given the protoplasm which is frequently necessary to demonstrate them. Although in some instances artifacts have been called Golgi bodies, there does not appear to be any question but that Golgi bodies are fairly consistent cytoplasmic inclusions. A valuable examination of our knowledge of the Golgi apparatus in the animal cell is given by Hibbard (1945).

The Golgi bodies are generally shown to appear as small segregation granules, spherules, globules, short rods, ovoid elements, osmiophilic nets, and specialized regions of protoplasm associated with contractile vacuoles. Their demonstration depends primarily upon their reduction of certain metallic compounds such as osmium or silver resulting in a blackening of the bodies. Even when such material is treated with turpentine or hydrogen peroxide, the bleaching process removes most cytoplasmic elements but leaves the Golgi bodies present. Overimpregnation, especially at high temperatures, is likely to show the discrete, isolated Golgi bodies as heavy black bands or rings. The commonly used cytological stains do not demonstrate them. Golgi material is generally believed to be composed largely of lipoid substance but it has been shown that lipoid and non-lipoid bodies react the same with Golgi techniques. A method generally used for differentiating Golgi material from mitochondria is the staining of the organism with a mixture of neutral red and Janus green. Golgi material is stained with neutral red and mitochondria with Janus green.

In ultracentrifuge methods with many kinds of metazoan cells, mitochondria have been found to be heavier than the Golgi apparatus which is said to be one of the lightest constituents of the cytoplasm. Neutral red granules and Golgi bodies can be distinctly separated showing that these are not identical substances.

The *vacuome* is a term that was originated to take the place of segregation granule. It is supposed to indicate the claimed homology between the neutral red bodies in animal cells and the vacuoles in plant cells. The Golgi apparatus and the vacuome are interpreted as being identical elements according to many investigators who believe that the granules stainable with neutral red are the same as granules demonstrated by Golgi methods. In contradiction, others have demonstrated that many cases of neutral red granules are not osmiophilic and hence believe the vacuome to be a reality. It has been shown that neutral red stains several different groups of granules and does not stain all stages of the same granule. One is therefore inclined to agree with MacLennan in that it is difficult to have these granules reveal fundamental homologies. In view of the lack of evidence of the true nature of such granules it appears unwise to place them in a category called the vacuome. For a further discussion of the "neutral red reaction" and the vacuome, the reader is referred to the work of Hall and Nigrelli (1937).

CONTRACTILE VACUOLES AND THE GOLGI APPARATUS

It was first suggested by Ramon y Cajal (1940) that the contractile vacuole is homologous with the Golgi apparatus of the Metazoa. Nassonov (1942) then demonstrated the presence of osmiophilic membranes around the radial canals in *P. caudatum* and many other ciliates. Like Nassonov, Gelei (1925, 1928) described similar structures in paramecium and called the region of specialized protoplasm in osmicated material, the "nephridial plasma." Nassonov ascribed a secretory function to the region and Gelei an excretory function. Gelei showed the Golgi material as short, deeply staining rods and as a net surrounding the distal portion of the radial canals. King (1935) also observed osmiophilic protoplasm in the walls of the distal part of the radial canals of *P. multimicronucleatum* after fixation with Schaudinn's fluid (plus acetic acid) and by other methods which do not usually preserve Golgi material (Fig. 32). He therefore rejected the theory that this specialized osmiophilic protoplasm is homologous with the Golgi material of metazoan cells. The fact remains that many ciliates are without radial canals, do not have osmiophilic protoplasm near the vacuoles or even in the vacuolar walls.

Much has been written about Golgi bodies and the vacuome but there is as yet no clear-cut evidence pertaining to their exact chemical composition or function.

3. Crystals.

Crystals and crystalline granules are commonly found in the cytoplasm of paramecia. If very small, such as the crystalline granules, they are likely to be distributed rather uniformly throughout the protoplasm. If large, they are likely to be concentrated at both ends of the animal, especially the posterior end (Figs. 10, 15, and 36). When large clusters of crystals are seen with transmitted light under low power of the microscope, the mass appears dark; under high power, it appears colorless and translucent. They are shown to be optically active when viewed with polarized light. That they are strongly birefringent in paramecium was first observed by Maupas (1883) and highly refractile by Schewiakoff (1893).

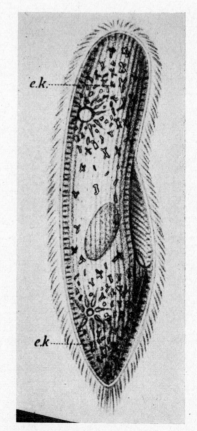

FIG. 36. *Paramecium caudatum* showing masses of crystals (e.k.) concentrated in the anterior and posterior ends of the body.

Crystals in Protozoa have been extensively studied by many investigators who have reported them to consist of uric acid, calcium oxalate, calcium carbonate, leucine, di- or tribasic calcium phosphate, amino acids, calcium chlorophosphate, a magnesium salt of a substituted glycine, sodium urate, and silica. Crystals have been considered by most as being excretory products although it has been suggested by some that they be regarded as reserve material in the cytoplasm.

Stein (1859) claimed to have observed the elimination of crystalline granules of *P. bursaria* with egested material. Maupas (1883) and Schewiakoff (1893) however, were unable to observe such a method of crystal elimination in *P. aurelia* and *P. caudatum* respectively. Schewiakoff reported that *P. caudatum* often contains crystals composed of calcium phosphate combined with some organic substance. He reported that when the paramecia were starved, the crystals disappeared completely in one to two days only to reappear when food was available to the paramecia.

Recently Bernheimer (1938) made a study of the crystals in paramecium and other Protozoa with special reference to their solubility in various reagents. He came to the conclusion that the crystalline bodies in *P. caudatum, P. multimicronucleatum, P. aurelia, P. bursaria, P. trichium,* and *P. woodruffi* are identical in all properties and that according to their shapes may be classified into crystals, granules, aggregates, and clusters (Fig. 37). They may range in maximum dimensions from 0.2 to 10 microns with aggregates measuring up to 25 microns. As reported by Bernheimer, forms of crystals frequently seen in paramecium are elongated, flattened prisms with singly or doubly truncated ends (Fig. 37 [4, 7]). Also present are crystals which appear to be contact twins (Fig. 37 [14]), interpenetrating twins (Fig. 37 [8 and 9]), and geniculated types (Fig. 37 [10]). Crystalline granules (Fig. 37 [5 and 15]) are usually less than 3 microns, in shape, are rounded to angular. The crystals are reported as being monoclinic, with a melting point of 400°–500° C and refractive indices of a 1.48 and γ 1.87. In studying the effects of a large number of reagents on crystals of paramecium, they are reported as dissolving immediately in HCl, HNO_3,

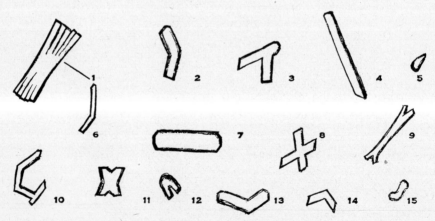

Fig. 37. The types of crystals found in *Paramecium multimicronucleatum*. For description, see text. (Bernheimer)

dilute H_2SO_4, NH_4OH, phosphoric acid and saturated aqueous solution of KOH. They dissolve within two minutes in hot water, strongly heated glycerol, and hot glacial acetic acid. The crystals are not dissolved in benzene, carbon disulfide, carbon tetrachloride, ether, chloroform, acetone, xylene, and solutions of melamine and sodium bicarbonate. They possibly may be related to melamine. The author (Wichterman, 1941) observed great numbers of optically active crystals in the posterior region of a race of zoöchlorellae-free *P. bursaria*. When races of normally green *P. bursaria* with few or no crystals were kept in complete darkness up to 25 days, the zoöchlorellae decreased in number but large numbers of crystals appeared. When restored to light again, the zoöchlorellae increased in number while the crystals were seen to disappear. When normally green *P. bursaria* which do not possess optically active crystals are irradiated with 300,000–600,000 roentgen units, the zoöchlorellae disappear but optically active crystals appear in the cytoplasm and persist indefinitely. This appearance of crystals appears to be correlated with a gradual change in the type of nutrition, i.e. from a holophytic to a holozoic one in which the colorless, x-rayed, crystal-bearing forms were dependent upon bacteria as a source of food (Wichterman, 1948 c).

That crystals in the Protozoa are not all alike chemically was shown by Bernheimer who reported that those in paramecium differ from the ones found in *Halteria* and several species of amebas under investigation.

4. Food Vacuoles and Digestive Granules.

The most conspicuous cytoplasmic elements of paramecium are the food vacuoles. These roughly spherical bodies vary in size and number. They contain ingested food particles—principally bacteria—and a small amount of fluid bounded by a thin but definite membrane. Because the term vacuole has been used to designate a number of other cytoplasmic elements and hence may result in confusion, Volkonsky (1934) proposed the term *gastriole* for this structure. Associated with the food vacuole are the *digestive granules*. These are fine cytoplasmic elements which are stainable with neutral red and are most readily observed in association with recently formed food vacuoles. These granules are found to cluster around the food vacuole and vary in size and shape in paramecium from small spherules to larger rods. According to MacLennan and Murer (1934) the digestive granules of paramecium are high in ash. A more detailed discussion of food vacuoles and digestive granules is given in Chapter 6.

5. Other Cytoplasmic Inclusions (Granules, Chromidia, Chromatophores, etc.).

Like most Protozoa, many cytoplasmic inclusions have been reported for paramecium in addition to those already described. (For "kappa" particles see Chapter 10.)

Lipoid granules can be demonstrated in recently fed paramecia. Zingher (1933) reported that paramecia in old cultures show larger amounts of fat than in new cultures. Zweibaum (1921) showed that fat is stored under conditions of low oxygen tension and then utilized and lost when the oxygen tension is restored; the rate of loss being dependent upon the temperature.

Small bodies called *segregation granules* can be made to concentrate and react positively with certain vital dyes.

Dunihue (1931) studied the relative amounts of segregation granules in paramecium and demonstrated a correlation of the decrease in numbers with starvation of the animals.

Diffuse masses of *carbohydrates* were demonstrated by Rammelmeyer (1925) as being visible in fixed paramecia.

Chromidia and *volutin granules* have been reported as protein reserves found in the cytoplasm. The exact definition and interpretation of each term is inconsistent according to the literature. Hertwig (1902) introduced the term chromidia for extranuclear chromatin found in the cytoplasm, but derived from the nucleus. Still others claim chromidia are mitochondria. Volutin granules are basophilic cytoplasmic elements which, because they are associated with chromatin and much like it in composition, are considered by some to be the same as chromidia.

Only one species, namely *P. bursaria,* possesses unicellular algae which belong to the genus *Chlorella.* They are small, green chlorophyll-bearing plants which may be discoid, spherical, ovoid or cup-like in shape and usually found in great numbers in the cytoplasm. For a more detailed account of these elements, see page 406.

G. The Nuclei

Paramecium, like all of the *Euciliata* and *Suctoria,* possesses dimorphic nuclei, i.e. one massive macronucleus and one or more very small micronuclei. Earlier, this condition was generally viewed as representing a segregation of two types of chromatin; one called the idiochromatin found only in the micronucleus and the other called the trophochromatin and found only in the macronucleus. However, the cytochemical investigation of Moses (1949, 1950) shows the two nuclei to be chemically identical.

1. The Macronucleus.

In general, the macronucleus is a conspicuous, ellipsoidal or kidney-shaped body found in the approximate center of paramecium (Figs. 5 and 6 and many other figures). Its contour is generally regular and smooth. It is of a compact type containing fine threads and tightly packed discrete chromatic granules of variable size and embedded in an achromatic matrix. The macronucleus of paramecium, like that of all ciliates, stains intensely with the usual chromatic dyes (particularly Feulgen) but does not undergo mitosis. Instead, it divides amitotically.

It is a matter of observation that in many ciliates, especially hypotrichs, there occurs a conspicuous change in structure and stainability of the macronucleus during fission. Such a change, although not nearly so great as in hypotrichs, is nevertheless found in paramecium. In a study of comparative ultraviolet absorption by the constituent parts of *P. caudatum,* Luyet and Gehenio (1935) reported that the absorption by the macronucleus is not uniform. They found that this nuclear body showed two distinct phases of absorption—an intense one at 2749 Å and another less intense at 2313 Å. They also reported the absence of a macronuclear membrane. By means of ultraviolet photography Luyet (1935) reported that material of the macronucleus consists of one phase which is transparent, more or less homogeneous, and forms the groundwork upon which is distributed the second or opaque phase. In their ultracentrifuge studies upon paramecium, King and Beams (1937) suggest that the opaque phase of Luyet probably corresponds to the chromatin while the transparent phase corresponds to the achromatic matrix. King and Beams were unable to find evidence of a macronuclear membrane in paramecia that had been fixed after centrifuging but they report that the peculiar manner in which the chromatin leaves the achromatic matrix in centrifuging leads them to believe that there may be a very thin membrane surrounding the macronucleus.

2. The Micronucleus.

The micronuclei of paramecium are comparatively small structures varying in length from a micron or less to approximately 8 microns (Figs. 5 and 6 and elsewhere in book). Although the number of micronuclei appears to vary in some species, they are nevertheless important aids in taxonomy (Table 3). Micronuclei may be classified into two groups as follows: a *compact* ("caudatum") type such as is found in *P.*

caudatum and a *vesicular* ("aurelia") type characteristic of *P. aurelia* (Fig. 4). In the compact type the micronucleus is commonly spherical, ellipsoidal, shield-shaped, or pyriform. It is generally found close to the macronucleus, often in a concavity or impressed into its surface. Occasionally fine chromatic granules and threads are uniformly distributed throughout the structure. More frequently it is definitely polarized, one end being differentiated into a clear polar cap which stains faintly with nuclear dyes.

CHEMICAL COMPOSITION OF THE NUCLEI

Moses (1949, 1950) made a comprehensive study on the quantitative cytochemical analyses of the macronuclei and micronuclei of *P. caudatum* and reported that although the macro-

Table 3

CHARACTERISTICS OF MICRONUCLEI IN SPECIES OF PARAMECIUM
(Wichterman)

Species	Number	Size	Shape	Structure	Position
P. caudatum*	1	8 μ	ellipsoidal	compact	Close to or in a cleft of macronucleus
P. aurelia*	2	3–5 μ	spherical	vesicular	Close to macronucleus
P. multimicro-nucleatum*	3–4 (up to 9)	0.7–2.5 μ	spherical	vesicular	Vicinity of macronucleus; arranged linearly
P. bursaria†	1	7 μ	ellipsoidal	compact	Close to or impressed into macronucleus
P. trichium†	1	4–7 μ	ellipsoidal	compact	Close to or impressed into macronucleus
P. calkinsi†	1–2	3–5 μ	spherical	vesicular	Not always close to macronucleus
P. polycaryum†	4(3–8)	3–4 μ	spherical	vesicular	Close to macronucleus
P. woodruffi†	3–4 (up to 8)	4–5 μ	spherical	vesicular	Not always close to macronucleus

* "Aurelia" group.
† "Bursaria" group.

nucleus is roughly more than 40 times the volume of the micronucleus, both nuclei are alike in their nucleic acid-protein composition. The fine threads and granules of the micronucleus, like those of the macronucleus, are Feulgen-positive hence contain desoxyribose nucleic acid. Since both types of nuclei contain similar concentrations of total protein, non-histone protein, desoxyribose nucleic acid, and ribose nucleic acid, Moses states that they must be actively metabolic nuclei, at least in their vegetative stages.

He noted further that there is about 20 times as much protein present as desoxyribose nucleic acid with little if any histone. His investigation revealed that large concentrations of ribose nucleic acid were found comprising roughly 10 per cent of the nucleoprotein content of both nuclei and he concluded that the two nuclei are not essentially different from each other except in size; also, that they are homologous to the metabolic nuclei of Metazoa.

Roskin (1945) and Shubnikova (1947) reported on the localization of ribose nucleic acid in the macronucleus and cytoplasm of P. caudatum and for the same species, Moses found that the great concentration of ribose nucleic acid in both nuclei is nearly as high as in the cytoplasm. It is concluded that protein is the major component of both nuclei comprising over 85 per cent of the nucleoprotein for each nucleus.

The micronucleus is surrounded by a thin nuclear membrane. The vesicular type, which is always spherical, possesses a deeply staining center of chromatin called the *endosome*. This is surrounded by a clear region which is also bounded by a nuclear membrane.

The micronucleus usually possesses granules of chromatin finer than those found in the macronucleus. Frequently, these micronuclear granules stain more or less intensely than those of the macronucleus. Luyet and Gehenio (1935) and the author (unpublished) were unable to observe micronuclei of *P. caudatum* in their photographs taken with different wave-lengths of ultraviolet light.

During mitotic division and in other micronuclear phenomena associated with reproduction, numerous, long, thin chromosomes are visible. The behavior of the nuclei during fission and other asexual and also sexual processes is considered in Chapter 9.

3. Nuclear Variation.

Variation in number of micronuclei has been established in a number of species of paramecium. We are aware of the variable numbers of small, vesicular micronuclei in *P. multimicronucleatum*. This species usually possesses three or four but may have up to seven. Variations in number of vesicular ("aurelia" type) micronuclei in certain other species of *Paramecium* are known and hence included in the species characterization. Thus *P. polycaryum* has usually four but may possess from three to eight; *P. woodruffi* usually possesses three or four but may contain as many as eight.

It is noteworthy that in all species of *Paramecium* possessing the large "caudatum" type of micronucleus, a single one is typically present. Nevertheless, reports have appeared dealing with variation from the characteristic unimicronuclear feature. In *P. caudatum*, Calkins (1906) reported a temporary bimicronucleate condition but this appears to have been *P. aurelia*. Diller (1940) reported nuclear variation in *P. caudatum* regarding size of micronucleus, its staining reaction at different periods, amicronucleate and bimicronucleate individuals, absence of the macronucleus, extrusion of macronuclear bodies, macronuclear fragmentation, and hypertrophy. Wenrich (1926) reported bimicronucleate specimens of *P. trichium* and the author has also observed them in rich cultures containing rapidly dividing individuals. Hamburger (1904) described *P. bursaria* with supernumerary micronuclei in her study of conjugation. In pedigreed cultures of *P. bursaria* maintained over seven years, Woodruff (1931) noted marked numerical micronuclear instability throughout the period. Individuals which were originally bimicronucleate later in cultures exhibited from one to four micronuclei and then finally became characteristically unimicronucleate although some became amicronucleate (Fig. 38). Chen (1946) reported abnormal nuclear changes during conjugation between certain Russian and American clones of *P. bursaria*. According to Chen, these abnormalities included: (1) the arresting of nuclear changes so that they do not proceed beyond the end of the first pregamic division; (2) the slowing

down of the rate of nuclear changes; (3) the deformation of chromosomes; (4) abnormal elongation of nuclei; (5) the formation of tripartite nuclei; and (6) the initiation of the process of nuclear exchange even though pronuclei are not formed and the exchange not consummated. Diller (1942) described a number of nuclear variations in the reconjugation phenomenon of *P. caudatum* and in ordinary conjugation of this species (1940). The writer (Wichterman, 1946 b) observed a pedigreed race of *P. caudatum* in which specimens were bimicronucleate although many were amicronucleate and others possessed variable numbers of micronuclei. By studying dividing forms in this unusual race, it was possible to observe the simultaneous production of bimicronucleate and amicronucleate daughters. Such a dividing unimicronucleate animal was shown to pass the dividing micronucleus with its two daughter micronuclei into only one daughter animal, leaving the remaining daughter amicronucleate. In addition, dividing bimicronucleate individuals

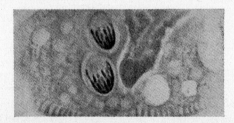

FIG. 38. *P. bursaria* showing bimicronucleate condition. (*Top*) Two micronuclei in an animal from a pedigree culture at the 22nd generation. (*Bottom*) Division of the two micronuclei in an animal from the same culture. (*Quart. J. Micro. Sci.*)

were observed which appeared to undergo fission normally with each daughter receiving two micronuclei. The race appeared to be parasitized with a microörganism which is believed by the writer to account for this micronuclear instability.

Amicronucleate races of paramecium have been encountered and cultured successfully. Woodruff (1921) maintained a race of *P. caudatum* in pureline culture and determined that it was definitely amicronucleate. The writer (Wichterman, 1946 b) reported amicronucleate *P. caudatum* in the race marked by unstable micronuclear behavior mentioned earlier and Diller (1940) reported amicronucleate specimens of the same species. Woodruff's originally bimicronucleate race of *P. bursaria* finally produced a race showing no micronucleus, yet this race appeared as healthy and vigorous as any micronucleate ones. Cytological investigations by Woodruff over a period of seven years revealed no evidence of endomixis or conjugation in the amicronucleate *P. bursaria* and it led him to state, "amicronucleate animals do not possess the power to undergo endomixis or conjugation—a significant and crucial proof that the micronucleus is not only morphologically but also actually functionally absent." He believed that this supported the view of the identification of the macronucleus and micronucleus as a segregation of somatic and generative elements into discrete bodies within the cell. Offsetting this interpretation of the inability of amicronucleate individuals to conjugate is the work of Chen (1940). He described conjugation between specimens of *P. bursaria* with and without micronuclei. Chen reported that the micronucleate conjugant showed normal behavior; the three pregamic divisions resulted in two pronuclei, one of

which passed over to the amicronucleate conjugant. Of course, the amicronucleate conjugant produced no pronuclei.

Concerning macronuclear variation, Diller (1936) made an extended cytological study, in this regard, of *P. aurelia* and later (1940) *P. caudatum*. He noted a number of different but simple fragmentations of the macronucleus not correlated with any special micronuclear activity. These phenomena he called *hemixis* to mean "a series of autonomous changes which the macronucleus undergoes in vegetative life, exclusive of those of normal binary fission" (see p. 262). Hemictic animals with the two micronuclei may show variable numbers and sizes of macronuclear fragments. In type "D" of hemixis, the animals are said to be devoid of macronuclear fragments and micronuclei. He did not definitely rule out the possibility that such animals are degenerative specimens near death.

Sonneborn (1940) reported that in certain races of *P. aurelia*, a small percentage of animals undergoing autogamy and conjugation develop a new macronucleus from each fragment of the old macronucleus instead of from products of the synkaryon (see p. 264). He further reported that viable animals with regenerate macronuclei commonly give rise to cultures in which micronuclei disappear. Sonneborn concluded that therefore hereditary characters (including mating type) cannot be directly determined by micronuclei for they persist in the absence of micronuclei, and that the mating type must be determined by the macronucleus.

Amicronucleate races of paramecium and other ciliates have been extensively reported in the literature. It has been demonstrated beyond a doubt that the micronucleus is not always essential for the maintenance of vital processes of the individual. Such amicronucleate races do not appear to differ structurally or physiologically from micronucleate ones. The fact remains that the vegetative life of paramecium and indeed other ciliates is little influenced by the multimicronucleate or amicronucleate condition. Amicronucleate specimens of *P. bursaria* have been shown to mate and conjugate with micronucleate ones (Chen, 1940).

The problem of the functional role of the micronucleus and the macronucleus has attracted the attention of many investigators. Although it is well established that the organism can live without a micronucleus, no races have ever been reported lacking the macronucleus, a structure which appears to be essential for the existence of the ciliate. There is also no record in which the micronucleus alone is adequate for the maintenance of vegetative functions. This structure warrants more detailed study than has been given it. It should be remembered that both the macronucleus and micronucleus have a common origin since they eventually develop from the synkaryon during the conjugation process. One would expect, therefore, that each receives a definite complement of genes. The micronucleus in its division shows well-formed chromosomes which are perpetuated (with their genes) in division and conjugation. On the other hand, the macronucleus, with its complement of genes can only divide amitotically during fission but this manner of division appears just as effectual in distributing its chromatic material. The fact that in conjugation of certain ciliates, namely *Metopus, Nyctotherus,* and *Euplotes,* a peculiar thread-like condition exists in the developing macronucleus (anlage) suggesting chromosomes, calls for a more detailed study of the structure. Beers (1946) aptly states, "The fact that the behavior of the macronucleus does not conform to the chromosome theory of heredity *in sensu strictu,* in that chromosomes are absent, may mean simply that a different mechanism for the distribution of the genes is involved."

A number of explanations has been given to account for micronuclear variation, and their origin is an interesting problem in protozoan cytology, and is of evolutionary significance. Hamburger (1904) in her study of conjugation in *P. bursaria* reported a number of irregularities in reorganized exconjugants which accounted for some specimens with supernumerary micronuclei. Precocious micronuclear division such as observed in *P. trichium* by Diller (1949 a) is another explanation. Parasitic microörganisms in *P. caudatum* have been shown to result in unstable micronuclear behavior (Wichterman, 1946 b). It is, of course, possible for a micronucleus to degenerate or disappear as was reported by Sonneborn (1940) for *P. aurelia.* It should be remembered that degeneration of some micronuclear products is a normal process for part of the conjugation process.

It has been known for many years that a large number of chemical and physical agents can alter mitosis and cell division in animals and induce abnormal division stages. Some of these agents are basic dyestuffs, narcotics, ether, alcohol, radium, ultraviolet and x-radiation, high and low temperatures, and hypotonic solutions (see Chapter 5). Burt (1945), using ethyl carbamate as well as other carbamates and chloral hydrate, induced aberrant nuclear behavior during the division of the ciliate, *Colpoda steinii.* Recently Lloyd (1947, 1948, 1949) induced uncoördinated growth, monster formation and nuclear variation in *P. caudatum* by using an isomer of hexachlorocyclohexane ("gammexane"). He reported that when extremely small concentrations of gammexane were placed in normal culture medium, macronuclei and micronuclei divided repeatedly while the cell threw out lobes in all directions. Up to 14 micronuclei were counted in one individual while the macronucleus was throwing off pieces of varying size. Buds became detached with varying nuclear combinations. While it is likely that all buds were not viable, some of them succeeded in starting fresh colonies of apparently normal individuals. (See monster formation, p. 355).

The behavior of the micronucleus and macronucleus during division and in other nuclear processes is considered in Chapter 9.

Chapter 4

Ecology, Collection, Cultivation, and Sterilization of Paramecium

A. Ecology
B. Collection Methods
C. General Culture Methods
D. Pure-line Mass Cultures
E. Methods of Sterilization of Paramecium

A. Ecology

During the past 200 years, paramecium has been reported from practically every part of the world. The best known species appear to be cosmopolitan, possibly only because of the fact that they are so well known and easily recognized. Lesser known species may eventually prove to be as widely distributed. Most of the species, namely *P. aurelia, P. caudatum, P. multimicronucleatum, P. bursaria, P. trichium,* and *P. polycaryum* are found naturally inhabiting the bodies of fresh water. In brackish water, i.e. where sea and fresh water mix, we find *P. woodruffi* and *P. calkinsi* although the latter has been reported from fresh water and water of greater salt concentration than sea water (Wichterman, 1950, 1951).

The fresh-water species are commonly found in lakes, ponds, pools, streams, rivers, reservoirs, aquaria, bodies of stagnant water, and occasionally drinking water. They may exist in large numbers in bodies of water where considerable putrefaction is occurring. As a matter of fact, large numbers of food bacteria in the water favor the presence of large numbers of paramecia. In a biologic survey of the Conestoga Creek basin of Pennsylvania, as edited by Patrick and Roberts (1949), the following species of *Paramecium* were reported as occurring in the following number of stations: *P. caudatum,* 20 stations; *P. aurelia,* 15 stations; *P. bursaria,* 12 stations; *P. multimicronucleatum,* eight stations; *P. trichium,* one station, and *P. polycaryum,* one station. Paramecia have been reported from inland salt ponds near salt mines and from sea water[1] but in most cases, it is possible to demonstrate that investigators observed instead of paramecium, other ciliated Protozoa. In other cases, it is impossible to be certain which ciliate was observed.

In a protozoölogical survey of the Savannah River made by the writer in 1952, *P. caudatum* and *P. bursaria* frequently were encountered together from widely scattered stations.

It is of interest to report that active specimens of *P. aurelia* and *P. caudatum* have been observed from the wet ground around Moscow by Nowikoff (1923). Fantham and Porter (1945) collected large numbers of *Sphagnum* and other mosses from many localities in Canada and the United States, then carefully washed the plants with aerated distilled water to obtain Protozoa. They

[1] Concerning the acclimatization of paramecium to varying concentrations of sea water, see page 207.

93

encountered many Protozoa including *P. caudatum,* "*P. putrinum,*" *P. aurelia,* and *P. bursaria;* the last two species being quite common.[2]

Gause, Smaragdova, and Alpatov (1942) studied the geographic variation in *P. aurelia, P. caudatum,* and *P. bursaria* and the stabilizing selection in the origin of geographic differences. Their investigations revealed the fact that the paramecia from northern Russia are larger than the southern ones and that these differences are hereditary. It is suggested by them that this result follows Bergmann's rule which predicts the increase of bodily dimensions to the north. Hereditary distinctions were also found in the optimal temperatures for multiplication of *P. bursaria* and *P. aurelia* from various localities. These investigators also showed that temperature optima for multiplication in northern, Moscow, and southern varieties of *P. bursaria* are closely correlated with the summer (July–August) temperatures of the localities from which they originated.

As reported by Kudo (1947), Uyemura and his co-workers found *P. caudatum* living in the thermal waters (36°–40° C.) of Japan. On the other hand, specimens have also been found in bodies of cold water during the winter months.

Juday (1908), Birge and Juday (1911), and Imel (1915) reported paramecium from the profundal, oxy-

genless region of lake bottoms where it was claimed that the specimens lived anaerobically. Lauterborn (1916) listed *P. aurelia* and *P. caudatum* as being found in a sapropelic habitat and Lackey (1932) and Lloyd (1946 a, b) reported paramecium from sewage tanks and filters. It can scarcely be claimed that all traces of oxygen are absent at such localities but the oxygen content of these environments is certainly lower than that to which paramecium is accustomed. In nature, paramecia live in bodies of water where there is at least some oxygen although the amount present may vary considerably. Paramecia are not anaerobic organisms; they require oxygen to live (see p. 218 for oxygen requirements). Table 4 shows the maximal survival time of paramecium under anaerobic conditions as given by many investigators. The wide discrepancy in results would suggest faulty technique on the part of some investigators, hence it may be questioned whether or not strictly anoxic conditions were obtained. Paramecia survive anaerobic conditions at 0°–5° C. better than at higher temperatures on account of the lower metabolic rate. In regard to nutritive conditions, starving paramecia die much more quickly than well-fed ones (Pütter, 1905; Barbarin, 1938).

It should be remembered that experiments designed in the laboratory to test survival of paramecium under anaerobic conditions differ considerably from the condition when it occurs in nature. In the laboratory, the change from an aerobic environment to an anaerobic one is abrupt while in nature there occurs a gradual decrease of the oxygen tension thereby permitting some adaptation through acclimatization.

The hydrogen-ion concentration of bodies of water in which paramecia are found in nature varies considerably

[2] Dr. K. W. Cooper has permitted the author to examine microscopically, a polished piece of amber from Canada of uncertain geologic age (possibly Cretaceous) which contained invertebrates of the sort expected in swamps or marshes. Among the inclusions is one specimen, along with *Alternaria*-like spores, which is very suggestive of a paramecium of the "caudatum" type in size, shape, general morphology, and surface markings. In a personal communication from Dr. J. A. Cushman, this authority reported that he knew of no account of a soft-bodied ciliate fossil such as this may prove to be.

Table 4

MAXIMAL SURVIVAL TIME OF PARAMECIUM UNDER ANAEROBIC CONDITIONS
(Modified after von Brand, 1946)

Species	Temperature	Strictly Anaerobic Conditions	Survival Time	References
P. aurelia	Room	?	Several hrs.	Budgett (1898)
P. aurelia	Room	?	4 days	Nikitinsky and Mudrezowa-Wyss (1930)
P. bursaria	Room	?	16 days	"
P. bursaria (cysts?)	35° C.	Yes	> 90 days	Becquerel (1936)
P. caudatum	Room	?	10 days	Pütter (1905)
P. caudatum	Room	?	18 days	Nikitinsky and Mudrezowa-Wyss (1930)
P. caudatum	Room	Yes	3 days	Liebmann (1936)
P. caudatum	Room	Yes	10 sec.	Gersch (1937)
P. caudatum (starved)	Room	?	1 hr.	Barbarin (1938)
P. caudatum (fed)	Room	?	> 2 days	"
P. caudatum	24°–28° C.	Yes	12 hrs.	Kitching (1939 a)
P. caudatum	0°–5° C .	Yes	> 30 days	Lindeman (1942)
P. caudatum	10° C.	Yes	< 30 days	"
P. multimicronucleatum	24°–28° C.	Yes	12 hrs.	Kitching (1939 a)
Paramecium sp.	4° C.	?	4 days	Fauré-Fremiet et al., (1929)
Paramecium sp.	15° C.	?	2 days	"
Paramecium sp.	25° C.	?	1 day	"
Paramecium sp.	Room	?	> 1 day	Loeb and Hardesty (1895)

between the clear, highly acid bog or swamp waters where *P. bursaria* is frequently encountered, to the slightly alkaline waters where it and other species may be found. In a pond, the bottom region is frequently acid because of the decomposing organic matter, while the surface water may be slightly alkaline. Paramecia grow best in an environment near neutrality (pH 7.0). Also *P. bursaria,* as well as others, is frequently found in the upper levels of the clear water which is, of course, rich in oxygen. It would appear that ecologically, the nature and extent of the food-bacteria present in a pond are of more significance than the hydrogen-ion concentration.

Factors such as the relation of paramecium to food, temperature, light, hydrogen-ion concentration, etc., are considered in detail in Chapter 6.

B. Collection Methods

Paramecium may be collected easily in nature and cultivated successfully in the laboratory. Since members of the various species are widely distributed, they are likely to be found in most ponds of fresh or stagnant water. The simplest and oldest method of collecting *Paramecium* consists merely of dipping a sample of water—about a liter or less—from the pond, to which is added some living or decayed vegetation such as algae, elodea, dead leaves, or other plant material. A wide-mouthed glass jar with screw lid serves as an ideal collecting container. After collecting, the container should be covered immediately, then labelled giving the date, location, and other information needed for the keeping of an adequate record. This may include the temperature of the water, its pH, color, etc. Special large pipettes (Banta, 1914) may be used to advantage in removing the water from selected places in the pond. The water may be examined immediately for *Paramecium* or it may be left in the container for several days to allow time for the paramecia to multiply. Generally upon standing for a period of three days, the paramecia, if they were present in the original sample, are likely to be found concentrated in a white ring slightly below the surface film; occasionally they are abundant in the surface scum. The more common species of *Paramecium* are likely to be found in the same locations in a pond and the directions as given above may apply to their collection. The reader is referred to the papers of Hyman (1925, 1931) for additional collecting and cultivating methods.

A concentration method may be used involving filtering in which a fine plankton net or filter paper is placed in a funnel and the sample of water is poured. When a small volume of water remains in the funnel after a large volume has passed through it, the residue is quickly poured into the collecting container and represents a concentrate. In this manner, many liters of water may be filtered and the concentrated sample be kept in a comparatively smaller container. This is a useful method where the number of paramecia may be few in a given location. Towing with a tow net is another concentration method and like all such methods, may be used to give quantitative results. The Sedgewick-Rafter method, although requiring special apparatus, is also designed to give quantitative results (McClung, pp. 390, 937).

To examine for paramecia, some of the pond water should be removed with a pipette, placed in a shallow dish and examined under a low-power binocular microscope. This simple culture, as described above, is likely to last for a week or even much longer in the laboratory. Such a culture is usually referred to as a "mixed" one because it will nearly always have growing in it many other kinds of microscopic animal life. But after a while, the paramecia in this culture will decline in number and eventually disappear unless freshly made infusions are used. Paramecia may be cultivated continuously and indefinitely if fresh organic infusions are prepared. Any one of the culture methods given immediately following, may be used for the cultivation of *Paramecium*. It should be remembered that the freshly boiled infusions be allowed to stand (ripen) for a day before being used.[3] Then a few cubic centimeters of the fluid containing the thriving culture of paramecia are removed with a pipette and inoculated into the newly made infusion. By repeating this general method of inoculating with paramecia into newly made infusions, the animals may be maintained indefinitely.

C. General Culture Methods

The number of different culture methods reported and described in the literature is enormous. Most of them are unnecessarily complicated; some are so unreliable that it would be exceedingly difficult to duplicate exactly a given cul-

[3] Until a culture is established, some prefer to leave the freshly prepared infusion in a container—preferably a large battery jar—uncovered. Bacteria from the air will fall into the infusion which will multiply and if suitable, serve as food for the paramecia. Cultures thereafter are kept covered. This method of course could not be used in the carefully controlled cultures which are considered later. In the preparation of the culture infusions, distilled water may not be available. Clear filtered water from ponds, lakes, or springs may be used. Bottled spring water is also satisfactory. If chlorinated tap water is used, it should first be allowed to stand for several days to a week in open containers.

ture technique. Since almost any food substance can be used in making cultures, methods have been reported for the culturing of *Paramecium* using the following: albumen, yeast, gelatin, pig's brain, sheep's brain, sheep's pancreas, dry serum, timothy hay, wheat grains, hay-wheat combination, rice-straw infusion, spinach, lily pads, lettuce, flour, grass, cabbage and cracker infusion, fresh-water mussels, beef extract, mushrooms, pond vegetation, beef suet, dilute milk, malted milk, dried skim milk powder, dried oats, cheese, mutton broth, apple juice, pineapple extract, bread crumbs, earth concoctions, lecithin, synthetic fluids and extracts, strains of living and dead bacteria, and cultures to which acids, salts and even nitroglycerin have been added. It is apparent that species of *Paramecium* will grow in a large variety of different cultural media in which organic decomposition is occurring. The bacteria in such an infusion serve as food for the paramecia which in turn increase in number.

Although a large number of different methods have been used by the author, there are presented here only those of proved merit in which the culture conditions can be continued with some degree of standardization (Wichterman, 1944, 1949).

All glassware that is to be used should be scrubbed clean, then allowed to dry thoroughly.[4] For the maintenance of pure or pedigreed cultures, glassware should be sterilized at 15 pounds pressure for 15–20 minutes in an autoclave or pressure cooker. Glassware that has been used previously for stains, fixing reagents, oils, etc. should never be used. If an autoclave is not available, glassware after it is scrubbed should be placed in a nitric acid bath which is made up of one part of commercial nitric acid to nine parts of water. The nitric acid oxidizes the organic matter without leaving an adsorbed residue on the glass. The glassware then should be rinsed well in distilled water and allowed to dry. The commonly used "cleaning solution" for cleaning laboratory glassware and made by saturating commercial sulfuric acid with commercial potassium dichromate should not be used. It has been found that as many as 15 rinsings with distilled water are necessary to remove the chromic acid adsorbed on the glass and chromic acid is toxic to living organisms (Richards, 1936).

Hay Infusion.

The most widely used culture method is that in which dry timothy hay (*Phleum pratense*) is boiled for a given length of time in a measured volume of water. This, known as a hay infusion, is allowed to stand (ripen) for a period of time before being inoculated with paramecia. A successful hay infusion can be made by placing 6 Gm. of cut timothy hay in 1 liter of distilled water and allowing the contents to boil for approximately 20 minutes or until the water becomes brown. A method used successfully by the author for many years is described as follows: The spikes and stems of clean dry timothy hay are cut into segments of approximately 1 inch in length. One and a half grams of the cut hay are placed in a 250 ml. wide-mouthed Erlenmeyer flask and dis-

[4] Efforts should be made to standardize the glassware used. Raffel (1930) made the important observation in his culture technique that drops of fluid which had been identical when placed on the depression culture slides were found to vary by a whole pH unit within 24 hours. This showed that the glass of the various culture slides—of French origin—differs in solubility. The French glass was made by a process involving the use of lead and the toxic effect of it killed the paramecia. Jennings (1945) found that quartz has no advantages over white (Pyrex) glass in the culturing of *Paramecium bursaria*.

tilled water added to the neck of the flask (240 cc.). The mouth of the flask is covered with an inverted, snugly fitting beaker and the contents of the flask allowed to boil for approximately 15 minutes. The covered flasks of cooked infusion are allowed to stand (ripen) for 24 hours or less, then inoculated with paramecia. For exacting experimental studies, these flasks should be plugged and autoclaved. This medium and method of preparation are excellent for the culturing of *Paramecium caudatum, P. multimicronucleatum,* and *P. trichium. Paramecium aurelia* and *P. bursaria* appear to grow better in less concentrated hay infusion. (1 Gm. of hay instead of 1½ Gm.) A culture of *Paramecium* in such a hay infusion may be kept for long periods of time by simply adding a few pieces of boiled hay about once a month.

Wheat Infusion.

Approximately 50 wheat grains are boiled in a liter of water for five minutes. The boiling prevents the grains from germinating in the culture. Upon standing for a day, paramecia from the original culture are then inoculated into this wheat infusion. A less concentrated infusion may be obtained by removing the boiled grains to a clean jar to which is added fresh water.

Oatmeal Infusion.

This method of Jennings is prepared by placing about 30 flakes of "Quaker Oats" (approximately ½ Gm.) in 100 cc. of boiling water (spring) and allowing the boiling to continue three minutes. The fluid is then filtered, kept for 24 hours, then inoculated with *Paramecium.* Jennings, et al. found that more satisfactory growth was possible by the addition of the green alga *Sticho-*

coccus. A method of growing the alga on agar in test tubes excluding bacteria, is described by Raffel (1930).

Lettuce Infusion.[5]

An excellent and widely used culture method is one that requires desiccated lettuce. The separated clean leaves of a head of lettuce are slowly dried in an oven until they are brown and crisp. Burned (black) leaves should be discarded. The leaves are next ground with a mortar and pestle, placed in a glass-stoppered bottle, and may be kept indefinitely. One and one-half grams of the desiccated lettuce are added to 1 liter of boiling distilled water and the mixture allowed to boil for five minutes. It is then filtered while hot into smaller (250 ml.) flasks, plugged with cotton, then autoclaved. Stoppered flasks of infusion so prepared are left standing overnight then are ready for use. After filtering and while the flasks are still warm (but not hot) small square sheets of thin paraffin ("Parafilm") are used to cover and seal the plugged mouths of the flasks—mainly to prevent evaporation. This sterile stock fluid in sealed flasks may be stored for long periods of time. One part of sterile distilled water and two parts of the stock lettuce infusion are mixed and when the mixture reaches room temperature the fluid is ready for use. For convenience, the author adds the distilled water to the stock solution before autoclaving then inoculates the flasks of infusion on the following day. An excess of $CaCO_3$ may be added to bring the pH up to 7.2. After inoculation, the flasks may be

[5] A powder called "Cerophyl" which is processed by Cerophyl Laboratories, Kansas City, Missouri, consists of the dehydrated leaves of young cereal plants, wheat, oats, barley, and rye. It can be used successfully in place of the desiccated lettuce in the cultivation of paramecium.

covered with sterile, inverted, snugly fitting beakers which serve as covers in place of the cotton. The stock fluid may also be used without dilution. All species of *Paramecium* mentioned as growing in the hay infusion also grow in the lettuce infusion, but it is the author's experience that the various species, with the exception of *P. bursaria,* do not live for as long a period of time in culture as in hay infusion, hence require additional subculturing. Lettuce infusion is especially recommended for the growing of *Paramecium bursaria.*

To establish cultures of *P. calkinsi* that have been found inhabiting fresh water, a hay or lettuce infusion should be used. Since this species is frequently encountered in brackish water, one should add a certain amount of sea water to the lettuce or hay infusions to match the salinity in order to obtain maximum growth. The author maintained a brackish race of *P. calkinsi* in pure hay infusion cultures for a year but was never able to obtain good growth. However, with the addition of two parts of filtered sea water to three parts of hay or lettuce infusion, enormous numbers of *P. calkinsi* were obtained readily. This ratio was found very satisfactory for cultivating a number of other races of *P. calkinsi,* especially in culture medium that was inoculated with the bacterium *Aerobacter aerogenes* although other species of bacteria served equally well as food. Unger (1926) successfully cultivated *P. calkinsi* in a medium consisting of one part 1.0 per cent salt solution (C.P. in glass-distilled water) and four parts of standard hay infusion.

P. woodruffi, which is found in brackish water, has been cultivated successfully by the author for many years by using filtered sea water and lettuce infusion in the same proportion as that used for *P. calkinsi.*

Frisch (personal communication) found that *P. calkinsi* and *P. woodruffi* grow well in a 1:1 ratio of sea water and hay infusion and will survive indefinitely (although not become abundant) in sea-water concentrations from 75–100 per cent.

The Food of Paramecium.

Normally, species of *Paramecium* depend upon bacteria as a source of food and the media described here naturally support bacterial growth. Apparently a large variety of different species of bacteria can be used. The use to which the paramecia are put will determine the type of cultural conditions necessary for the investigator. One who wishes to use species of *Paramecium* for cytological or teaching purposes might be concerned with having only a single species of *Paramecium* present with little regard for the bacteria in the culture. Here one may place into the culture medium a given species of *Paramecium* without strict regard for bacteriologic technique and the paramecia are likely to grow in this chance combination of mixed bacteria.

But the exacting needs of many experimenters require in addition to a standardizable culture medium an exact knowledge of the microörganisms—generally carefully selected strains of bacteria, algae, or yeast—that live in the culture medium and support growth of *Paramecium.* Recently there have come into use, precision[6] methods for cultur-

[6] In the lettuce culture medium described by Jennings (1939) for growing *P. bursaria,* 20 cc. of the medium is inoculated with a 1 mm. loop of the bacterium, *Flavobacterium brunneum,* taken from a three- to five-day-old agar slant and also three 2 mm. loops of the alga *Stichococcus bacillaris* from an 18-day-old agar slant. He reported that the paramecia divided rapidly and became large and plump. Later however, Jennings (1942) found that the alga was unnecessary for good growth.

ing ciliates upon a single known species of bacterium or other microörganism or in media free of all microörganisms. Such carefully controlled cultures have yielded important knowledge on food requirements, growth, reproduction (fission-rates) and other physiologic problems in cellular biology.

Hargitt and Fray (1917) cultivated ciliates on pure cultures of bacteria. After sterilizing *Paramecium,* they cultivated *P. aurelia* and *P. caudatum* on 11 species of bacteria from 30 species they had isolated. Since then, cultures of *Paramecium* maintained under strict bacteriologic control and using only a single species of bacterium have been reported by Oehler (1920 a), Phillips (1922), Philpott (1928), Giese and Taylor (1935), Parpart (1928), Raffel (1930), Glaser and Coria (1930), Giese (1938), Phelps (1934), and others. It has been shown that *P. aurelia* grow on a diet of *Pseudomonas ovalis, Flavobacterium brunneum, Bacillus niger, Bacillus cereus, Serratia marcescens (Bacillus prodigiosus), Achromobacter candicans, Bacillus subtilis (Bacillus flavescens), Pseudomonas fluorescens, Aerobacter aerogenes, Bacillus pyocyaneus, Bacillus candicans, Erythrobacillus prodigiosus, Aerobacter cloacae.*

P. caudatum will grow on *Aerobacter aerogenes, Pseudomonas ovalis, Bacillus subtilis, Achromobacter pinnatum, Serratia marcescens (Bacillus prodigiosus), Ps. fluorescens, Esch. coli, Bacillus proteus, Bacillus coli.*

P. multimicronucleatum has been cultivated on *Pseudomonas ovalis, Ps. fluorescens, Bacillus subtilis, B. cereus, B. megatherium, Aerobacter cloacae, Aerobacter aerogenes, Erwinia carotovora, Alcaligenes faecalis, Proteus vulgaris, Serratia marcescens, Phytomonas tumefaciens, Bacillus pyosepticus.*

P. bursaria has been grown on *Flavobacterium brunneum* (and the alga *Stichococcus bacillaris*) and *P. calkinsi* has been grown on *Aerobacter aerogenes.*

For additional information on factors dealing with bacteria and their suitability refer to p. 156.

There is a possibility that a large number of single strains of species of bacteria not yet tried may serve as food for paramecia. In his thorough study of food requirements in ciliates, Oehler (1920 a) reported that flourishing growths of *P. aurelia* were obtained on pure cultures of *Ps. fluorescens* in contradiction to the work of Hargitt and Fray (1917) who found that *P. aurelia* would not grow on this species of bacterium. While it is true that Oehler used mass cultures only and Hargitt and Fray used isolation cultures, there is the possibility of racial differences in the ciliates or the bacteria involved. Hetherington (1934 a), Leslie (1940 b), and others showed that a *particular strain* of a given species of bacterium is important in the growth of the ciliate. In the literature, we find what appear to be conflicting claims on the success or failure to obtain favorable growth of *Paramecium* with a definite species of bacterium. Examples, in addition to the conflicting accounts of Oehler and Hargitt and Fray cited above, are those of Phelps (1934) who found *Serratia marcescens (B. prodigiosus)* favorable to the growth of *P. aurelia* but DeLamater (1939) who reported it as being unfavorable. In the growth of *P. caudatum, Esch. coli* was reported as favorable by Chejfec (1920) but unfavorable by Johnson (1936). Hardin (1944) reported two strains of *Ps. ovalis* derived from a common parent stock which, after numerous transfers under probably different conditions, yielded different results. One strain of *Ps. ovalis* was shown to be immediately toxic for *Paramecium multimicronucle-*

atum while the other strain of the bacterium was not. Since it has been shown (Hargitt and Fray) that all species of bacteria in an infusion may not be used as food for paramecia, the fact that certain species of bacteria may have deleterious effects on the growth of the ciliates should be taken into account. DeLamater (1939) studied the relation of different species of bacteria in culture to induction of endomixis in *P. aurelia*. She reported that certain species of bacteria (*Flavobacterium brunneum, Bacillus niger, B. cereus,* and *S. marcescens*) ably supported growth of the paramecia. However, when the paramecia were cultured on *Bacillus* (*Esch.*) *coli,* little change was noted in the fission-rate but there were produced abnormal divisions (chains), an increase in deaths, and irregular fragmentation of the macronucleus.

In addition to single strains of bacteria and algae used in cultures, Loefer (1936 a) described a method for maintaining pure-line mass cultures of *P. caudatum* on a single species of yeast, *Saccharomyces ellipsoideus;* Glaser and Coria (1930) used the yeast, *Saccharomyces cerevisiae* for the cultivation of the same species and Gause and his associates (1939–1942) used as a food source the yeast *Torula utilis* in a saline medium for several species of *Paramecium*.

A strictly pure culture is one that contains a single species of *Paramecium* free of all bacteria or other microorganisms. Lwoff (1932, 1938) pioneered in pure-culture studies with his successful culture of the ciliate, *Glaucoma piriformis,* as reported in his treatise on the nutrition of Protozoa. This type of investigation has been continued by Kidder and his associates, as well as by a few others.

After many unsuccessful attempts to culture other holozoic forms in sterile media, Lwoff concluded them to be obligatory particulate feeders. He pointed out that along with morphologic evolution of microörganisms, there has been a physiologic evolution involving a successive loss of function concerning their abilities to utilize different compounds as nutrient materials. Accordingly, the simplest protozoöns can synthesize their proteins with nitrates as the only source of nitrogen, another group requires ammonium salts, another amino acids, still another requires peptones and finally the more specialized forms require a more complex type of particulate food. He has carried this idea over to the vitamin requirements of Protozoa. It is known that certain of the simpler Protozoa can synthesize both parts of the vitamin B_1 molecule. Others in a different group are unable to synthesize the parts but can fuse them if they are furnished in the medium. Still a higher group of Protozoa must be supplied with the intact vitamin molecule to live and grow.

The idea of physiologic evolution has been questioned by others and only continued experimentation upon pure-culture methods will provide the answer to this fertile field of research.

Since species of *Paramecium* are normally bacteria-eaters and are associated with many different kinds of bacteria in nature, the establishment of a pure culture first involves the complete sterilization of *Paramecium* followed by the substitution of an adequate diet in place of bacteria. This has been done for a number of flagellates and ciliates in various laboratories (Hall, 1941) but very few cases of such cultures have been reported for *Paramecium*. The establishment of monobacterial and pure cultures requires rigorous care and scrupulously clean methods such as are employed in bacteriologic technique where the utmost precautions must be taken to maintain sterile conditions.

All experiments upon growth, feeding, fission-rate, division, and other physiologic problems must take into account that the paramecia in a culture with even a single species of bacterium add another important variable to the experiment. Since the paramecia are not primarily dependent upon the culture fluid for food but the bacteria growing in it, there is presented in effect an ecologic problem. In reality, the paramecia are dependent in turn upon the nutrition, growth, and reproduction of the associated bacteria. In a typical monobacterial culture in which a single species of *Paramecium* is growing with a single species of bacterium, factors such as oxygen or carbon dioxide tension, oxidation-reduction potentials, temperature, hydrogen-ion concentration, etc. should always be considered in relation to bacteria in the culture. It therefore becomes apparent that in order to obtain accurate knowledge in certain fields of general physiology such as growth, we should try to eliminate all bacteria in a culture by substituting the principal source of food, i.e. bacteria, with another substance. If this is not possible for the species of *Paramecium* under investigation, carefully selected strains of species of bacteria should be used in which growth and other physiologic factors of the bacteria are well known to the investigator. Bacteria-free culturing of *Paramecium* presents the most difficult task for the experimenter but one which is likely to yield very profitable results and discoveries in the basic field of nutrition, growth, and growth requirements.

Oehler (1919) was the first investigator to report cultivation of ciliates in the complete absence of other living organisms. He obtained pure cultures of *Colpoda steini* by using dead *Bacillus*

coli, powdered dry egg white, and pulverized fish as well as other substances. However, Kidder and Stuart (1939) were unable to confirm this. The reader is referred to the works of Peters (1921), Oehler (1921–1924 b), the Chattons (1923), Lwoff (1924, 1925, 1929, 1932), and finally Hall (1941) for a comprehensive bibliography of papers dealing with the cultivation of ciliates free of other living microörganisms. Bacteria-free cultures have been described by Glaser (1932), Glaser and Coria (1933, 1935) for *P. caudatum* and *P. multimicronucleatum;* by Loefer (1936) for *P. bursaria,* and by Johnson and Baker (1942), Johnson and Tatum (1945), and Johnson (1950) for *P. multimicronucleatum.* Glaser and Coria (1933) reported the successful cultivation of *P. caudatum* in a bacteria-free medium containing heat-killed yeast cells, liver extract, and fresh rabbit kidney. The medium was prepared as follows: To 100 cc. of sterile tap water, 0.5 Gm. of Lilly liver extract (※343) was added. Filtering was done first through paper, then through a sterile Berkefeld "N" filter. The filtered extract had a reaction of pH 6.2–6.4 and was tested for sterility. Finely ground liver may be used instead. The cells of baker's yeast were grown, washed, and heat-killed. Using aseptic precautions, small pieces of kidney were removed from a freshly killed rabbit and were added to the test tubes containing liver extract and heat-killed yeast. Sterile paramecia were added to the medium and grown at room temperature. They reported that *P. caudatum* had been carried through 22 subcultures for six months in the Lilly liver extract medium and 12 subcultures over a three-month period for fresh liver extracts. Hetherington (1934) repeated the experiments of Glaser and Coria (1933) but came to the conclusion that these workers probably did not work with sterile paramecia, although methods were employed by them to obtain sterility. On the other hand, Brown (1934) made a detailed bacteriologic examination of the cultures of *P. caudatum* and *P. multimicronucleatum* used by Glaser and Coria. This investigator reported the cultures to be free of bacteria. Glaser and Coria (1935) emphasized their use of sterile paramecia in the cultivation of *P. caudatum* when they reported on the culturing of *P. multimicronucleatum* in bacteria-free media.

P. bursaria was grown bacteria-free in a synthetic medium by Loefer (1936) who reported that concentrations of both salts and organic substances are important factors in determining successful cultures. The growth cycle was studied in a tryptone medium containing various mineral bases.

Johnson and Baker (1942) reported successful bacteriologically sterile cultures of *P. multimicronucleatum* in a medium of pressed yeast juice devoid of all other cells. They maintained five strains of paramecia through 18 transfers for a period of 10 months. In these sterile cultures, they were unable to obtain a rate of growth equal to the highest rates obtainable in bacterial cultures of paramecium. In the pressed yeast juice medium 0.5 divisions per day occurred while in their bacterial cultures a division rate of 1.0–2.0 is reported.

The method of preparing the pressed yeast juice by Johnson and Baker is given by them as follows: One pound of Fleischmann's baker's yeast is ground with an equal weight of washed, fine white sand, which is then mixed with 125 Gm. of diatomaceous earth and reground to a sticky dough. This dough is next wrapped in two layers of diaper cloth previously dampened with distilled water and placed in a 4½-inch perforated screw press cylinder. Pressure is applied as rapidly as possible by a manually operated screw press. The yield of about 120 cc. of yeast juice is collected in an ice-packed flask and refrigerated overnight after which it is forced by pressure through a Seitz bacteriologic filter. For the cultures, 5 cc. of triple distilled water is sterilized and upon cooling 0.5 cc. of yeast juice added, then inoculated with sterile paramecia. Transfers and sterility tests are made at 14-day intervals. Later Johnson and Tatum (1945) used the same method for obtaining fractions of the yeast juice. Instead of using all the juice, the first half is discarded. The remainder which has greater activity and is more easily filtered is collected and sterilized by means of bacteriologic filters. They report fair growth of paramecia when the pressed yeast juice was added to triple distilled water or Osterhout's medium but better growth when added to 0.5 per cent proteose-peptone (one part of pressed yeast juice to 20 parts of 0.5 per cent proteose-peptone). Their cultures were grown in an incubator at 25° ± 1° C. The pressed yeast juice contains two fractions essential for the growth of *P. multimicronucleatum*. One is a heat-labile, non-dialyzable fraction which can be precipitated with ammonium sulfate and the other is a heat-stable dialyzable fraction. Johnson (1946) reported a higher daily fission-rate of 1.0 and over for paramecium cultured in a sterile medium of pressed yeast juice containing in addition, the flagellate, *Polytomella*. Later, Johnson (1950) was able to substitute hydrolyzed nucleic acid in one per cent proteose-peptone medium for the heat-labile fraction and to obtain growth equal to that obtained with pressed yeast juice. The growth properties of pressed yeast juice and the chemical composition of the synthetic media in the bacteria-free cultures are discussed in the section on growth (Chapter 6).

In investigations upon the cultivation of *P. aurelia* in the absence of other living organisms, van Wagtendonk and Hackett (1949) found the following to sustain growth: (a) equal volumes of 0.5 per cent yeast autolyzate sterilized by filtration through a sintered glass bacterial filter and autoclaved lettuce infusion which had been inoculated with the bacterium *Aerobacter aerogenes* 24 hours before autoclaving; (b) equal volumes of 0.5 per cent yeast autolyzate sterilized by autoclaving and an autoclaved lettuce infusion which had been inoculated with *A. aerogenes* 24 hours previously. Sterility tests were employed by them to check the culture medium. In either of these two media, *P. aurelia* showed 1.7 fissions per day as compared to 4–6 fissions per day in the lettuce-bacterium medium. The Burbancks (1950, 1951) also reported growth (and autogamy) of *P. aurelia* in non-living media using methods comparable to those of van Wagtendonk and Hackett.

D. Pure-line Mass Cultures[7]

In nature, *Paramecium* lives in association with bacteria as well as other microörganisms including algae and Protozoa and a typical random collection of paramecia will yield in the sample, other forms of microscopic animal and plant life. Such a collection represents a "wild" or mixed culture. In beginning a species-pure culture, that is, one in which all the progeny of paramecia are derived from a single specimen, it is necessary to first isolate an individual. This is done by using a simple capillary pipette which has been

[7] The technique for maintaining daily isolation cultures for determining fission-rates is described on page 361.

drawn out to a fine tip of about 200 microns diameter. The tip can be fire-polished by quickly drawing it through a Bunsen flame. A sample of the wild culture is placed in a shallow dish (Petri) under a low-power dissecting microscope and a typically active specimen removed and placed in a very small volume of uncontaminated culture medium. The author uses a small sample of the original wild culture and adds a large volume of the sterile culture medium before isolation. This serves to acclimatize the paramecia to the medium and lessens the concentration of organisms present thereby making the task less difficult. A single specimen of *Paramecium* thus isolated is placed in a depression slide or small culture dish to which has been added about five drops of sterile culture fluid.[8] Isolation cultures are then kept in moist chambers. Such a chamber may be constructed easily by using a large culture dish which consists of a lid and bottom with overall dimensions of 20 × 8 cm. Using the lid as a bottom, line it with several sheets of filter paper of the same diameter as the dish, then add distilled water to wet the paper. On the filter paper place glass supports such as Syracuse watch glasses, upon which are laid the depression slides. When the depression slides with isolated paramecia have been placed in the dish, cover with the original bottom which will now serve as a lid. Petri dishes may also be used in the same manner for moist chambers. Ob-

[8] Depression slides are microscope slides having one or two concavities for fluid. Special dishes convenient for isolation cultures and holding more fluid than depression slides are Columbia culture dishes measuring 42 × 42 × 7 mm. with a single deep ground glass depression; pyrex "spot-plates" of clear glass and each possessing nine wells; Bureau of Plant Industry Syracuse watch glasses which hold about 1 cc. of fluid and can be stacked; and thin watch glasses, about 50 mm. wide.

servations of the cultures should be made daily, and when the isolation cultures contain a dozen or more specimens, the progeny[9] may be transferred to a larger volume of infusion with a sterile pipette. A Syracuse watch glass to which new fluid has been added is very satisfactory for this second transfer and it need not be kept in the moist chamber but should be covered. At room temperature a fairly rich species-pure mass culture should appear in a few days. The author has found it convenient to keep all species-pure cultures in 250 ml. wide-mouthed flasks. An inverted tightly fitting beaker is used as a cover. However, stoppered test tubes or other closed glass containers will serve equally well.

Unless bacteriologic precautions are taken, it is practically certain that in the transfers made above, bacteria also have been transferred in the isolation of *Paramecium*. A flask containing *Paramecium* (and their associated bacteria) may, in a day or so after standing, acquire a thin surface scum of bacteria. It has been the author's experience that the success of obtaining a good rich mass culture may depend upon the periodic breaking of this film by the gentle shaking of the flask. With such a culture now established, it is possible to maintain the species of *Paramecium* indefinitely by simply inoculating new flasks of culture media with a small amount of paramecium-culture. Without varying this technique, the author has maintained without interruption, species-pure cultures of many races of all species of *Paramecium* in permanent culture up to 16 years.

[9] If a single specimen of paramecium is isolated and allowed to multiply so that eventually a large number of individuals are produced by vegetative fission alone, the group so derived from the one member is called a *clone*.

Cultures for Demonstrating the Mating Reaction and Conjugation.

Perhaps the most spectacular phenomenon associated with sexuality in *Paramecium* is the mating reaction leading to conjugation. This was discovered by Sonneborn (1937 c) for *P. aurelia* and reported later by Jennings (1938) for *P. bursaria*. Additional information bearing upon the mating reaction with subsequent conjugation is to be found in Chapter 10. Beside *P. aurelia* and *P. bursaria*, the mating reaction with resulting conjugation has been seen to occur in *P. caudatum*, *P. multimicronucleatum*, *P. trichium*, and *P. calkinsi*.

If a single specimen of paramecium is isolated and allowed to multiply by vegetative fission alone so that eventually a great number of individuals are produced, the clone represents a pure-line mass culture which is of a given mating type. It is suggested that one begin with diverse clones of *P. bursaria* —the green paramecium—because this species is widely distributed in nature, is easily cultivated, and rather stable in regard to mating type. It is recognized by its green color due to the presence of symbiotic algal cells called zoöchlorellae which are contained in the body of the ciliate.

Ordinarily members of a clone do not mate and conjugate but when members of two diverse clones of opposite mating types are mixed, they almost immediately agglutinate or clump together and subsequently conjugate. After mixing, a few are seen to begin sticking together, then the clumps become larger and larger until, five or six minutes later, immense clumps of a hundred or less individuals are formed. These clumps then become smaller resulting in joined pairs of paramecia undergoing conjugation. In cultures of *P. bursaria* growing in this laboratory, the most favorable time for demonstrating the mating reaction is from approximately 10 A.M. to 2 P.M. with the reaction strongest nearer noon.

E. Methods of Sterilization of Paramecium

Paramecium is one of the animals that has been sterilized internally as well as externally. In the method of isolation described for obtaining mass cultures of species-pure *Paramecium,* it is obvious that wild bacteria of unknown species are also transferred and become established in the culture and serve as food. Such a culture is sometimes called a "pure-line" or "pure-mixed" one, meaning pure regarding the species of *Paramecium* but mixed and unknown regarding the species of bacteria. The exacting requirements of modern research, however, often call for culturing methods in which the food requirements or species of bacteria are known. All methods in which paramecia are cultivated either in carefully controlled bacterial infusions or bacteria-free media first require internal sterilization. There are available precise methods for eliminating all bacteria or other associated microörganisms and it is possible to begin a culture with a sterile protozoön. There have been reported in the literature a number of methods for the sterilization of paramecium and other Protozoa in which a basic principle or combination of principles are employed. All the methods call for rigid bacteriologic technique. Bacteria are known to be present not only externally on the paramecium but internally as well. Some bacteria, in the form of bacterial spores, may be defecated only to germinate and contaminate the sterile fluid.

Paramecia may be sterilized by repeated washing and dilution with sterile fluid, by migration of the paramecia through previously sterilized fluid, by combinations of these methods, or by the use of bactericidal agents.

Washing and Dilution Methods.

The essential feature involved in these methods which are the most widely used for sterilization of paramecium, is washing the specimen free of its associated bacteria by dilution with sterile fluid. The washing may involve the handling of single isolated specimens or large numbers of individuals requiring centrifugation.

Hargitt and Fray (1917) attacked the problem of sterilization of *P. aurelia* and *P. caudatum* with some measuse of success and laid the groundwork for more precise methods. They attempted sterilization by the use of isolated specimens and mass paramecia requiring centrifugation. Single specimens of paramecium were isolated with capillary pipettes and washed through a series of five sterile wash fluids in watch crystals and depression slides which were then placed in Petri dishes. They found that a single *Paramecium aurelia* swimming in a large volume of fluid in such a watch crystal was often lost or that much time was needed in recovering it. The use of larger volumes of media was abandoned by them and they next used only small amounts of sterile tap water in ordinary depression slides enclosed in Petri dishes. Here, the animal was isolated with a sterile pipette, the cover of the dish being raised only enough to permit the entrance of the capillary tip. The specimen was then deposited in the sterile medium of the slide and the cover of the dish dropped. Using a sterile pipette each time, the paramecium was washed in five different culture fluids. They were careful not to introduce bacteria into the sterile washing fluids that might have resided in the rubber bulb of the pipette.[10] To determine the reduction in number of bacteria by this washing and whether or not sterilization was accomplished, they examined each wash fluid bacteriologically by inoculation of nutrient agar plates with the wash water. Table 5 shows the results of the washings of

failed to sterilize a paramecium although the bacteria were reduced in number in the subsequent washing stages. Parpart concluded that the paramecia had ingested spores of bacteria and these, in food vacuoles, had not been removed by the rapid washings. With this end in view, he allowed the animals to remain in the fifth wash fluid for five hours during which time the

Table 5

SHOWING REDUCTION IN NUMBER OF THE BACTERIA BY WASHING PARAMECIUM THROUGH FIVE WASH FLUIDS IN WATCH CRYSTALS AND IN DEPRESSION SLIDES WHEN ENCLOSED IN STERILE PETRI DISHES (HARGITT AND FRAY, 1917)

Plated Wash Water	*Washed in Watch Crystal*	*Washed in Depression Slide*
Plate I first wash water	2500 colonies per drop	2000 colonies per drop
Plate II second wash water	1500 " " "	49 " " "
Plate III third wash water	1000 " "	3 " " "
Plate IV fourth wash water	8 " " "	Sterile
Plate V fifth wash water (Paramecium also on plate)	1 colony " "	Sterile

paramecium in watch crystals and depression slides.

Thus they reported that while sterilization of paramecium was not accomplished in watch crystals, the paramecia were sterilized after the fourth wash in the depression slides.

Perhaps the most reliable method used for the sterilization of paramecium that is now standard procedure in many laboratories is the method developed by Parpart (1928). He found that in the majority of cases five and even 10 uninterrupted washings in sterile media

[10] They did not sterilize the rubber bulb since heat destroys its elasticity. To prevent contamination of bacteria that might have been in the bulb, they placed a small roll of cotton in a narrow constricted portion of the pipette and sterilized the whole. The plug filtered all bacteria which might have been present in the bulb.

paramecia egested their bacterial spores. This was followed by five final washings resulting in sterile animals. Certain fundamental features were introduced into this method which are worthy of note. The entire wash fluid was transferred to agar plates instead of random samples in checking for sterility. A separate, sterile pipette is used for each transfer and all transferring is done under a special hood in which strict attention is given to rigorous bacteriologic technique. The hood consisted of a wooden frame measuring 36 x 15 x 11 inches with a glass top and cloth sides. The front cloth served as an entrance and was loose at the bottom. A binocular microscope was placed at one end with sufficient focal length to permit its ocular lenses to extend through and above the top. Cloth, with slits for the ocular lenses, was glued to the edges of the glass surrounding the lenses. For the handling of the animals, the transfer pipettes were attached to a rubber tube plugged with cotton and operated by means of mouth suction. The transfer pipettes were made of soft glass having an inner diameter of 2 mm. and a wall of 1 mm. thickness and drawn out to capillary fineness with an average inner diameter of 213 microns. The wide end was plugged with cotton, each pipette plugged in a separate test tube and sterilized with the depression slides in Petri dishes in a dry oven between 160° and 170° C. for 45 minutes.

The steps in the washing are as follows:

(a) Three piles of 5 Petri dishes were placed under the hood and 6 drops (⅓ cc.) of sterile fluid was put into each of the enclosed depression slides. The lowermost Petri dishes served as moist chambers for the 5th wash and hence contained slides under the depression slides.

(b) The 15 pipettes necessary for the transfers were placed under the hood with the sterile wash fluid.

(c) The culture containing the paramecia was placed on the microscope stand and a single individual transferred to the uppermost slide in each stack of dishes. Each animal was transferred successively to the depression slide in the Petri dish immediately beneath. By working in rotation from stack to stack, the animals remained in each wash about one minute.

(d) When all three animals were in the 5th wash, from 3–4 cc. of sterile distilled water was added to the lowermost dishes. This prevented excess evaporation from the depression slide while the 5th wash fluid and the animals were being incubated for 5 hours at 25° C.

(e) At the end of 5 hours, each animal was again transferred through 4 washes.

(f) From the last of these, the 9th, the animal was transferred to the desired culture media. The Petri dish of this, the 10th wash, was converted into a moist chamber as in "d."

The fact that all animals of the many tested were sterile in the 10th wash shows the efficiency of the Parpart method as indicated in Table 6.

Kidder, Lilly, and Claff (1940) used a modification of this method in which a single Syracuse watch glass is enclosed in a cellophane bag. The ends of the bag are folded and the bag containing the dish sterilized in the autoclave. Upon cooling, the bag is carefully opened and 5 ml. of sterile wash fluid are placed in each watch glass with a sterile serologic pipette. The protozoön is placed in the wash fluid with a micropipette inserted through the open end

Table 6

SUCCESS OF 10 WASHINGS; THE ANIMAL REMAINING IN THE FIFTH WASH
FIVE HOURS (PARPART, 1928)

Total Number of Animals Tested	Broth Cultures of the Fifth Wash Fluids		Broth Cultures of the Tenth Wash Fluid plus the Animals	
	Sterile	Infected	Sterile	Infected
50	17	33	50	0

of the bag. They claim three important advantages in their modification which are as follows:

1. The top of the dish and the fluid is never exposed to the outside air (and its bacteria) from above, hence obviates the use of a hood.
2. Since water does not condense on the cellophane (but does so when glass covers are used) the protozoön can be seen clearly at all times.
3. The use of so large a container as a Syracuse watch glass with its 5 ml. of fluid increases the dilution factor.

An improvement in their method would be the use of smooth, concavity dishes, in place of the straight-walled Syracuse watch glasses.

ful in appreciably reducing the number of bacteria in their five washes but unsuccessful in obtaining sterile animals.

However, Kidder and Stuart (1939) modified this method by two important additions as follows: the use of additional washes (at least 15) and the use of plugged centrifuge tubes for the washes.[11] They were successful in the sterilization of *Colpoda steinii* by centrifugation and the method can be modified easily to sterilize *Paramecium*.

Rosser (1941) reported a method of freeing paramecia from bacterial contamination by passing them through drops of sterile media. Wash-drops were pipetted into the center of Petri dishes, then paramecia were transferred from drop to drop in a small loop of 0.127

Table 7

SHOWING REDUCTION IN NUMBERS OF BACTERIA BY WASHING *Paramecium* THROUGH FIVE WASH FLUIDS IN A CENTRIFUGE WHEN WASHES ARE PLATED (HARGITT AND FRAY, 1917)

Number of wash water	I	II	III	IV	V
Number of colonies of bacteria on the plate	500	200	50	11	3

In addition to the wash-dilution method in which single paramecia are isolated and sterilized, there is another modification of the same principle using centrifugation. This was attempted by Hargitt and Fray (1917) who described their method as follows, "A pipette-ful of *Paramecium* was placed in a sterile centrifuge tube and considerable sterile hay infusion added to dilute the bacteria. The paramecia were thrown to the bottom of the tube and as soon as the tubes came to rest, the wash fluid was removed and fresh fluid added with a sterile pipette. This was repeated five times. Samples of the five wash fluids were then plated in Petri dishes and the following number of bacteria found present" (Table 7).

One observes that they were success-

mm. platinum wire. Paramecia were allowed to remain for at least 15 minutes in each wash-drop with the exception of the third in which the specimens were kept for about four hours. An improvement in this method would be the addition and utilization of more wash-drops. A simple apparatus for washing Protozoa was also described by Turner (1931).

Migration Methods.

The principle of migration involves the introduction of paramecium con-

[11] The centrifuge tubes were stoppered with large cotton plugs and autoclaved. To prevent the cotton plug from being drawn into the tube during centrifugation, the cotton was folded over the tube and held in place with a rubber band.

taminated with bacteria into a sterile fluid. The specimen is then allowed to swim or migrate across a comparatively large distance of sterile medium. Its feature resides in the fact that the contaminating bacteria are left behind in the migration and the sterile ciliate may be removed from another end of the container with a sterile pipette. This is accomplished by using a Petri dish or long box-like glass trough of sterile fluid with lid. The dish of fluid is placed on the stage of a dissecting microscope and when the fluid has come to rest, the cover is lifted carefully and "wild" paramecia introduced into one side of the dish. The dish is quickly covered and some of the paramecia migrate to the opposite edge of the dish where they are removed with a sterile pipette and placed in fresh culture fluid. Time of removal of the paramecia is an important factor in migration methods. Since paramecia move faster than even the motile bacteria which may be present, one should remove the ciliates as soon as they reach the opposite point of the container. A shield or hood should be placed over the dishes during the operation to prevent contamination of bacteria from the air. Frequently this method results in a great reduction of bacteria but may not always result in sterile paramecia. However, several migration dishes may be employed so that when a number of paramecia have migrated across the dish they may be removed and placed into another one. This procedure may be repeated as in the washing methods. Since migration across a dish takes a comparatively short time, bacterial spores in food vacuoles of paramecium may still be present. These viable spores when egested into sterile culture media are likely to germinate, reproduce, and contaminate the culture again. An interruption of several hours between migrations (as in

the Parpart wash method) would enable the paramecia to defecate their food vaculoes after which the migration could be continued. Hetherington (1934) made use of the dilution-migration principle in which he alternated dilution with migration. He placed a drop of culture rich with *Colpidium colpoda* in the left side of a Columbia culture dish containing 1 ml. of sterile fluid. Upon migrating to the right side of the dish, a number of the ciliates were removed with a sterile micropipette and transferred to the left side of a second dish where the process was repeated a second and finally a third time. The selected ciliates were left in the third dish for three hours (to defecate bacterial spores) then the specimens were allowed to migrate by the same method in a fourth and fifth dish where they were left to stand again for three hours to complete the defecation. Specimens were again removed and allowed to migrate in a sixth and seventh dish after which the specimens, now sterile, were placed in the culture medium.

A large number of variations of the migration method have been reported in the literature for the sterilization of Protozoa. An ingenious capillary tube was described by Stone and Reynolds (1939) for the sterilization of the parasitic flagellate, *Trichomonas hominis,* but with modifications it appears to have practical application for the sterilization of certain species of *Paramecium.* The capillary tube is about 20 cm. in length and consists of a series of sharp bends or loops which serve as bacterial traps. Purdy and Butterfield (1918) claim to have sterilized *Paramecium* by allowing an organism to swim through 30 feet of sterile water!

The fact that species of *Paramecium* are negatively geotrophic has practical advantages for sterilization purposes.

Glaser and Coria (1930, 1935) made use of this feature in the migration of paramecia through a large tube about 13 inches long and possessing a ¼-inch bore and a fine tapering point (Fig. 39). The wide end is plugged with cotton and the unit sterilized. Sterile fluid is drawn up to within two inches of the top by suction applied to the wide end through a rubber tube. About 2 ml. of a wild culture containing paramecia are drawn up the migration tube and this forms a layer beneath the sterile fluid. The tapered end is sealed in a flame and the tube then mounted upright in a rack as shown in the figure. After a time, samples from the top of the tube contained some paramecia which had eliminated themselves of most microörganisms in their migration. To render the paramecia sterile, the process was repeated.

Glaser and Coria (1930) described another migration method in which a V-shaped tube is filled with Noguchi's semisolid medium. The larger arm of the tube measures 12 cm. in length with an inner diameter of 28 mm. The smaller arm is 9 cm. in length with an inside diameter of 8 mm. (Fig. 40). After sterilization, the tube is filled with 15 ml. of the sterile melted medium and allowed to partially solidify. The wild culture is then placed at the bottom of

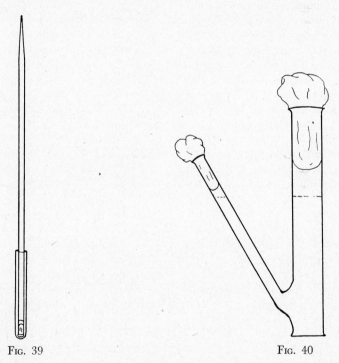

FIG. 39   FIG. 40

FIG. 39. Migration tube. The tube is filled with a sterile fluid to within about 2 inches from the top. Protozoa to be sterilized are drawn up under the sterile fluid and the small tip sealed in a flame. The Protozoa migrate upward and are taken off at the top.

FIG. 40. V migration tube for semisolid media. The Protozoa to be sterilized are injected through the small arm and deposited at the bottom of the V. They migrate up through the semisolid medium and are removed at the top of the large arm. (Glaser and Coria)

the tube with a long, fine pipette by injection through the small arm. One is to avoid the introduction of air bubbles. The tube is left to stand at room temperature to allow paramecia to reach the top of the large arm where specimens are then recovered from the surface.

Claff (1940) designed an ingenious apparatus utilizing the principle of migration-dilution for the sterilization of paramecium and other ciliates that are negatively geotrophic. The apparatus is a rather compact unit and can be sterilized readily. It consists of a series of six flasks each holding approximately 145 ml. of fluid and joined together so that the top of one flask is connected by small-bore rubber tubing to the bottom of the next succeeding flask (Fig. 41). A 1-liter reservoir (flask) capable of being raised to a level higher than the other flasks is joined to the bottom of flask No. 1 by rubber tubing. A test tube into which a glass tube is inserted through a cotton plug is connected by rubber tubing to the top of flask No. 6. The fluid in the raised reservoir is held in check by a clamp. By releasing the clamp it is possible to transfer a small

volume of fluid from the reservoir into the bottom of flask No. 1. This effects a transfer of the uppermost fluid from each flask into the bottom of the next succeeding flask in line. Finally this results in the uppermost fluid in flask No. 6 to be transferred into the test tube. A vaccine port near the bottom of flask No. 1 is fitted with a vaccine cap and paramecia to be sterilized are injected through it with a sterile hypodermic needle. The paramecia are then allowed to migrate to the narrow top of the flask and after large numbers are collected, they are forced over into the bottom of flask No. 2 by about 1½ ml. of sterile medium from the reservoir. The process is repeated after each migration and the fluid drained from the system at the top of flask No. 6 is kept in five sterile test tubes and used as bacteriologic controls on the fluid preceding the paramecia. The migration of the paramecia may be followed macroscopically or with a hand lens. Paramecia are obtained in the sixth test tube and after careful handling, placed in the appropriate sterile culture medium. Claff gives evidence for the sterilization of *Paramecium caudatum*. Better success is accomplished

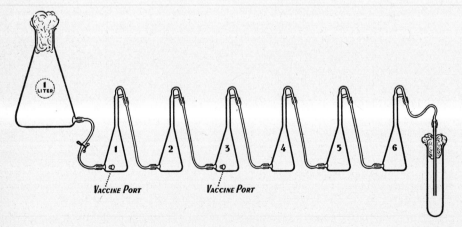

Fɪɢ. 41. Diagram of migration-dilution apparatus drawn to show connections (explanation in text). (Claff)

by first feeding the paramecia for several days on a diet of yeast or non-spore-forming bacteria before the sterilization is attempted. Advantages claimed for the apparatus and method are that it is a closed system, that there is present a large dilution factor of the order of 10^6, and that there is little chance of contamination through faulty manipulative technique. Also a large number of sterile organisms may be obtained in as short a time as an hour or less. A number of precautionary steps are necessary for successful operation of the apparatus and the reader is referred to the original paper for them. Paramecia are relatively powerful swimmers so that migration through the flasks might be too rapid to effect internal sterilization. Attention should be called to the fact that bacterial spores present in the food vacuoles may subsequently contaminate the culture of paramecia with bacteria. This can be remedied however by either allowing some of the paramecia to remain in one of the sterile flasks of infusion for several hours, or by permitting them to migrate through the apparatus more than once to eliminate all bacterial spores.

Electrophoresis (Cataphoresis).

The migration of paramecia under the influence of an electric current is called *electrophoresis* (see also Chapter 8). When an electric current is sent through a suspension of paramecia, they typically migrate toward the cathode or negative terminal.

Making use of this principle as a method of sterilization, Amster (1922) was able to free ciliates of bacteria by placing non-polarizable electrodes in a culture of ciliates to which a 0.05 per cent sodium chloride solution was added for conductivity of a current of electricity. The ciliates (*Balantiophorus*)

moved to the cathode and bacteria to the anode. He found that six repetitions were sufficient to sterilize the ciliates.

Sawano (1938) took advantage of this phenomenon to collect and concentrate paramecia in a "pure" state from mass culture. The equipment is shown diagrammatically in Fig. 42 in which the lower end of a glass tube 6 cm. in diameter and 24 cm. in length is covered with a cellophane membrane, fastened with petroleum jelly and secured tightly with a rubber band to prevent leakage. The covered end is placed in a glass dish 6 cm. in depth and 12 cm. in diameter. A siphon about 22 cm. in length is connected to a suction pump to remove culture media. A 0.5 per cent sodium chloride solution is poured in the dish; the tube is then filled with the culture of paramecia and fastened to a stand in such a position that the surface of the cellophane membrane is sufficiently soaked in the salt solution. A platinum foil electrode (anode, $+$) is inserted in the surface of the tube and the cathode ($-$) in the bottom of the dish, then connected to the plus and minus poles respectively of

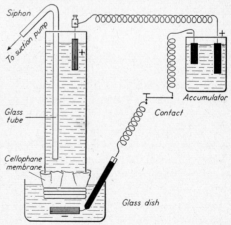

FIG. 42. Diagram showing the equipment for the collection of paramecia by means of electrophoresis. (Sawano)

an accumulator. When the circuit is closed, the paramecia in the tube migrate at once toward the cathode and remain on the surface of the cellophane membrane forming a conspicuous white sediment. When almost all of the paramecia have been deposited, the water in the tube is drawn off carefully through the siphon leaving fluid to a depth of 2 cm. The membrane is then removed and the concentrated paramecia now on the cellophane surface are washed thoroughly and transferred to a beaker. The suspension of paramecia is then diluted with twice the volume of pure water and, according to Sawano, the specimens are allowed to consume as large a quantity as possible of the bacteria remaining in the suspension. He reported that when left overnight, paramecia evacuated the ingested bacteria.

While the method of Sawano may be effective for quickly concentrating paramecia in a relatively "pure" state, it must be admitted that the concentrated specimens are not bacteria-free. However, by repeating the operation several times and by using sterile media each time, it should be possible with the addition of further refinements to obtain sterile specimens. Humphrey and Humphrey (1948) described a method for reducing the numbers of bacteria in a suspension of *P. caudatum* by an electrically directed migration through a sterile column of liquid which they claimed to be suitable for experiments in respiratory metabolism.

By a more exacting method, van Wagtendonk and Hackett (1949) successfully sterilized *P. aurelia* by the use of an electromigration tube in which there was an electric transport of the paramecia against a continuous flow of sterile agar solution.

The author (Wichterman, 1952) has described a method and device for con-

centrating abundant dividing stages of paramecium (see page 411).

Bactericidal Agents (Antibiotics).

A number of cases have been reported in which various agents such as chemicals, heat, roentgen rays, and the like have been utilized in killing bacteria associated with the protozoön thereby effecting sterility. The use of chemicals and heat for sterilization has achieved a small measure of success upon encysted amebas and ciliates. Oehler (1924 a, b) reported that he obtained sterile *Colpoda* from cysts by using heat. This seems a rather unreliable method since we are aware of bacteria that are able to withstand even very high temperatures. Generally the higher temperatures will kill the encysted ciliates before the bacteria are destroyed. The fact remains that the use of heat for sterilization of species of *Paramecium* is of little value since one is obliged to work with trophic forms of the species.

Recently Seaman (1947) reported sterilization of *Colpidium campylum* and *P. multimicronucleatum* using penicillin and it has been successfully used by van Wagtendonk and Hackett (1949) in the sterilization of *P. aurelia*. In addition to penicillin antibiotics such as aureomycin, streptomycin, bacitracin, terramycin, and others may prove to be successful in the sterilization of paramecium.

Effecting sterilization of ciliates by the use of roentgen rays was reported by Brown, Luck, Sheets, and Taylor (1933) on *Euplotes*. They reported sterilization of ciliates by exposure of trophic and encysted stages of *Euplotes* to x-rays of 2110 roentgen units per second. When applied to *Paramecium*, it should be kept in mind that the x-rays be of such a strength that they destroy the bacterial spores within the food vacuoles but not cause irreparable damage to the endoplasm where the food vacuoles are found.

Sterility Tests.

It has been emphasized in the preceding account that rigorous bacteriologic technique must be employed in obtaining sterile paramecium. All dilution, wash, and migration methods or combinations of them result in almost immediate reduction of bacteria. But it is possible that very small numbers of bacteria be present in the final washes or that bacterial spores in the food vacuoles contaminate the organism internally even though the external surface be free of all bacteria. Such bacterial spores when defecated are capable of germination only to contaminate the sterile fluid. Therefore the fluid and the paramecia should be checked thoroughly for contaminating bacteria. Tests for sterility are generally made by inoculating the medium and paramecia to be tested into a culture medium of broth or by distribution on a nutrient solid medium such as extract agar. If the broth becomes turbid, it is contaminated with bacteria (turbidity test). But as pointed out by a number of investigators, lack of turbidity does not always indicate sterility. Certain bacteria settle to the bottom of a tube, grow slowly, and form small masses while the broth is left clear. More certain than the turbidity test is the use of the plate method. This consists of streaking or spreading the fluid and paramecia to be tested over a solid nutrient agar medium. A number of such plates should be prepared, kept at various temperatures from room temperature to 37° C., and examined for contaminating bacterial colonies at periods from 24 hours up to 14 days. The standard,

sterility testing media employed by Johnson and Baker (1942) and Johnson and Tatum (1945) consist of the following: 0.5 per cent Difco yeast extract plus 0.5 per cent dextrose; 0.03 per cent beef extract; Difco nutrient agar plus 0.5 per cent dextrose; and Brewer's thioglycollate anaerobic medium.

Although one is unlikely to find anaerobic bacteria in paramecia from wild cultures, a test for them should be made. A simple test to determine their presence is given by Kidder (1941) in which tubes containing not over 3 ml. of nutrient broth and a 2- to 2½-inch layer of paraffin oil, are plugged with cotton and autoclaved for 20 minutes at 15 pounds' pressure. Rubber stoppers may be sterilized at the same time. Immediately after sterilization the rubber stoppers are fitted tightly into the tubes, which are then allowed to cool. When the broth is cool, inoculations are made by injecting the material to be tested through the paraffin oil into the broth after which the rubber stopper is immediately replaced. Some investigators see little value in the use of anaerobic sterility tests since the animals are only grown aerobically. Within the limits of the experiment, the system is sterile if all aerobic bacteria are removed. The testing of the wash fluids for bacteria with Gram stains is another valuable check.

The object of sterilization of paramecium is of course making the specimens bacteria-free. But in so doing, we remove the chief source of food of paramecium since these ciliates are principally obligate bacteria-feeders. However, in starting with a sterile paramecium, the experimenter is now in a position to keep his culture truly pure by attempting to substitute a synthetic culture medium containing among other substances, tryptone, proteose peptone, yeast extract, dead yeast cells, or perhaps dead bacterial cells instead of the usual diet of living bacteria. Notable attempts have been made to maintain vigorous cultures of paramecia for fairly long periods of time in such media. The question arises: can such cultures be kept indefinitely as in cultures where paramecia utilize bacteria for food? If it can be accomplished successfully, we are in a position to attack some of the basic problems concerned with nutrition, growth, and reproduction. On the other hand, with sterile paramecia available, one is able to select as a source of food, a single known species of bacterium whose growth and other characteristics are well known to the experimenter.

Concluding Remarks.

The location of cultures in a room is of considerable importance. They should not be placed near a window in the presence of direct sunlight. Windows facing to the north are ideal for the cultivation of *P. bursaria*—the green *Paramecium*—which requires light. Cultures of other species of *Paramecium* should be placed on tables or shelves where they may receive some indirect light.

The writer has suspended from the ceiling of the laboratory, a 30-inch ultraviolet "Sterilamp" which is left on throughout the night. It appears to be very successful in reducing the number of foreign microörganisms whose spores may be present in the air. Eighty per cent of the ultraviolet radiation generated by the lamp is in the region of 2537 Angstroms. Radiations of this wave-length are the most destructive to bacteria and molds which may contaminate cultures.

All species of *Paramecium* grow well at ordinary room temperature. It has been shown by a number of investigators that paramecia will divide at temperatures between 5° and 35° C. but that the optimum temperature yielding greatest growth is between 24° and 28° C. (see p. 191)

Cultures with a pH of approximately 7.0 or slightly higher yield the best growth in paramecium. The relation of pH to temperature factors, food and growth requirements of paramecium, and the utilization of disolved proteins are considered in Chapter 6.

Additional references to methods or modifications of existing ones for the washing, isolating, sterilizing, or cultivation of paramecium and closely related forms are as follows: Hansen (1927), Luck, Sheets, and Thomas (1931), Turner (1931), Rosenberg (1932), Rosser (1941), Medes and Stimson (1942), Stanley (1945), Wichterman (1949), and Sonneborn (1950 b).

Chapter 5

Physical and Chemical Properties of the Protoplasm of Paramecium

A. **Protoplasmic Structure and Optical Properties**
B. **Chemical Structure of Paramecium**
C. **Surface Area and Volume**
D. **Specific Gravity of Paramecium and Its Contents**
E. **Viscosity**
F. **Permeability**
G. **Effects of Various Agents on Protoplasm of Paramecium**
    1. Effects of Water and Different Anisotonic Media
    2. Effects of Temperature
    3. Effects of Salts, Acids, Alkalies, and Electrolytes
    4. Effects of Hydrogen-ion Concentration
    5. Effects of Mechanical Agitation Including Centrifugation
    6. Effects of Electric Current, Magnetic Fields, and Sound Waves
    7. Effects of Ultraviolet Light, Fluorescence, X-rays, Radium, Radioactive Substances and Isotopes
    8. Effects of Alcohol, Narcotics, Vital Stains, Photodynamic Action and Sensitization, Carcinogenic Agents, Oligodynamic Action, Alkaloids and Related Drugs; Other Inorganic and Organic Compounds Including Metallic Salts, Organic Extracts, Enzymes, Hormones, Vitamins, Sera, Toxins, etc.
    9. Effects of Hydrostatic Pressure

A. Protoplasmic Structure and Optical Properties

It is evident that our knowledge of the general morphology and special cytology of *Paramecium* is far greater than our knowledge of its special physiologic functions.

In the Metazoa, specialized cells which are grouped into tissues, organs, and organ-systems perform the sum total of the various physiologic processes which we recognize as nutrition, respiration, excretion, irritability, and reproduction. In *Paramecium,* as they are for all Protozoa, the physiologic processes are performed by, and occur in, the single cell.

Long before von Mohl in 1846 introduced the term *protoplasm* for the "slimy, granular semifluid" in plant cells, early workers recognized the existence of a viscid substance within living cells. Several years before von Mohl, Dujardin made first recognition of the living substance as such when he proposed the term *sarcode* for the living substance of a foraminiferan protozoön and of other lower animals. According to Dujardin, the sarcode was considered as a living jelly; sticky, like mucus, glutinous and transparent. It was an easy step for investigators to identify Dujardin's animal sarcode with von Mohl's plant protoplasm. Max Schultze,

117

in 1863, demonstrated the universal occurrence and fundamental similarity of protoplasm in the cells of all living organisms. From that time up to the present, protoplasm has been observed and investigated with a variety of methods by an ever increasing number of workers in the many disciplines of science.

Concepts of the visible structure of protoplasm led to theories that defined the living substance as reticular, fibrillar, alveolar, etc. A shift by many investigators, who placed great reliance upon fixing agents which coagulated the protoplasm, had much to do with these views.

When viewed through the microscope with ordinary transmitted light, the protoplasm of *Paramecium* appears as a colorless, translucent substance in which are embedded small granules, crystals, food, and other vacuoles, all of which vary in size and number. Kite (1913) and others believed that the protoplasm of *Paramecium* was a soft, elastic and glutinous gel, the surface of which seemed to be more viscous than the interior. In dark-field illumination and polarized light, crystals, food particles, and other granules shine brilliantly. Cytoplasmic inclusions vary considerably with the functional behavior of the organism. For a detailed discussion of these protoplasmic inclusions of *Paramecium,* the reader is referred to pages 80–86.

The translucency of protoplasm in *Paramecium* varies with the shape, size, and number of granules and vacuoles present and, to a certain extent, with the species. It is the writer's experience that *P. multimicronucleatum* is rarely as transparent as *P. caudatum* and *P. aurelia,* but this observation may be due to the larger size of *P. multimicronucleatum.* In all species, the matrix or hyaloplasm is completely translucent. It is claimed by some that photography with ultraviolet light discloses structure in the hyaloplasm. Since ultraviolet rays have a coagulative effect on protoplasm, the method is open to criticism. The active, streaming movement of the protoplasm, termed *cyclosis* and readily seen in all species of *Paramecium,* is most rapid in *P. bursaria.*

The macronucleus of *Paramecium* can be seen easily in the living condition if slight pressure is applied to the organism. The structure appears uniformly granular and pale yellow in color. The micronucleus in *P. caudatum,* while more difficult to demonstrate in living specimens, can be seen by the same method.

An important constituent of the chromatin found in the nuclei of *Paramecium* is nucleic acid of the thymus type and generally called DNA or thymonucleic acid (zoönucleic acid, chromonucleic acid, desoxypentose nucleic acid or desoxyribose nucleic acid).

The so-called yeast type of nucleic acid (phytonucleic acid, plasmonucleic acid, pentose nucleic acid, or d-ribose or ribose nucleic acid) is found chiefly in the cytoplasm (see also Chapter 3).

All species of *Paramecium* have well-defined polarity; an anterior-posterior axis being a persistent feature in all stages of vegetative and reproductive life. The same may be said about the dorsoventral surfaces.

According to Child and Deviney (1926) a physiologic (metabolic) gradient exists since evidence is available showing quantitative differences decreasing from the anterior end to the posterior end of the organism. Thus for *P. caudatum* these workers show a differential susceptibility, decreasing from the anterior end posteriorly, to ultraviolet radiation; visible light after sensitization by eosin, methylene blue, or neutral red; the weak bases (NH_4OH

and NH_4Cl) ; the strong base NaOH; the weak acids, acetic, CO_2; KCN; the strong acids, HCl, H_2SO_4; the dyes, neutral red, and methylene blue.

While the persistence of a fairly rigid shape is characteristic of the various species of *Paramecium,* the author wishes to emphasize that certain stages in the life of this protozoön appear less rigid than in other stages. For instance, exconjugants appear more flexible than typically vegetative stages. Nevertheless, vegetative stages exhibit considerable elasticity and flexibility of the body. This is apparent in the ultracentrifuge studies of King and Beams (1937) and readily can be demonstrated when paramecia are placed on semisolid agar or even by watching them enmeshed in cotton fibers.

It would seem likely that the cortex of *Paramecium* with its ectoplasmic structures serves to give the organism support and shape, accounts for its rigidity and perhaps gives some measure of protection. Chambers (1924) tore the ectoplasm with a microneedle and observed that the fluid endoplasm emptied into the external culture medium. This was followed by a disintegration of the ectoplasm. Although it is somewhat inconclusive, he noted that the endoplasm occasionally formed a surface film which bounded the extruded mass. Merton (1928) investigated *Paramecium* deprived of its ectoplasm which included the pellicle, cilia, and trichocyst layer. The remaining portion of such a specimen resembled the fanshaped appearance of certain amebas. Some of these fan-shaped forms were not only able to survive and move but also divided. One is reminded of the work of Nadler (1929) who was able to remove the entire pellicle from the ciliate *Blepharisma* by immersing the organism in weak solutions of strychnine sulfate or morphine. The "denuded" specimens retained their normal shapes and within a few days formed new pellicles.[1] The entire process could then be repeated. If one were able to do the same with *Paramecium,* it might throw some light upon the mechanism responsible for causing the mating reaction.

Fine protoplasmic structures, such as minute granules otherwise imperceptible, are revealed with the polarizing microscope. Certain other structural properties of *Paramecium* are more readily visible with the phase and electron microscope. X-ray diffraction studies should reveal added information on the structure of protoplasm.

B. Chemical Structure of Paramecium

The chemical composition and structure of the protoplasm of *Paramecium* have been investigated in considerable detail by a number of workers, namely Sosnowski (1899), Greeley (1904), Grobicka and Wasilewska (1925), and MacLennan and Murer (1934). Grobicka and Wasilewska reported that fresh protoplasm from well-fed *P. caudatum* contained the following:

87.40–91.60% average 89.00% water*
 8.40–12.60%   "   11.00% dry substance
 0.55– 0.59%   "   0.57% ash
 0.86– 1.59%   "   1.14% glycogen
 0.84– 1.19%   "   1.05% fatty acid
 0.72– 1.54%   "   1.06% total nitrogen

*Iida (1940) reported the water content of *Paramecium* to be 79 per cent.

The dry substance contained the following:

 4.95– 5.19% average 5.07% ash
12.70–17.20%   "   14.90% glycogen
 7.83–10.94%   "   9.18% fatty acid
56.10%   "   56.10% protein
12.35–13.74%   "   12.81% total nitrogen

[1] Giese (1946) was unable to repeat this experiment in his strain of *Blepharisma.*

Of the 12.81 per cent nitrogen, 8.97 per cent consisted of protein nitrogen and 3.84 per cent of non-protein nitrogen; the 12.81 per cent nitrogen of the dry substance corresponded to 1.06 per cent nitrogen in the living paramecia. The glycogen content of the paramecia in terms of fresh weight amounted to 0.86 to 1.59 with an average of 1.14 per cent but after five days of starvation fell to an average of 0.68 per cent. Only 0.35 per cent cholesterol was found in the dry substance.

Bayer and Wense (1936 a, b) have extracted acetylcholine from *Paramecium* and reported that extracts contained Adrenalin or a substance resembling it since the extracts produced an Adrenalin-like effect on rabbit intestine.

MacLennan and Murer (1934) studied *P. caudatum* by microincineration, stained control sections, vital staining and osmic impregnation. They reported that most of the cytoplasmic components and organelles, including vacuome, chondriome, cilia, basal granules, trichocysts, food vacuoles, and the nuclei could be identified by their ash (Fig. 43). No ash was found in the pellicle and only a small amount in the hyaloplasm. According to these investigators, the larger amount of ash in the endoplasm as compared with the ectoplasm was due to the large number of granules in the former region. Nuclei and some of the granules near them showed traces of iron.

By means of the microincineration technique and two specific colorimetric tests, Lansing (1938) demonstrated the presence of calcium in *P. caudatum*. It was shown that calcium salts are concentrated in the cortex of the cell and that calcium is a constituent of the cortical ash.

Kruszynski (1939) also reported the presence of calcium as well as iron in microchemical investigations of micro-incinerated *P. caudatum*. According to him, varying amounts of inorganic substance are encountered in different specimens. This is perhaps due to the physiologic state of the organisms. He observed that the ash of extruded trichocysts was composed mainly of a calcium salt.

C. Surface Area and Volume

In estimating the surface area and volume of *P. caudatum,* Fortner (1925) compared the cell-body as equal to the sum of two half-rotating ellipsoids and presented the following formulas in which

m	represents	the length of *Paramecium*
b	"	the radius of its greatest diameter
ψ	"	the variable coefficient $\dfrac{b}{m}$
4.5	"	the coefficient of ψ and m^2
2.01	"	the constant for the volume

Surface Area.

The following formulas[2] may be used to determine the surface (O) of *P. caudatum.*

Formula a, $O = m^2 \cdot \psi \cdot 4.5$ or
Formula b, $O = m^2 \cdot \psi \cdot 4.9$

If, as suggested by Fortner, the sculpturing of the surface of paramecium results in doubling the value, the formulas should be as follows:

Formula c, $O = m^2 \cdot \psi \cdot 9$ or
Formula d, $O = m^2 \cdot \psi \cdot 9.8$

For example if we take a typical *P. aurelia* which measures, say, 130 microns in length and 40 microns in width and use formula "a," the surface area is found to be 11,712 μ^2. For a typical *P.*

[2] According to Fortner, the coefficient of 4.9 is not very accurate but it gives comparable values. The value of 4.5 as a maximum of Ψ gives more accurate values, but at smaller Ψ values, relatively larger errors.

caudatum which measures 250 microns in length and 70 microns in width, the surface area is found to be 39,375 μ^2.

Volume.

The following formula may be used to determine the volume of *P. caudatum*.

$$V = m^3 \cdot \psi\, 2 \cdot 2.01$$

Using the same typical specimen of *P. aurelia* which measured 130 microns in length and 40 microns in width, the formula gives the volume as 104,182 μ^3. Calculating for the same typical specimen of *P. caudatum* which measured 250 microns in length and 70 microns in width, the volume is found to be 628,125 μ^3.

While Fortner originally devised his formulas of surface area and volume for *P. caudatum*, they may be applied equally well to other members of the "aurelia" group (*P. aurelia* and *P. multimicronucleatum*) because of close similarity in body form.

Popoff (1908) assumed the body of *P. caudatum* to be an ellipsoid and computed the volume by using the following formula:

$$V = 4\,\pi\, LBT/24$$

in which L is the length, B the breadth and T the thickness.

Using the same measurements for the typical *P. aurelia* and *P. caudatum* employed in Fortner's formula, we find the former species to have a volume of 108,853 μ^3 and the latter species to have a volume of 641,083 μ^3. It is thus seen that Popoff's formula yields fairly comparable but slightly higher values.

In determining the volume of *P. multimicronucleatum*, Jones (1933) used plasticine models in water. He noted that the anterior third of the body was approximately equal (in volume) to the buccal groove of the middle third of the body. Jones concluded that the result was a cylinder having a height equal to one third of the total length and a base equal to the diameter. He considered the posterior portion to be a cone of similar height and base to the cylinder referred to above. By adding together the volumes represented by a cone and a cylinder each having a height equal to one-third the length of the animal and both having the same base, Jones reported that the volume of *P. multimicronucleatum* was as follows:

$$V = \frac{Lb^2}{3}$$

in which L is the length and b the breadth.

It will be found that this formula yields values greatly lower than those of Fortner and Popoff.

D. Specific Gravity of Paramecium and Its Contents

Jensen (1893) attempted to determine the weight of paramecium by observing floating specimens in a potassium carbonate solution of a given density. His method, which induced shrinkage of the paramecia, gave the rather high value of 1.25 as the specific weight. Platt (1899) studied paramecia which had first been killed with acetic acid or osmic acid fumes and then placed in gum arabic solution of known specific gravity. From her experiments she concluded that the specific weight of paramecium is approximately 1.017. Lyon (1905) obtained specific gravities of 1.048–1.049 by centrifuging living paramecia in gum arabic solutions. Kanda (1914 a, b, 1918) repeated the procedure and obtained values of 1.0382–1.0393.

Fetter obtained the specific gravity by placing paramecia that averaged 260 microns in length in sugar solutions of

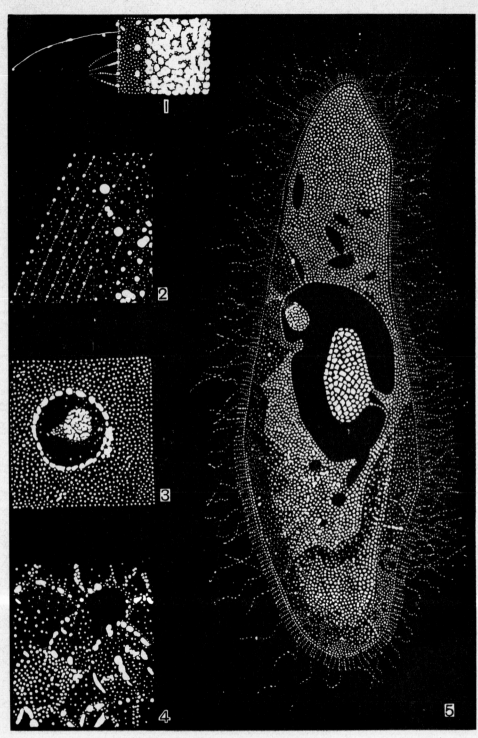

Fig. 43. (*For legend see facing page.*)

varying percentages and then centrifuging. She reported that with certain concentrations of the sugar solution, the paramecia were found at the bottom of the tube; with heavier concentrations, they were found at the top of the tube. She then obtained the specific gravity of the solution between the highest concentration at which they were on the bottom of the tube and the lowest concentration at which they were found on the top of the tube. The result gave a specific gravity of 1.038.

Leontjew (1923, 1924, 1926), who also used as his method the rate of sinking of *Paramecium* in a solution, gave the specific weight as 1.045–1.05. Koehler (1922) reports that workers give a specific gravity of 1.038 and 1.039 which is in agreement with the findings of Fetter and Kanda.

If, as mentioned by Kalmus (1931), we consider the specific weight and volume of a typical specimen of *P. caudatum* with a length of 200 microns and a width of 65 microns, its weight is given as follows:

(Weight of *Paramecium*) $P = 0.46 \cdot 10^{-6}$ gr.

McClendon (1909) fixed and stained *Paramecium* after centrifuging and reported that the crystals and nuclei were displaced centrifugally. This is in agreement with the findings of Harvey (1931) as determined by the microscope-centrifuge in which he reported that the crystals and nuclei were thrown down rapidly.

By means of the ultracentrifuge, King and Beams (1937) found that cell contents of various species of *Paramecium* were stratified in the following order from centrifugal to centripetal regions: crystals in vacuoles, compact macronuclear chromatin, food vacuoles and neutral red inclusions, achromatic matrix of macronucleus, endoplasm, large clear vacuoles, and finally fat.

E. Viscosity

The early investigators were content to describe living protoplasm as a viscous fluid. Not satisfied with such an inadequate description of the physical behavior of living material, the modern biologist is interested in knowing how viscous it is or its viscosity when compared with water.

As in all living cells, the protoplasm of paramecium is a colloid. Colloids generally show changes from a comparatively fluid consistency to a firm, jelly-like one or vice versa. The result is change in the consistency or viscosity of protoplasm. Viscosity is now recognized as a fundamental attribute of protoplasm. According to Heilbrunn (1943), viscosity[3] is the force which tends to hold the particles of a substance together when a shearing force acting

[3] The term *poise* is the common unit of viscosity measurement and is derived from the name of the great French physiologist Poiseuille who pioneered in viscosity measurements. At room temperature, the viscosity of water is approximately one hundredth of a *poise,* or a *centipoise.*

FIG. 43. Camera lucida drawings of sections of *Paramecium caudatum* viewed with darkfield illumination after incineration. Figs. 1, 2, and 4, × 2480; figs. 3 and 5, × 1240. Figs. 1, 10 μ thick; fig. 2, 3, and 5, 5 μ thick; fig. 4, 3 μ thick. (1) Small portion of a section showing cilia, basal granules, a discharged trichocyst, and a few vacuome granules in the endoplasm. The contrast between ectoplasm and endoplasm is clearly shown. (2) Tangential section showing ciliary rows and longitudinal neuromotor fibrils. (3) Food vacuole showing a relatively light concentration of ash around it. (4) Detailed drawing of the endoplasmic ash, showing granules, rods, and filaments. (5) Entire longitudinal section. The micronucleus lies at the left of the macronucleus. (MacLennan and Murer)

on the substance tends to pull it apart. Thus when a substance is more viscous, it flows less readily. Methods have been devised for determining the protoplasmic viscosity of cells in which a quantitative result may be obtained. Viscosity has been studied electromagnetically, by observations of granules in Brownian movement, by observations on rate of diffusion of dyes in protoplasm, by the methods of microdissection, and by centrifugation. The centrifuge method has been used with considerable success in the study of many animal cells—especially isolated ones. It has also been applied to paramecium. With the use of a formula, the viscosity is measured by recording the number of seconds that it is necessary for granules to move through a given part of the cell. The centrifugal force varies as the square of the number of turns of the centrifuge per second. With the increase of speed of centrifugal turn, the speed of granular movement is found to increase as the square of the number of turns of the centrifuge per second.

Fetter (1926) made a determination of the protoplasmic viscosity of paramecium by the centrifuge method using the following form of Stokes formula:

$$V = \frac{2 \, cg \, (\sigma - s) \, a^2}{9 \, \eta}$$

V is the velocity of movement of the granules in the protoplasm; c, the centrifugal force; g, the gravity constant; σ, the specific gravity of the granule; s, the specific gravity of the animal; a, the radius of the granule, and η, the viscosity of the liquid. In the formula, the specific gravity of the entire animal (s) is used instead of the specific gravity of the fluid between the granules. This is due to the fact that in paramecium it is the movement of the ingested particles through the entire protoplasm that is

being measured, rather than the movement of protoplasmic granules through the rest of the protoplasm. For a comprehensive account of the centrifugal method and the application of the Stokes formula, the reader is referred to the work of Heilbrunn (1921, 1926).

With the viscosity of water represented as 0.01 poise or a centipoise, Fetter reported that the absolute viscosity of the protoplasm of paramecium is 8726 times as great as that of water or 8726 centipoises with iron-fed paramecia and 8027 centipoises with starch-fed paramecia.

It is suggested by Fetter that this admittedly high viscosity value of the protoplasm (endoplasm) of paramecium may be due to the "complicated fibrillar system which ramifies through all parts of the cell." The interpretation is incorrect because this system is ectoplasmic. Since she measured the endoplasmic and not the ectoplasmic viscosity of paramecium, another explanation is needed to account for this high viscosity.

Brown (1940), using a specially designed direct vision microscope centrifuge (which he described in detail), also studied the protoplasmic viscosity of *Paramecium* sp. Results were calculated after using Cunningham's (1910) modification of Stokes' formula. He reported possible errors that appear to have been made by Fetter and improvements in the technique. Brown stated that the viscosity of protoplasm in paramecium is not greater than 50 centipoises, and that it may be much less.

There is a need for more extended studies of the absolute protoplasmic viscosity of paramecium.

Changes in viscosity of paramecium can be produced experimentally by a large number of different agents such as temperature, acids, alkalies, electricity, narcotics, etc. These changes are frequently so great that easily recognizable effects may be seen in locomotion, the behavior of contractile vacuoles, cyclosis, body shape, trichocyst extrusion, ciliary behavior, and Brownian movement. Agents which affect the viscosity and special structures to the extent of influencing behavior in paramecium are discussed later.

F. Permeability

Paramecium has been studied rather extensively in respect to its permeability to water, salts, acids, drugs, and a wide variety of agents. While much has been discovered in investigations upon membranes at the outer surface of other Protozoa, little evidence is available concerning the structural part responsible for functioning as a selective membrane in paramecium.

The external covering of paramecium, called the cortex, contains the more viscous ectoplasm in which is found the trichocyst layer, fibrillar system and pellicle with cilia. The outer surface is generally not sticky or adhesive but it is so prior to, during, and immediately after certain sexual processes. The cortex or a constituent of it has the properties of a membrane with great selectivity. *Paramecium* may possess a submicroscopic membrane in addition to the pellicle. Additional information, possibly discoverable by microdissection studies, may be of help.

The surface membrane, though it normally may not be visible, is a fundamental feature of the protozoön. It regulates in various degrees, the passage of oxygen into the body and the diffusion of waste materials from the protoplasm. Water, dissolved organic materials and chemicals also pass through the membrane. Experiments have shown that the membrane functions as a precise, osmotic regulator with the protoplasm of paramecium possessing a much higher osmotic concentration that that of its environment.

Large amounts of water are taken into the protoplasm of paramecium. Adolph (1943) reports that *P. caudatum* eliminates its own body volume of water in 15–20 minutes. Water intake is due in part to the water content of ingested and formed food vacuoles. It should be remembered that particles of food—mainly bacteria—are always taken into the protoplasm in a tiny droplet of water. But this method does not account for all of the water taken into the protoplasm. Much of it must come through the surface membrane.[4] This is especially noticeable during sexual union between two paramecia since the contractile vacuoles continue to function and eliminate liquid material which cannot now come from ingested food vacuoles.

The function of the contractile vacuoles in relation to permeability is discussed on page 226.

Nirenstein (1920) found that there was no general correlation of ability of acidic and basic dyes to penetrate paramecium with their partition coefficients in either acidic, basic, or neutral oils. But a close correlation existed with the partition coefficients in an oil which contained both an organic acid and an organic base.

Packard (1925) reported that light increases permeability to NH_4OH in paramecium and this has been confirmed by Child and Deviney (1926).

[4] When paramecia are joined together in conjugation or cytogamy feeding is stopped and of course food vacuoles with water do not pass into the protoplasm. Indeed there occurs dedifferentiation of ingestatory structures when the ciliates become tightly joined in the sexual process.

G. Effects of Various Agents on Protoplasm of Paramecium

As a test animal for the evaluation of the action of drugs and other agents, paramecium has decided advantages over most organisms. Starting with a single specimen, it is soon easily possible to obtain a pure-line pedigreed culture containing enormous numbers of paramecia. Larger animals are generally costly, and the method of rearing them is time-consuming. With paramecium, one may obtain as many as four or more generations a day, and maintaining their progeny requires comparatively little space.

With paramecium it is possible to deal with a completely isolated cell which is at the same time an entire organism. A reagent can be administered more uniformly and efficiently than with most other types of organisms since each and every specimen can be immersed uniformly in a solution. Contact with the test reagent can be maintained as long as is desirable then broken off at will by simply transferring the organisms to original culture fluid or another test reagent. Generally the action of the reagent can be observed quickly under the low powers of the microscope and the effect of the reagent can be followed individually on a single cell even while in the medium. This allows for speed and precision of observation generally impossible with other test organisms.

As a result of the reagent, the division rate, which is an index of vitality, can be determined easily and expressed in quantitative terms. On the other hand, this action of the reagent may be manifested in loss of motility which may include a change in ciliary action or its complete cessation, dysfunction of contractile vacuoles, blistering of the pellicle, swelling or vacuolation of the body, and disintegration of the body finally ending in death.

Recently, precise microchemical tests have been devised and other methods are being developed to the point where extremely accurate physiologic experiments upon paramecium and other Protozoa can be carried out to successful completion. See also Andrejawa (1931).

The literature dealing with the effects of various agents upon paramecium is enormous. Hence, in a book of this kind, the results of investigators must of necessity be brief but references are listed for more extensive information.

1. Effects of Water and Different Anisotonic Media.

By far the largest volumetric constituent of protoplasm is water which composes the continuous or dispersing phase of the colloidal system. It is evident that any immediate change in the water content of paramecium results in a change of viscosity. Water loss, especially in the case of hypertonic solutions, causes shrinkage of the body with an increase of protoplasmic viscosity. On the other hand, when water is added to the protoplasm as in the case of a hypotonic solution, paramecium tends to swell and results in a decrease of viscosity.

Paramecia can live in pure distilled water, but they will not reproduce because of lack of food (Daniel, 1908; Estabrook, 1910 b).

The effects of heavy water on *Paramecium* and other Protozoa have been studied by Barnes and Gaw (1935), Gaw (1936), Taylor, Swingle, Eyring, and Frost (1933), and Harvey (1934). According to Harvey, paramecia were killed in six to 10 hours by 80–100 per cent heavy water; in 0.2 per cent they were not appreciably affected. Lethal effects resulted in reduced locomotion, swelling of body, dysfunctioning of contractile vacuoles, and blistering of the body finally resulting in disintegration.

It has been shown that the contractile vacuoles in paramecium prove to be favorable indicators of the effects of concentrated heavy water. The vacuoles become greatly enlarged and their pulsation ceases in 95 per cent D_2O. When in 30 per cent D_2O for 24 hours, the rate of pulsation may be only one-half as fast as in the controls.

REFERENCES ON THE EFFECT OF DISTILLED WATER.

Bokorny (1905 a)
Daniel (1908)
Estabrook (1910 b)
Kitching (1939 a)
Ludwig (1927)
Peters (1908)
Towle (1904)
Wallengren (1902)

REFERENCES ON THE EFFECTS OF OTHER DIFFERENT ANISOTONIC MEDIA.

Balbiani (1898) glycerin
Behrend (1916) glycerin
Enriques (1902) cane sugar
Estabrook (1910 b) cane sugar
Finley (1930) sea water
Fortner (1925) cane sugar
Fortunato (1940) solid agar medium
Frisch (1935, 1939) sea water
Hayes (1930) sea water
Herfs (1922) sea water
Jennings (1931) sea water
Kalmus (1931) sea water
Kölsch (1902) sugar, glycerin
Massart (1889) sugar, glycerin
Schlieper (1930) sugar, glycerin
Spek (1919, 1920) sugar, glycerin
Wenrich (1928 a) sea water
Yasuda (1900) sugar, glycerin
Yocom (1934) sea water
Züelzer (1910) sea water

2. Effects of Temperature.

As is generally well known, high temperatures increase the velocity of chemical reactions and increase the rate of molecular action in liquids. The result is somewhat the same upon protoplasm.

Too high a temperature results in irreversible coagulation of the protoplasm and death. Cytoplasmic vacuolization as a result of heating beyond the optimum occurs in paramecium (Khainsky, 1910 a). A gradually increased temperature effect is shown in a relative fashion as follows:

Normal environment > increase in viscosity > reversible   coagulation > irreversible coagulation (death)

Low temperature tends to increase the viscosity and at the same time decrease the rate of locomotion in paramecium.

Like so many of the Protozoa that have been investigated, paramecium shows remarkable resistance to low as well as high temperatures. Jacobs (1919) has shown that paramecia have the ability to acclimate themselves to increasingly high temperatures, Efimoff (1924) to cold, and Wolfson (1935) to subzero temperatures.

In his work on *Paramecium caudatum,* Greeley (1904) reported that temperatures slightly above normal resulted in absorption of water by the protoplasm which markedly increased the fluidity and motility of the organism.

Oliphant (1938) reported that changes in temperature which are not lethal do not induce reversal in direction of the effective beat in the cilia of *P. multimicronucleatum,* but that lethal temperatures induce reversal in ciliary action if death is not immediate. He reported that the rate at which cilia beat in reverse varies directly with temperature and that reversal is associated with increased viscosity in response to temperature.

Chalkley (1930 a, b) reported that the resistance of *P. caudatum* to a temperature of 40° C. varies with the hydrogen-ion concentration of the medium and exhibits two maxima of resistance: one on the alkaline and one on the acid side with a region of minimal resistance at neutrality. In saline solutions the mechanism of death by heat appears to vary with different hydrogen-ion concentrations. At pH 6 or less the cell coagulates, at pH 8 or more the organism disintegrates; between these two extremes death occurs from rupture by swelling. Chalkley also found an increase in thermal resistance of paramecium on the addition of Ca and a decrease on addition of K.

Woodruff and Baitsell (1911 a, b) and others have made studies of the effect of constant temperatures on cell division in paramecia with special reference to the temperature coefficient.

Starved paramecia have been found to be more resistant to heat and cold than so-called "physiologically young" or well-fed specimens. Giese and Reed (1940) reported that in tests with cold and heat (5° C. and 39° C.), *P. caudatum* was found to be more resistant when starving than when well fed and dividing, and the same general effect was found to be the case in *P. multimicronucleatum* by Doudoroff (1936). Garner (1934) studied the relation of numbers of *P. caudatum* to their ability to withstand high temperatures and Olifant (1936), Cole (1925), Giese (1952), the effect of different temperatures.

Giese and Crossman (1945) found that *P. caudatum* and *P. multimicronucleatum* are much more readily killed by heat at 42.3° C. if they are first exposed to ultraviolet light. It was noted that almost complete recovery from ultraviolet light as judged by heat sensitivity occurred within four to five days. They concluded that heat does not sensitize paramecia to ultraviolet light but ultraviolet light sensitizes them to heat.

The acclimatization of paramecia to changes in temperature is discussed on p. 203. The effects of temperature on locomotion, growth, reproduction, and other physiologic functions are considered under their respective functions.

3. Effects of Salts, Acids, Alkalies, and Electrolytes.

The action of various salts and salt antagonisms upon the viscosity and behavior of Paramecium has been investigated by many (see especially Jennings, 1931).

According to Greeley (1903, 1904) KCl increases the viscosity or coagulates protoplasm of *Paramecium caudatum* while NaCl decreases the viscosity. Woodruff and Bunzel (1909) found a general parallelism between the ionic potential of a series of 16 cations and their toxicity on paramecium. The series ranged from the most toxic in Ag (ionic potential $+ 1.163$) to the least toxic ion K (ionic potential $- 2.92$).

Spek (1920) found that LiBr, LiCl, and KOH cause paramecium to increase in volume, and he attributed this to a swelling action of the colloid resulting in increased viscosity. Heilbrunn (1928) reported that lithium salts coagulate or increase the viscosity of *Paramecium* and *Stentor*.

Mast and Nadler (1926) by observing the direction of locomotion and ciliary activity in *P. caudatum* in relation to many chemicals reported that all the univalent salts and hydrates tested induced reversal of cilia except $(NH_4)_2SO_4$ and $(NH_4)C_2H_3O_2$, but that the bivalent and trivalent cation salts and hydrates, except $CaHPO_4$, $MgHPO_4$, and $Ba(OH)_2$, did not. According to these investigators, polyvalent cation salts were found to neutralize the effect of monovalent cation salts and they concluded that ciliary reversal in paramecium is associated with differential absorption of cations as well as subsequent changes in electric potential. In addition, they believed that other factors were involved.

Oliphant (1938) found that the chlorides, bromides, and iodides of potassium, lithium, sodium, and ammonium induce reversal in the direction of the effective beat of the cilia in *P. multimicronucleatum,* but that the chlorides, bromides, and iodides of calcium and magnesium do not. He found also that the duration of reversed ciliary action varies directly with the concentration of the salt which induces it. He has shown that paramecium reacts to increase in monovalent cation salt concentration in the environment by reversal in action of the cilia but does not react to increase in bivalent cation salt concentration. It is his belief that this difference in response must be due to some difference in the effect of monovalent and bivalent cations upon the protoplasm of paramecium. Evidence indicates that monovalent cations induce increase in the viscosity and bivalent cations decrease viscosity of the protoplasm. Oliphant is of the opinion that reversal in ciliary activity is due to or associated with an increase in protoplasmic viscosity.

Jacobs (1912, 1922) bubbled CO_2 through culture media containing paramecia and found that a short exposure to it caused a decrease while a longer exposure an increase in the viscosity. Both of these effects are reversible although the second one, according to Jacobs, tends to pass into an irreversible coagulation if the exposure is continued for a long enough time. He suggested that CO_2 may be an important factor in producing many of the natural changes in the consistency of protoplasm which have hitherto been unexplained.

In experiments dealing with toxic potassium chlorate solutions, Fortner (1926 a, b) reported that the duration of life for 300 *P. caudatum* having the same relative body surface is inversely proportional to the concentration raised to a certain power. The numerical value of the exponent indicating this power is suggested as a measure of the toxicity of the substance tested.

The reader is referred to the section on movement, ciliary activity and reversal of cilia (Chapter 8) for further information on reaction of paramecium to salts.

Additional references on the effects of salts, acids, alkalies, and electrolytes on paramecium or closely related forms follow:

Alpatov (1937) salts (and heat resistance)
Balbiani (1898) salts
Bancroft (1906 a) salts
Barratt (1904 a, b, 1905 a, b) acids and bases
Baskina (1924) HCl
Boell (1946) sodium azide
Borowski (1922) calcium carbonate, calcium phosphate, bases and acids
Burge and Estes (1928) many inorganic salts
Calkins (1904) inorganic salts
Calkins and Lieb (1902) inorganic salts
Chatton and Tellier (1927) various chloride solutions
Child and Deviney (1926) salts, bases
Collett (1919–21) acids and salts
Dale (1913) various electrolytes
Daniel (1908) salts
Dogiel and Issakowa (1927) magnesium salts
Eisenberg (1925, 1926, 1929, 1930) various ions
Eisenberg-Hamburg (1932) calcium and other salts
Estabrook (1910 a, b) sodium chloride
Fortner (1925) potassium chlorate
Hirsch (1914) salts
Hutchison (1915) salts
Jennings (1915) acids and salts
Jennings and Moore (1902) carbonic and other acids
Klokaciova (1927) salts
Loeb and Wasteneys (1911) acids and salts
McCleland and Peters (1920) salts
Middleton (1922) inorganic salts
Middleton (1928) ammonium tartrate, ammonium sulfate, potassium nitrate
Motolese (1920) picric acid
Nagai (1907) salts

Nikitinsky (1928) carbonic acid
Okazaki (1927 a, b) acids, cations, and various drugs
Park (1929) osmic acid
Peters (1904) salts
Peters and Rees (1906) ions
Port (1927, 1928) salts
Prowazek (1910 a, b, c) acids, bases
Sand (1901) inorganic salts
Scharrer (1933) carbolic acid
Spek (1919, 1920) LiBr, LiCl, KSCN, and other inorganic salts
Stempell (1924) ions
Sun (1912) inorganic salts
Towle (1904) salts
Vieweger (1912) salts, acids, and bases
Woodruff (1905) inorganic salts
Yasuda (1900) salts
Zirkle (1936) acids and bases and radiosensitivity
Zweibaum (1912) aluminum, mercuric and sodium chloride, and many others

It is of course well known that the protoplasm of paramecium like all other cells can be irreversibly coagulated resulting in death. Agents such as acids, certain salts as mercuric chloride, alcohols, high temperature, etc., can produce this phenomenon either singly or in combination with others. Because of this fact they are generally used as fixing agents for the detailed study of structure in paramecium.

4. Effects of Hydrogen-ion Concentration (see also pp. 40 and 188).

It has been shown that paramecium as well as a large number of other Protozoa can live in rather wide ranges of hydrogen-ion concentrations. Loefer (1938) studied the effect of hydrogen-ion concentration on growth and morphology of *P. bursaria*. Organisms lived within pH ranges of 4.2–8.4 with growth best at pH 6.8. Chase and Glaser (1930) found an acceleration in the ciliary locomotion of Paramecium when the pH was shifted to either side of the neutral point.

Pollack (1928) reports that studies on

the effects of pH concentration on protoplasmic viscosity are complicated by the rate of entrance of the ions and by the fact that they may be neutralized by the buffers of the protoplasm.

The effects of hydrogen-ion concentration upon ciliary activity and locomotion, growth, and food-vacuole formation are discussed elsewhere in the book.

Additional references dealing with effects of the hydrogen-ion concentration follow:

Beck and Nichols (1937) (with fluorescent dyes)
Bodine (1921)
Chalkley (1930 a, b)
Crane (1921)
Darby (1929)
Gaw (1936)
Johnson (1929)
Jones (1930)
Lee (1942 b)
Mayeda (1928)
Mayeda and Date (1929)
Phelps (1931)
Shapiro (1927)
Wichterman (1948 d, 1949)

5. Effects of Mechanical Agitation Including Centrifugation.

Morse (1909–10) and more recently King and Beams (1941) have investigated certain effects of agitation of *Paramecium caudatum*. When paramecia are shaken by hand or put in an agitating device for a short time, rate of movement and ingestion are markedly decreased. Mechanical agitation also caused a decrease in the viscosity of the endoplasm which was associated with an increase in the rate of cyclosis. King and Beams also noted that the macronucleus and micronucleus undergo a change in refractive index and color upon mechanical agitation and become clearly visible.

Mechanical agitation of paramecium especially by centrifugation, has thrown a great deal of light on the nuclear apparatus, cytoplasmic inclusions and the protoplasm in general. Centrifugation permits intracellular inclusions to be thrown into particular stratification levels in accordance with their relative specific gravities. Also it is a method of estimating protoplasmic viscosity.

It so happens that paramecium proves to be an exceedingly useful organism upon which to make such studies. It has been shown that many centrifuged paramecia, even after the point of stratification of intracellular inclusions, will, when removed from the centrifuge, re-cover their normal shape, acquire typical distribution of inclusions, and undergo fission.

In his study of centrifuged paramecia, McClendon (1909) found that the entire macronucleus moved centrifugally after the macronuclear chromatin had been precipitated centrifugally against the nuclear wall. He further added that the division rate was increased after centrifuging but this report may be criticized because of too few experiments. Contrariwise, Yancey (1931) reported that centrifuging one to 45 minutes at

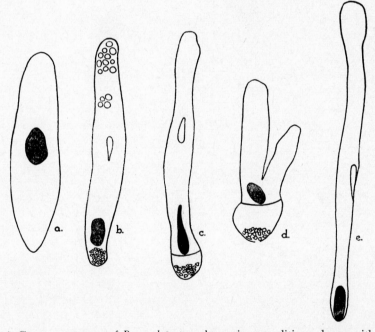

Fig. 44. Gross appearance of *P. caudatum* under various conditions, drawn with camera lucida. ($\times$ 235). (a) Normal uncentrifuged, with centrally located macronucleus. (b) Centrifuged for 5 minutes at 21,000 times gravity; crystals and chromatic part of the macronucleus at the centrifugal pole; recovered. (c) Centrifuged for 10 minutes at 21,000 times gravity; a layer of fluid has appeared between the chromatic part of the macronucleus and the crystals at the centrifugal pole; cilia still beating when drawn, died later. (d) As in (c), but the animal was so oriented during centrifuging that the crystals were forced down against the pellicle at the center of the body, instead of, as usual, at one end; cilia still beating when drawn, died later. (e) Centrifuged for 15 minutes at 21,000 times gravity; here the crystals have been forced through the pellicle, the chromatic part of the macronucleus about to be extruded; dead when drawn. (King and Beams)

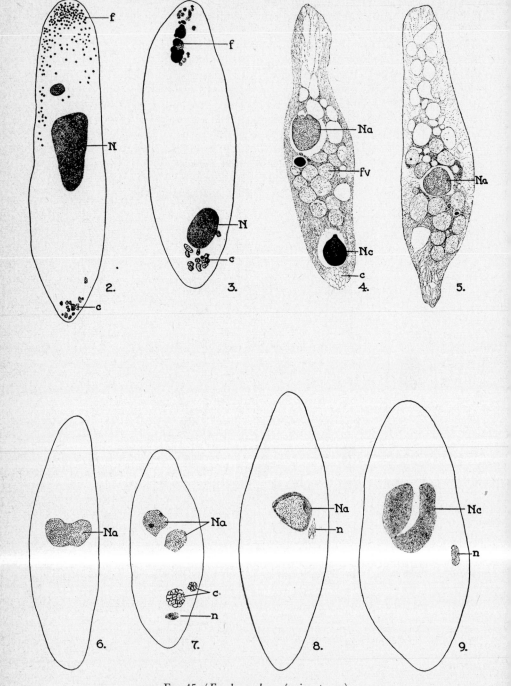

FIG. 45. (*For legend see facing page.*)

21,000–32,000 times gravity lowered the division rate in P. multimicronucleatum. King and Beams (1937) however state that neither P. caudatum nor P. multimicronucleatum are able to survive centrifuging at 21,000 times gravity for 15 minutes but that when paramecia are centrifuged at slower rates, a lowering of the division rate occurs.

In an extensive study, King and Beams (1937) have shown in their investigations of the effect of ultracentrifuging P. caudatum, P. aurelia, and P. multimicronucleatum that it is possible to bring about a redistribution of the various nuclear and cytoplasmic components. This was made possible by the use of the enormous forces of the air-driven ultracentrifuge developed in 1930 by Dr. J. W. Beams. In one model, an air pressure of 50 pounds produces a speed of approximately 180,000 revolutions per minute or a centrifugal force equal to 400,000 times the force of gravity (Figs. 44 and 45).

King and Beams report that upon centrifuging P. caudatum at 21,000 times gravity, the organisms become greatly elongated and the chromatin is occasionally forced from the achromatic matrix of the macronucleus. Depending upon their relative specific gravities, the materials in Paramecium are redistributed so that one finds at the centrifugal end crystals, a layer of fluid, micronucleus and macronuclear chromatin, food vacuoles and neutral red inclusions, achromatic matrix of the macronucleus, endoplasm with large clear alveoli, and fat, at the centripetal. They report instances when the crystals, micronucleus and macronuclear chromatin are extruded from the body. Occasionally, paramecia were observed dragging the extruded macronucleus behind and connected to the body by a "long streamer of chromatin" which throws some light on the viscosity of macronuclear chromatin.

In animals which survived centrifugation, displaced cytoplasmic components returned to their normal positions but

---

Fig. 45. All figures from P. caudatum ($\times$ 425), centrifuged for 7½ minutes at 21,000 times gravity, unless otherwise indicated.
(c) Crystals. (f) Fat. (f.v.) Food vacuoles. (n) Micronucleus. (N) Macronucleus. (Na) Achromatic portion (matrix) of macronucleus. (Nc) Chromatic portion of macronucleus.
(2) Centrifuged at 21,000 times gravity for 1 minute; fixed in Champy and post-osmified. Macronucleus in normal position, fine globules of fat at centripetal pole, crystals at centrifugal. (3) Technique as in 2. The fat has collected in larger masses centripetally, crystals and compact chromatic portion of macronucleus centrifugal. Matrix of macronucleus not visible. Figs. 4 to 9 fixed in Schaudinn and stained in hemalum. (4) One hour after centrifuging. Macronucleus separated into a compact mass of chromatin, centrifugally located with the crystals, and an achromatic matrix, in the usual position of the macronucleus. Food vacuoles between the chromatic portion of the macronucleus and the achromatic matrix. (5) As in 4 the crystals and the chromatic portion of the macronucleus have been forced out of the cell. (6) Five hours after centrifuging. Micronucleus and chromatic portion of macronucleus lacking; achromatic matrix present. (7) Thirty hours after centrifuging. Crystals in large, compact globules; micronucleus and achromatic matrix of macronucleus present. (8 and 9) The two daughter cells resulting from fission of a single Paramecium, which occurred within 24 hours after centrifuging. In (8), micronucleus and the achromatic matrix of the macronucleus, a narrow rim of chromatin material surrounding the achromatic matrix. In (9), micronucleus and the chromatic portion of the macronucleus which apparently has prepared for, but failed to accomplish, division. (King and Beams)

occasionally large masses of crystals were passed to daughter cells in fission. The two components of the macronucleus, namely the chromatic portion (Feulgen positive) and the achromatic matrix, did not fuse after centrifugation, but they were often connected by a thin strand of chromatin. Animals deprived of the micronucleus in a number of cases survived and divided but amicronucleate races could not be established. The study also revealed that the membrane of the contractile vacuole appears to be a temporary one.

6. Effects of Electric Current (see also Chapter 8), Magnetic Fields, and Sound Waves.

When an electric current is passed through a small vessel of water containing paramecia, the organisms first swim toward the cathode or negative electrode. With an increase in electric current, the paramecia swim backward toward the anode and the organisms become short and thick. When the current is further increased, the paramecia burst, sometimes at one end (Fig. 46).

Statkewitsch (1903 a) experimented with paramecia which had been stained in the living condition with certain chemical indicators. To such organisms, an electric current was applied. He reported that the current caused chemical changes within the protoplasm as shown by the indicators. It was noted that endoplasmic granules and vacuoles became more alkaline in reaction.

It appears that the action of electric currents on protoplasm and protoplasmic structures is at first local on a particular area of the body, then the current appears to have a polarizing effect on the organism.

Paramecia and closely related ciliates have been investigated in relation to their behavior in high frequency electric

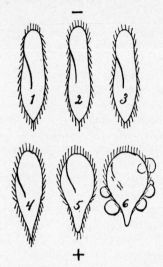

Fig. 46. Progressive cathodic reversal of the cilia and change of form in Paramecium as the constant electric current is made stronger. The cathode is supposed to lie at the upper end. The current is weakest at 1, where only a few cilia are reversed. (2–6) Successive changes as the current is gradually increased. (Jennings)

and magnetic fields and sound waves. Kahler, Chalkley, and Voegtlin (1929) exposed suspensions of *P. caudatum* in electric and magnetic fields at 10 Mc (megacycles) and in electric fields at 75 Mc for periods of one to four hours. Gale (1935) used the electric field at 15, 30, and 50 Mc for one to six hours on *P. caudatum* and *Chilomonas*. These investigators found that the Protozoa were destroyed at 41°–43° C. when treated without temperature control. Slow direct heating with a water bath produced identical results, but when the suspensions were kept at sublethal temperatures by frequent chilling or by suspending the organisms in a non-conducting, non-heating medium no after-effects were detected.

Working with *Paramecium* and *Vorticella,* Jellinek (1936) attempted to exclude large-order heating effects by working with an extremely weak field. In single-drop cultures, the ciliates were immobilized or destroyed immediately at 100 Mc.

In a comprehensive study, Summers and Hughes (1940) investigated *Colpidium* in high-frequency electric and magnetic fields and came to the conclusion that there is no large-order effect on the physiologic organization of the ciliate which could be attributed to a non-thermal mode of energy transfer in the 13–60 Mc range of frequencies.

Ackerman (1950, 1951, a) studied the optimum frequencies necessary for sonic disintegration of five species of *Paramecium*. After using intense sound waves of audible frequencies, different maxima were found for the species of *Paramecium* with the optimum frequencies increasing as the size of the specimens decreased (Ackerman, 1952).

Additional references dealing with the effects of electric current follow:

Andrejewa (1930)
Bancroft (1905, 1906 b)
Luyet (1933, 1936)

Response of *Paramecium* to electric currents in respect to behavior stimulation, etc. is considered in Chapter 8.

7. Effects of Ultraviolet Light, Fluorescence, X-rays, Radium, Radioactive Substances and Isotopes.

EFFECTS OF ULTRAVIOLET LIGHT. All types of cells or organisms are affected by ultraviolet radiation providing the radiation can penetrate to the protoplasm. The ultraviolet ranges from approximately 150 Å at the short end of the spectrum to 3900 Å at the long end of the spectrum.[5] From 150–2000 Å constitutes the Millikan, Lyman and Schumann regions which are absorbed by most materials including air and

[5] Å represents the angstrom unit which is 10^{-8} cm.

water hence are little used in biologic experimentation. The portion of the spectrum which is generally used in biologic research is from 2000–3900 Å with 2000–3000 Å usually referred to as the short, abiotic, or lethal ultraviolet and 3000–3900 Å as the long, biotic, or non-lethal region. It has been demonstrated that the lethal effect of quartz ultraviolet radiation on Protozoa is strongest at the shorter wave-lengths and weakest at wave-lengths approaching 3130 Å. While inorganic salts, carbohydrates, and fatty materials of protoplasm absorb ultraviolet, the simple proteins and nucleoproteins show a highly specific absorption. On the other hand, proteins show little absorption for visible light.

Hertel (1905) using wave-lengths of 2100, 2320, 2800, 3340, and 3830 Å showed that at equal intensities, the shorter the wave-length the greater the destructive action upon *Paramecium* and that 3340 and 3830 Å had no effect. Hughes and Bovie (1918) studied the effects of fluorite ultraviolet light on the rate of division in *P. caudatum*. They noted that when irradiated with Schumann rays of about 1600 Å, the paramecia became sticky but divided at the same rate as the controls.

Harris and Hoyt (1919) reported that since the absorption bands of proteins are due to absorption by tyrosine and phenylalanine, these amino acids might act as sensitizers, hence absorbing the radiation and transferring it to other materials in the cell. They demonstrated that a solution of tyrosine screened paramecia effectively against lethal ultraviolet radiations.

In an attempt to determine which of the wave-lengths of the quartz mercury arc are most destructive to paramecium, Sonne (1929) and Weinstein (1930) made comparisons on the basis of the total energy delivered per mm.[2] in order to kill the ciliates. Sonne noticed different visible effects of different wave-lengths on paramecium and believed that different substances in the proto-plasm may show a preferential absorp-tion and different reactions at the various wave-lengths. As an example, shorter wave-lengths may be so com-pletely absorbed by the lipoids on the surface that the main effects might be surface phenomena. On the other hand, longer wave-lengths which may pene-trate to the interior of the cell would act internally. Sonne found 2804 Å the most effective while Weinstein reported that 2654 Å was the most efficient wave-length. The wave-lengths, with their percentages of effectivity accord-ing to Weinstein, follow: 2537 Å (96.4 per cent) ; 2654 Å (100 per cent) ; 2804 Å (87.4 per cent) ; 3020 Å (20 per cent) and 3130 Å (11 per cent). The incident energy necessary to kill the paramecia varied from 2162 ergs/mm.^{2} at 2654 Å to 19,629 ergs/mm.^{2} at 3130 Å.

One is referred to the valuable texts by Duggar (1936), Lea (1947), and Nickson (1952) which deal with the general biologic effects of radiations on protoplasm and particularly the com-prehensive studies made by Giese and his associates on the effect of ultraviolet radiations upon paramecium. In studies dealing with ultraviolet light, the choos-ing of the end-point is of great impor-tance. The general effect of ultraviolet radiations upon paramecium and a con-sideration of the end-point was reported by Giese and Leighton (1935 a, b) who noted that upon first being irradiated at wave-lengths 2537, 2654, 2804, and 3025 Å, the paramecia showed slight stimulation,[6] then a gradual decrease

in activity, accompanied with a shorten-ing and broadening of the cell. Con-tracile vacuoles became huge spheres with the canals enlarged, seemingly jelled, and conspicuous. Cilia of the body then became slow moving and un-coördinated. Vesiculation often occurred before the oral cilia became inactivated; in some cases even while some of the body cilia were still active. The vesicu-lation usually occurred with the forma-tion of a clear vesicle at the posterior end, although at wave-lengths 2537 and 3025 Å, many vesicles occasionally formed. When the internal contents of the animal were forcibly ejected into a vesicle, it burst and the organism dis-integrated leaving a mass of scattered granules in Brownian movement. Vesic-ulation was chosen as the criterion of death since it is an end-point easily ob-served and hence not subject to con-siderable individual interpretation. Also this end-point is proportional to dosage of irradiation (Fig. 47).

According to Child and Deviney (1926) the most extreme effects on *P. caudatum* consist of complete disinte-gration and coagulation extending a greater or lesser distance posteriorly from the anterior end. Approximately 35 per cent of paramecia under inves-tigation showed a second region of dis-integration at the posterior end or dis-integration of the vesicle which was extruded from the anus.

Bovie and Hughes (1918) studied the effects of quartz ultraviolet light on the rate of division of *P. caudatum,* and Hinrichs (1927, 1928), using ultraviolet radiation, prevented the division of *P. caudatum* and produced axial double animals. When repeated with double animals, chains of three or four indi-viduals were formed. Retardation of division in normal and starved para-mecia and recovery from sublethal dos-ages were studied also by Giese (1939,

---

[6] The first effect of ultraviolet light on paramecia is a stimulation to greater activity possibly corresponding to an initial decrease in viscosity (Giese, 1945 a, b).

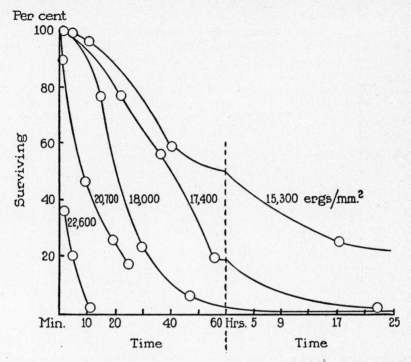

FIG. 47. Rate of vesiculation after ultraviolet irradiation (λ 2537 Å) in *P. multimicro-nucleatum*. (Giese and Leighton)

1945) (Fig. 48). Alpatov and Nastju-kova (1932) investigated the effects of various ultraviolet-ray exposures on the division rate in *Paramecium*.

Giese and Reed (1940) studied the resistance to ultraviolet radiation in three species of *Paramecium*, and comparisons were made on the basis of the effects of the radiations on the division rate. *P. multimicronucleatum* was found to be least resistant, *P. aurelia* intermediate, and *P. caudatum* most resistant. Three stocks of *P. multimicronucleatum* obtained from diverse localities showed widely different resistances, but three stocks of *P. caudatum* were fairly similar in resistance. Alpatov and Nastjukova (1934) also noticed differences in susceptibility to ultraviolet radiation between *P. caudatum* and *P. bursaria*. It appears that starved paramecia are

more susceptible to ultraviolet radiations than well-fed individuals, not only for the sublethal doses, but also for lethal ones as shown in *P. multimicronucleatum*, *P. aurelia*, and *P. caudatum* by Giese and Reed (1940) and *P. bursaria* by Tang and Gaw (1937). Giese and Reed could find no clear correlation between susceptibility to ultraviolet radiation and size, nuclear constitution, general vigor and resistance to other environmental agents. Giese (1945) also observed that ultraviolet radiation prolongs the time of reversal of ciliary activity. The increased time of ciliary reversal is produced by a relatively small dosage as compared to the dosage which results in immobilization, disruption, and vesiculation of the protoplasm of paramecium.

Of interest is the discovery by Giese

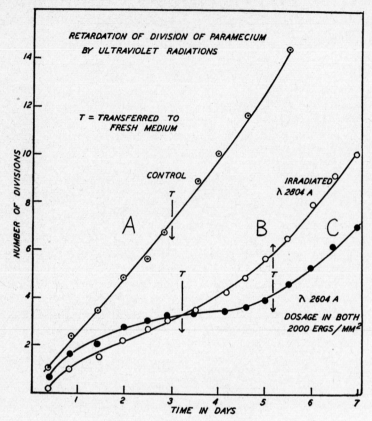

Fig. 48. Retardation of division of *Paramecium caudatum,* from the logarithmic phase of a culture, by ultraviolet light. Each point is the average of 10 cultures. In the first 3 days are shown the immediate effects of the radiation. The rate of recovery is shown in the remaining 4 days. Transfers to fresh medium were made at "T." (Giese)

and Crossman (1945) that heat does not sensitize paramecia to ultraviolet light but ultraviolet light sensitizes them to heat. *P. caudatum* and *P. multimicronucleatum* were found to be more readily killed by heat at 42.3° C. if they were first exposed to ultraviolet light.

Giese and Leighton (1935 a) found that when *P. multimicronucleatum* cultivated under controlled conditions were irradiated at known intensities of light of wave-lengths 2537, 2654, 2804, 3025, and 3130 Å, the approximate absorption of the light by the organisms

was greatest and of the same order of magnitude at the three shortest wave-lengths, considerably less at 3025 Å, and least at 3130 Å. They reported that paramecia did not die when irradiated with high dosages of intense light at 3130 Å. At the other wave-lengths, 50 per cent vesiculation occurred when between 10^{12} and 10^{13} quanta had been absorbed by a *Paramecium,* and they believed that a very large number of molecules in the ciliate are affected before vesiculation occurs.

Additional references dealing with the effects of ultraviolet light upon

Paramecium and closely related forms follow:

Boland (1928)
Gale (1935)
Giese (1942)
Kimball (1949)
Kimball and Gaither (1950, 1951)
Luyet and Gehenio (1935)
MacDougall (1931)
Monod (1933)
Roskin and Romanowa (1929)

FLUORESCENCE. The emission of light from matter under the influence of an exciting radiation agent is termed *fluorescence*. It is of interest here as one of the secondary processes following excitation of the electrons in the molecules of protoplasm by absorption of radiation, abiotic wave-lengths of 3000 Å and shorter being strongly fluorescent. Rentschler (1931) reported that paramecia killed by the arc light fail to fluoresce. Giese and Leighton (1933) described the fluorescence of *P. multimicronucleatum* in the monochromatic light of the wave-lengths ranging from 3660 Å to 2537 Å of the quartz mercury arc. Beck and Nichols (1937) studied the action of fluorescent dyes on *Paramecium* as affected by pH.

EFFECTS OF X-RAYS. It has been known for a long time that paramecia are able to survive exceedingly high dosages of Roentgen rays; with low, sublethal dosages they become perceptibly accelerated (Wichterman, 1947 b, 1948 b). Dosages of 200,000 *r* and above noticeably retard motility in *P. bursaria* and there are generally no survivors above 600,000 *r*. Irradiation with x-rays markedly increases the viscosity of the protoplasm of paramecia; greater dosages lead to irreversible coagulation. Giese and Heath (1948) reported that after dosages of 280,000 to 410,000 *r*, *P. caudatum* becomes especially sticky.

Earlier workers merely exposed organisms to the action of x-rays, and information upon unit dosage is absent. At the present time with modern x-ray apparatus it is possible to deliver rays of constant quality and intensity at a given number of Roentgen units (*r*) per second or minute. Schaudinn (1899) appears to have been the first to observe the effects of x-rays upon *Protozoa*.

In what appears to be an exploratory investigation, Bardeen (1906–08) exposed paramecia to x-rays and reported that after a 12-hour exposure, no apparent effect was noted upon the ciliates. He is the first to have observed the effect of x-rays upon conjugants and reported that there was no effect in joined pairs undergoing conjugation or upon their offspring. It is likely that he worked with minimal dosages, but this information is lacking from his account.

Schneider (1926), Hance and Clark (1926), Hance (1931), and Power and Shefner (1950) investigated the effect of x-rays upon vegetative and/or dividing forms of *Paramecium*. Hance and Clark found the division rate to suffer a slight initial depression lasting two to five days following the exposure in *P. caudatum* and *P. multimicronucleatum*. This is in agreement with the work of Giese and Heath (1948) on *P. caudatum* and the author's findings in *P. bursaria* where irradiation was found to temporarily inhibit division. The greater the dosage, the greater the delay in fission. However, if there is survival of paramecia after exposure to x-rays, the fission-rate slowly increases until it is normal. (See also Hance, 1927–1928.)

According to Hance and Clark, dividing specimens showed no different effects than typical vegetative ones when irradiated. They reported that doses repeated at various intervals generally failed to interfere more markedly with the division rate than a single dose. Repeated irradiation caused the paramecia to become slightly swollen without apparent interference with their vitality. According to these investigators, treatments lasting from 10 minutes to three to four hours depressed division, and longer or repeated exposures under some conditions raised the reproductive rate.[7]

Back (1939) irradiated *Paramecium caudatum* to 2/3–5/6 of the lethal dose and found that division of the ciliates could be suspended for several weeks. Accordingly, he considered this lack of division a type of injury inflicted upon the organism.

With *Colpidium colpoda,* a ciliate closely related to *Paramecium,* Crowther (1926) found that a considerable exposure to x-rays produced little visible alteration either in the appearance or motion of the organisms. He observed that with a given sublethal exposure, the ciliates became perceptibly accelerated with rapid and excited appearing behavior followed by a zigzag motion. When removed at this stage, the majority of ciliates survived. Lethal effects were noted when the forward progression stopped, vacuoles became distended and with the formation of a bubble of clear substance which was extruded from the body, death followed. It was found that the protoplasm of irradiated *Colpidium* appeared less transparent and distinctly more granular than in normal specimens.

With *P. bursaria,* somewhat similar effects were noted with sublethal dosages of 100,000 r in that the paramecia became more active in their swimming behavior. Also similar blebs or vacuoles of a clear, structureless substance were

formed upon the pellicle of *P. bursaria* after longer irradiation and active cyclosis was considerably decreased or even stopped (Wichterman, 1948 b).

Taylor, Thomas, and Brown (1933) found that dosages of 46 x 10[4] r produced death in the ciliate, *Colpidium campylum* within 15 minutes following exposure. They reported that x-radiation with 35 x 10[4] r of a sterile culture medium of 10 per cent yeast extract, tap, and distilled water rendered the medium highly toxic or lethal to the ciliates. Their tests showed hydrogen peroxide in the irradiated water in concentrations above 1:100,000. They found that de-oxygenated water following the irradiation gave negative tests for H_2O_2 and was not toxic to colpidia. Organic materials such as sheep's blood, agar, gelatin, and bacteria added to the tap water before or after its x-radiation protected the colpidia against its toxic action. They believe that hydrogen peroxide was in large measure responsible for toxic and lethal effects but suggest that other toxic agents may be produced.

Lethality of x-rays in paramecium and other microorganisms is related to the gradual accumulation of a toxic concentration of hydrogen peroxide due to ionizing radiations—a fact established long ago by many investigators using a variety of microorganisms. An excellent review of the subject of radiation chemistry is to be found in the symposium report by Allsopp, Burton, et al (1951). The role of externally produced hydrogen peroxide in damage to *Paramecium aurelia* by x-rays has been studied by Kimball and Gaither (1952). Taylor (1935) reported that in addition to hydrogen peroxide, the air adjacent to the medium during irradiation of ciliates in enclosed chambers was lethal to the specimens. He believed the formed gas to be ozone. In a study of

---

[7] The x-rays were produced at 30 kilovolt peak and 22 milliamperes, filtered through very thin cardboard and applied at a target distance of 25.5 cm. Such rays produced about 6 x 10[12] pairs of ions per gm. per second in air.

environmental factors concerned with the lethality of x-rays in paramecium, Wichterman and Figge (1952) attempted to analyze the cause of variability as shown in survival curves. It was found, for instance, that under certain conditions, the LD 50 may vary from 75 to 350 kiloroentgens. The ratio of the number of paramecia to volume of culture fluid was found to be not nearly so important as the ratio of volume of air to the volume of culture fluid in plastic irradiation chambers of various sizes and shapes. Evidence indicates that this toxic gaseous substance comes from the airspace of the irradiation chamber and that the rate of diffusion into the culture fluid containing the paramecia depends upon the surface area of the medium and its depth and volume. Delicate tests employed did not reveal the toxic gas to be ozone. To eliminate factors which cause variation in survival curves, a 2 cc. Nylon hypodermic syringe was used as an irradiation chamber. This device absorbs very little radiation, eliminates air from the irradiation chamber and permits the introduction of various substances to be tested upon paramecia during irradiation. Accurate sampling of specimens after intervals of irradiation without changing the depth of the medium is also a desirable feature. It is thus ideal for the study of lethality of x-rays in paramecium and may prove useful for similar studies with other microorganisms.

With the Nylon syringe method, Figge and Wichterman (1952) irradiated *Paramecium caudatum* in control media and in media containing hematoporphyrin, sodium nitrite and sodium pentobarbital. Paramecia in control media require very high dosages of irradiation to produce lethal effects (LD 50, 24 hours, 350 kiloroentgens). Paramecia irradiated in chemically non-toxic porphyrin concentrations of 1:10,-000 to 1:40,000 caused 100 per cent mortality with 100 kr within 12 hours. It is known that hematoporphyrin sensitizes paramecia to visible and near ultraviolet light. The above and other experiments demonstrate in a conclusive way that porphyrin sensitizes paramecia to short wave-length radiations such as x-rays. Preliminary experiments with sodium pentobarbital and sodium nitrite in the concentrations employed (1:1000 – 1:4000) indicate that these substances do not protect paramecia from lethal effects of x-radiation.

In an inconclusive paper Baldwin (1920) reported the combined action of x-rays and of vital stains upon paramecia, and more recently Halberstaedter and Back (1943) observed the effect of the combined action of colchicine and x-rays and colchicine alone upon *Paramecium caudatum*. The concentrations of crystalline colchicine in the basic culture media were 0.0005 per cent, 0.00025 per cent, and 0.000125 per cent with the x-ray intensity at the distance of the irradiated object being 80,000 r/m. In colchicine combined with the x-rays, the paramecia were maintained for 48 hours in the colchicine solution then transferred to normal culture media for irradiation.

These workers showed that the lethal dose for normal paramecia fluctuated between 200,000 and 700,000 r, the half-value doses lying between 300,000 and 250,000 r. Their experiments demonstrated that the immediate lethal dose is considerably smaller—about 50 per cent—for paramecia which received treatment earlier with colchicine (Fig. 49).

Brown, Luck, Sheets, and Taylor (1933) studied the action of x-rays upon another ciliate, *Euplotes taylori* in culture fluid and on culture fluid alone. They found that culture fluid exposed to x-rays did not induce death or produce toxic effects on *Euplotes*. The minimum lethal dose to kill the protozoön was found to be 2110 r/sec. but much less exposure was sufficient to kill the associated bacteria in the culture fluid. Irradiated bacteria were found to be unsatisfactory as a source of food for *Euplotes*. This difference in resistance to x-ray exposure made it easily possible to sterilize this ciliate. Their important observation adds another method for the sterilization of Protozoa, and it is very likely that many others, including *Paramecium,* can be sterilized in the same manner.

In a comprehensive study of the effect of x-rays upon *P. caudatum* and its culture fluid, Back and Halberstaedter (1945) reported that doses of 100,000 r were used without visible effects. However, increased irradiation of paramecia became visible in the

alteration of the manner of their movement. This change was noticeable when one half of the immediate lethal dose had been applied. Cilia failed to behave normally and the paramecia swam irregularly backward and forward. With greater dosage, movement ceased and cytolysis soon occurred. They reported that complete cessation of movement served as an indication of the impending death of the organism and defined the immediate lethal dose for *P. caudatum* as that dose which produced complete cessation of motility within 10–15 minutes after irradiation. When one million r were applied to the culture media alone, no toxic or secondary effect was observed on the paramecia in contrast to the work of Piffault (1939) who reported that irradiation of culture fluid in which *P. aurelia* were swimming may produce secondary or toxic effects which should be taken into

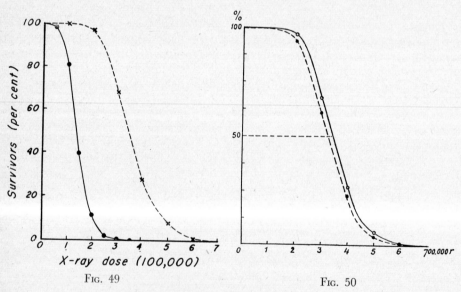

Fig. 49   Fig. 50

Fig. 49. Full line, treated in advance with colchicine; broken line, normal paramecia.

Fig. 50. X-ray survival curve of *P. caudatum*. "Family" curve represented by broken line; "clone" curve by unbroken line. (Courtesy, Back and Halberstaedter. Influence of biological factors on the form of Roentgen-ray survival curves, *Am. J. Roentgenol.*)

account in this type of investigation. The author (Wichterman, 1948 b) found that culture fluid irradiated with one million r has no lethal or toxic effect on *P. bursaria*.

Back and Halberstaedter performed experiments to determine the difference in x-ray susceptibility between immediate (family) and remote (clone) descendants of a single individual. The family represented four to eight descendants of a single organism. They reported that the lethal dose for different individuals of a single clone was found to be variable. Application of 200,000 r caused death in a small percentage of the irradiated population. With increase in the dosage, this percentage rises. At 600,000 r, almost all the irradiated specimens die. All survivors of this treatment died after the dose was increased to 700,000 r. The dose which produces a 50 per cent mortality rate lies between 350,000 and 400,000 r (Table 8).

When paramecia were irradiated in the same drop, members of a clone died at different x-ray doses. However, under the same conditions, members of a family always died at the same dose.

The results of the above-mentioned experiments are plotted as a survival curve in Fig. 50. The shape of the curve (unbroken line) is typically sigmoid and corresponds to the sigma curve obtained by different investigators for biologic materials. These investigators believe that the experimentally derived mortality curve could also be adequately explained in terms of biologic variation alone, hence the curve might be regarded as one of biologic variation.

The author (Wichterman, 1951) irradiated late dividing stages of *P. caudatum* with 300,000 r with the result that such dividers remained suspended in the division process for six to 48 hours, occasionally longer. Generally the x-rayed dividing cells do not live to produce daughters but if they are produced, they persist for a day or longer, then die. Occasionally one of the daughters produced by an irradiated divider dies while the remaining daughter lives to divide later. In some instances, an irradiated divider, after gradual recovery from x-ray effects, may produce viable daughter cells which produce flourishing cultures. Irradiation of dividers can produce giant C-shaped monsters (Fig. 129) from which normal appearing specimens may originate. Vegetative stages of paramecium appear more resistant to irradiation than dividing cells. Roentgen-ray sensitization and action has been investigated by Joseph and Prowazek (1902) and Nikitin (1930), and Kimball (1949 a) has reported on the inheritance of mutational changes in *P. aurelia* (using vigor as a gauge) after treatment with x-rays. Recently Giese and Heath

Table 8

PERCENTAGE MORTALITY AFTER IRRADIATION OF *P. caudatum**
(BACK AND HALBERSTAEDTER, 1945)
(M.C. = MASS CULTURE, TOTAL 750 SPECIMENS;
F = FAMILIES, 4–8 SPECIMENS OF 107 FAMILIES)

100,000 r		200,000 r		300,000 r		400,000 r		500,000 r		600,000 r		700,000 r	
M.C.	F	M.C.	F	M.C.	F	M.C.	F	M.C.	F	M.C.	F	M.C.	F
0	0	3.0	6.6	32.2	39.4	73.5	77.7	93.8	96.3	99.0	99.1	100	100

* Data for 112 typical irradiation trials.

(1948) sensitized *P. caudatum* to heat by using sublethal dosages of x-rays. They found that if paramecia were irradiated then exposed to a sublethal dosage of heat, the organisms were killed; but if the same heat exposure preceded the same dosage of radiations, they were not. It was found that sensitivity to both heat and x-rays was much greater in paramecia from the log-growth phase than in those from the stationary phase of a culture.

The author (Wichterman 1947 b, 1948 b) irradiated opposite mating types of *Paramecium bursaria* with dosages ranging from 100,000 *r* to 1,000,000 *r* in steps of 100,000 *r* (Table 9). It was found that normally green *P. bursaria* could be made permanently colorless as a result of irradiation with 300,000 to 600,000 *r*. Such colorless forms as produced by irradiation had the same sex type as before irradiation and readily mated and conjugated with members of the normally green opposite sex type. However, irradiated conjugants remained joined together in the sexual process much longer than unirradiated conjugants; some were unable to complete conjugation and died (Table 10). The same is true with irradiated conjugants of *P. calkinsi* (Wichterman, 1951 a). It is of interest to report that Lee (1949) described change of mating type in his clones of *P. bursaria* following exposure to x-rays. In the clones irradiated with 600,000 *r* by the author, progeny are still colorless and of the same mating type as before irradiation three years after the experiment (Wichterman, 1947 b).

EFFECTS OF RADIUM AND RADIOACTIVE SUBSTANCES AND ISOTOPES. In studies upon the effect of radium rays, Züelzer (1905) reported that multiplication of *P. caudatum* was stopped by 24-hour exposure but that *P. bursaria* was much

more resistant. In this regard, see also Spencer and Melroy (1943).

Markovits (1928) found that specimens of *P. caudatum* were most susceptible to radium radiations during their reorganization period when their division rate is low and most resistant when it is high. According to him this relation is the reverse of what is found in cells with mitotic division, and he regarded nuclear division in paramecium as amitotic.

Kimura (1935) studied the biologic action of rays from radioactive substances and his account is concerned with the effects of small-dose radiation upon the reproductive activity of unicellular organisms (see also Roskin, 1929). The author (unpublished, 1949) exposed mating types of *P. bursaria* to radioactive phosphorus (P^{32}). This isotope emitted beta particles or rays, delivering 10 *reps* (roentgen equivalent physicals) per minute. Even when irradiated with 14,000 *reps*, no unusual effects could be detected in locomotion or in the mating phenomena of paramecium. On the other hand, Kimball (1947 b, 1949 b) reported on the induction of mutations in *P. aurelia* by beta radiation, also P^{32}, when using vigor of paramecia as a criterion for mutation in irradiated exautogamous clones.

Powers (1948) exposed *P. aurelia* to the radioactive isotopes P^{32} and $Sr^{89,90}$ Y^{90} all of which are beta emitters. In his study of death following irradiated exautogamous specimens, he noted that in the low-dose range, the effect of radioactive P is much more pronounced than that of equivalent levels of radioactive Sr, Y.

Not only entire specimens of *P. aurelia* grown in a medium containing tritium but sections cut at one micron were found to produce well-defined radioautographs (Fitzgerald, Eidinoff, Knoll, and Simmel, 1951).

Table 9

EFFECT OF ROENTGEN RAYS UPON LOCOMOTION, THE MATING REACTION AND
CONJUGATION OF *Paramecium bursaria*
(Wichterman)

Dosage in roentgens (r) of two irradiated mating types of opposite sex (C and D)	Effect on locomotion	Effect on mating reaction and conjugation		
		Result when two mating types were mixed directly after irradiation	Day following irradiation when mating reaction first occurred	Day following irradiation when conjugation preceded by the mating reaction was seen to occur
100,000 r	More active than normal	Mild mating reaction with small clumps of only 10 or less paramecia	1st (immediately followed by conjugation)	1st
200,000 r	Slower than normal	Mild mating reaction with small clumps of only 10 or less paramecia	1st (immediately followed by conjugation)	1st
300,000 r	Slower than normal	Mild mating reaction with small clumps of only 10 or less paramecia	1st (no conjugation)	2nd
400,000 r	Slower than normal	As above but 45 minutes after mixing	3rd, 4th and 5th (no data beyond 5th day)	
500,000 r	Slower than normal	No mating reaction	3rd day, then daily	Few pairs on 7th day and more on following days but most on 12–14th
600,000 r	Slower than normal	No mating reaction	3rd day, then daily; clumps increasing in size	Few pairs on 7th day and more on following days but most on 12th–14th
700,000 r	Very slow moving	No mating reaction	Never	Never
800,000 r	Very slow moving	No mating reaction	Never	Never
900,000 r	Very slow moving	No mating reaction	Never	Never
1,000,000 r	No locomotion but ciliary activity	No mating reaction	Never	Never

Table 10

LENGTH OF TIME CONTROL AND X-RAYED CONJUGANTS OF *Paramecium bursaria* REMAINED JOINED TOGETHER IN CONJUGATION PROCESS* (Wichterman)

Hours after mixing opposite mating types	Number of conjugants separated in controls	Number of conjugants remaining in controls	Number of conjugants separated in irradiated set	Number of living conjugants remaining in irradiated set	Remarks
20	0	50	0	50	
20½	0	50	0	50	
21½	9	41	0	50	
22¼	16	25	0	50	
22½	6	19	0	50	
23	5	14	0	50	
23½	5	9	0	50	
24	3	6	2	48	
24½	6	0	4	44 ←	*Note:* All of controls separated at this time while irradiated conjugants are just beginning to separate.
25	—	—	5	39	
25½	—	—	0	39	
26	—	—	2	37	
26½	—	—	2	35	
27½	—	—	4	31	
28	—	—	2	29	
29	—	—	4	25	
29½	—	—	1	24	
30	—	—	0	24	
30½	—	—	0	24	
31	—	—	0	24	
31½	—	—	0	24	
32	—	—	0	24	
32½	—	—	0	24	
33	—	—	0	24	
33½	—	—	0	24	
42½	—	—	2	22	
43½	—	—	2	20	
44½	—	—	0	20	*Note:* Nineteen irradiated pairs of conjugants died during the process.
45½	—	—	0	20	↓
47	—	—	1	19	
48	—	—	0	19	
49	—	—	0	18	1 pair died and discarded
53	—	—	0	18	
55	—	—	0	16	2 pairs died and discarded
57½	—	—	0	15	1 pair died and discarded
66	—	—	0	15	
68	—	—	0	14	1 pair died and discarded
70	—	—	0	13	1 pair died and discarded
75	—	—	0	13	
78	—	—	0	11	2 pairs died and discarded
81	—	—	0	4	7 pairs died and discarded
82	—	—	0	0	4 pairs died and discarded

* In this experiment there were used 50 pairs of control conjugants and 50 pairs of conjugants irradiated with 300,000 *r* 19–20 hours after mixing members of each mating type. All paramecia of one mating type came from one culture, while all members of the opposite mating type came from another culture. Temperature during experiment: 24.5–27° C.

In conclusion it may be said that in general, paramecia as with many Protozoa, respond to roentgen rays, ultraviolet rays, and the rays from radium and radioactive substances in such a way that small dosages accelerate locomotion. However, with greatly increased dosages, locomotion ceases, and there occurs vacuolation and coagulation of the protoplasm resulting in death. One is impressed with the extraordinarily high dosages of roentgen units required to kill paramecia.

8. Effects of Alcohol, Narcotics, Vital Stains, Photodynamic Action and Sensitization, Carcinogenic Agents, Oligodynamic Action, Alkaloids and Related Drugs; Other Inorganic and Organic Compounds Including Metallic Salts, Organic Extracts, Enzymes, Hormones, Vitamins, Sera, Toxins, etc.

ALCOHOL. Calkins and Lieb (1902) found that alcohol in medium doses, e.g. one part in 2500 parts of culture medium, acted as a continued stimulus to the division rate of *Paramecium*. However, the experiments of Woodruff (1908 a, b, c) failed to show such a marked uniformity of effect from alcoholic treatment. Woodruff reported that alcohol produced opposite effects on the division rate at different periods in the life-cycle. Minute doses of alcohol were found to decrease the rate of division at one period of the life-cycle and increase it at another period. When the division rate is increased with alcohol, the effect is not continuous. The rate of division gradually falls below that of the controls followed by a fluctuation above and below the controls. With repeated increases of dosage (even when the amount is doubled three times) the fission-rate soon becomes more rapid but then falls again below that of the controls.

Treatment with alcohol lowers the resistance of *Paramecium* to copper sulfate. Woodruff (1908 c) and Loefer and Hall (1936) studied the effect of ethyl alcohol on growth of Protozoa in bacteria-free cultures. They found that no acceleration of growth by alcohol occurred and that growth was stopped in 2 per cent alcohol. Too high a concentration of alcohol will, of course, irreversibly coagulate the protoplasm.

Additional references dealing with the effects of alcohol follow:

Bills (1922, 1923, 1924)
Daniel (1909)
Estabrook (1910 b)
Hunt (1907)
Kisch (1913)
Loefer and Hall (1936)
Matheny (1910)
Metalnikow (1912)
Sand (1901)
Schürmayer (1890)
Tsukamoto (1895)

NARCOTICS. The great physiologist, Claude Bernard, observed that certain drugs caused anesthesia or depression of the central nervous system in higher animals and that when the drugs were allowed to react for longer periods of time, they caused a similar reversible depression in all living cells.

Narcosis is a general phenomenon demonstrated by all living animals. In their specific action, narcotics generally leave little trace of their presence either physiologically or morphologically after the effect has disappeared. The typical narcotics are those frequently classified as the general anesthetics and the hypnotics.

Many theories have been expressed concerning their mode of action on protoplasm and the organism. For them, one is referred to the comprehensive reviews of Henderson (1930) and Winterstein (1926).

Alverdes (1922 a) observed that *Paramecium* in 0.1 per cent solution of chloral hydrate lost its body cilia in 48 hours and became motionless. When removed from this solution and placed in normal culture medium, the organism regenerated its cilia.

Additional references on narcotics pertaining to *Paramecium* and some closely related forms follow:

Alpatov (1937) narcotics, heat resistance, and electrical stimulation

Alverdes (1922 a) chloral hydrate

Baskina (1924) Chloretone

Burt (1945) chloral hydrate, ethyl carbamate, and other carbamates

Cole (1925) Chloretone

Figge and Wichterman (1952) sodium pentobarbital and x-rays

Fortner (1925, 1926 c) osmo-narcosis and potassium chlorate

Galina (1914) chloroform

Goldschmeid-Hermann (1935) narcotics

Ishiwawa (1911) chloroform

Kisch (1913) ether, chloroform

Kissa (1914) chloroform and chloral hydrate

Leichsenring (1925) ethylene, nitrous oxide, ether, chloroform

Loeb and Wasteneys (1911) chloroform and chloral hydrate

Löhner (1913) chloroform and chloral hydrate

Nagai (1907) chloroform and chloral hydrate

Okazaki (1927 a, b) lethal doses of various drugs in paramecia and mice

Peters (1904) chloroform

Pütter (1911) chloroform and chloral hydrate

Rothert (1904) ether and chloroform

Schürmayer (1890) chloroform and chloral hydrate

Stefanowska (1902) nitrous oxide, acetylene

Szücs and Kisch (1912) nitrous oxide and acetylene

Verworn (1889) nitrous oxide and acetylene

Wieland (1922) nitrous oxide and acetylene

VITAL STAINS. It is well known that certain dyes slightly or not at all toxic to protoplasm have an affinity for specific structures inside living cells. As a result of the selective affinities of the dyes, conclusions can, in some cases, be drawn as to the chemical constitution of structures which react to the dye. These dyes are generally used in greatly diluted solutions. When Congo red is used as an indicator, its red color changes to blue in weak acids. Neutral red as an indicator may be yellowish red (alkaline), cherry red (weak acid), and blue (strong acid); it may also stain the nucleus slightly as well as Golgi bodies. Janus green B stains chondriosomes and methylene blue stains nuclei and many cytoplasmic constituents. Certain colloidal dyes (Victoria blue, night blue) fail to penetrate plant cells but penetrate *Paramecium* and certain animal cells (Barnes, 1937). References pertaining to dyes employed on *Paramecium* follow:

Baldwin (1920) methylene blue, eosin alone and in conjunction with x-rays

Ball (1927) methylene blue, eosin

Barbarin (1940) neutral red

Becker (1926) methylene blue, eosin

Bragg and Hulpieu (1925) methylene blue, eosin

Brandt (1881) methylene blue, eosin

Carben (1914) methylene blue, eosin

Certes (1885) neutral red

Child (1934 a, b) methylene blue

Child and Deviney (1926) neutral red

Cole (1934) acid fuchsin

Costamagna (1899) neutral red

Day (1930) alizarin blue

Dunihue (1931) neutral red

Efimoff (1922, 1925) neutral red

Fortner (1928 c) neutral red

Fyg (1929) various vital stains

Gersch (1937) basic vital stains

Greenwood and Saunders (1894) Janus green

Halter (1925) Janus green

Heinrich (1903) Congo red, rhodamin III, safranin, fluorescein sodium

Horning (1926) Janus green

Howland (1924 a) alizarin blue

Kalmus (1928 a, b) methylene blue

Kite (1915) basic vital stains

Koehring (1930) neutral red reaction

Kōno (1930) basic vital stains

Ledoux-Lebard (1902 a) eosin and other substances

Makarov (1940) vital stains in relation to microscopic structure
Mast (1947) many indicator dyes
Neuschlosz (1919, 1920) methylene blue, trypan blue, fuchsin
Nirenstein (1920) basic vital stains
Prowazek (1897, 1899, 1901 a, 1902) basic vital stains
Przemycki (1894, 1897) basic vital stains
Rohde (1917) basic vital stains
Rumjantzew and Kedrowsky (1927) basic vital stains
Serrano (1879) basic vital stains
Spencer and Melroy (1940, 1941, 1943) eosin, fluorescein, and phenanthrene
Strelnikow (1924) basic vital stains
Vonwiller (1921) basic vital stains
Wankell (1922) basic vital stains
Wilson (1927) neutral red

PHOTODYNAMIC ACTION AND SENSITIZATION. When visible light is used in conjunction with certain dyes such as eosin or rose bengal, a marked effect occurs in protoplasm and the resulting action is called *photodynamic action.* These dyes, which are generally fluorescent ones, are not effective to the same extent in darkness.

Paramecia treated with eosin were subjected to the intermittent light of an electric lamp by the Efimoffs (1925). They concluded that the toxic effects of eosin and light are irreversible and can be used as a proof of the mass-action law. One is referred to the works of Tappeiner and Jodlbauer (1904, 1907) and Blum (1941) for a monographic treatment of photodynamic action.

Recently Giese (1946) reported that specimens of *P. caudatum* are readily killed even at low intensities of light in the presence of photodynamic dyes. He found that in 1:20,000 and 1:40,000 eosin solutions, paramecia are killed after 12 hours and in 1:60,000 after 36 hours exposure to the fluorescent lamp. In 1:100,000 and 1:400,000 eosin solutions, the paramecia divided at a retarded rate while controls in the dark divided at the same rate with or without

dye. Giese and Crossman (1946) observed that visible light of high intensity does not injure *P. caudatum* or sensitize them to heat but if photodynamic dyes are added, paramecia are readily killed by visible light of high intensity and they are sensitized to heat by sublethal dosages of light. Paramecia so sensitized are killed when subjected to a sublethal exposure to heat. When the light and heat were applied in the reverse order, namely heat then light, no ill effects were observed. It was noted that when the concentration of the dye was reduced, a larger light dosage was required. Sublethal dosages of light in the presence of dyes did not affect the division rate even when three-fourths of the lethal dosage had been used.

Photodynamic action and sensitization of *Paramecium* have been investigated by the following:

Beck and Nichols (1937) fluorescent dyes and pH
Carben (1914) methylene blue, eosin
Danielsohn (1899) acridine derivatives
Dognon (1927, 1928) fluorescence and various sensitizing agents; light intensity
Efimoff (1922, 1925) various compounds
Feiler (1928 a, b) quinine
Figge and Wichterman (1952) hematoporphyrin and x-rays
Fischer and Kemnitz (1916) porphyrin
Hausmann (1911) hematoporphyrin
Hausmann and Kolmer (1908) blood and bile
Henri and Henri (1912) selenium
Hertel (1904, 1905) eosin in magnesium light
Jodlbauer (1904, 1908, 1913) various compounds
Lampiris (1915) gold and platinum salts
Metzner (1927 b) various compounds
Osthelder (1907) acridine
Prát (1917) acridine
Raab (1900) acridine
Salomonsen (1903) eosin
Szücs and Kisch (1912) various compounds
Tappeiner (1896, 1909) phenylquinoline and Phosphine (acridine derivatives)
Tappeiner and Jodlbauer (1904, 1907) fluorescent compounds

CARCINOGENIC AGENTS. Wolman (1939) reported a proliferative effect of the carcinogenic hydrocarbons, 3,4-benzpyrene, methylcholanthrene, and 1,2,5,6-dibenzanthracene on cell division in paramecium. Mottram (1940, 1941, 1942, 1944) obtained monstrosity in amicronucleate clones of *Colpidium* sp. (which he earlier reported as *P. aurelia*) upon exposure of the ciliates to the carcinogenic hydrocarbons, 3,4-benzpyrene, methylcholanthrene, and 1,2,5,-6-dibenzanthracene. Using paramecium, Tittler and Kobrin (1941, 1942) were unable to obtain similar results.

The action of methylcholanthrene upon *Paramecium multimicronucleatum* has been studied by Spencer and Melroy (1940, 1941, 1943), Daniel, Spencer and Calnan (1945), and Spencer and Calnan (1945). Spencer and Melroy reported that the immediate effect of the carcinogen was to stimulate cell division. They claimed that besides an immediate effect, there was apparent a cumulative one and that methylcholanthrene-adapted paramecia survived longer than the control animals. After continuous exposure of strains of *P. multimicronucleatum* to methylcholanthrene as well as to crystal violet, neutral red, eosin, radium, fluorescein, and phenanthrene (a non-carcinogenic polycyclic hydrocarbon) over long periods of active multiplication, Spencer and Calnan concluded that an environment which is apparently harmless to individuals may have a cumulative effect on the species resulting in its eventual destruction.

A serious objection to most of these experiments dealing with the effects of carcinogens is that the cultures of paramecia were not pure ones since they contained various kinds of other microorganisms. It is thus difficult to ascertain whether or not the effect is due directly to the carcinogen or indirectly to the bacteria which are used as food. It is possible that the carcinogens may have served as an added carbon source of food for some of the bacteria.

Tittler (1948) studied the effect of 1,2,5,6-dibenzanthracene, 3,4-benzpyrene, and methylcholanthrene upon growth of *Tetrahymena geleii* in pure culture. The carcinogens did not affect growth in this ciliate, which is closely related to paramecium.

In regard to the photodynamic action of carcinogenic agents, Mottram and Doniach (1947) reported that paramecia which had been subjected to benzpyrene overnight and repeatedly washed in tap water retained their photosensitivity. Similar reactions occurred when dibenzanthracene, shale oil, and coal tar were used to sensitize paramecia. Doniach (1939) also made a comparison of photodynamic activity of some carcinogenic compounds with non-carcinogenic ones.

The work of Lloyd (1947, 1949) concerning exposure of *P. caudatum* to benzene hexachloride or "gammexane" is discussed on page 358.

See also sections on acclimatization (p. 203) and monsters (p. 335).

OLIGODYNAMIC ACTION.[8] Minute traces of compounds of certain metals, especially copper, have been shown to confer a certain toxic property upon water and in turn upon protoplasm. Brass has a somewhat similar toxic effect while tin does not.

References dealing with oligodynamic action upon *Paramecium* follow:

Eckert and Feiler (1931) quinine and silver nitrate
Israel and Klingmann (1897) copper
Junker (1925) copper

[8] Nägeli (1893) discovered that a very small quantity of copper in water which emanated from copper coins was extremely toxic to *Spirogyra*. He created the name *oligodynamic* for the action.

Löhner and Markovits (1922) copper
Ludwig (1927) silver nitrate

ALKALOIDS AND RELATED DRUGS. The alkaloids are generally considered to be nitrogenous substances—usually of plant origin—which produce pronounced physiologic effects upon organisms. Better known examples of alkaloids are quinine, morphine, cocaine, strychnine, and colchicine.

According to Giemsa and Prowazek (1908) and Acton (1921), quinine lowers the division rate but Sand (1901) reported the rate to be increased with appropriate concentrations. Strychnine produced a temporary increase of fission-rate as reported by Calkins and Lieb (1902) for paramecium.

Colchicine ($C_{22}H_{25}O_6N$) is an alkaloid poison which is extracted from the seeds of a plant known as meadow saffron. Its action has been investigated recently by a number of workers. It has been shown that colchicine in very weak concentrations results in doubling of the number of chromosomes although occasionally the mitotic figure is destroyed. The action of the drug may at first inhibit typical cell division while the chromosome splitting goes on as usual. Subsequently the effect of the drug apparently disappears, and the cells continue to divide with the doubled number of chromosomes. While colchicine has the effect of stopping or freezing mitosis in cell division, some few

workers have maintained that it stimulates mitosis. Halberstaedter and Back (1943) studied the effect of colchicine alone and combined with x-rays upon *P. caudatum*. They used three concentrations of crystalline colchicine in basic culture media as follows: (1) 0.0005 per cent, (2) 0.00025 per cent and (3) 0.000125 per cent (Table 11). The 0.0005 per cent solution was lethal but the 0.00025 and to a more marked degree the 0.000125 per cent solutions exerted a stimulating effect upon cell division. In fact the last two concentrations gave a higher division rate than in the controls. The results of 50 experiments are summarized in the following table.

For the combined effect of colchicine and x-rays, see page 141.

Atebrin (quinacrine) and Plasmochin (pamaquine) are two of the compounds that have been synthesized to supplement quinine in the treatment of malaria. Atebrin is a derivative of acridine, and Plasmochin is a derivative of quinoline, the latter occurring in coal tar and bone oil.

References dealing with the effects of alkaloids and related drugs on paramecium follow:

Acton (1921) cinchona alkaloids
Alexandrowa and Istomina (1903) "salamander poison" (steroidal alkaloids)
Barros (1940) colchicine
Baskina (1924) quinine

Table 11

INFLUENCE OF COLCHICINE ON THE DIVISION RATE OF *P. caudatum*
(Modified after Halberstaedter and Back, 1943)

Numbers of Individuals in the Different Solutions

Counted after	0.0005% Sol.	0.00025% Sol.	0.000125% Sol.	Control Culture
0 hours	50	50	50	50
24 hours	0	114	130	74
48 hours	0	450	609	275

Bichniewicz (1913) quinine

Brahmachari, et al. (1932, 1933) quinoline compounds

Calkins and Lieb (1902) strychnine

Cantacuzène (1925) quinine derivatives

Chejfec (1937) quinine

Crane (1921) alkaloids

Dehorne and Morvillez (1926) various alkaloids

Eckert and Feiler (1931) quinine

Epshtein and Babikova (1936) arichin and plasmocide types

Estabrook (1910 b) strychnine

Feiler (1927, 1928 a, b, 1929 a, b, 1931) Plasmochin (pamaquine), quinine, and other alkaloids

Gause (1937) cinchonine isomers

Giemsa and Prowazek (1908) quinine, atropin, strychnine

Goyal (1935) the alkaloid, conissine

Grethe (1895) quinine derivatives

Hausman and Kolmer (1907) colchicine

King and Beams (1940) colchicine

Kriz (1924) strychnine and nicotine

Lehmann (1927) phenol

Malm (1930) alkaloids in relation to pH

Neuschlosz (1919, 1920) quinine

Okazaki (1927 a, b) caffeine, cocaine, strychnine, saponin

Pick and Wasicky (1915) emetin, papaverin

Potts (1944) quinine and Atebrine (quinacrine)

Prowazek (1910 b) alkaloids (strychnine, pilocarpine); saponin, bile

Roskin and Dune (1929) quinine

Roskin and Romanowa (1929) quinine and ultraviolet light

Sand (1901) quinine

Sarkar (1936) quinine

Schürmayer (1890) strychnine, cocaine

Shishyaeva and Muratova (1936) quinine and ultra short waves

Šmelev (1928) quinine

Weyland (1914) papaverine, "saffrol" (colchicine)

OTHER INORGANIC AND ORGANIC COMPOUNDS INCLUDING METALLIC SALTS, ORGANIC EXTRACTS, ENZYMES, HORMONES, VITAMINS, SERA, TOXINS, ETC.:

Abderhalden and Schiffmann (1922) Optone (thyroid)

Ball (1925) thyroid, liver, hypophysis, and glycogen

Barbarin (1940) KCN, CuSO₄, KCl, CaCl₂

Baskina (1924) mercuric chloride

Bauer (1926) histamine, Adrenaline (epinephrine)

Bayer and Wense (1936 a, b, 1937) choline, acetylcholine, Adrenaline (sympathin) hormones

Bramstedt (1937) posterior pituitary hormone

Browder (1915) lecithin, cholesterol

Budington and Harvey (1915) thyroid

Burge and Estes (1926, 1928) insulin, thyroxin

Burge, Estes, Wickwire and Williams (1927) insulin, thyroxin, Adrenaline, and posterior pituitary injections

Calkins and Eddy (1916–17) pancreatic vitamin

Carben (1914) copper salts

Chambers (1919) hormones and glandular products

Chopra and Chowhan (1931) Indian cobra venom

Cori (1923) thyroid extracts and thyroxin

Davenport and Neal (1896) mercuric chloride

Donatelli and Pratesi (1937) sodium glycerophosphate and tetramethylammonium compounds

Essex and Markowitz (1930) rattlesnake venom

Figge and Wichterman (1952) sodium nitrite and x-rays

Flather (1919 a, b) polished and unpolished rice, glandular extracts

Fühner (1917) arsenic

Geckler (1949, 1950 a) nitrogen mustard

Geppert and Shaw (1937) antiseptic agents including mercuric chloride, m-hydroxybenzoic acid, m-chlorophenol, m-nitrophenol, m-cresol, resorcinol monomethyl ether, phenol, resorcinol, and m-aminophenol

Gerard and Hyman (1931) cyanide sensitivity

Grover and Shaw (1947) thiourea

Hall (1944) vitamins

Holter and Doyle (1938) catalase

Hamilton (1904) human serum (scarlet fever and pneumonia serum)

Hammett (1929) cystine, cysteine, glutathione, insulin, etc.

Harnisch (1926) arsenic acid

Harrison and Fowler (1944, 1945) rabbit serum

Harrison, Sano, Fowler, Shellhamer, and

Bocher (1948) sera from cancerous and non-cancerous persons

Heubner (1925) cantharidin, histamine

Humphrey and Humphrey (1947, 1948) succinic dehydrogenase; cyanide sensitivity

Israel and Klingman (1897) copper

Johnson and Tatum (1945) pressed yeast juice

Jollos (1913 a, 1921) arsenious acid

Kalmus (1928 a) potassium cyanide

Kidder (1946) folic acid

Kidder and Dewey (1945) riboflavin, pantothenic acid, biotin, niacin, pyridoxine

Lacaillade (1933) bee venom

Lampiris (1915) gold and silver salts

Ledoux-Lebard (1902) blood serum

Little, Oleson, and Schaefer (1951) aureomycin

Loefer (1936) many peptones and carbohydrates

Löhner and Markovits (1922) copper

Ludwig (1927) silver nitrate

Lund, B. L. (1918) potassium cyanide

Lund, E. J. (1918 a, b, c, 1921) potassium cyanide

Lwoff (1925) amino acids, organic extracts

McCleland and Peters (1920) mercuric chloride

Masugi (1927) blood serum

Melnikov, Avetesyan, and Rokitskaya (1941) phenols (soluble sodium phenolates: 24 compounds)

Metz (1947) formalin

Metz and Butterfield (1951) proteolytic and non-proteolytic enzymes

Metz and Fusco (1948) antiserum

Metz and Fusco (1949) lyophilized paramecia

Middleton (1940) inactivated liver extract

Middleton and Wakerlin (1939) liver extract

Mitrophanowa (1925) mercuric chloride

Mukerji, Dutta, and Ganguly (1942) dextro-rotatory hydrocupreidine derivatives

Nelson and Krueger (1950) respiratory inhibitors: thiourea and thiouracil with urea and uracil

Neuhaus (1910) arsenic, antimony, mercury, copper

Neuschlosz (1919, 1920) arsenic, antimony

Nowikoff (1908) thyroid extracts and others

Okazaki (1927) Adrenalin (epinephrine)

Pace (1945) potassium cyanide

Pacinotti (1914) glycogen

Paneth (1890–91) hydrogen peroxide

Pecker (1915) human serum

Peters and Opal (1909) copper sulfate

Philpott (1925, 1926, 1927, 1928, 1930, 1931 a, b, 1932) toxins of diphtheria and others, pathogenic bacteria, cobra antiserum and other snake venoms

Powers and Raper (1950) nitrogen mustard

Prowazek (1908) lecithin

Pütter (1905) glycogen

Rakieten (1928) antiserum and antigen of *Staphylococcus aureus*

Rayl and Shaw (1941) surface active toxicants, sodium laurylsulfonate, and hexyl-resorcinol

Rees (1922 b) Ascaron

Riddle and Torrey (1923) thyroxin

Rohdenburg (1930) adrenal, insulin, parathyroid, thyroid, ovary, testis, spleen, thymus, pituitary, liver

Roskin (1946) serum from cancerous persons

Rössle (1905) guinea-pig serum

Sand (1901) arsenic

Sato and Tamiya (1937) potassium cyanide, carbon monoxide, and pyridin hydrosulphite

Sawano (1938) cathepsin

Schürmayer (1890) antipyrine and acetanilid

Seaman (1947) penicillin

Sharpe (1930) hydrogen sulfide, sodium sulfide

Shaw and Geppert (1937 a, b) various toxic agents and meta-substituted phenols: phenol, mercuric chloride, m-hydroxy-benzoic acid and saponin

Shaw, Ordal, and Wingfield (1949) uranium

Shaw-Mackenzie (1916) copper salts

Shoup and Boykin (1931) potassium cyanide

Shumway (1914, 1917, 1929) thyroid

Sickels and Shaw (1935) purified diphtheria toxin

Šmelev (1928) arsenic

Strauss (1923) thyroid

Sun (1912) uric acid

Surber and Meehan (1931) sodium arsenite, arsenious oxide

Takenouchi (1918) filtrates from diphtheria, tetanus, staphylococci and streptococci cultures

Tappeiner (1896) phenylquinoline, phosphin

Tittler and Kobrin (1941, 1942) carcinogenic agents (see also p. 150)

Torrey (1934) thyroxin

Torrey, Riddle, and Brodie (1925) thyroxin

Tunnicliff (1928, 1929) toxins and antitoxins of measles, scarlet fever, diphtheria; urine

Ugata (1926) amino acids, nucleine acids, beef extract

van Wagtendonk (1948 a, 1949) proteolytic enzymes

Wieland (1922) nitrous oxide, acetylene, hydrogen

Woodruff (1908 d) copper sulfate

Woodruff and Baitsell (1911 a) beef extract

Woodruff and Swingle (1922, 1923, 1924) thyroxin

Woodward (1928) Arbacia egg-secretion

Ferguson, Holmes, and Lavor (1942) studied the effect of the following sulfonamide drugs in solutions of distilled water: sulfanilamide, sulfathiazole, sulfapyridine, and sulfaguanidine. They observed that *P. caudatum* was more resistant to these compounds than *Hydra*. The order of toxicity of these compounds is as follows: sulfapyridine (most toxic), sulfathiazole, sulfanilamide, and sulfaguanidine (least toxic.)

Propionic acid has been shown to inhibit the growth of mold on bread. Potts (1943 a) reported that propionic acid and its sodium salt have an appreciable action on *P. caudatum*, the acid however being many times as active as its sodium salt. Later, he investigated the action of a number of recently synthesized arsenicals, namely benzene, pyridine, and pyridone compounds. The pentavalent arsenicals were found to be only weakly active in killing *P. caudatum* and among the trivalent arsenicals, the pyridones were conspicuously inactive. The most active killing compounds were the arsinoxides and chloroarsines derived from benzene and pyridines (1943 b). The same investigator (1944) tested the action of 23 compounds—some related and some not—on *P. caudatum*. Sulfanilamide was found to be conspicuously ineffective. Benzenestibonic acid and tartar emetic appeared to be more active than the pentavalent benzene arsenicals. Also tested were Atebrine (quinacrine), fluorescein, quinine plus urea, HCl, and others.

9. Effects of Hydrostatic Pressure.

Regnard (1891), who was the first to study the effects of hydrostatic pressure on organisms, subjected paramecium and other ciliates to pressures up to 300 atmospheres. At this pressure no effect was detected but at 400 to 600 atmospheres, the ciliates became immobile and the cilia swollen. Pressure release brought recovery unless the pressure had been maintained for an hour or so.

Ebbecke (1935, 1936) obtained similar results with paramecium and found that they could recover even after exposure to 800 atmospheres. He reported that under high pressure, there occurred a decrease or at times a complete cessation of locomotion followed by a rounding up of paramecia.

Syngajewskaja (1929) studied the growth of *P. caudatum* during its life cycle in respect to the osmotic pressure.

Chapter 6

Metabolism: Nutrition, Secretion, and Growth of Paramecium and Factors Influencing Growth

Like every living cell, paramecium is in a state of continuous chemical and physical change. From the chemical point of view growth consists essentially in the transformation of simple, unorganized foodstuffs such as inorganic salts, fats, carbohydrates, amino acids, etc. into new, more complicated substances which form a part of the organized protoplasm.

The sum total of all the chemical activities involved in the living protoplasm is termed *metabolism*. Frequently those activities concerned with the building-up processes are referred to as *anabolism* while those concerned with the breakdown or production of energy are listed under *catabolism*. However, the two processes are closely interrelated and it is difficult at times to distinguish between anabolic and catabolic activities. Metabolism covers such topics as nutrition, growth, secretion, respiration, and excretion.

A. Nutrition and Secretion

1. Types of Nutrition and the Food of Paramecium.

Protozoa obtain nourishment by a number of different methods which, briefly stated, are as follows: *holozoic* (heterotrophic, zoötrophic), *holophytic* (autophytic, phototrophic, phytotrophic), and *saprozoic* (saprophytic).[1] Holozoic nutrition is the common method by which all higher animals obtain nourishment. The animal requires as a food source materials such as proteins, carbohydrates, and fats which are commonly found in the protoplasm of other organisms hence supplying the requirement of organic carbon. In this type of nutrition there is involved ingestion, secretion, digestion, absorption, assimilation, and the final egestion of indigestible materials. Most of the Protozoa and all species of paramecium except the green *P. bursaria* belong to this class.

Holophytic nutrition is characteristic of chlorophyll-bearing organisms. An example is shown in the contained green zoöchlorellae of *P. bursaria*. In the presence of sunlight, oxygen is liberated from the chlorophyll in the zoöchlorellae and this combines with carbon and other elements from water and inorganic salts. The result is the formation of complex, energy-holding compounds which Pringsheim has shown to pass from the zoöchlorellae to the ciliate for the utilization as food.

Saprozoic nutrition is generally found

[1] Lwoff (1938, 1951), Pringsheim (1937), and Hall (1939, 1941) on the basis of nitrogen requirements describe a more extended classification of types of protozoan nutrition. Useful reference works dealing with the biochemistry and physiology of paramecium and other *Protozoa* are Lwoff (1951) and Prosser, Brown, et al (1951).

in those Protozoa which lack ingestatory structures. The animals are nourished by the diffusion of decomposed and dissolved organic material found in molecular size in the environment. This type of nutrition is well exemplified in *Opalina* and other astomatous ciliates.

In certain instances, some Protozoa are able to utilize more than one of the aforementioned types of nutrition and the type is called *mixotrophic*.

Paramecia feed naturally upon different kinds of bacteria. When brought into the laboratory and cultivated, it is important that suitable bacteria in sufficient numbers be available as a source of food (see p. 99). In addition to bacteria, paramecia have been grown in the laboratory on yeast cells, other microorganisms, and in bacteria-free media as described in Chapter 4.

Different species of bacteria furnish the principal source of food for paramecia. In the laboratory paramecia are generally cultivated on one or more species of bacteria grown in the same culture with the ciliates (see p. 105). Ludwig (1928 a) cultivated *P. caudatum* on *B. subtilis* and estimated that with a division rate of two per day, a single specimen required 1500 bacilli per hour. On the other hand, Chejfec (1929) estimated that *P. caudatum* ingested and digested from two to five million *Bacterium coli* in 24 hours. Cutler and Crump (1923 a, 1925) and others have shown that the rate of division in Protozoa is a function of the quantity of food.

Hargitt and Fray (1917) reported that cultures of mixed bacteria as a rule are far superior as a diet for paramecium to any one kind of bacteria. From the work of many it appears that *B. subtilis* is probably the principal food of paramecium in the typical hay infusion. Phillips (1922) found that *P. aurelia* grew fairly well on her "C" strain of bacteria (a streptothrix) with a daily fission-rate of 1.03 but that the rate was increased to 1.79 on a mixture of strains "A" and "C." Strain "A" alone was not capable of supporting the growth of the paramecia.

As postulated by Oehler (1916), a protozoön may not feed on a given species of bacterium because of excessive formation of carbon dioxide, ammonia, trimethylamine, or some other substance deleterious to the growth of the ciliate as a direct result of bacterial metabolism. This unquestionably holds for the species of *Paramecium*. Also, a given species of bacterium may be inadequate for nutrition because of the establishment of unfavorable hydrogen-ion cencentration in the medium as a result of bacterial growth, size or shape of the bacteria, and the fact that a given bacterium may not be sufficiently constituted chemically to supply the nutritive requirements of the ciliate (Luck, Sheets, and Thomas, 1931; De-Lamater, 1939).

Oehler (1922 b), Loefer (1936 a), Pringsheim (1928), Mast (1947), and others have cultivated paramecia successfully upon certain species of living yeast cells. Loefer (1936 a) successfully maintained pure-line mass cultures of *P. caudatum* on the yeast *Saccharomyces ellipsoideus* at a pH of 6.7–6.8. Pringsheim claimed that while yeasts are quite satisfactory for *P. caudatum* and *P. bursaria*, the bacteria are more suitable for the former and algae for the latter. This may have some relationship to the fact that *P. bursaria* will digest some of its symbiotic zoöchlorellae when other food is scant. Lund (1918) cultivated paramecia in tap water to which compressed yeast was added. Bozler (1924 a) maintained that cooked yeast cells leave the body of paramecium wholly unchanged, but paramecia have also been cultivated on dead yeast cells

provided the latter are not decomposing.

Certain algae, namely *Chlorella* spp., *Prototheca zopfi*, small *Ulotricaceae*, *Stichococcus* spp., and *Hormidium* sp. have been described as being satisfactory as a food source. Raffel (1930) used *Stichococcus bacillaris* as food for *P. aurelia* and Pringsheim reported that larger organisms, including *Polystoma uvella* and *Chlorogonium elongatum*, produced optimum growth while *Cosmarium* spp. and *Euglena* spp. were less satisfactory as a food source for *P. bursaria*.

Powers and Mitchell (1910) were able to induce paramecium to feed first on the flagellate *Chilomonas* and later on small ciliates.

That paramecia may ingest many different kinds of microörganisms is clear but it is not evidence that they can be digested and utilized as food. Wenrich (1926) observed the ingestion by *P. trichium* of small green flagellate cysts in considerable numbers which became lodged in the ectoplasm of the ciliate.

Bragg (1939 a, b) observed that *P. caudatum*, *P. aurelia*, and *P. multimicronucleatum* ingested green, chlorophyll-bearing *Euglena* spp., *Chlamydomonas* sp., *Chlorotolys regularis*, and that *P. trichium* ingested *Chilomonas*. He noted however that the green organisms were egested while still green in color indicating that digestion of them failed to occur. Bragg (1939 a) also reported that Diller has seen *P. trichium* packed with *Chilomonas paramecium* and Wichterman (1941) has fed zoöchlorellae to colorless *P. bursaria*. The fact remains that reliable information is lacking concerning the actual digestion of these chlorophyll-bearing organisms by paramecium.

It is generally known that Protozoa which normally feed on other living microörganisms require a heat-labile substance for growth; heat-killed bacteria therefore will not support continued growth of paramecium. Emery (1928) claimed that a measurable quantity of amino acids is utilized in *P. caudatum* in place of bacteria as food. He reported that in a mixture of equal parts of 10 amino acids, 100,000 paramecia would use 48.3 per cent of a 0.1 per cent solution in 12 hours.

Recently Johnson and Baker (1942) and Johnson and Tatum (1945) have reported successful growth of *P. multimicronucleatum* with fractions of sterile, unheated, pressed yeast juice (baker's yeast) (see p. 103). They found that the pressed yeast juice supplies two fractions which are essential for the growth of paramecium. One is heat-stable and the other heat-labile and protein in nature.

Loefer (1936) studied the effect of a large number of desiccated peptones and carbohydrates for their effect in promoting the growth of bacteria-free *P. bursaria*. The following were found to be most suitable in the order named: Proteose-peptone (Difco), Bacto-tryptone (Difco), Bacto-veal (Difco), and Seidenpepton (Hoffman-LaRoche).

For the carbohydrates tested, dextrose, mannose, maltose, dextrin, and melezitose were best in the order named. Levulose, galactose, mannitol, lactose, sucrose, soluble starch, and salicin were without much effect while arabinose, xylose, rhamnose, and inulin were unfavorable.

2. The Feeding Apparatus.

For a description of detailed structures relating to the feeding apparatus refer to Chapter 3. The feeding apparatus of paramecium is essentially the same in all species. It consists of the shallow ciliated *oral groove* which extends from the anterior end of the body to a region slightly below the middle of the body. A ciliated funnel-shaped depression called the *vestibulum* is found at the posterior end of the groove which leads into the *mouth*. Nirenstein (1905) and others describe the mouth as a fixed oval opening.[2] In typical vegetative specimens the mouth can be observed as a persistent opening broadly oval in outline. The mouth leads into the curved tapering tube commonly called the *cytopharynx* which runs directly toward the center of the body, turns backward sharply, proceeds parallel with the body surface then turns sharply again at nearly right angles to the first turn (Figs. 16 and 51). As described elsewhere in detail, the cytopharynx contains numerous cilia on its inner surface existing in two bands. One is called the penniculus (Gelei, 1934 a; Lund, 1933, 1941) and extends from the anterior left edge of the cytopharynx almost to the posterior oral edge. The other called the "Vierermembran" extends from the anterior aboral edge to the posterior oral edge. The penniculus consists of eight rows of short cilia and the "Vierermembran" of four rows of long cilia. Mast (1947) is in general agreement with the findings of Gelei and Lund although there is some disagreement among the investigators in regard to certain details.

Bozler (1924 a, b) reported that approximately 10 long fibers are found attached to the right wall of the pharynx about half-way up its length and which extend nearly to the posterior end of the body. These fibers, called by him "Schlundfaden," are said to be fairly rigid, fixed in position and in the same

---

[2] Frisch (1937) described the mouth as a narrow slit bounded by a raised thickened border in the form of an elongated oval that is sometimes closed. It should be remembered that the observation of Frisch was on an exconjugant and it is possible that redifferentiation had not been completed.

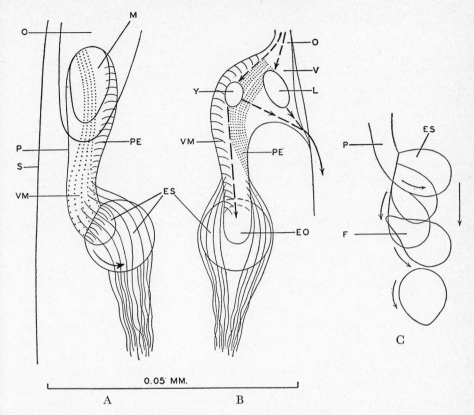

0.05 MM.

A   B

FIG. 51. Camera outline showing the structure of the feeding apparatus in *Paramecium aurelia*. (A and B) Two views, one at right angles to the other. (*O*) Oral groove; (*V*) vestibulum; (*M*) mouth; (*P*) pharynx; (*EO*) esophageal opening; (*ES*) esophageal sac and esophageal fibers; (*PE*) penniculus; (*VM*) "Vierermembran"; (*Y*) yeast cell; (*L*) a large particle; (*S*) surface of the body; broken lines, paths taken by particles during the process of feeding; arrows, direction of movement. The cilia which project into the esophageal sac from the end of the pharynx are doubtless part of the "Vierermembran." There are, of course, many more than are represented. (C) Camera outline of a portion of the feeding apparatus and some sketches illustrating the movement of the food vacuole during its separation from the pharynx in *Paramecium aurelia*. (*P*) pharynx; (*ES*) esophageal sac; (*F*) food vacuole; arrows, direction of movement. (Mast)

plane. Gelei reported the fibers near the proximal end of the cytopharynx and not in one plane while Lund, who called them postesophageal fibers, maintains that they are not fixed and attached near the distal end on all sides of the cytopharynx. Mast observed the fibers in the living condition but failed to observe their manner of attachment.

The cytopharynx finally leads into an elliptic opening designated the "esophageal sac" by Mast which develops into the food vacuole. Cilia in the cytopharynx force fluid with particles in suspension against the membrane over the distal opening of this structure to produce this sac. According to Mast the sac consists of a thin elastic membrane which separates the content of the sac from the cytoplasm and changes greatly in form and size as the food vacuole develops. Earlier Lund (1941) reported that the posterior extremity of the pharyngo-esophageal cavity (cytopharynx) consists of a blind pouch heavily ciliated on its ventral surface which he called the "esophageal process." On the other hand Mast reported that he could not find such a structure at the end of the cytopharynx in his study of living paramecia.

3. Selection of Particles, Ingestion, and the Formation of the Food Vacuole.

To test food selection in paramecium one may use nutritious or digestible substances such as the starches of corn, potato, rice, arrowroot, etc., dried or boiled egg white or yolk, milk powders, finely ground or powdered spices, etc., or indigestible substances such as carmine,[3] India ink, Chinese ink, talc, powdered charcoal, etc.

Indicators[4] are useful to render the food vacuoles more visible and to test their pH. Bragg (1939 b) reports that the following amounts of indicators dissolved in 100 cc. of distilled water give

good results: (a) neutral red, 50 mg.; (b) phenol red, 100 mg.; (c) litmus, 150 mg.; and (d) Congo red, 50 mg. In using, a drop of the indicator solution is mixed in a drop of culture on a slide containing paramecia, covered with a cover-glass, then observed under the microscope.

Bragg (1939 b), Wichterman (1941), Mast (1947), and others have seen paramecia ingest single-celled green algae or flagellates and Bragg has raised the question if such forms are digested and utilized by the paramecia.

Metalnikow (1907, 1912), Bozler (1924 a, b), Lozina-Lozinsky (1929 a, b, c, d, 1931), Bragg (1936 a, 1939 a), Mast (1947), and others are in agreement concerning the fact that while paramecia ingest many kinds of particles, more are taken into the body which are digestible than those which are indigestible. Metalnikow (1912, 1915, 1917) placed *P. caudatum* in suspensions of carmine and other substances for given lengths of time and determined the content, circulation and number of vacuoles thus formed. He found that paramecia formed fewer food vacuoles containing indigestible particles than they did in media with digestible particles. It appears from his work that *P. caudatum* not only selects certain types of particles from others but that it can "learn" to do so.

Wladimirsky (1916) explained Metalnikow's results differently. He believed that a kind of depression occurred in the physiologic state of the organism and that *P. caudatum* is unable to select or learn to select food.

The work of Lozina-Lozinsky, Bragg, and others would indicate that paramecia demonstrate differential selection of particles to be ingested and that a certain degree of learning is shown thereby confirming the earlier results of Day and Bentley (1911) (see p. 248).

---

[3] Gleichen (1778) appears to have been the first to introduce the use of colored particles into the body of ciliates in order to study the food vacuoles. First called gastric vacuoles, the food vacuoles were clearly recognized as centers of digestion. It led to Ehrenberg's (1833, 1838) complicated, but erroneous, conception of the Polygastrica theory in ciliates. Carmine particles are commonly used to demonstrate ingestion in paramecium. Bragg (1937) found that toxic properties of certain carmines could be removed by first washing 1 Gm. of carmine in 1 liter of hot or cold tap water.

[4] Hance (1925) reported that an infusion of red cabbage leaves (30 Gm. to 1 liter water) makes a combined culture medium and indicator. Purple color indicates a basic condition and red an acid one. Under certain conditions this may also be used as an acid-indicator in the food vacuoles but it is not always reliable (Bragg and Hulpieu, 1925).

Lozina-Lozinsky presented evidence to show that selection of food particles is not due to differences in form, size, or specific gravity of the particles.

According to Bozler selection is made at or near the entrance to the pharynx by the response of individual cilia to contact with the particles. Thus when a particle comes into contact with a cilium, the cilium responds—depending upon the physical characteristics of the particle—by throwing the particle into the cytopharynx or out of it. Lozina-Lozinsky (1931) is of the opinion that particle discrimination is not dependent upon an individual cilium but rather upon a central coördinating system and that the acceptance of a particle is the result of a chemical rather than a physical stimulus.

Dembowski (1922) was able to demonstrate that *P. caudatum* could ingest sulfur, which it usually ignores, as well as glass, porcelain, barium sulfate, and calcium carbonate by rubbing a small amount of egg yolk or starch into these substances.

Bragg (1936 b) observed that *P. trichium* showed a greater degree of selection of bacteria from carmine than *P. caudatum*. He reported that hemoglobin powder even when present in large amounts was completely rejected by *P. trichium* although the organisms continued to ingest bacteria from the medium and form bacterial vacuoles. Bragg also demonstrated that the addition of neutral red to a culture of *P. trichium* inhibited the selection of carmine thus supporting the suggestion regarding the involvement of a chemical factor.[5]

[5] Nelson (1933) found that the parasitic ciliate, *Balantidium coli*, readily ingested starch and blood cells but that inulin, yeast, carmine, Fuller's earth, sulfur, and certain small Protozoa were ingested in only small quantities. However, when inorganic particles were first coated with colloidal starch, they were ingested as readily as starch grains.

Lund (1941) holds that particles of all sizes are trapped in the paraesophageal fibrils with most of the fluid circulating back into the pharyngeal cavity. The particles of food then slip individually through the paraesophageal fibrils into the growing bulge on the dorsal side of the cytopharynx. This bulge which develops into the future food vacuole is bounded by a plasma membrane and appears to be retained by about 10 long postesophageal fibrils which are capable of undulatory movements (Fig. 23). More particles become concentrated by the paraesophageal fibrils as they slip through to increase the volume of the food vacuole. Lund observed that with the growth of the vacuole, the postesophageal fibrils constrict about its base then the vacuole is pressed posteriorly. Once the vacuole is released, the fibrils conduct it rapidly into the cytoplasm.

We are indebted to Mast (1947) for his comprehensive study of the food vacuole in four species of living paramecia, namely *P. caudatum, P. multimicronucleatum, P. aurelia,* and *P. trichium.* He observed that when feeding, many particles of all sorts from the surrounding medium are carried into the vestibulum but that only a small proportion of these are passed on through the mouth and into the cytopharynx. Some of the particles leave the vestibulum immediately or else are circulated for a while in the vestibulum before being discharged. Mainly the smaller particles and some larger ones leave the vestibulum, pass through the mouth, enter the cytopharynx where they pass rapidly downward to the esophageal sac. Some of the particles however are found to pass the mouth, come into contact with the ciliary band ("Vierermembran") on the aboral wall of the cytopharynx, and are passed out of the mouth again. Bills (1922) and

Bozler (1924 a, b) held that particles could be rejected even after they had passed down the cytopharynx. Mast however reports that once the particles have passed the first bend in the cytopharynx, they are not rejected but pass directly down into the esophageal sac. Both Bozler and Mast report the ingestion of starch grains and other large particles measuring 11–18 microns. Since the distal end of the cytopharynx measures only 5–7 microns in diameter at its smallest region, this structure must be capable of considerable expansion to permit the passage of such large particles.

Mast believed that selection among the larger particles takes place in the vestibulum and selection among the smaller ones in the proximal region of the cytopharynx and vestibulum. He was unable to explain the nature of the stimulating agent involved in the selection.

Mast and Lashley (1916) report that when swimming actively, paramecia ordinarily do not ingest anything. The writer finds them to be most active in feeding when they are least active in swimming (Fig. 52).

Under high magnification, one can observe easily that actively beating cilia of the oral groove and vestibulum sweep bacteria and other minute particles to the cytopharynx. Some of these particles are carried out again while others are swept down the ciliated cytopharynx and into the esophageal sac. Particles in the fluid of the sac rotate rapidly. Finally the sac increases in size due to the accumulation of ingested particles with fluid, then the sac separates from the cytopharynx to become a food vacuole or gastriole.

Nirenstein (1905) studied in detail the initial movements of the food vacuole of *P. caudatum* and his observations have been applied by many to all members of the genus. The work of Bragg (1935 a, b) has shown this interpreta-

FIG. 52. Individuals from a pure-line mass culture of *Paramecium caudatum* feeding on bacteria. When eating in this manner, they stop active swimming and "browse" around the food. (Wichterman)

tion to be erroneous; indeed not only are they different but he has shown that one is able to distinguish a species of paramecium by the initial movements of the food vacuole alone.

After a food vacuole is formed by constriction of the esophageal sac, a portion of the sac remains as a membrane over the distal opening of the cytopharynx. It is this membrane which bulges slightly into the cytoplasm to form a new esophageal sac. At first it contains relatively much fluid[6] and few particles but with increase in size the concentration of particles increases until the sac is almost filled with them. However the relation between amount of particles and of fluid in a sac varies. Long cilia which extend from the end of the cytopharynx and into the sac beat vigorously causing the particles in the sac to rotate (Mast, 1947).

Bütschli (1889) and Horning (1926) reported that no membrane is found at the end of the cytopharynx but that the droplets which pass into the cytoplasm from this structure are surrounded only by a surface film while Gelei (1934) believed a membrane to be present at the end of the cytopharynx, the membrane persisting as the esophageal sac enlarges.

[6] Eisenberg-Hamburg (1925), Frisch (1937), Mast (1947), and others maintain that all water which is excreted by the contractile vacuoles enters the body through the cytopharynx; Bozler (1924 a, b), Fortner (1926), and Müller (1932) believe that some of it enters through the pellicle, and Kitching (1934, 1936, 1938) in a number of experiments contends that all enters through the pellicle. In this regard the writer has for years made extended observations upon the behavior of living joined pairs of paramecia in conjugation and cytogamy (Wichterman, 1940, 1946 f). Relatively soon after the paramecia are joined in sexual process dedifferentiation of each cytopharynx occurs, then the structure disappears. However without a mouth and cytopharynx in each member of a pair, the contractile vacuoles are seen to function in a normal manner so that excreted water from the contractile vacuoles must now come through the pellicle.

Nirenstein (1905) believed the cytoplasm to exert a suction on the esophageal sac to cause this body to enlarge rapidly to its maximal size while Bütschli (1889) and Bozler (1924) believed the enlargement due to pressure of fluid forced into the sac by action of the pharyngeal cilia. They all agree that fluid is first observed in the sac which is followed by solid particles which are forced into the sac by action of cilia lining the pharynx.

4. Size, Shape, Structure, and Rate of Formation of the Food Vacuole.

SIZE. One who observes the process of ingestion in paramecium cannot help but notice the variation in size of food vacuole formation. Metalnikow (1912) found that when paramecia (probably *P. aurelia*) are transferred from a medium poor in digestible materials to one that is rich containing bacteria, milk, egg yolk, etc., the first vacuole formed is always huge. On the other hand paramecia in a poor medium with indigestible particles such as carmine, Chinese ink, etc. form abnormally small food vacuoles. Mast (1947) repeated the experiments of Metalnikow and found that the size of the vacuoles is not at all closely correlated with the composition of the medium. Dogiel and Issakowa-Keo (1927) reported that in salt solutions of $MgCl_2$, $MgSO_4$, and $FeSO_4$ with Chinese ink the food vacuoles formed in *P. caudatum* are long, tubular, and coiled but in solutions of $BaCl_2$ they are small and spindle-shaped. Mast repeated these experiments but was unable to confirm them.

By dissolving rice starch in saliva and placing *P. aurelia* in the dextrin thus formed Cosmovici (1931, 1933) stained the specimens with iodine. He found that a canal often convoluted or enlarged stained blue with the iodine which extended from the mouth to the anus. Cosmovici concluded from this and from investigations upon other ciliates that these organisms possessed a closed capillary digestive system through which food material is moved by waves of cytoplasmic contraction. Mast repeated these experiments but could not confirm them.

Frisch (1937) found that in flourishing cultures of *P. multimicronucleatum* the average diameter of food vacuoles measured 17.25 microns to 25 microns with 65.55 microns for the largest. As the cultures declined, the vacuoles became progressively smaller reaching finally 3.45 microns. The size of the food vacuoles so formed varied directly with the size of the paramecia.

When paramecia are placed in viscous substances such as polyvinyl alcohol or methyl cellulose, unusually large food vacuoles are formed. Lee (1942 a, b) maintains that the size of the food vacuoles in paramecium is independent of the hydrogen-ion concentration of the medium. The size of the food vacuoles appears to be caused by the following environmental factors namely, the quantity and quality of the food particles in suspension, their chemical composition, and the viscosity of the surrounding medium (Mast, 1947). To this must be added the physiologic state of the organism (Unger, 1926).

SHAPE. Newly formed food vacuoles in paramecium are usually nearly spherical, occasionally spindle-shaped (Nirenstein, 1905; Bozler, 1924 a, b; Dunihue, 1931; Gelei, 1934 a; Bragg, 1935 b, 1936 a; and Mast, 1947). Bragg (1936 a) reports that the food vacuole upon leaving the cytopharynx is usually spindle-shaped in *P. caudatum,* rarely so in *P. multimicronucleatum,* and never in *P. aurelia* and *P. trichium* and that its form is a species character. According to Mast the shape of the food

vacuoles depends largely if not entirely upon the viscosity of their content.[7]

STRUCTURE. The food vacuole consists of a fine elastic membrane which is capable of considerable stretching. This membrane bounds the vacuolar contents consisting of a small amount of the culture medium—mainly water—bacteria and other particles suspended in the culture. Mast (1947) reports that the membrane is impermeable to acid. King and Beams (1937) have shown that food vacuoles removed from paramecium and in water retain their form for over one-half hour. This is followed by a wrinkling of the vacuolar membrane then the breakdown of the vacuole. Investigations by many have shown that the membrane performs in much the same manner as the selective, semipermeable membranes of animal cells. It is shown to be permeable to water, certain dyes, enzymes, diffusible and digested food materials.

RATE OF FORMATION. Metalnikow (1912) and Lee (1942 a, b) contend that the rate of formation of food vacuoles is correlated with the hydrogen-ion concentration and the temperature of the medium. Bozler reported that the formation of a food vacuole requires four to five minutes if the particles in suspension are scarce but only one to two minutes if abundant; this is in general agreement with the findings of Frisch (1937).

5. Separation of Food Vacuole from the Cytopharynx.

In his detailed study of the food vacuole, Mast (1947) reported that the

---

[7] Dogiel and Issakowa-Keo (1927) immersed paramecium in salt solutions of Mg-SO₄, MgCl₂, and FeSO₄ and noticed the food vacuoles to become greatly elongated or sausage-shaped. At times these became swollen and extruded through the gullet. When placed in BaCl₂, the food vacuoles became small and spindle-shaped.

esophageal sac enlarges then becomes nearly spherical with the major axis directed backward. Next, the substance in the sac rotates strongly and the particles in suspension vibrate vigorously and become much concentrated. After the sac has become about twice as wide as the pharynx, the sac slides posteriorly from the diagonal end of the pharynx and becomes pear-shaped, a nipple being drawn out on the vacuole as it leaves the pharynx then a small portion of the esophageal sac remains as a membrane over the opening in the pharynx which bulges out into the cytoplasm slightly, forming a shallow sac which soon enlarges to form a new vacuole (Fig. 51C). After the nipple which connects the newly formed food-vacuole with the tip of the diagonal end of the pharynx has broken, the vacuole moves rapidly through the cytoplasm nearly to the posterior end of the body. On the way it creates currents in the cytoplasm, turns through approximately 270° and becomes spherical. When it has reached the end of this course, it stops momentarily, then slowly passes forward near the aboral surface toward the anterior end of the body, and then back along the opposite surface, to the anus. Mast observed considerable variation in these phenomena. At times he noticed that the cytoplasm around the vacuole flowed slowly backward and appeared to elongate the vacuole, occasionally with such force as to seemingly pull it away from the pharynx as reported earlier by Gelei (1934).

Stein (1859-78), Bütschli (1887-89) and Bragg (1935 b, 1936 a) reported that contraction of the distal end of the cytopharynx and cyclosis are involved in separation of the food vacuole. Cyclosis is the characteristic circulatory streaming of the protoplasm. Mast failed to observe any change in the cytopharynx and has noted the food vacuole to leave the cytopharynx in the total absence of cyclosis. Nirenstein (1905) contends that cyclosis carries the food vacuole off the cytopharynx and into the cytoplasm; Bozler (1924 a, b) that it is not carried off by cyclosis but that there is a periodic backward streaming of the cytoplasm adjoining the "Schlundfaden." More recently cyclosis in paramecium has been studied by Luboska and Dembowski (1950). Kalmus (1931) and others believe that surface tension plays an important role in the formation of the food vacuoles but this is discounted by Mast. Lund (1941) states that the food vacuole is separated from the cytopharynx and forced posteriad by the action of the postesophageal fibrils and Mast lends support to this opinion.

Bozler believed that the separation of the food vacuole from the cytopharynx is initiated by contact of particles with the inner surface of the vacuolar membrane and that paramecia will not form food vacuoles in a medium devoid of particles. On the other hand Schewiakoff (1893) maintained that particles are not necessary for food vacuolar formation. For P. trichium, Bragg (1935 b) stated that contact of large particles with the inner surface of the vacuolar membrane is always immediately followed by separation of the vacuole from the cytopharynx but that many of the vacuoles formed had no large particles. For P. caudatum he reported that many of the food vacuoles formed had no large particles and concluded that large particles are not necessary. Mast (1947) had observed in four species of Paramecium, food vacuole formation without large particles entering into their formation.

The fact remains that little evidence is available concerning the processes involved in the causes of separation of the food vacuole from the cytopharynx.

6. Behavior and Fate of the Food Vacuoles in the Cytoplasm.

Upon leaving the cytopharynx the food vacuole passes rapidly posteriad on a fixed course which appears to be due to the action of the esophageal fibers. It then proceeds slowly—due to cyclosis—on a varied course to the anus. The general direction of the path taken by the food vacuole is first along the aboral side toward the middle or anterior region of the organism. Fortner demonstrated that the circulation of the food vacuole may be limited first to the cyclosis in the lower half of the organism before the food vacuole is passed into the larger stream of endoplasm until finally egested (Fig. 53).

According to the physiologic state of the organism and the type of food that has been ingested, the food vacuole remains in the body for one to three hours at room temperature. It appears that food vacuoles which contain indigestible materials remain in the organism for a shorter length of time than those with digestible food substances.

According to Nirenstein (1905) and others the food vacuole upon leaving the cytopharynx becomes spherical, decreases in size to about one tenth of its original size, then increases until it is somewhat larger than at first. This feature has been followed closely by Mast (1947) (Fig. 54) who observed that the vacuole, after it became free from the pharynx, rapidly became spherical and slowly decreased in size to nearly one-twentieth and then very rapidly increased until it was nearly as large as originally. As the vacuole decreased in size, the bacteria in it died and aggregated in numerous small clumps. Fluid passed from the vacuole into the surrounding cytoplasm until little was left leaving the vacuole so packed with solid particles that the surface was very irregular. According to Mast the numerous "neutral-red granules" at the surface of the vacuole when it left the pharynx, disappeared during its rapid enlargement. He reported that all these phenomena varied greatly in extent and time; for example, under some conditions there was no perceptible change in the size of the vacuoles at all.

Mast believed the phenomena in respect to change in size of the food vacuole of paramecium similar to those occurring in the *Peritricha* as reported by Mast and Bowen (1944). Accordingly it is maintained that the loss of fluid which results in the decreased size of the food vacuole is due in part to difference in the osmotic concentration of fluids in the vacuoles and the cytoplasm: the osmotic concentration of the cytoplasm being higher than that of the food vacuole.[8] Also the inward pressures of the stretched elastic membrane of the vacuole is in part a factor. It is claimed that the increase in viscosity of the food vacuole is then correlated with an increase in its acidity.

7. Hydrogen-ion Concentration of the Food Vacuole.

To determine the pH in a food vacuole, chemical substances called indicators are used which change in color according to the acid or alkaline nature of the vacuolar contents.

As early as 1879, Engelmann observed the ingestion of blue litmus particles in *P. aurelia* and noted that the particles became red. He concluded that the observed change demonstrated that the protoplasm is acid; he failed to recog-

[8] Fortner (1933) added numerous crushed bacteria to a culture of *P. caudatum* and observed the retention of water in the food vacuoles with no appreciable decrease in size. It is likely that the osmotic concentration of the substance in the food vacuoles was as high or even higher than that of the surrounding cytoplasm.

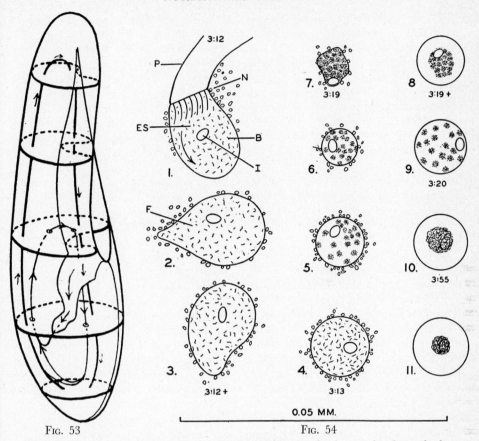

Fig. 53

Fig. 54

Fig. 53. Cyclosis in Paramecium and paths taken by food vacuoles in endoplasm. (Kalmus)

Fig. 54. Camera outlines of a food vacuole in a specimen of *Paramecium aurelia,* showing changes in form, size, and content, and changes in the position of neutral-red granules at the surface. (*P*) Pharynx; (*ES*) esophageal sac; (*F*) food vacuole; (*B*) bacteria; (*I*) indigestible body; (*N*) neutral-red granules; 3:12–3:55, time.

Note that the vacuole rapidly became spherical then decreased slowly but greatly in size, then increased rapidly and greatly. Note also that the numerous neutral-red granules at the surface of the vacuole all disappeared during its rapid enlargement and that the bacteria appear to have died and agglutinated during the decrease in size. (Mast)

nize that this color change was correlated with the content of the food vacuoles. Metchnikoff (1889), LeDantec (1890), Greenwood (1894), Greenwood and Saunders (1894), Nirenstein (1905), and many others have investigated the food vacuole and found that the pH of its contents increases greatly, then decreases again.

In their studies upon *P. aurelia* and *P. caudatum,* Shipley and DeGaris (1925) reported that the forming food vacuole is filled with an alkaline fluid (the so-called "preliminary alkaline phase") which later becomes acid and then alkaline again. Their work suggested that an alkaline substance is secreted by the cytopharynx. Shapiro

(1927) repeated and extended these experiments and declared verification of the "preliminary alkaline phase"; contrariwise, Dunihue (1931) was unable to confirm it. Under high magnification (1200 x) Mast (1947), after numerous observations and the use of several indicator dyes, failed to obtain evidence of alkalinity in the forming food vacuole and stated that the purplish pink color observed in the lower magnifications was due to refractive phenomena, instead of change in acidity. Mast therefore emphatically declared there is no preliminary alkaline phase of the food vacuole in paramecium.

As reported by Mast, the food vacuole on its course through the body usually decreases greatly in size while the acidity of its content increases greatly. The vacuole then enlarges very rapidly while the acidity of its content decreases greatly. He found considerable variation at times with no perceptible changes but in others the acidity in some vacuoles increased to a maximum at least as high as pH 1.4 and then decreased in some to approximately pH 7.8. The change in acidity was found to be definitely correlated with the change in size of the food vacuole.

Bozler made observations on the food vacuole of paramecium containing yeast cells stained with Congo red and concluded that the acidity of the food vacuole contents increased approximately to pH 3. On the other hand Nirenstein (1925) reported the hydrogen-ion concentration to approach pH 1 and this was confirmed by Kalmus (1931). Shapiro (1927) studied paramecia in culture fluid containing litmus, Congo red, and phenol red and concluded that at first while forming, the food vacuole has a pH of about 7.6, then increases to a maximum of pH 4 then decreases to pH 7.

Mast performed numerous experiments dealing with the hydrogen-ion concentration of the food vacuole. He first used neutral red crystals in acetate buffers and later the indicator neutral red (pH 6.8, red–pH 8, auburn). He noted that as the vacuole decreased in size, the hydrogen-ion concentration of its content increased in some cases to more than pH 4 and then as it increased in size decreased to approximately pH 8. When paramecia were placed in culture fluid containing numerous bacteria and an excess of Congo red (pH 3, blue–pH 5, orange) the fluid in the newly formed food vacuoles became light yellow, the undissolved particles of Congo red, yellowish brown and the bacteria colorless. As the vacuoles decreased in size after they had left the pharynx, fluid in them decreased greatly in quantity and became light blue, then the bacteria died and became dark blue and the particles of Congo red became dark blue. As the vacuoles increased in size, the fluid in them increased greatly in quantity and became orange, then the bacteria and the particles of Congo red soon became orange and remained so until they were eliminated. When the food vacuoles had become smallest, the color of their content resembled that of the buffer pH 3.2, hence Mast concluded that the maximum acidity with Congo red to be pH 3.2 and this was reached when the vacuoles had decreased to minimum size.

In addition, Mast used the indicators thymol blue (pH 1.2, red–pH 2.7, yellow) and meta cresol purple (pH 1.2, red–pH 2.8, yellow). In thymol blue buffered solutions yeast cells to be ingested became lemon yellow at pH 3.5. After the yellow yeast cells left the pharynx in food vacuoles, the yeast cells in some became pink like those in buffer pH 1.4, then as the vacuole increased in size became yellow again. The results with meta cresol purple were found to be the same as with thymol blue. Mast asserts that the relatively high concentration of acid in the food vacuole without injury to the neighboring cytoplasm is convincing evidence in support of the contention that the vacuolar membrane is impermeable to acid. It should be remembered that culture fluid even as low as pH 3.5 is fatal to paramecia.

The increase in acidity is believed to be due to secretion of acid by the cytoplasm adjoining the vestibulum and the pharynx and to impermeability of the vacuolar membrane to hydrogen-ions and loss of water while the decrease in acidity is due to entrance of alkaline fluid from the cytoplasm. This increase in acidity probably causes hydrolysis thereby increasing the osmotic concentration which results in an inflow of fluid containing digestive enzymes (Mast).

To test the maximum alkalinity of the content of the food vacuole in paramecium, Mast used Nile blue (pH 7, blue–pH 8, purple); phenol red (pH 6.8, yellow–pH 7.4, pink); bromthymol blue (pH 6, yellow–pH 7.6, blue) and cresol red (pH 7.2, yellow–pH 8.8, red). The indicators were dissolved in culture fluid which contained bacteria and yeast cells and the color of the food vacuole contents compared to the color of the buffer solutions. It was found that the maximum alkalinity of some of the food vacuoles may be nearly pH 7.8 but much lower (pH 7.2–7.4) in most of them. The maximum alkalinity varied somewhat with the acidity of the fluid ingested.

According to Mast the increase in alkalinity in the food vacuole in paramecium is brought about by entrance of alkaline fluid from the cytoplasm; the cytoplasm adjoining the food vacuole secreting neither acid nor base. On the other hand a large number of investigators hold that the acid and the base in the food vacuole are secreted by the adjoining cytoplasm.

8. Secretion, Digestion, and Reserve Foods.

Much has been written about digestion in paramecium but accounts are contradictory and certain aspects of the process are based upon broad assumptions. That so little is known is somewhat surprising since in recent years the research upon enzymes has been vigorous. Microdissection, newer centrifugal and critical microchemical methods are now available which permit the investigation of chemical reactions in a single cell or even a constituent of it.

Enzyme production and digestion in paramecium have been correlated with discrete granules, globules, or other fine elements in the protoplasm. Accordingly these have been referred to as neutral red granules or globules, digestive granules, segregation granules, mitochondria, chondriosomes, etc.

Prowazek (1897) discovered the neutral red granules and their characteristic manner of gathering near the surface of the forming food vacuole in paramecium. He maintained that they contained digestive enzymes ("Fermentträger") but failed to explain how the enzymes in the neutral red granules enter the vacuole. Nirenstein (1905) accepted the interpretation of Prowazek but in addition asserted that the granules came from the cytoplasm and passed through the vacuolar membrane thereby carrying the enzyme directly into the vacuole. He regarded the granules as carriers of a tryptic ferment; Roskin and Levinsohn (1926) demonstrated the oxidase reaction in them. Nirenstein's explanation is accepted by many (Rees, 1922 b; Bozler, 1924 a, b; Fortner, 1926 b; Volkonsky, 1929; Müller, 1932, and others).

In contradiction, Khainsky (1910) obtained no evidence to indicate that granules or globules passed from the cytoplasm into the food vacuoles of *P. caudatum*. Instead he maintained that neutral red globules formed in the vacuole and passed out through the vacuolar membrane into the cytoplasm. He also stated (1910) that the food vacuole in paramecium is acid during the entire period of protein digestion and becomes neutral to finally alkaline when the solution of the food substance is terminated.

Also in *P. caudatum* Koehring (1930) concluded that the granules do not enter the food vacuole but "bombard" the newly formed food vacuole. In the same species Dunihue (1931) asserted that the surface of the forming food vacuole becomes closely packed with neutral red globules throughout most of the digestive period but do not enter the vacuole. According to Fortner (1926 b, 1928 c) the neutral red granules in paramecium originate in the cytoplasm adjoining the macronucleus and enzymes are closely associated with them. He reported the granules to contain protein and fat and to decrease greatly in number during starvation. Horning (1926) maintained that the mitochondria carried digestive enzymes into the food vacuole.

Mast (1947) studied the neutral red granules at the surface of the food vacuole and obtained no evidence that granules or globules pass in or out of the vacuolar membrane. He contended that the granules originate in substance ingested by paramecium and that those in the cytoplasm do not enter the food vacuole. It is suggested by some that "digestive granules" or "segregation granules"—certain types of neutral red granules—may synthesize proteins and similar materials.

Hartog and Dixon (1893) appear to have been the first to study the digestive enzymes of Protozoa (*Pelomyxa*) *in vitro;* their extract contained an enzyme considered by them to be a pepsin. Mouton (1902) studied the enzymes of *Amoeba zymophila* in a glycerin extract and concluded that the protease was very close to the trypsin of the vertebrates; in reality it was a cathepsin. Mesnil and Mouton (1903) obtained an enzyme extract from *P. aurelia* and considered the enzyme also to be proteolytic and similar to trypsin (cathepsin).

As a result of the investigations of Sawano (1938), and Hollis and Doyle (1937) on proteolytic enzyme systems in paramecium, it has been shown that the ciliate is capable of hydrolyzing certain proteins and peptides to amino acids. Holter and Doyle (1938) reported the presence of the enzyme catalase in paramecium.

Using the methods of the Willstätter school, Sawano (1938) obtained enzyme preparations from *P. caudatum*. He reported that the proteinase of paramecium is a cathepsin or cathepsin-like one and not trypsin or pepsin. Cathepsin, which is found in the tissue cells of vertebrate animals, acts most effectively at the isoelectric points of proteins hence its optimal pH is between that of pepsin and trypsin. Sawano also reported that amino-polypeptidase and dipeptidase which are commonly found in organisms are demonstrated abundantly in paramecium. Hollis and Doyle (1937) found that extracts of paramecium contained a dipeptidase. The amylase found by them was activated by sodium chloride and had a pH optimum of 6.0.

It is of interest to report that no one

has observed any evidence of digestion in the food vacuole[9] during its acid phase. It is possible however that invisible preliminary changes occur in this stage as suggested by Nirenstein and that the acid in the food vacuoles marks a beginning of the digestive process although it has been clearly demonstrated that digestion takes place principally in the food vacuoles during the later alkaline phase.

Meissner (1888) and others have demonstrated with Protozoa that starch granules in the food vacuoles show corrosive effects, lose their double refraction in polarized light, and stain red with iodine indicating a change of starch to dextrin. Fabre-Domergue (1888) and Cosmovici (1931 a) with special techniques reported the presence of several different dextrins in paramecium. According to Burge, Estes, Wickwire, and Williams (1927) paramecia use the three simple sugars, dextrose, levulose, and less rapidly, galactose as shown in their experiments. Insulin increased the utilization of all three sugars in paramecium just as in higher animals. Pituitrin also increased utilization of these sugars while thyroxin decreased it. Small amounts of Adrenalin increased and large amounts decreased the sugar metabolism of the paramecia. Burge and Estes (1928) reported that the inorganic salt, dipotassium phos-

phate, was found to increase greatly sugar utilization in *P. caudatum*.

Barbarin (1939) found that in an atmosphere of hydrogen, the glycogen reserve of *P. caudatum* was reduced to one-half in 10 hours and to an insignificant amount in 20 hours. According to Zweibaum (1921) the glycogen content of conjugants of paramecium was markedly different from that of nonconjugating individuals.

Pringsheim (1928) found that paramecium utilized wheat starch better than rice starch. Leichsenring (1925) in observing that the addition of dextrose, galactose, sucrose, lactose, maltose, dextrin, and soluble starch to starved *P. caudatum* increased the oxygen consumption, showed that the organisms could utilize these polysaccharides, mono- and disaccharides in solution. It would appear that the monosaccharides are stored as glycogen or other complex polysaccharides, are converted to fat, or are oxidized. Cunningham and Kirk (1941) found that both starch and glucose exert a protein-sparing action and concluded that protein is used as a source of energy by the ciliates under conditions of starvation.

The work of Kidder and his associates dealing with nutrition in the ciliates *Tetrahymena geleii* and *T. vorax* is of considerable importance. In the six strains of *T. geleii* and two strains of *T. vorax* that were investigated by Kidder and Dewey (1945) for their ability to ferment carbohydrates, all produced acids from glucose, fructose, mannose, dextrin, glycogen, and starch. None fermented pentoses, disaccharides other than maltose, inulin, or polyhydric alcohols while three produced acid from galactose and two from lactose.

Carbohydrate reserves in the form of soluble glycogen are dissolved in the protoplasm; their existence in fixed specimens of paramecium has been re-

[9] Living microörganisms such as bacteria when ingested by paramecia die soon after the food vacuole is formed. Metalnikow (1904, 1912) observed that bacteria were killed in about 30 seconds after the food vacuole was detached from the cytopharynx of *P. caudatum*. Fortner (1933) believes that the ingested microörganisms are killed by a toxic substance originating from the neutralred granules in *P. caudatum*. Mast reports that there is considerable evidence showing that the pharynx may secrete a toxic substance which passes into the forming food vacuole. It is very likely that this so-called killing substance is in reality the acid in the food vacuoles as well as the digestive enzymes.

ported by Rammelmeyer (1925). Glycogen granules of the cytoplasm can be stained with iodine but with starvation, the granules are utilized and disappear.

There is still difference of opinion concerning the digestion of fat. Paramecia can ingest fat and its presence demonstrated by Sudan III and osmium tetroxide. Many observations show the close relationships between carbohydrate and fat metabolism in paramecium. According to Pütter (1905) fat is produced from glycogen by paramecium under anoxybiotic conditions and this belief of transformation from carbohydrate to fat is shared by von Brand (1934, 1946) and Barbarin (1939). Nirenstein (1910) held that starch fed to paramecium was converted to fat then stored as such in the body and this was supported by the work of Pringsheim (1928).

It would appear that fat can form as an end product of carbohydrate metabolism in paramecium under anaerobic conditions but not necessarily under aerobic ones.

Nirenstein (1910) believed that paramecia richly supplied with fat show a high fat composition and that the fat inside is hydrolyzed in the food vacuole and resynthesized thereby serving as food and reserve material. After making detailed observations on bacteria, yolk granules, fat globules, and starch grains in the food vacuoles of *P. caudatum,* Nirenstein (1905) found no indication of digestion in any of these substances until after the fluid in the vacuoles had become alkaline. After this had occurred the bacteria and yolk granules gradually decreased in number and finally disappeared entirely while some of the starch grains corroded. Nirenstein's observations upon paramecium have been confirmed by Pringsheim (1928), Zingher (1933), and Mast (1947).

Nutritional studies have been made on the ciliate, *Tetrahymena* to determine its amino acid requirements. Thus Peterson (1942) has shown that only peptones exceeding a minimum molecular size are adequate. Others have grown this ciliate successfully on amino acid mixtures. Klein (1943) obtained satisfactory growth on a mixture of 15 amino acids of which the omission of any one reduced the growth rate.

In the investigations upon *Tetrahymena* and other ciliates, it has been shown that the nutritive requirements vary not only from one species to the next but from one strain or race to the next within a species. It is more than likely that the same conditions will be found in paramecium.

From the nutritional work done so far on ciliates, it would appear that while substances in the food vacuole are digested to form monosaccharides, amino acids and other breakdown products, protein digestion need not possibly proceed beyond the peptone stage in certain cases.

Lipoid reserves are those lipoid granules which are formed during active feeding and utilized by the organism. They are frequently lost during hunger but Zingher (1933) reported that old cultures of paramecium show larger amounts of fat than fresh cultures. Weatherby (1927) believed the presence of unusually large amounts of visible fats in such cultures to be due possibly to the alkalinization of the protoplasm in *P. caudatum.*

Zweibaum (1921) reported that fat in paramecium is stored under conditions of low oxygen tension only to be lost when the oxygen tension is restored; the rate of loss being dependent upon the temperature. He also observed an increase in the amount of fatty acid just prior to conjugation. Parasitic microörganisms in the macronucleus of paramecium result in the production of large amounts of crystals and visible lipoids in the cytoplasm according to Fiveiskaja (1929).

Paramecia can ingest fat globules but it is maintained by Meissner (1888) and Staniewicz (1910) that they leave the body undigested. Mesnil and Mouton (1903) were unable to find lipoid enzymes in paramecium. Cunningham and Kirk (1941) report that olive oil

The work of Nirenstein (1910) indicates that protein may be utilized in the formation of fat reserves as observed when paramecia are fed with heat-coagulated egg albumin.

Small bodies or granules—known as protein reserve bodies—are found in the protoplasm of paramecium and have been referred to in the literature as chromidia, protein granules, basophilic granules, volutin, etc. Mitochondria,

Table 12

THE DEGREE OF ABSORPTION OF SPECIFIC AMINO ACIDS IN *Paramecium caudatum* (Emery, 1928)

	Per cent
Mixture of different amino acids (except arginine)	48.3
Glutamic acid hydrochloride	45.6
Cysteine hydrochloride	26.3
Aspartic acid	25.1
Tyrosine	17.7
Arginine	15.9
Alanine	15.5
Glutamic acid	13.2
Leucine	12.0
Glycocoll	9.6
Tryptophane	9.6
Phenylalanine	7.7

did play some role in the metabolism of *P. caudatum* as shown by their experiments.

Proteins are digested by the action of hydrolases which attack either the protein molecule or intermediate products of hydrolysis finally producing amino acids.

Burge, Estes, Wickwire, and Williams (1927) found that the amino acids glycocoll, tyrosine, norleucine, isoleucine, cystine, and a mixture of naturally occurring amino acids were utilized by paramecium in their experiments. As mentioned earlier, Emery (1928) found that a measurable quantity of amino acids is utilized in place of the normal bacterial food in *P. caudatum* (Table 12).

which are commonly found in paramecium, are claimed by many to be composed of both lipoid and protein material. Since mitochondria are found in close proximity to the food vacuole, some express the opinion that they are related to digestion. It is generally agreed that mitochondria possess enzymes, especially oxidizing ones.

Cytoplasmic elements called "segregation granules" are bodies able to concentrate and store proteins. Dunihue (1931) correlated the decrease in numbers of these granules with starvation. Other cytoplasmic granules in *P. caudatum* called "digestive granules" and stainable with neutral red are reported to be high in ash by MacLennan and Murer (1934) (Fig. 43).

9. Inanition.

If paramecia are deprived of all food in the environment, soon all the reserve materials of the cytoplasm are exhausted and the inanition phenomena as described by Wallengren (1902) appear. Upon utilization of the contained reserved food materials, the paramecia become transparent and there follow volume changes of the macronucleus, vacuolization of the cytoplasm, loss of cilia, and degeneration of specialized structures. When richly supplied with food in the form of suitable bacteria, the starved paramecia may be restored to normal. As reported earlier, Dunihue (1931) correlated the decrease of cytoplasmic elements called segregation granules with starvation.

Wallengren (1902), Enriques (1909), and Lipska (1910, 1910 a) reported that the sizes of paramecia which were undergoing starvation changed. Jones (1933) noted that the average length of *P. multimicronucleatum* decreased as the division rate increased and vice versa, prior to the inanition period (Fig. 55).

Under conditions of starvation, Cunningham and Kirk (1941) concluded that in the presence of the utilizable substrates, glycine, fibrin, and olive oil, the principal nitrogenous end product of the metabolism of *P. caudatum* is ammonia. Barbarin (1940) found that susceptibility of *P. caudatum* to lethal concentrations of KCN, $CuSO_4$, KCl, $CaCl_2$, and neutral red was increased during a protracted period of starvation.

The role of starvation in the conjugation of paramecium is discussed on page 369.

10. Egestion.

After the digestion of food materials in the food vacuoles, egestion or defecation occurs very rapidly at the anus (cytopyge, cytoproct, anal spot) which is located terminally or subterminally on the oral surface of the body. While some small food vacuoles may be egested in their entirety usually one observes a fast stream of fine particles to be passed out of the anus during egestion. In life, the anus can be seen only while the organism is actively egesting material.

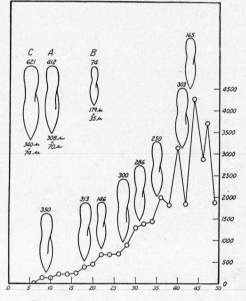

FIG. 55. The "average animal size" and the "population per cc." in culture 3. The abscissa represents time in days. The ordinate represents the population in a cubic centimeter of culture. The paramecia directly above the population curve are drawn to scale to represent the "average animal" on the days pointed to by the posterior tips of the animals. The average volume of one animal in thousands of micra is printed immediately above. (A) Represents the largest paramecium which was measured in Culture 3. (B) The smallest measured in Culture 3 or any of the series of cultures. (C) The largest individual measured in this clone. (A, B, and C) Members of one clone of *P. multimicronucleatum*. (Jones)

B. Growth and Factors Influencing Growth

1. Measurement and Growth of Individual Paramecia.

Measurements of individual paramecia are generally made by placing either living or fixed specimens in a drop of fluid on a microscope slide. A cover-glass placed over the drop should be supported by small pieces of glass, such as finely drawn glass capillary rods, to prevent the cover-glass from crushing the specimens. With the use of a calibrated ocular micrometer which is placed in the eyepiece of the microscope, it is comparatively simple to make the measurements of length and breadth.

Most investigators have used fixed and preserved specimens to make their measurements. Care must be exercised in the killing and fixing process since improper fixation frequently results in greater shrinkage. Worcester's fluid was preferred by Jennings for his detailed measurements while Gause, the present writer, and many others used Schaudinn's fluid with success. Generally, even so-called reliable fixing agents introduce a shrinkage factor but this may be so slight as to be considered negligible.

It is indeed surprising that comparatively few investigators have measured living specimens.[10] When living paramecia are first placed on a slide, the organisms generally move too quickly to be measured accurately. However, if a small amount of bacterial sediment from the culture is placed on the slide with the drop containing paramecia, one observes that after about five min-

utes or so, the organisms become fairly inactive as they feed on the bacteria (Fig. 52). It is at this time that they can be measured with considerable accuracy.

In his study of growth in paramecium, Jennings (1908) noted that whenever a large and small specimen (belonging to a given group) were isolated at the same time, the large specimen as a rule divided first. This suggested that size differences in a culture may be largely matters of growth and that small specimens may be young ones that had recently been produced by fission; variations in size such as shown in his frequency polygons may be largely individual growth differences.

Jennings therefore studied growth by following the changes of form and size in living specimens and by a comprehensive statistical examination of the dimensions of individual paramecia of approximately known age. In his classical study entitled "Heredity, Variation and Evolution in Protozoa," Jennings (1908 a, b) is responsible for presenting the first detailed measurements upon paramecium (*P. aurelia* and *P. caudatum*) and for treating the data statistically. Although this work was done over 40 years ago, his methodology, precision of observation and reasoning make it worth careful study even today. His statistical methods and analyses, which form the foundation of his study of growth and variation, are given below in brief; the original work should be consulted for additional information.

For most of the tables, the constants computed were the following: the mean, standard deviation, and coefficient of variation, for length and for breadth; the mean index or ratio of breadth to length; and the coefficient of correlation.[11] Computations were

---

[10] Adolph (1931) has shown that the growth in size of microörganisms even smaller than paramecium can be measured from motion pictures. The method is worthy of application to paramecium.

[11] For measurement techniques and the making of statistical analyses, the book en-

made by the aid of seven-place logarithms and of Crelle's and Barlow's tables.

The standard deviation was computed using the formula:

$$\sigma = \sqrt{\frac{\Sigma\ (V - V_\circ)^2}{n}\ - r_1{}^2\ - \frac{1}{12}}$$

The *mean index* is the mean of the quotient $\frac{\text{breadth}}{\text{length}}$: it shows essentially what percentage the breadth is of the length. This mean was found without computing the index for each individual by the following formula:

$$i = \frac{A_B}{A_L}\ (1 + C_L{}^2 - rC_B\ C_L)$$

Where i is the mean index, A_B is the mean breadth, A_L the mean length, C_B the coefficient of variation for breadth,

titled "Quantitative Zoology" by Simpson and Roe (1939) will be found useful.

C_L the same for length, and r is the coefficient of correlation between length and breadth.

In his final analysis, Jennings was able to show not only that his calculations disclosed a size difference existing between *P. aurelia* and *P. caudatum* at a time when both species were considered by many to be variations of the one species, but also that paramecia rapidly multiplying and growing in culture should be more variable than those which are stationary. He proceeded to test this last-mentioned theory and using the calculations just described, measured *P. caudatum* from a given culture showing rapid multiplication and from a culture of the same clone in which multiplication was not in rapid progress. His results, which are shown in Tables 13 and 14, were then graphically represented in the polygon of Fig. 56. The tables and figures are also useful in showing Jennings' general method for assembling his data and arriving at his calculations.

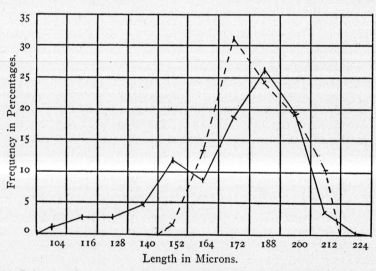

FIG. 56. Polygons of variation in length for a culture of descendants of *P. caudatum* that is rapidly multiplying and one that is not. The continuous line represents the rapidly multiplying culture of Table 14; the broken line the stationary culture of Table 13.

Table 13

MEASUREMENTS OF *P. caudatum* FROM A CULTURE IN WHICH MULTIPLICATION IS NOT RAPID (STATIONARY CULTURE) (JENNINGS, 1908)

Length in Microns

Breadth in Microns	148	152	156	160	164	168	172	176	180	184	188	192	196	200	204	208	212	Total
32	1			2		1	1	3	1									9
36					2	6	5	1	2	2	2							20
40		1		1		3	8	4	4	4	3	1	3	3				35
44				1		1	2	2	2	3	3	4	1	3				22
48					1		1		3	2	1	3	4	3	1	4	2	25
52									1	1		1	2	1	2	3	4	15
56									1			2				1		4
60									1	1				1	2			5
	1	1	0	4	3	11	17	10	15	13	9	11	10	11	5	8	6	135

Length—Mean, $185.008 \pm .836\mu$ Breadth—Mean, $43.556 \pm .392\mu$
St. Dev., $14.420 \pm .592\mu$ St. Dev., $6.748 \pm .276\mu$
Coef. Var., $7.794 \pm .324$ Coef. Var., $15.490 \pm .651$

Mean Index, 23.517 per cent.; Coef. Cor., $.5955 \pm .0375$.

Table 14

MEASUREMENTS OF *P. caudatum* FROM THE SAME CULTURE AS SHOWN IN TABLE 13 IN WHICH MULTIPLICATION IS RAPID (JENNINGS, 1908)

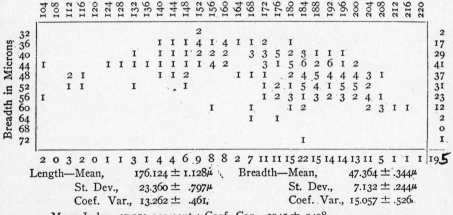

Breadth in Microns	104	108	112	116	120	124	128	132	136	140	144	148	152	156	160	164	168	172	176	180	184	188	192	196	200	204	208	212	216	220	Total
32													2																		2
36									1	1	1	4	1	4	1	1	2		1												17
40								1	1	1	1	2	2	2		3	3	5	2	3	1	1	1								29
44	1						1	1	1	1	1	1	1	1		4	2		3	1	5	6	2	6	1	2					41
48					2	1				1	1	2				1	1	1		2	4	5	4	4	4	3	1				37
52					1	1			1					1					1		2	1	5	5	1	5	5	2			31
56	1														1				1		1	2	3	3	3	2	3	2	1		23
60																	1		1			1	2				2	3	1	1	12
64																			1	1											2
68																															0
72																							1								1
	2	0	3	2	0	1	1	3	1	4	4	6	9	8	8	2	7	11	11	15	22	15	14	14	13	11	5	1	1	1	195

Length—Mean, $176.124 \pm 1.128\mu$ Breadth—Mean, $47.364 \pm .344\mu$
St. Dev., $23.360 \pm .797\mu$ St. Dev., $7.132 \pm .244\mu$
Coef. Var., $13.262 \pm .461$ Coef. Var., $15.057 \pm .526$

Mean Index, 27.153 per cent.; Coef. Cor., $.3945 \pm .0408$.

From a study of the tables and figure it is evident that the variability has become much greater in the rapidly growing culture with the range of variation of length in the stationary culture being from 142 to 212 microns while in the rapidly multiplying culture being from 104 to 220 microns. It will be noted that in the latter, the range has almost doubled in extent; the coefficient of variation in length has also almost doubled changing from 7.794 when the culture was stationary to 13.262 when it was rapidly multiplying. The range of variability for breadth has likewise increased considerably although the coefficient of variability shows little change. The correlation between length and breadth has become considerably less in the rapidly multiplying culture decreasing from .5955 to .3945. The mean length has slightly decreased and the mean breadth slightly increased in the growing culture.

GROWTH OF A SINGLE PARAMECIUM. In the production of two young daughters, the dividing *P. caudatum* can be first recognized by being widest in diameter at the approximate center of the body as the organism takes on the appearance of a fat spindle. The dividing mother appears to be a less forceful swimmer[12] than a mature, non-dividing specimen. In the early stage of division one can recognize a slight constriction encircling the center of the body which gradually becomes more pronounced. Concomitant with this constriction is the swing toward dedifferentiation of the oral groove and ingestatory structures. Prior to this constriction, the macronucleus has begun to elongate for its amitotic division as the micronucleus prepares to divide mitotically. Other specialized cytosomal changes occur which with the behavior of nuclear structures are considered in greater detail on page 254.

Jennings reported that the constriction does not pass squarely across the body but obliquely, being farther back on the oral surface. Therefore, when the two daughters are measured separately, they will appear to differ in length according to the place where the measurement is taken. In reality, in early stages of fission, the unseparated daughters are equal to one-half the length of the dividing mother. The posterior half of the dividing mother is usually slightly broader than the anterior half and with progressive deepening of the constriction, the two halves lengthen until they separate.

[12] It is the writer's experience that one finds a greater percentage of dividing specimens on the bottom of the culture container than elsewhere.

At the beginning of fission, before the constriction reaches a depth of about 10 microns, there is little relation between the length of the body and depth of constriction, indicating that the halves have not as yet begun to lengthen. The length of the youngest at this period may be taken as that characteristic for the young individuals in their earliest recognizable condition before growth has begun.

In the unseparated halves (potential daughters) after lengthening had begun, the length of the halves increased as the constriction deepened. As the length thus increased, the breadth decreased. As reported by Jennings, with each increase of 10 microns in the depth of the constriction, the breadth of the body decreased 2.630 microns and a large part of the decrease in breadth occurred in the first stages of constriction.

While the length is increasing, the breadth is decreasing and growth tends to decrease the correlation between length and breadth or even to make it negative.

In summation, it may be stated that before fission, the body thickens and becomes shorter. When the constriction first appears the anterior and posterior halves differ in form. Then as the constriction deepens, the two halves become longer. The anterior half becomes more pointed and slightly more slender than the posterior half. Near the end of fission, the two halves are about the same length with the posterior half about 4 microns longer. Upon separation, the two recently separated daughters can be distinguished from each other by the slight size difference and easily by the fact that the anterior one has a pointed, pear-shaped form (Figs. 78 and 79).

It generally requires slightly more than a half hour from the earliest appearance of the constriction until separation into two daughters. As would be expected, temperature is a factor in this time relationship.

Unseparated young may be divided into two classes: one consisting of those specimens before lengthening has begun and the other consisting of specimens after lengthening has begun. It was shown by Jennings that in the study of many hundreds of dividing specimens, the mean length of the youngest stages of the new individuals was considerably *less* than one half of the mean length of the individuals not undergoing fission. The mean length of the young was found to be 87.848 microns while that of the non-dividing specimens was 199.960 microns, or 24.264 microns more than twice the mean length of the young individuals. The mean breadth of the youngest stages was found to be slightly greater than that of non-dividing specimens—55.48 microns, in place of 50.22 microns. The mean index, or ratio of breadth to length, was consider-

ably more than twice as great in the young as in the adults; in the former it was 63.136 per cent; in the latter 25.114 per cent.

On the basis of his many measurements, Jennings reported that the mean length of *P. caudatum* beginning fission was 175.696 microns while the mean length of specimens not undergoing fission in a random sample of the same culture was 199.960 microns. It is apparent therefore that paramecia beginning to divide become shorter but increase in breadth. However, immediately prior to separation of the dividing specimen, the two unseparated young have somewhat more than half the original length of the dividing mother, indicating an increase in length just before separation. Jennings' data and Table 15 show that the recently separated daughters grow rapidly in length. Even those five minutes old are considerably longer than those which have just separated. As the young specimens

Table 15

GROWTH IN MICRONS OF DEVELOPING *Paramecium caudatum* FROM TIME OF FIRST RECOGNIZABLE STAGE IN FISSION TO THE NEXT FISSION PERIOD (Modified after Jennings)

Stage	Age	Mean Length in Microns	Mean Breadth in Microns
1	Beginning of constriction in fission delimiting two unseparated halves (totaling 16.2 microns), each	82.600	65.893
2	Fifteen minutes after beginning of constriction of two unseparated halves in fission	85.774	64.493
3	Recent daughter 2½ minutes after separation	107.660	59.355
4	9½ minutes after separation	128.000	60.168
5	23 minutes	143.348	54.284
6	40 minutes	149.920	55.840
7	82½ minutes	161.524	54.192
8	4 hours	176.560	58.922
9	12 hours	188.988	62.796
10	18 hours	199.048	56.496
11	Beginning of constriction in fission delimiting two unseparated halves totaling	165.200	64.893

grow in length, they become more slender. Breadth reaches its minimum then begins to increase but more slowly than the length. According to Jennings, these later changes showing marked fluctuations in breadth are due to environmental conditions while the growth in length continues steadily for at least 18 hours or prior to the next fission stage, at which time the individual becomes shorter and thicker as described earlier. It is significant that there is much less variation in dimensions at the beginning of specimens undergoing fission than in adult speci-

mens measured at random from the same culture.

Simpson (1901 a, b, 1902) studied binary fission in respect to variation and made careful measurements of length and breadth of *P. caudatum* at stated intervals after fission. Popoff (1909 a) measured the growth of *P. caudatum* after determining the length, breadth, thickness, and volume by killing one sister cell immediately after fission and the other at a specific time after fission. Richards (1941) shows Popoff's results of micrometer units converted into microns and then plotted (Figs. 57, 58,

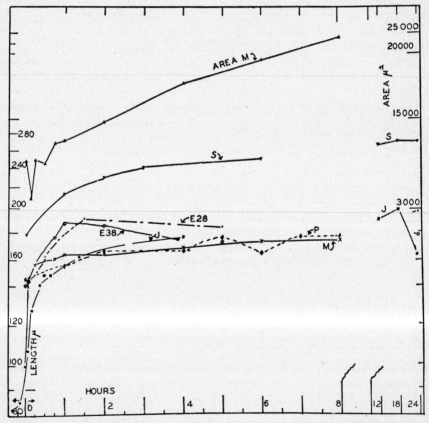

FIG. 57. Growth in length and in area of *Paramecium caudatum*. (M) From Mizuno (1927). (S) From Schmalhausen and Syngajewskaja (1925). (E28 and E38) From Estabrook (1910). (J) From Jennings (1908). (P) From Popoff (1909). (Richards)

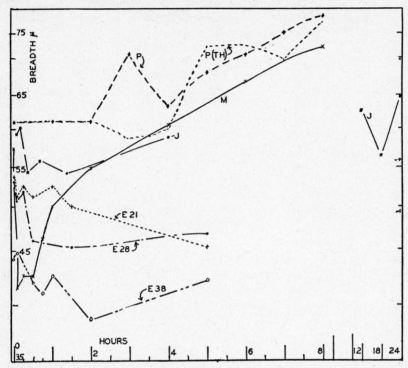

FIG. 58. Growth in breadth and thickness (TH) of *Paramecium caudatum*. (P) From Popoff (1909). (M) From Mizuno (1927). (J) From Jennings (1908). (E21, E28, and E38) From Estabrook (1910). (Richards)

and 59). Standard size at zero time is the average of the cells killed at fission and the average differences for the intervals were added successively by Richards to obtain the data shown in the figures. In his study, Popoff noted that the volume of the macronucleus was proportionally reduced until immediately prior to fission at which time the macronucleus rapidly increased in volume, then divided.

In addition to the studies by Jennings and Popoff upon growth of *P. caudatum*, Estabrook (1910 a), Erdmann (1920), Schmalhausen and Syngajewskaja (1925), Syngajewskaja (1926), and Mizuno (1927) have made valuable contributions using the same species. Schmalhausen and Syngajewskaja (1925) measured 44 living *P. caudatum*

from the time of division and found that 78 per cent of the total growth in length occurred within the first hour. The graphs presented by Adolph (1931) as drawn from various sources show essentially the same feature; his excellent monograph on the regulation of size in unicellular organisms will be found helpful to those interested in growth studies. Data of the aforementioned investigators have been analyzed by Richards (1941) and the results depicting growth changes in length, area, breadth, thickness, and volume are shown graphically in the figures. The data of growth as shown in the figures are based upon averages from many specimens, not upon the growth of a single one.

Fig. 57 shows that the growth curves

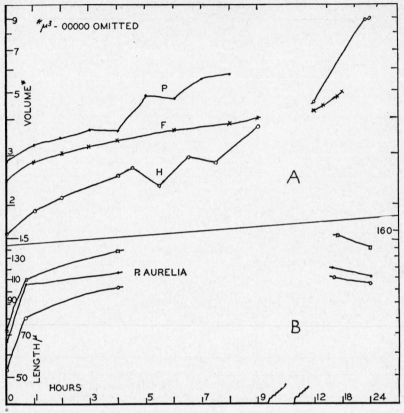

FIG. 59. (A) Growth in volume of *Paramecium caudatum* (P), *Frontonia leucas* (F), from Popoff (1907, 1908), and of *Hartmanella hyalina* (H) from Cutler and Crump (1927). (B) Growth in length of *P. aurelia*. Data from Erdmann (1920). (Richards)

in length are similar with only minor variations. The size differences are very likely racial characters since the paramecia were cultivated by the various investigators in essentially the same manner. The curves for growth of three races of *P. aurelia* as studied by Erdmann (1920) have been plotted by Richards (Fig. 59) and are seen to be similar to those of *P. caudatum*.

As would be expected, the curves of growth for paramecium as determined mathematically are characteristic logarithmic growth curves. They are of the same form and character as the sigmoid growth curves which have been obtained from similar studies made upon plants and higher, multicellular animals including man.

If a single specimen of paramecium[13] be isolated in a small amount of culture fluid containing suitable bacteria as food, division will produce two daugh-

[13] Smith (1934) noted that when specimens of *P. caudatum* were isolated from old cultures and then cultivated, an appreciably long lag period occurred before fission, as compared with specimens similarly isolated from young cultures. He reported that the older the paramecia from such old cultures, the longer the lag period. The lag period was attributed to the physiologic condition of the animal and changes in the environment.

ter cells. If the daughters are isolated and placed into a fresh volume of medium, growth and reproduction appear to be potentially unlimited as demonstrated by many, especially Woodruff (1943) with his well-known pedigree isolation cultures. When the rate of growth is practically constant, the growth curve for the sum of the paramecia may be plotted on arithlog paper and the result will be a straight line because the growth (y) is exponential, $y = Y_o e^{kt}$, when Y_o is the amount of seeding or the growth at time, $t = 0$ and e, is the Naperian base. The proportionality constant $k = (\ln 2)/G.T.$, or $0.639/G.T.$ The generation time (G.T.) is the time between divisions (Richards, 1941).

The study of pedigree isolation cultures, histograms, and the life-cycle is considered in Chapter 11.

2. Measurement and Growth of Populations.

Populations of paramecia may be measured by counting the number of organisms in a sample of the population with a hemocytometer (Woodruff, 1912) or with a Sedgewick-Rafter apparatus (Hall, Johnson, and Loefer, 1935). Two important parts of this latter apparatus consist of the counting cell[14] which is simply a brass ring ce-

mented to a microscope slide to make a cell exactly 1 mm. deep and secondly, an ocular disk. The ocular disk (Whipple micrometer) contains a large square ruled into 100 smaller squares. With the proper combination of lenses in the microscope, one side of the large square corresponds to 1 mm. on the stage of the microscope as determined by the use of a stage micrometer.

With the counting cell in place on the stage of the microscope, one simply counts all the organisms within the large square of the ocular disk. This will give the number of organisms in 1 cubic millimeter since the cell is exactly 1 mm. deep. Before counting, the paramecia should be killed by heat or a fixing reagent. For obtaining their data, Jennings preferred. Worcester's fluid and Gause, Schaudinn's fluid both of which yielded good results. According to Hall, Johnson, and Loefer the probable error of their counts is usually well below 5 per cent.

The sample may be diluted in the counting chamber if the paramecia are too greatly concentrated. At least two samples—generally more—should be taken from a well-shaken culture.

From a number of samples of a given culture, the average count is determined for each point on the graph. Such results may be expressed in terms of x/x_o which is the ratio of final to initial concentration of paramecia per cubic centimeter. As an example, Fig. 60 shows Loefer's data obtained by this method in studying the growth of bacteria-free cultures of *P. bursaria* in synthetic media containing both salts and organic substances.

A simple method for making population counts as used by Jones (1933) is given as follows: the number of paramecia in one half of 1 cc. was determined by removing individuals from a well-shaken culture with a foot-oper-

---

[14] Loefer (1936 b) used a counting cell which was a modification of the Sedgewick-Rafter cell in his growth studies of *P. bursaria*. The cell consisted of a circular metal ring having an inside diameter of 35.7 mm. and a thickness of 1 mm., mounted with balsam on a 4 x 7.5 cm. glass slide. For counting, it was put on the stage of a binocular microscope, 1 ml. of the fixed culture placed in the cell and a direct count then made of all paramecia in the sample. The slide was divided into 32 spaces with a diamond pencil by means of five vertical and five horizontal lines on the lower surface, thereby facilitating counting of the paramecia with a 48 mm. objective.

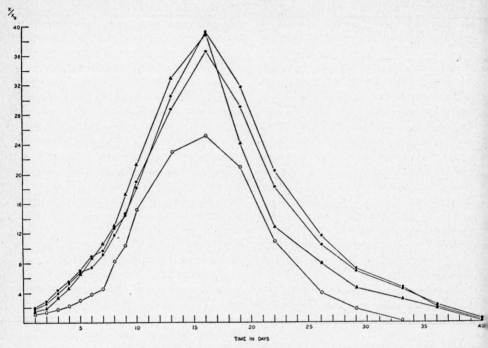

Fig. 60. Growth cycle of *Paramecium bursaria* in 0.5 per cent tryptone, made up with, (1) tap water (● ———— ●), (2) medium B (■ ———— ■), (3) medium C (⊙———⊙), (4) medium D (▲———▲). x/x₀ equals ratio of final to initial concentration of ciliates per cubic centimeter. (Loefer)

ated pipette and counting the specimens as picked up in the pipette.

At the present time little work has been done on the detailed history of ciliate populations in pure culture. The review, dealing with experimental populations of *Paramecium* and other ciliates by Johnson (1937), should be consulted. Loefer's (1936 b) study is the only one showing complete growth curves of paramecium in pure culture. It will be seen that his curves showing the growth of *P. bursaria* (Fig. 60) are comparable to the numbers curve in the hypothetical population of Fig. 61.

However, the work of Loefer and that of Phelps (1934, 1936) on *Glaucoma pyriformis* indicate that in general, the growth of ciliates in pure culture

follows the general trends observed in populations of bacteria and yeasts. Allee (1941) may be perfectly correct in stating that it makes little difference to the student of populations whether his beasts feed on microscopic fodder or the kind that comes in bales since there may be hay and weeds in both.

As described by Buchanan and Fulmer (1928) and reported by Hall (1941) it is convenient to characterize seven phases of growth of microörganisms in culture (Fig. 61). These phases are as follows:

1. An *initial stationary phase* during which there is no increase in numbers of organisms in the population.

2. A *lag phase* during which there is

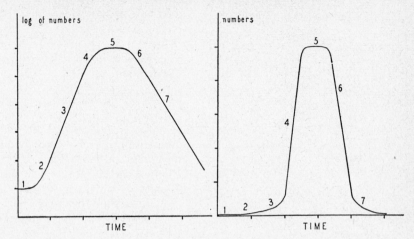

FIG. 61. Growth phases in a hypothetical population. In the curve at the left, logarithms of numbers are plotted against time; on the right, numbers are plotted against time for a comparable population. Successive growth phases are numbered from 1 to 7. (Hall)

positive growth acceleration where the division rate increases to a maximum.

3. The *logarithmic growth phase* during which the maximal division rate is maintained.

4. The *negative growth acceleration phase* in which the division rate steadily decreases.

5. The *maximum stationary phase* during which the population of individuals remains practically constant.

6. The *accelerated death phase* during which the total population begins to decrease.

7. The so-called *logarithmic death phase* in which the population decreases at a fairly constant rate.

Conditions such as food, pH, temperature, etc. which are likely to alter the population growth curve are considered later in this section. As reported by Hall (1941) (Fig. 62) certain environmental conditions may produce any one of the following effects on growth curves:

1. An increase in the maximal population without an appreciable effect upon the growth rate. In other words, the length of the logarithmic phase might be increased without a change in the division rate as shown in curve B as compared to the control medium shown in curve A.

2. An increase in the growth rate without a change in density of the maximal stationary phase as shown in curve C.

3. An increase in growth rate and maximal density of population as shown in curve D.

4. A decrease in maximal density of population with no appreciable effect on the early growth rate as shown in curve E.

5. A decrease in the growth rate without any effect on the maximal density of population as shown in curve F.

6. A decrease in growth rate and maximal density of population as shown in curve G.

Curves A, B, and E would show no significant differences in the early his-

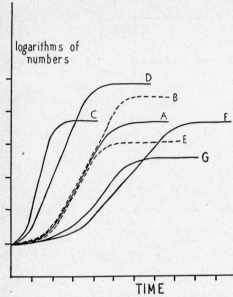

logarithms of numbers

D

B

C A F

E

G

TIME

Fig. 62. Hypothetical modifications (curves B-G) of the normal growth of a population (curve A), from the initial stationary to the maximal stationary phase. (Hall)

tories of the cultures. Each, however, reaches a different maximal stationary phase, one higher and one lower than that of the control (curve A). Early examination of cultures A, C, and F would show significant differences in population density and in growth rate, although each culture eventually reaches the same maximal density (Hall, 1941).

Phelps (1934, 1936) has shown that age and other qualities of the inoculum are factors that must be considered in comparative studies on population growth. Jones (1933) studied population and size changes within a pure-line of *P. multimicronucleatum* and observed that the length increased when the population count decreased and vice versa prior to the thirty-fourth day of the culture (Fig. 55). After that time, the population declined in number and specimens in length until the

paramecia had practically disappeared from the culture.

3. Growth in Relation to Food Concentration.

As would be expected, a factor of paramount importance in growth studies is the concentration and distribution of available food in the culture medium. A fairly great concentration of suitable bacteria will, within certain limits, provide an adequate source of supply of food for paramecia.[15]

Unless it be a single species of paramecium with no other microörganisms present in the culture, the typical paramecium-culture with its associated bacteria presents in reality an ecologic association in which bacteria are growing while serving as food for the Protozoa. The medium must therefore support not only the growth of paramecium but the bacterium as well. It is easy to understand that conditions in such a culture are never likely to be exactly the same from day to day. To control and standardize the amount of bacteria ingested by paramecia is a major problem in growth studies. Leslie (1940 a) in studies dealing with nutrition of *P. multimicronucleatum* reported that quantity of food (bacteria) was regulated by the use of a photoelectric turbidimeter with measured bacterial suspensions with a relatively high degree of accuracy.

It is possible to start not only with sterile culture fluid but sterile paramecia as well. Then bacteria of given age and concentration in pure culture

[15] It has been the writer's experience that a day or so after inoculation of culture medium, greater growth is obtained by a mild agitation of the container than if it is left standing. Certain species of bacteria used as food will grow on or slightly below the surface of the medium forming a more or less continuous film, with the ciliates concentrated below it.

may be added to the media containing paramecium. However, once in ecologic combination the bacteria in the paramecium-culture also have a growth curve which is generally not taken into account in dealing with growth curves of the ciliate.

The growth of paramecium in sterile, pure culture without the presence of bacteria or other microörganisms is discussed on page 101.

QUALITY OF AVAILABLE FOOD. Calkins held that *Bacillus subtilis* was the principal source of food of paramecium in the typical hay infusion. He probably had in mind species of *Paramecium* belonging to the "aurelia" group. Hargitt and Fray (1917) found that more rapid growth occurred in their cultures of *P. aurelia* and *P. caudatum* when fed with *B. subtilis* than any of eight other pure strains of hay-infusion bacteria. Attempts were made by Johnson (1936) to grow *P. caudatum* on seven species of bacteria separately and on mixtures of these but only suspensions of *Bacillus subtilis* alone supported good growth.

Phillips (1922) found that *P. aurelia* grew fairly well on her "C" strain of bacteria (a streptothrix) but grew faster on a mixture of bacterium "A" and "C" despite the fact that "A" alone was incapable of supporting growth of paramecia.

The extensive investigations of Oehler (1916, 1920 a, b, 1921, 1924 a, b), Luck, Sheets, and Thomas (1931), Lwoff (1932, 1951), Sandon (1932), Leslie (1940 a, b), and Cunningham and Kirk (1941) upon nutrition of these organisms should be consulted for detailed information.

A species of bacterium may be unsuitable for food because of its size, shape, age, or chemical composition. Dead bacteria appear to be unsuitable for sustained growth. It has been shown conclusively that many species of bac-

teria provide an excellent source of food while others are deleterious. Also, certain races or strains of a given species of bacterium may provide adequate food while other races of the same species may be inadequate, even toxic. The species of bacteria which have been reported in the literature as being suitable for food of paramecium are listed on page 100.

From their experiments, Hargitt and Fray (1917) concluded that cultures of mixed bacteria are, as a rule, far superior as a food source for paramecium to any one species of bacterium. On the other hand some investigators have cultivated a given species of *Paramecium* on a single species of bacterium for years.

Leslie (1940 b) determined the relative suitability of some 30 species of bacteria as food organisms for *P. multimicronucleatum* when grown in an inorganic medium. Of these non-pathogenic representatives of several different families, he found about one third of the number suitable as a food source.

Species of *B. coli, B. subtilis,* and *Ps. fluorescens* have been reported in the literature as being both suitable and unsuitable food organisms for the cultivation and growth of paramecium. Using several strains of *B. coli* and *B. subtilis,* Leslie reported that certain strains of the same species were satisfactory as a food source and others were not. There was shown conclusively a racial difference existing in a given species of bacterium in respect to suitability as food.

While attempting to free a stock culture of *P. multimicronucleatum* from the contaminating colorless flagellate, *Oikomonas termo,* Hardin (1944) discovered that such bacteria-containing cultures nearly always flourished. However, cultures of the paramecia without the flagellate but with bacteria universally perished in a few days. He believed this due to a beneficial effect of *Oikomonas* which resided in its ability to detoxify bacteria used for nutrition by paramecium.

Oehler found that bacteria from young cultures served as a better source of food than bacteria from old cultures; certain rather resistant strains proved to be edible if eaten when young. Leslie (1940 b) stated that aging of *Ps. fluorescens* four days before use so improved its food qualities that it was found to be very satisfactory food, but if the bacterium was taken from a 24-hour-old slant culture, it was definitely unsuitable.

From the foregoing account, it is obvious that the only way of testing the nutritional value of a given species or race of bacterium is to attempt using it as a source of food for paramecium.

4. Growth in Relation to pH of the Culture Medium.

The acidity or the alkalinity of the culture medium is related to the cultivation, growth, and size of paramecium. The hydrogen-ion concentration, designated by the symbol pH, is the negative logarithm of the hydrogen-ion concentration ($pH = \log 1/c_H + = - \log c_H +$). With the use of simplified methods, the pH of a culture may be determined quickly and easily by the use of chemical indicators or it may be done electrometrically. At neutrality the hydrogen-ion concentration is 10^{-7} normal (7.0) being equal to the hydroxyl-ion concentration.

Studies have been made upon bacterized cultures and bacteria-free or pure cultures of paramecium in an effort to determine the pH range in which the ciliates can live and the optimum pH which supports growth best. The result has been the accumulation of a large body of evidence showing that the pH relationships of Protozoa are complex, varying not only with the particular species but also with the composition of the culture medium. Crane (1921) found that the toxicity of quinine for paramecium was much greater at pH 8.0 than in more acid solutions. Roskin and Dune (1929) reported that the effect was reversed in the ciliates which had been growing in a hay infusion containing saccharated iron oxide. Paramecia which had been grown in saccharated iron oxide were much more resistant to quinine above pH 8.0 and much less resistant below pH 8.0 than were ciliates grown in hay infusion without colloidal iron. Thus the usual pH range is altered by the presence of quinine and this quinine effect is reversed by the past history of the organism.

Typical hay infusions at first usually become quite acid reaching as low as pH 4.7 and then become alkaline reaching a pH of 8.0 or more (Bodine, 1921; Jones, 1930). The same is true of freshly made lettuce infusion which has a pH of 5.0 only to become alkaline later (Wichterman, 1948 d). According to Jahn (1934), as the pH of the infusion changes, toxic substances may become beneficial or beneficial substances may become toxic. Besides having a direct influence upon growth rate, the pH of the medium can modify the effects of other factors.

Saunders (1924) and Pruthi (1926–27) reported that *Paramecium* sp. have an optimum pH of 7.8–8.0. Darby (1929, 1930) and data of Raffel as reported by Jahn (1934) present evidence to show that *P. aurelia* grows in a range from 5.7–7.8 with an optimum of 6.8–7.0; and Phelps (1931, 1934) for the same species reports a range of 5.9–8.2, adding that the range 5.9–7.7 is favorable. According to Darby (1930) 7.0 is optimum for *P. caudatum,* and Jones

(1930) found that the range is 4.8–8.3 for *P. multimicronucleatum* with 7.0 being optimum for greatest multiplication. The fact that these investigators used different media may explain their somewhat divergent results which may indicate that the growth may not be due so much to pH but to the chemical and bacterial composition of their different media.

From the foregoing accounts and the author's observations (Wichterman, 1948 d, 1949), it appears that cultures with a pH of approximately 7.0 yield the best growth in paramecium (Table 16). Before inoculation with paramecia an excess of $CaCO_3$ added to a 250 cc. flask culture of infusion brings the pH to 7.2 and this yields excellent growth.

The effect of hydrogen-ion concentration on growth and morphology has been studied by Loefer (1938) in which he used bacteria-free strains of green *P. bursaria*. Without the complicating factor of other microörganisms being present, such a pure culture is ideally suited for pH investigations. Growth in relation to pH of the medium was determined in a Bacto-tryptone and in a Proteose-peptone medium. According

Table 16

GROWTH OF PARAMECIUM IN RELATION TO THE HYDROGEN-ION CONCENTRATION
OF THE MEDIUM

Species	pH Range of Medium	Optimum pH for Growth	Reference
Paramecium sp.	—	7.8 – 8.0	Saunders (1924)
Paramecium sp.	7.0 – 8.5	7.8 – 8.0	Pruthi (1926–27)
P. aurelia	5.7 – 7.8	6.7	Morea (1927)
P. aurelia	5.7 – 7.8	7.0	Darby (1929, 1930)
P. aurelia	5.9 – 8.2	5.9 – 7.7	Phelps (1931, 1934)
P. aurelia	5.8*– 8.9	6.4	Gaw (1936)
P. aurelia	—	7.2	Sonneborn (1936)
P. aurelia	—	6.8 – 7.0	Pace and Kimura (1944)
P. aurelia	—	7.0	Giese (1945)
P. aurelia	6.2 – 7.3	7.0 – 7.2	Wichterman (1948 d)
P. caudatum	6.0 – 9.5	7.0	Morea (1927)
P. caudatum	5.3 – 8.2	7.0	Darby (1930)
P. caudatum	—	6.7 – 6.8	Loefer (1936 a)
P. caudatum	5.8*– 8.9	7.0	Gaw (1936)
P. caudatum	—	6.8 – 7.0	Pace and Kimura (1944)
P. caudatum	—	7.0	Giese (1945)
P. caudatum	6.2 – 7.2	6.9 – 7.1	Wichterman (1948 d)
P. multimicronucleatum	4.8 – 8.3	7.0	Jones (1930)
P. multimicronucleatum	5.8*– 8.9	7.5	Gaw (1936)
P. multimicronucleatum	—	7.2	Johnson and Hardin (1938)
P. multimicronucleatum	—	7.0	Giese (1945)
P. multimicronucleatum	6.2 – 7.5	6.5 – 7.0	Wichterman (1948 d)
P. bursaria	5.3 – 8.0	6.7 – 6.8	Loefer (1938)
P. bursaria	5.0 – 7.4	7.1 – 7.3	Wichterman (1948 d)
P. polycaryum	5.8*– 8.9	7.0	Gaw (1936)
P. polycaryum	5.0 – 7.5	6.9 – 7.3	Wichterman (1948 d)
P. trichium	6.2 – 7.1	6.7 – 7.1	Wichterman (1948 d)
P. calkinsi	6.5 – 7.8	7.1 – 7.4	Wichterman (1948 d)
P. woodruffi	6.5 – 7.5	7.0 – 7.5	Wichterman (1948 d)

* Lived only short time at these concentrations.

to Loefer growth was observed from pH 4.9–8.0 with a maximum at pH 6.8 (Fig. 63). Measurements of the paramecia in the tryptone media showed an average length of 86.5 microns at pH 7.6–8.0 to 129 microns at pH 6.0–6.3. Greatest variation occurred in cultures at pH 4.6 in which no growth occurred. The shortest organisms at the acid and alkaline extremes of growth were the widest; those in cultures from pH 5.7–7.4 were narrowest with an average width of 44 microns (Fig. 14). Loefer showed that optimum for size does not correspond with the pH most favorable for multiplication (Table 1).

Jones (1928, 1930) followed the population growth of an inoculation of 200 specimens of *P. multimicronucleatum* in 70 cultures at 80° F., and made counts of 0.5 ml. samples periodically. The pH of the medium was also measured. A maximum concentration of 10^6 paramecia was obtained in 700 ml. cultures. Growth stopped when the pH decreased to 5.0. With hay-flour infusions, two cycles of growth were found; the first was terminated by the high acidity and when the acidity returned to about pH 7.0, the second growth cycle started. During a three-day period, Jones found that the number having died at the close of the first growth cycle exceeded the maximum number present during the second period. He believed the death of the paramecia was due to toxic excretion products which were neutralized by the materials liberated from the cytolysis of the dead animals. The decline of the population was related to

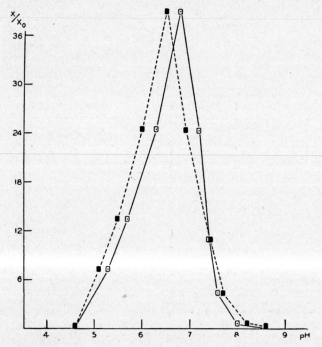

F<small>IG</small>. 63. Growth of *Paramecium bursaria* at different hydrogen-ion concentrations in a Bactotryptone medium. x/x_0 = ratio of final to initial concentration of ciliates per ml. pH of cultures at the beginning of incubation (■ - - - - - ■); pH after 21 days growth (⊡———⊡). (Loefer)

the decline of the food and by periodically renewing the food, cultures were maintained for four years.

5. Growth in Relation to Temperature
(see also Acclimatization).

Within limits, the higher the temperature the greater the metabolic activities of the organism. The result is a more rapid growth and a greater reproductive rate. Temperature is a factor which can be controlled with ease. There are of course temperature limits between which a species of *Paramecium* can live and there appears to be an optimum temperature range for greatest growth. In the typical bacterized paramecium-culture it is necessary to consider not only the temperature effect upon the ciliate but upon the bacteria as well. Indeed temperature will not only directly affect metabolic activities of the ciliate but it can and does modify the action of substances in the medium. As an example, Collett (1919–21) reported that the toxicity to paramecium of the organic acids acetate and butyrate in acid solutions was increased with rise in temperature from 10° to 20° and to 30° C.

It has been shown that temperature characteristics vary in the same species under different conditions. Chalkley (1930 a) placed specimens of *P. caudatum* in four culture media as follows: balanced saline, saline with potassium excess, saline with calcium excess, and saline with sodium excess all with pH from 5.8 or 6 to 8.4 or 8.6 at 40° C. for two to 16 minutes. He discovered that the resistance to heat death varied with the hydrogen-ion concentration with maxima appearing in the alkaline and acid ranges and a minimum close to 7.0. Acidity decreased and alkalinity increased resistance and between pH 6.6 and 7.6 excess of potassium decreased resistance and excess of calcium increased resistance.

In their experiments dealing with fission-rate and geographical variation in species of *Paramecium* from different parts of Russia, those collected from the north (especially *P. bursaria*) were found to possess lower temperature optima of division rate than those clones obtained from the south according to Gause, Smaragdova, and Alpatov (1942). These fission-rate characteristics were also found to be hereditary.

From the studies of Schürmayer (1890), Mendelssohn (1902), Rautmann (1909), Popoff (1909 a), and others it appears that paramecia will divide at temperatures between 5° and 35° C. and that the optimum temperature is between 24° and 28° C. For *P. aurelia*, Woodruff and Baitsell (1911 c) reported that the optimum temperature zone for reproduction is between 24° and 28.5° C. Johnson and Hardin (1938) found that cultures of *P. multimicronucleatum* grown at 26° C. had a consistently higher fission-rate than those grown at 20° C.

Giese (1952) found that when *P. multimicronucleatum* is cultivated on *Pseudomonas ovalis* in lettuce infusion, the division rate increased only slightly above 30° C. and at 34.5° C. multiplication continued for only 30 hours. At 36.5° C. the paramecia divided more slowly than at 30° C. and died after 20 hours. Specimens did not divide at 38° C. but when they were transferred to fresh culture medium at 26° C. they divided at a rate comparable to the control specimens. Survivors of paramecia briefly subjected to 42° C. divided, after a brief lag, at a rate comparable to controls when they were placed in culture at 26° C. Giese noted that starved paramecia were no more sensitive to high temperature than specimens from the logarithmic phase of a culture. He obtained similar results with *P. caudatum* and *P. aurelia* but these species appeared to be less resistant than *P. multimicronucleatum*. He reported that while all three species of *Paramecium* tested survived at 10° C., specimens failed to divide. Well fed paramecia from the logarithmic phase of the culture died at 5° C. but starved specimens lived for many days. The resistance of starved paramecia to low temperatures probably has ecological significance since specimens present in bodies of water during cold periods are likely to be in a relatively starved condition, hence survive.

Jollos (1913 a) observed that in races of paramecium, smaller animals were produced when the temperature of their environment was raised. When the temperature was lowered to that of the usual environment, the length of all animals returned to their original size.

P. caudatum was cultivated by Glaser and Coria (1933) in a medium free from other living microörganisms (dead yeast cells). They reported that the organisms were grown at temperatures ranging from 20° to 28° C. and that the optimum was 25° C.; at 30° C. the ciliates were killed. Doudoroff (1936) found that for *P. multimicronucleatum* its resistance to raised temperature was low in the presence of food but its resistance rose to a maximum when the food was exhausted. In a study of Protozoa living in various thermal waters of Japan, Uyemura and his associates (Kudo, 1947, p. 18) discovered that *P. caudatum* was found in water having a temperature of 36°–40° C.

Efimoff (1924) reported that *Paramecium* divided once in approximately 13 days at 0° C., was able to withstand freezing at −1° C. for 30 minutes but died when kept for 50–60 minutes at the same temperature. *P. caudatum* died in less than 30 minutes when exposed below −4° C.; quick and short cooling not below −9° C. produced no injury but when prolonged, the ciliates became spherical and swollen to four to five times their normal size.

In studying *Paramecium* sp. in gradually descending subzero temperature, Wolfson (1935) observed that as the temperature decreased, the ciliate often swam backward and bodily movements ceased at −14.2° C. but cilia continued to beat for a time. Paramecium recovered completely from a momentary exposure to −16° C. but long cooling at −16° C. brought about degeneration. No survival was observed when water containing paramecia was frozen.

The temperature coefficient of biologic processes is sometimes expressed in terms of Q_{10}. This is the rate of the speed or velocity constant of a process or reaction at a given temperature to the velocity constant at a temperature 10 degrees lower; it is held (van't Hoff's rule) that the Q_{10} of a chemical reaction lies between 2 and 3. The Arrhenius formula has received considerable usage of late in biologic reactions. The van't Hoff temperature coefficient, Q_{10} which gives the ratio of the velocities for 10-degree intervals is claimed to be less precise than the critical thermal increment (μ) in the Arrhenius formula. Since the latter is a simple multiple of the former, a μ of 13,200 corresponds to a Q_{10} of 2 for the 10-degree interval between 30° and 40° C. For the interval between 10° and 30° C., the Q_{10} corresponding to the same value of μ would be 2.1 (Heilbrunn, 1943; see p. 433 for a detailed account of the van't Hoff, Arrhenius, and other formulas as applied to temperature coefficients).

Woodruff and Baitsell (1911 c) studied the temperature coefficient of the rate of reproduction in *P. aurelia*. They concluded that the temperature coefficient (Q_{10}) of the average rate of reproduction is approximately 2.70 and therefore the rate of cell division is influenced by temperature at a velocity similar to that for a chemical reaction.

Mitchell (1929) in a reëxamination of the temperature coefficient in paramecium found that if the change of division rate with temperature were stated in terms of the van't Hoff-Arrhenius equation, the value of μ would equal 23,000 between 12° and 25° C. If applied to the findings of Woodruff and Baitsell, the average Q_{10} found by Mitchell is 3.6 which differs considerably with that reported by Woodruff and Baitsell.

Glaser (1924) in observing the forward movement of paramecium in respect to temperature found μ equal to 16,000 between 6° and 15° C., and 8,000 between 16° and 40° C. This would result in a Q_{10} of approximately 2.7 and 1.6 respectively.

6. Growth in Relation to Light and Darkness.

Recent investigations have shown that light of definite wave-lengths has specific effects upon the growth of organisms. Packard (1925) observed that permeability of paramecia to NH_4OH is greater when the organisms are exposed to light than when they are in darkness. This change in permeability was also demonstrated in paramecia exposed to monochromatic red light, becoming greater as the wave-length shortened and reaching a maximum in the near ultraviolet. He also showed that the permeability of paramecia varies with the division rate; rapidly dividing cells have a high permeability and those with a slow division rate a low permeability. Since permeability increases as the duration to light is prolonged, this factor has a bearing on growth espe-cially in pure cultures which contain diffusible substances in the medium.

Hutchinson and Ashton (1929, 1931) reported that when *P. caudatum* was irradiated with monochromatic light certain frequencies in the red, yellow, and the near ultraviolet parts of the spectrum were found to stimulate growth. Those in the green and in the far ultraviolet beyond 3000 Å retarded growth and even killed the organisms. According to these investigators, there was a marked similarity between the curves showing the increase in rate of plasmolysis and those showing the effect of irradiation on the growth rate of *P. caudatum*. Generally, visible and near ultraviolet light caused an increase in the rate of plasmolysis of less magnitude than that caused by the far ultraviolet. All visible and near ultraviolet lines used with the exception of 4960 Å, 4078 Å, and 3022 Å caused stimulation of division rates and those lines had the least effect on the rate of plasmolysis.

The importance of light is well known in the growth of the green paramecium, *P. bursaria,* and is discussed in regard to symbiosis on page 406. However, comparatively little is known concerning the relationship between light and growth in other species of *Paramecium* or for that matter in other colorless Protozoa.

By keeping infusions in light and in darkness, Eddy (1928) observed no difference in growth of ciliates which could be interpreted as a result of the direct effect of light. This is in agreement with the findings of Maupas (1888) and Woodruff (1905). Woodruff and Baitsell (1911 b) performed experiments in which cultures were kept in total darkness except for the short time daily when they were examined—usually not more than three minutes. A control culture carried in the light disclosed that light did not influence the rate of reproduction in paramecium. Richards (1929) found a seasonal rhythm of growth in *P. aurelia* with a high point in July which is the month of maximum sunlight. He referred to the ecologic study of Wang (1928) in which the sunlight and temperature maxima do not coincide but called attention to the fact that there is a closer correlation of sunlight than of temperature. Richards suggested that the slowing down of ciliate division rates might be due to an accumulative deficiency of light of shorter wave-lengths.

Zhalkovskii (1938) found that reflected light stimulated multiplication of *P. caudatum* in the red, but had a depressing action in the violet. Filtered, transmitted light had a greater depressing effect than reflected light and the difference was believed to be due to the polarizing effect of the reflected light.

That light appears to be of little significance in at least a number of colorless species of paramecium is shown in the carefully controlled experiments of a large number of investigators where maximum growth and highest fission-rates are obtained in the darkness of a constant-temperature incubator.

For the effects of ultraviolet light, roentgen rays, and radium rays refer to page 135.

7. Growth in Relation to the Waste Products of Metabolism.

When paramecia grow in culture, products of metabolism are formed which pass directly into the culture medium. The same is true of the bacteria growing in the typical bacterized paramecium-culture. The effects of the metabolic products as a result of ciliate and bacterial growth in culture have led to extensive literature on the subject with conflicting claims.

In a series of publications, Woodruff (1911–13 a) proposed the theory that the accumulation of excretory materials in a protozoan culture produced inhibitory effects on division rates and growth. This was questioned by many others who maintained that the significant inhibiting influence in a culture was not the accumulation of excretory products but lack of food. Woodruff concluded that while waste products of metabolism inhibit growth of homologous species, growth of a different species may be relatively unaffected by the same substances. It is of interest to report that Dimitrowa (1932) found growth of *P. caudatum* to be accelerated by the addition of small amounts of old medium to fresh medium in her experimental cultures. She might have introduced bacteria from the old cultures into the new medium and observed instead the effects of bacterial feeding by paramecia.

Gause (1934 f) in his extensive experiments with protozoan populations accepted the conclusions of Woodruff concerning the effects of waste products on growth and attempted to eliminate these substances by changing the medium in his cultures every 24 hours. According to Lwoff and Roukelman (1926) growth stops in bacteria-free cultures of the ciliate *Glaucoma piriformis* long before the nutrient materials are exhausted as measured by total N, amino N, amide N, and peptone N. That the supply of N compounds is not greatly decreased before growth stops indicates that cessation of growth is caused by the accumulation of waste products in the medium.

Phelps (1936) reported that waste products of bacteria might have a depressing effect on the protozoan division rate and suggested that different kinds of food (bacteria) might cause the same protozoön to produce excretory products quite different in nature and effects.

On the other hand Taylor and Strickland (1938) cultivated the ciliate, *Colpoda duodenaria* in balanced salt solution with a single species of bacterium as a source of food. They found that excretion products accumulating in the medium for four months did not affect their growth curves. Jones (1933) for *P. multimicronucleatum* concluded that excretory products were insignificant in his study but that food was the principal growth factor.

In bacterized balanced salt medium and in bacterized hay medium, Johnson and Hardin (1938) reported no significant inhibiting effect on the reproductive rates of *P. multimicronucleatum* grown in old culture medium with the waste products supposedly accumulating. However, culture medium conditioned with the bacterium (*P. fluorescens*) for 15–30 days when used for the culture of paramecium definitely inhibited reproduction.

In the ciliates *Colpidium* and *Glaucoma,* Hall (1941) suggested a vitamin deficiency as one explanation for inhibition of protozoan growth by conditioned medium.

Information has been presented by several workers to show that media with so-called waste products of metabolism —sometimes called conditioned media —may have a decidedly beneficial effect on Protozoa. It has been found that growth of certain ciliates would occur in bacteria-free medium only after the culture medium had previously been acted upon by living bacteria. Leslie (1940 a, b) postulated that living *Pseudomonas fluorescens* furnished a growth-promoting substance or food factor for paramecium the lack of which in suspensions of dead *Ps. fluorescens* might account for its unsuitability.

Additional suport to the idea of a conditioning effect in populations of paramecium has been given by Ludwig and Boost (1939). They made a reanalysis of the results of several investigators by plotting the relative growth rate against the amount of population produced in the medium. Their graphs show a change of relative growth rate as the population increases by the addition of new individuals.

In an analysis of the work of Gause, Nastukova, and Alpatov (1934), Ludwig and Boost show in Fig. 64 that there was an initial rise in the relative growth rate and then a decrease. The initial rise is attributed to a beneficial effect of autoexcretory products. It is suggested that the beneficial operation of the autoexcretory products may be caused possibly by a change in pH, redox potential, or some other factor or factors of the medium. (See also section on allelocatalysis.)

THE WASTE PRODUCTS. Howland (1924) demonstrated the presence of uric acid in her cultures of paramecium. Later, Weatherby (1929) found that substances generally used in the preparation of culture media—hay, wheat, oats, etc. contain considerable quantities of uric acid. On the basis of tests, he concluded that the bulk of nitrogen excreted by paramecium is in the form of urea; no excretion of ammonia, uric acid, creatine, or creatinin was found.

Jones (1933) reported that the greatest concentration of uric acid was found in cultures of *P. multimicronucleatum* when the population of paramecia was greatest. He also made repeated tests for urea and ammonia but the tests proved negative. Adolph (1922) came to the conclusion that either waste-products are not deleterious through their retarding of metabolism or urea is not an excretory product in paramecium.

The effects of waste products of bac-

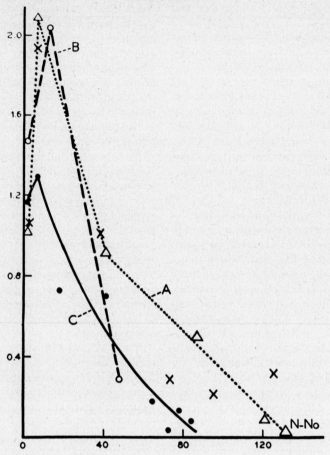

Fig. 64. Growth of Paramecium. [From Ludwig and Boost (1939), after Gause, Nastukova and Alpatov (1934)]. Abscissa = number of animals, ordinate = dN/Ndt. (Curve A) (△) *P. caudatum* in *caudatum*-medium. (B) (o) *P. aurelia* in *caudatum*-medium. (C) (●) *P. caudatum* in *aurelia*-medium. (D) (×, not drawn) *P. aurelia* in *aurelia*-medium.

terial metabolism such as CO_2, NH_3, trimethylamine, etc. (Oehler, 1916) are also factors which must be taken into account in the typical bacterized paramecium-culture.

As far as the writer is aware, no one has even considered the waste material evacuated from the anus of paramecium. In a rich culture of actively feeding paramecia, the amount of discrete material egested by the ciliates can be enormous.

Refer to Chapter 7 for a more detailed account of excretion and other metabolic functions.

8. Growth and Oxygen and Carbon Dioxide Relationships—the Redox Potential.

Increased growth of paramecia directly below the surface of mass cultures indicates in part their sensitivity to oxygen. Growth can be extended below the surface by aeration. The oxygen tension of a culture medium is an important factor upon which the reduction potential depends. It appears that oxidation-reduction potential can be meaningful only if we know and can control the oxygen consumption in experiments.

When applied to paramecia in culture medium, the redox potential is considered an indication of the oxidizing or reducing power of an oxidation-reduction system. Thus the more positive the redox potential the more highly oxidized is the culture medium. When the potential is more negative, the culture medium will be found to be more highly reduced. The components of the medium determining the reduction potential and which affect the ciliate are varied and complex. It is known that the redox potential varies with the pH of the culture medium but information at the present time is still lacking concerning detailed relationships between growth and the redox potential of culture media with Protozoa; concerning this refer to Efimoff, Nekrassow, and Efimoff (1928).

For Protozoa in general, we have little or no quantitative data in regard to the oxygen tension and the relative degrees of aerobiosis and anaerobiosis upon growth of Protozoa. One is referred to the review on this subject by Jahn (1934) who reports that several investigators have found that the oxidation-reduction potential of hay infusions shortly after inoculation are low but that during the first few weeks rise, gradually reaching a still higher value after three months. It is pointed out that the low initial readings were probably due to removal of free oxygen as a result of rapid bacterial growth at which time the potential is rapidly changing. Jahn reports that the reduction-potential of natural waters is quite high due to the presence of appreciable oxygen and that low reduction potentials are probably found only in waters containing little oxygen and considerable organic matter. This fact should be taken into account in problems dealing with ecology of paramecium.

9. Autocatalysis, Allelocatalysis, and Related Phenomena.

Increased crowding of organisms ("overcrowding") has been shown to reduce the rate of increase in population growth while Allee (1931, 1934, 1938) and others have presented evidence to show that a situation of "undercrowding" may exist in a population. That is to say, a population of a few paramecia may not show a rate of increase as great as a population somewhat larger but otherwise under similar conditions.

Relationship of density or concentration of paramecia to volume of medium is a factor that should be investigated in more detail. With precision culture methods such as have been employed recently to study nutritional requirements in ciliates, a statistical treatment of this problem is likely to yield profitable results in our knowledge of growth. Richards (1941) makes the important suggestion that organisms swimming in larger volumes of media may expend more energy in locomotion hence leave less energy for growth and reproduction.

If a bacteria-free paramecium is introduced into a drop of sterile culture medium there is represented a physicochemical habitat reduced to simplest terms. Ecologically speaking when an organism is affected by its habitat we have an *action;* when it influences its habitat the process is called a *reaction* and the whole represents an ecologic action-system. Concerning the existence or well-being of the protozoön, the action of the habitat on the introduced organism or its reaction on the drop of culture, or both, may be neutral, beneficial, or harmful (Allee, 1941).

If to this simplified microcosm we place a second organism like the first, the resulting changes in the ecologic action-system consist principally of the effects—direct or indirect—of one paramecium upon the other. The direct effects are called coactions and some of the many possible types of interactions are illustrated in Fig. 65 in which B, C, E, and F represent coactions in part and D less so. These coactions may be essentially neutral, beneficial, or harmful for one or both paramecia. Mating is one good example of a coaction.

The paramecia, in their reaction to the habitat, may fix some toxic material such as waste products of metabolism (a disoperative effect) more efficiently than either can do singly. On the other hand, the two paramecia may coöperatively condition their environment more effectively in some peculiar way. An example of this is the "Robertson effect" or allelocatalysis and this would represent a coöperative effect.

In his studies upon the ciliate *Enchelys*, Robertson (1923) proposed the theories called *autocatalysis* and *allelocatalysis* respectively which he believed must be considered in growth investigations of ciliates and other microörganisms.

For autocatalysis Robertson stated that some necessary substance is dissipated by the cells into the surrounding medium and that the reproductive rate of the ciliates is nearly proportional to

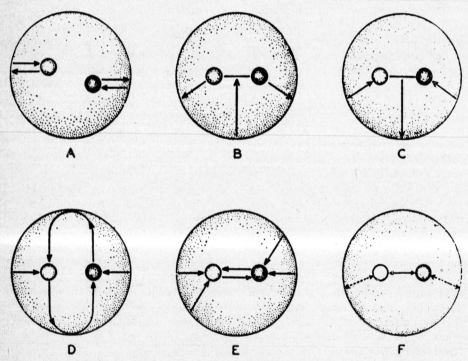

FIG. 65. Some of the possible types of ecological action systems when two organisms are living in a common environment. (Allee)

its concentration. When this substance is too dilute, (dissipated into too large a volume of medium) then reproduction, indeed maintenance, become impossible. He claimed that the growth of one organism or a population was autocatalytic because the growth curves were sigmoid and hence could be fitted by equation for a typical autocatalytic chemical reaction. For additional detailed information his works should be consulted (1908, 1921, 1922, 1924 a, b, c).

Subsequent work by a number of investigators working with Protozoa and other microörganisms has shown that the sigmoid growth curve is simply the result of the environment. When the environment is favorable, there is the regular geometric increase; when unfavorable, due to the diminution of food (bacteria) and accumulation of metabolic waste products, there is the decrease. The rate of growth is fairly constant when the environment is maintained fairly constant. In the main, the theory of autocatalysis has not been accepted by students of growth.

According to Robertson (1924 b) allelocatalysis is the acceleration of multiplication by the nearness of a second organism in a restricted volume of the medium. He reported (1921) that two Enchelys or two Colpidium in a drop of culture medium divided more than twice as rapidly as the division rate of one individual in an environment of equal volume. He was impressed with the fact that a ciliate may divide not at equal but at progressively shorter intervals. In other words, a culture begun with one ciliate may after a "lag-period" of 19 hours after inoculation produce only four individuals; four hours later, eight individuals; three hours later, 16 individuals; 48 hours after the original isolation, over 1000 individuals. In a consideration of certain other factors involved in growth, Robertson concluded that the only possible inference that can be drawn from this phenomenon is that the ciliates discharge into the culture medium some substance which accelerates their own multiplication. He believed this evidence of the existence not only of an internal supply of autocatalyst but also an external supply of catalyzer which is communicated to the surrounding medium by each of the ciliates which inhabit it.

Cutler and Crump (1923, 1925) using Colpidium were unable to confirm allelocatalysis in carefully controlled experiments and Greenleaf (1924, 1926) also failed to demonstrate it using P. aurelia and P. caudatum.

According to Petersen (1929) the division rate of P. caudatum was accelerated in volumes of 0.83 ml. culture fluid but not in volumes of less than 0.21 ml. Dimitrowa (1932) and others have claimed better growth in so-called "conditioned" media—media which originally contained actively growing paramecia—than in non-conditioned media. On the other hand, Meyers (1927) failed to demonstrate allelocatalysis with conditioned medium upon P. caudatum and Johnson and Hardin (1938) found the same with P. multimicronucleatum. Indeed Johnson and Hardin found that the conditioned medium had an inhibitory effect upon growth, as did Di Tomo (1932) with P. caudatum.

In working with P. caudatum, Darby (1930) proposed that allelocatalysis might be merely a pH effect. After carefully controlling the food supply of P. caudatum by feeding specimens measured quantities of bacteria each day, Chejfec (1929) was unable to obtain an accelerative (allelocatalytic) effect independent of the food supply.

Jahn (1934) has suggested that allelocatalysis might be explained in terms of the oxidation-reduction potential and CO_2 tension of the medium. He also presented mathematical evidence for defects of autocatalysis.

While Robertson's theories and deductions have been the subject of much adverse criticism, they at least have the merit of having stimulated much fruitful investigation in growth studies of Protozoa.

10. Growth in Relation to Vitamins, Hormones, and Other Substances (see also p. 175).

Although the flagellates have been studied rather extensively in respect to vitamins and other essential substances needed for growth and metabolism, comparatively little is known about the conditions in ciliates generally and practically nothing about the requirements in paramecium. This is easy to understand since many of the flagellate Protozoa are able to exist upon very simple compounds; indeed it has been shown that some can synthesize their own vitamins. On the other hand, paramecium, like other holozoic ciliates, depends principally upon bacteria as food hence we must deal with a more complex nutritional requirement unless the bacterial food supply can be substituted with simple compounds like the amino acids, etc.

EFFECTS OF VITAMINS. As would be expected, the earlier studies were, in the light of modern research, crude attempts to discover the role of vitamins, hormones, and other factors in respect to growth. At present a distinction is usually made between a *growth factor* and a *growth stimulant*. A *growth factor* is generally interpreted as an essential substance required by the organism for growth which it itself is unable to synthesize for its needs, i.e. a specific amino acid. On the other hand, a *growth stimulant,* while it may produce noticeable effects upon growth, is not essential for growth, i.e. certain accessory vitamins.

Calkins and Eddy (1916–1917) stated that pancreatic vitamin did not alter the division rate in paramecium. Flather (1919 a) in studying the effects of a diet of polished and unpolished rice upon paramecium concluded that the former was less favorable as a food but neither gave as high a fission-rate as the malted-milk controls. Orange juice likewise did not alter the fission-rate. It should be kept in mind that in these early studies, no serious attempt was made to control carefully the bacterial food supply hence the conclusions drawn are open to question.

Recently, comprehensive nutritional studies have been made on the ciliate *Tetrahymena* by Kidder and Dewey and a bibliography of their work giving technical methods, amino acid, and vitamin requirements may be found in their 1947 paper. One is also referred to the papers of Elliott (1935, 1939, 1950), Hall and Cosgrove (1944), and Hall (1944) for additional information and references.

Because this ciliate is a fairly close relative of paramecium, a study of the thorough investigations upon *Tetrahymena* should be profitable for those planning similar research upon paramecium.

Tetrahymena geleii has been maintained in bacteria-free culture in amino acid media. Without the contaminating effects of bacteria in the medium, investigations upon growth and biochemistry can be made with considerable reliability and results will reflect the ciliates' metabolism only. Johnson and Baker (1943) have investigated certain vitamin requirements in relation to populations of *Tetrahymena geleii* and found that thiamine (B_1) when added to fresh proteose-peptone medium produced a higher maximum population; when added to long-time cultures there was a secondary increase in population almost to the original peaks. Cultures with a mixture of thiamine, riboflavin (B_2), and pyridoxine maintained the highest population level and cultures with para-aminobenzoic acid had higher maximum populations than those under other conditions but died out earlier than other types of cultures. However, this was prevented by the addition of thiamine.

The growth-factor requirements of *Tetrahymena geleii* have been determined rather precisely in recent years although early work resulted in conflicting reports. For example, Kidder and Dewey (1942, 1944, 1945) showed that strain *W* did not require thiamine, pantothenic acid, nicotinic acid, riboflavin, and pyridoxine but later (1949) found all of them essential. With the purification of the unknown growth factor *protogen* by Stokstad et al. (1949), much of the confusion in the earlier work disappeared. Elliott (1935, 1939), employing a crude medium, showed the growth-stimulating properties of pantothenic acid and the essentiality of thiamine for another strain of *Tetrahymena geleii, E* (*Colpidium striatum*). More recently (1950) using a highly purified medium, he has demonstrated the essentiality of thiamine, pantothenic acid, nicotinic acid, pyridoxine, and the unknown growth factor, *protogen*. Riboflavin was shown to be stimulatory but not essential.

Thiamine is a requirement for growth of *Glaucoma pyriformis* according to Lwoff and Lwoff (1937, 1938). In his nutritional studies of *Colpidium striatum*, Klein (1943) used eight vitamins (vitamin A ester concentrate, thiamine hydrochloride, riboflavin, nicotinic acid, calcium pantothenate, pyridoxine, ascorbic acid, and viosterol in oil) with amino acids to obtain continued growth and permanent maintenance of cultures. Thiamine and riboflavin are required for *Colpidium campylum* (Hall, 1942) and it has been shown that *Colpoda duodenaria* requires large amounts of thiamine, pantothen, riboflavin, nicotinamide, and pyridoxine. It does not need p-aminobenzoic acid, biotin, or inositol (Tatum, Garnjobst, and Taylor, 1942; Garnjobst, Tatum, and Taylor, 1943). Hall (1944) found that an adequate supply of thiamine is essential to growth of *Glaucoma pyriformis* for the production of normal population densities and that it exerts an important stabilizing effect upon the population. Riboflavin was also found to be a significant requirement for production of normal densities of this ciliate.

EFFECTS OF HORMONES AND OTHER GLANDULAR SUBSTANCES. It has been known for a long time that endocrine glands and their products, the hormones, greatly influence metabolic activities and growth in metazoan organisms. With this fact in mind, many investigators have attempted to discover the effects of glandular substances upon Protozoa.

Nowikoff (1908) appears to have been the first to investigate the effect of thyroid on the division rate of Protozoa. Using desiccated sheep thyroid in distilled water containing *P. caudatum,* he reported an increase of numbers of paramecia when compared with the ordinary hay infusion culture. He failed to take into account the bacterial food factor in his mass cultures so that his conclusions are open to question. Making use of daily isolation cultures, Shumway (1914, 1917, 1929) reported the reproductive rate to be greater (65 per cent) than in ordinary hay infusions when *P. aurelia* and *P. caudatum* were cultivated in emulsions of fresh, boiled, and commercially powdered thyroid. The thyroid-fed paramecia were smaller, more active, and more transparent than the controls. Also the former possessed a more vacuolated protoplasm with a tendency toward the formation of three contractile vacuoles. The work of Abderhalden and Schiffmann (1922) in the main supported the investigations of Shumway. Budington and Harvey (1915) used freshly prepared thyroid extracts of the cat, bird, frog, fish, and turtle on paramecium and *Stylonychia* in culture and observed that all the extracts increased the fission-rate. Cori (1923) claimed that when desiccated thyroid extract is placed in hay infusions containing paramecia (*"P. putrinum"*) the division rate was accelerated by about 12 per cent.

On the other hand Woodruff and Swingle (1922–23, 1924) reported that thyroid or thyroxin failed to accelerate division in *P. aurelia;* actually the latter was found to depress the fission-rate. This is essentially in agreement with the findings of Riddle and Torrey (1923) and Torrey (1924).

Using nine different clones of *P. aurelia* and *P. caudatum,* Ball (1925) noted that the clones reacted differently under similar conditions. He discovered that with an uncontrolled bacterial food supply, members of the same clone of paramecium may divide at a significantly higher rate in solutions of desiccated thyroid than in control hay infusion cultures but by increasing the bacterial food supply of the control, it was possible to produce a fission-rate greater than in the thyroid-fed culture. When approximately equal numbers of bacteria were provided for both lines no significant increase in division rate was observed. Ball concluded that his evidence points to the fact that thyroid

accelerates the division rate in paramecium by providing a favorable bacterial food supply and not by the specific action of the thyroid hormone.

In addition to the observations made upon thyroid substances as affecting reproduction and growth in paramecium, other glands have been similarly investigated.

According to Shumway, suspensions of thymus, spleen, ovary, suprarenal, and pituitary have no appreciable effects on growth of paramecium. Woodruff and Swingle came to the same conclusions using pineal and pituitary material upon *P. aurelia.* Desiccated hypophysis produced no significant increase in fission-rate (Nowikoff, 1908; Shumway, 1917) but a slight increase was noted by Chambers (1919) while Abderhalden and Schiffmann (1922) reported a decrease followed by a marked increase when transferred to a fresh medium. Calkins and Eddy (1916–17) reported that pancreatic vitamin did not alter the division rate. Woodruff and Swingle (1924) observed either an increase or no effect on fission while Flather (1919 b) noted that pineal extract and Adrenaline accelerated the action of the contractile vacuoles. Woodruff and Swingle found a heightened division rate with fresh frog pituitary over the beef extract controls. Nowikoff and Shumway claimed desiccated suprarenal failed to increase the division rate while Chambers (1919) found a slight increase. Tethelin is claimed by Robertson (1923) to increase the fission-rate.

Besides using thyroid, Ball (1925) studied the effect of desiccated liver and hypophysis upon many clones of *P. aurelia* and *P. caudatum.* Like his observation made upon thyroid, he found no effect of these substances upon the multiplication of paramecia. Rohdenburg (1930) reported that extracts of adrenal gland, spleen, testis, thymus, and liver stimulated growth in paramecium while insulin, parathyroid, thyroid, pituitary depressed division rate and ovary extract was not significant. He raises the question as to whether or not these effects are due to some specific property of the gland or to extrinsic forces. Since no mention at all was made of bacterial food in his work, the findings are of dubious value.

For the most part, the conflicting claims of these investigators are due to the inadequate measures to insure standardized cultural conditions. For their results to have meaning, factors which must be taken into account in growth studies of this kind are given in Chapter 4. All work reported to date has been done upon cultures with bacteria present as the source of food. This in itself is a variable which must be considered of paramount importance. It is of course true that the task before the experimenter is difficult since paramecium is essentially a bacterial feeder but we can hope for some clarification of this problem by the use of pure cultures of paramecium which may be grown without the contaminating presence of bacteria as the only source of food.

OTHER CONSIDERATIONS RELATING TO GROWTH. In the bacteria-free culture medium, where specified amino acids, vitamins, etc., are added to glass-distilled water, it is of course possible to determine with considerable accuracy the exact chemical composition of the milieu. What also should be taken into account is the possible interaction of these amino acids not only among themselves but among vitamins and other substances in the medium. In organic media such as lettuce and hay infusions, it would be of considerable value to know a great deal more about their chemical and biologic characteristics.

There is comparatively little informa-tion available concerning the mineral requirements of Protozoa but calcium and iron appear to be essential for growth in a number of species that have been investigated.

According to Packard (1926) small amounts of NaCl accelerate the rate of fission in paramecium but larger amounts retard it and Herfs (1922) reported that from 0.25 to 0.5 per cent NaCl increased the fission-rate. Finley (1930) found that *P. aurelia* increased in numbers from 132 individuals per cc. in 10 per cent sea water to 550 individuals per cc. in 103 per cent sea water. *P. caudatum* was found to increase from 22 individuals per cc. in 13 per cent sea water to 242 individuals per cc. in 74 per cent sea water but decreased to 22 individuals per cc. in 103 per cent sea water. Frisch (1939) reported that the rate of fission increased in sea-water cultures of *P. caudatum* and *P. multimicronucleatum* until a concentration of approximately 6 per cent is reached. After that it decreased and was inhibited for one to four days. However, addition of more food increased the fission-rate.

According to Hammett (1930) the —SH or sulfhydryl group is the essential chemical stimulus to growth. He reported that increase of paramecia is accelerated under rigidly controlled conditions by extremely low concentrations of —SH; i.e. one part of sulfur as —SH in ten million parts of culture medium.

11. Acclimatization.

Although the term acclimatization originally meant an adjustment toward or adaptation to climate, it has come to mean any adjustment to physical or chemical environment. Paramecium, like every organism, is able to make some adjustment to meet varying or unfavorable conditions of its environment and grow. The change of environment may be an abrupt one in which case paramecia frequently die. Certainly most cases of successful acclimatization have been accomplished by slow, gradual changes in environment.

There are many types of acclimatization. Those which principally concern us are acclimatization to temperatures, chemicals, and osmotic concentrations.

TEMPERATURE. It was noted by Mendelssohn (1902) that paramecia maintained at the usual temperatures showed an optimum of approximately 24°–28° C. but if they were kept at 36°–38° C. for some hours, the optimum rose to 30°–32° C. Jollos (1913 a, 1914, 1921) performed numerous experiments upon the acclimatization of paramecium. Over a period of many years he studied in detail 58 clones of *P. caudatum* and 46 clones of *P. aurelia* not only in respect to temperature but chemicals as well. In observing the effects of resistance of *P. caudatum* to gradually increasing temperatures, he found that the thermal death point could be raised one or two degrees, occasionally three, in long-continued experiments.

In his experiments Jollos paid especial attention to the study of the permanence of acclimatization effects that could be passed on to succeeding daughter progeny. To them Jollos gave the name *Dauermodifikationen* which means long-lasting modifications that are inherited for many generations but finally disappear. This important feature in inheritance is treated more fully in Chapter 10.

With high temperatures, he noticed that no lasting effect was produced unless the paramecia were subjected to diverse temperatures for long periods of time. In one clone of *P. aurelia* which had been subjected to a temperature of 31° C. for two and one half to three years, there resulted an increased growth rate when compared with the controls. The experimental lines of paramecia showed three divisions per day while the controls divided not more than two times per day. The tolerance to high temperatures was increased and the increased tolerance lasted in some cases to six months after removal from the high temperatures, but the tolerance finally disappeared.

Jollos noted that the modifications due to high or low temperatures decreased only at endomixis or conjugation and not at the intervening periods of fission. Thus if subjected to high temperatures for a short time, the induced modification was said to disappear at the first endomixis after restoration to normal environment. If the high temperature had continued for a long time, the modification lasted through the first or second or later endomictic period but finally disappeared at one of these periods. The same was found to be true for conjugation.

It should be remembered that nutritive conditions such as the quantity and quality of bacterial foods were not considered in great detail by Jollos and may have been a factor of importance in his results. In this regard Doudoroff (1936) demonstrated that resistance is low to raised temperature in *P. multimicronucleatum* when food is present, but rose to a maximum when the food was exhausted. He also reported that there was no significant difference in the resistance between vegetative and conjugating individuals.

Jacobs (1919) studied acclimatization as a factor affecting the upper thermal death points in races of *P. caudatum* and in races of a large, three-vacuolate species which appears to have been *P. multimicronucleatum*. His results are summarized in Table 17.

One observes from the table, consid-

Table 17

Times Required to Kill Approximately One Half of the Individuals of Races
of *P. caudatum* and *P. multimicronucleatum* at Different Temperatures
(Modified after Jacobs, 1919)

Temperature C.	Paramecium caudatum				Paramecium multimicronucleatum		
	Race 2	Race 3	Race 4	Race 5	Race 1	Race 6	Race 7
43°	—	—	—	—	30 sec.	15 sec.	30 sec.
42°	15 sec.	20 sec.	20 sec.	20 sec.	1.5 min.	1 min.	2 min.
41°	45 sec.	45 sec.	1 min.	1 min.	4.5 min.	8 min.	4 min.
40°	2.5 min.	2.5 min.	2.5 min.	5 min.	13 min.	20 min.	7 min.
39°	3 min.	3 min.	4 min.	9 min.	18 min.	—	18 min.
38°	—	3.5 min.	7 min.	—	—	—	—
37°	—	4 min.	—	—	—	—	—
36°	—	6 min.	—	—	—	—	—

erable variation in the length of life at a given temperature. Shown also is the fact that *P. multimicronucleatum* is remarkably resistant to high temperatures as compared to *P. caudatum*. Jacobs remarked that in a number of different experiments, the degree to which a slow change increased the final resistance varied considerably with different races and under different experimental conditions but that in all cases it was very appreciable.

Efimoff (1924) studied acclimatization to cold and this is discussed on page 191.

Chemicals. The literature on acclimatization of Protozoa to various chemicals, drugs, poisons, etc. is voluminous (see p. 147). In the parasitic forms, this aspect has been studied in considerable detail. Preparations such as quinine and arsenic are used widely to treat diseases caused by pathogenic Protozoa. When repeatedly introduced into the infected host, it is often observed that the Protozoa become acclimatized to their action in the usual doses so well shown by the work of Ehrlich. It is not known whether this acclimatization to the action of drugs is due to natural selection of the more resistant strains of the protozoan parasite or whether it depends upon the formation of adaptive modifi-

cations on the part of the protozoön. Of significance is the fact that possibly these are the modifications ("Dauermodifikationen") or inherited changes by which the Protozoa become more resistant to chemical agents.

In the ciliate, *Stentor*, Davenport, and Neal (1896) reported acclimatization to a 0.00005 per cent solution of mercuric chloride. Death followed when the concentration was raised to 0.001 per cent after two days. These workers also demonstrated acclimatization to quinine. *Stentor* and *Spirostomum* were cultivated successfully and acclimatized in a 1 per cent solution of ethyl alcohol by Daniel (1909). He observed racial differences in *Stentor* in respect to acclimatibility. One is also referred to the works of Hafkine (1890) and Neuhaus (1910).

Neuschlosz (1919, 1920) discovered that by subjecting *P. caudatum* in weak solutions for about a month, the paramecia became acclimatized to greater concentrations of quinine, methylene blue, trypan blue, fuchsin, arsenic, and antimony. He then proceeded to make a chemical analysis of control animals that were immediately killed by these agents and those that had become acclimatized. In the latter it was found that the paramecia were in some way able to destroy the injurious compound while the non-acclimatized ones were not. This investigator concluded that by acclimatization, the metabolism of paramecium apparently had been changed in some manner so that it was able to secrete a substance that destroys the compound to which it was acclimatized. It was shown that the change in metabolism is of a characteristic nature for each substance since acclimatization to one fails to increase resistance to other substances.

When in compounds where they are trivalent, arsenic and antimony are highly poisonous to paramecia but in compounds which are pentavalent, the compounds are only mildly toxic if at all. Neuschlosz found that the slowly acclimatized paramecia transform the toxic trivalent compounds into non-toxic pentavalent compounds.

Harnisch (1926) was of the opinion that in the case of acclimatization to arsenic, it is not the ciliates but the accompanying bacteria that are so altered in their metabolism as to transfer the substance. Chejfec (1937) also acclimatized paramecium to quinine but his results failed to support the contention of the mechanism of acclimatization as described by Neuschlosz.

The work of Jollos (1913 a–34) is of considerable importance not only in regard to acclimatization but also in its relation to inheritance, hence it will be treated more fully under that division. He tested the effects on resistance by subjecting *P. caudatum* gradually to increasing concentrations of arsenic and reported that resistance to this compound was increased within two months. His basic method in selection consisted essentially of cultivating paramecia in a weak, non-lethal concentration then subjecting them to a stronger concentration which was lethal to most. Those that survived were returned to

the weak solution and allowed to multiply. This procedure was again repeated and it was found that the animals acquired the ability to resist higher and higher concentrations. His method greatly increased the resistance to arsenic and this increased resistance was inherited for long periods of time ("Dauermodifikationen") even after the paramecia were restored to control culture fluid without arsenic. In one instance the resistance was inherited in fission for approximately nine months through 600 fission periods then gradually lessened in resistance until the characteristic finally disappeared.

While gradual increases of the toxic substance are usually continued for a long period of time in most successful cases of acclimatization, Jollos also was able in a few cases to show that a marked resistance was acquired in a very short time in arsenic solutions. He also reported racial differences in respect to acclimatization and that conjugation of *P. caudatum* resulted in a loss of acquired resistance to arsenic.

The findings of Jollos dealing with the effects of salts of potassium, calcium, lithium, and other common chemicals are significant in acclimatization and growth studies. Using *P. aurelia,* he noted that a N/200 solution of calcium nitrate decreased the fission-rate while potassium chloride increased it. When the paramecia were cultivated in the potassium chloride solution the fission-rate rose above the normal even after the organisms were placed in the control media.

In conclusion, one may raise the question of the part played by bacteria or other factors that may have had some bearing in respect to growth in these studies.

OSMOTIC CONCENTRATION. There appears to be considerable tolerance shown by certain species of *Paramecium* to changes in osmotic concentration of the environment. In some species these changes must be made in a gradual manner; in others they may be made abruptly. Attempts have been made to acclimatize fresh-water inhabiting species to increasing concentrations of sea water and brackish-water forms to fresh water.

In the early literature beginning with 1786, O. F. Müller and many others described species which they called *Paramecium* as inhabiting brackish, salt, or sea water. From these early accounts one cannot at all be certain that the species described are valid ones. Indeed many of these early descriptions of so-called species of *Paramecium* have been relegated to other genera. For a more extended account of these early reports refer to page 1 and the review by Frisch (1939). More recently Lepsi (1926) and Kalmus (1931) erroneously state that Calkins (1902 a) found *P. caudatum* in the sea water of Woods Hole, Mass., and Kalmus is incorrect in reporting that Parona (1880) reported *Paramecium* from sea water.

It can be stated with certainty that one species, *P. calkinsi* in nature, is found in fresh and brackish water and sea water of great salt concentration. *P. calkinsi* was originally obtained and described by Woodruff (1921) from fresh water but Wenrich (1928 a) and others have obtained this species from the brackish water of Stuart's Pond, Woods Hole, Massachusetts. The author has reported *P. calkinsi* from "brine" pools of Cape Cod (Wichterman, 1951).

In a simple experiment, Wenrich reported on the acclimatization of a brackish-water race of *P. calkinsi*. He found that in all concentrations of sea water, this brackish-water race of *P. calkinsi* lived and behaved normally. However, when the paramecia were transferred directly from the brackish pond water to pure sea water, they became shrivelled and died. When the paramecia were transferred directly from the brackish water to fresh water, no harmful results could be detected.

In the race of *P. calkinsi* studied by Bullington (1930) the paramecia could be transferred directly from both fresh water and sea water. This was done by transferring a piece of the boiled hay or a grain of boiled wheat with paramecia from the culture from which the transfer was to be made to a clean dish and adding pure sea water or fresh water as the case might be. It was noted that during acclimatization, the paramecia at first moved less actively. Bullington found a slight shrinkage of the body in salt-water transfers and a slight swelling in fresh-water transfers but these effects soon disappeared in cultures.

The writer (unpublished) studied several brackish-water races which were cultivated successfully in equal parts of hay infusion and filtered sea water. A gradual change throughout the period of a year was made in an effort to acclimate the paramecia to pure hay or lettuce infusion. Organisms lived but grew poorly in the hay and lettuce infusions during that time. However, at the end of the year filtered sea water was again added to the hay or lettuce infusion to the amount of one part of the former to two parts of the latter and excellent growth resulted. After a period of four years, these cultures are still thriving in the laboratory in this combination.

P. woodruffi, the largest of the "bursaria" group, was discovered and described by Wenrich in brackish water at the Marine Biological Laboratory, Woods Hole, Mass., in company with *P. calkinsi* and has been cultivated in fresh water and sea water (Wenrich,

1928 a, b; Bullington, 1930). Wenrich noted that this species from brackish water was slightly yellow but when the organisms were placed in fresh-water cultures, the color disappeared. The yellow color was partly restored when a small amount of sea water was added.

Bullington also reported that this species is able to live and flourish in either fresh or sea water equally well and may be transferred from one to the other in a very short time. Successful acclimatization usually required four days from fresh water to pure sea water but the individuals could be transferred directly back to fresh water.

From the preceding account, it is evident that P. calkinsi and P. woodruffi can acclimatize themselves to fresh or sea water with comparative ease and grow. Although all other species of Paramecium are commonly found in fresh water,[16] attempts have been made to acclimatize them to saline or sea water of various concentrations. This aspect has been investigated by Balbiani (1898), Yasuda (1900), Züelzer (1910), Herfs (1922), Adolph (1925), Chatton and Tellier (1927), Finley (1930), Bullington (1930), Yocom (1934), and Frisch (1935, 1939). Recently, Gause and his associates have studied this problem in great detail in regard to acclimatization and natural selection.

When P. caudatum and P. aurelia are transferred directly from fresh water to sea water or water containing salt, it has been reported that the paramecia adapt themselves in maximum salt concentrations from 0.3–0.75 per cent depending upon the author's accounts. When the salt concentration[17] is gradually increased, however, many investigators report that P. caudatum can survive even when the salt concentrations of the culture reaches approximately 1 per cent. In the earlier references in the literature in which species of fresh-water paramecium have been described as living in sea or salt water in nature, the salt concentration, when given, was usually less than 1 per cent.

Finley (1930) reported that P. aurelia and P. caudatum can tolerate up to 10–20 per cent sea water but in gradual adaptation can survive a maximum concentration of more than 3.03 per cent—the concentration of Woods Hole, Mass., sea water—even retaining normal shape and locomotion in 100 per cent sea water. This latter phase of the investigation needs confirmation. He also claimed that during the period of acclimatization, the paramecia increased greatly in number.

Yocom (1934) reported that when introduced into 10 per cent sea water, P. caudatum showed little if any effects while a change to 20 per cent sea water resulted in death of the organisms after five minutes. However, when the sea water was added gradually in small amounts to fresh-water cultures of P. caudatum, the specimens lived in concentrations higher than 20 per cent sea water. Above 20 per cent concentration of sea water there was a marked decline in the population and he reported that

---

[16] Very early reports in the literature are excluded because of uncertainty in species identification by the investigators but Quennerstedt (1865), Florentin (1889), and Levander (1894) reported paramecia in saline environments. Smith (1904) reported P. caudatum from the Gulf of Mexico, New Orleans, and Edmonson (1920) reported P. caudatum and P. trichium from the salty Devil's Lake Complex of N. Dakota.

[17] Sea water is carefully and thoroughly filtered or it may be boiled for 10 minutes to kill all microörganisms and upon cooling, restored to its original volume by the addition of glass-distilled water. Also, artificial sea water can be used successfully. Finley (1930) described a simple weighing apparatus for determining the gradually increased salinity of the water as a result of evaporation. Gause and his associates used Osterhout's saline medium.

in cultures of 60 per cent or more, few if any paramecia survived.

In a comprehensive study of the experimental adaptation of *P. caudatum* and *P. multimicronucleatum* to sea water, Frisch (1939) reported that the concentration of sea water in which the last individual died varied from 33–52 per cent and required 21–42 days to reach the maximum concentration. It was found that the average maximum concentration of sea water for seven cultures of *P. multimicronucleatum* was 44.7 per cent and for five cultures of *P. caudatum* 44.6 per cent showing the two species to be quite equal in their adaptability.

Frisch presented tables showing that the length, width, and volume of paramecia decreased progressively with increased sea-water concentrations in addition to changes in shape of the specimens (Table 18). However, with the addition of ripe nutrient medium to the cultures, size and volume of the paramecia were increased and the specimens assumed their normal shape.

Growth of the population as shown in Table 19 increased in sea-water cultures until a concentration of approximately 6 per cent was reached. The table discloses that culture VIII reached the maximum number of 17,000 individuals on the fifteenth day in 20 per cent sea water but that later there was a steady decline in numbers until the last specimen died on the twenty-second day in 48 per cent sea water. Also shown is that the rate of fission during the first day was 1.5 (from 300 to 900 individuals); that it increased to 2.5 (from 900 to 5000 individuals) during the second day at which time the sea-water concentration had increased to 6 per cent and then decreased to 0.4 (from 5000 to 8000 individuals) during the third day. Frisch concluded that at least up to 6 per cent, sea water increases the rate of fission beyond the rate observed in fresh-water cultures from 1.4 to 2.5 and that higher percentages of sea water markedly decreased the rate.

The size of the contractile vacuoles and food vacuoles decreased in sea-water cultures and the average rate of pulsation of the contractile vacuoles and average rate of formation of food vacuoles were found to be lower in such cultures than in fresh-water cultures.

Frisch concluded that the inability of

Table 18

RELATION BETWEEN THE SIZE OF *Paramecium* AND THE CONCENTRATION OF SEA WATER IN THE CULTURE MEDIUM (Frisch, 1939)

Per Cent of Sea Water	Average Length in Micra	Average Width in Micra
Paramecium caudatum (VII)		
0	189	46
10	163	40
20	153	42
40	145	40
Paramecium multimicronucleatum (VI)		
0	300	76
10	241	66
20	250	63
40	218	62

Table 19

Day of culture	% of sea water	Total number of individuals in 10^{-3}	
		Culture VIII Fed only at the 10, 20 and 40% concentrations	Culture IX Fed on additional days marked by an asterisk
Start	5.0	0.3	0.3
1	5.5	0.9	0.9
2	6.0	5.0	4.0
3	6.5	8.0	6.0
4	7.0	11.0	9.0*
5	7.5	13.0	9.0*
6	8.0	13.0	12.0
7	8.5	13.0	15.0
8	9.0	14.0	15.0*
9	9.5	14.0	19.0
10	10.0	16.0	19.0
11	12.0	15.0	21.0
12	14.0	15.0	23.0
13	16.0	15.0	20.0*
14	18.0	15.0	19.0*
15	20.0	17.0	22.0
16	24.0	14.0	21.0
17	28.0	14.0	24.0*
18	32.0	12.0	22.0*
19	36.0	10.0	19.0*
20	40.0	8.0	18.0
21	44.0	5.0	10.0
22	48.0	.07	1.0
23	52.0	.0	.01

P. caudatum and *P. multimicronucleatum* to become acclimatized to sea water is due partly to a shortage of food in sea cultures and to the lethal changes produced by the salts of sea water upon the protoplasm and organelles.

In the course of his investigations in spiraling of the eight well-defined species of *Paramecium*, Bullington (1930) briefly reported upon the acclimatization of seven of the species to sea water. *P. bursaria* was found to be extremely sensitive to sea water; specimens were unable to live in concentrations which did not exceed 5–6 per cent. With *P. aurelia*, body size decreased visibly at 25 per cent sea water, half normal size at 35 per cent with death occurring at 37 per cent concentration 23 days after the

beginning of the experiment. *P. caudatum* appeared to be less adaptable to sea water than the preceding species since specimens began to flatten at 20 per cent concentration and more so at 30 per cent. All individuals were dead at 32 per cent concentration 20 days after the test. *P. multimicronucleatum* and *P. polycaryum* were able to live in concentrations of sea water varying from 40–42 per cent but did not survive in percentages above those given.

Finally the reports of variations in tolerance shown when fresh-water species of *Paramecium* are introduced into increasing concentrations of sea water are due to a number of factors. Some are differences in technique, salinity of the sea water, racial differences within the species, and nutrition. In most cases the investigators merely reported a tolerance to a given concentration of sea water; it would be more desirable to have information dealing with the ability of a species to reproduce and grow in such a concentration.

The food factor[18] is an important item generally overlooked in this type of investigation. Bacteria which normally grow and supply food in the typical fresh-water infusion may not be able to acclimatize themselves to such increased concentrations of sea water, hence the food supply would soon be exhausted while the acclimatized paramecia starve. Information bearing upon suitable bacteria that are able to acclimatize themselves to sea water would be of value in such experiments.

To avoid repetition, the important work of Gause and his associates dealing with acclimatization to increased saline concentrations and natural selection is considered in the following section relating to the struggle for existence.

12. The Struggle for Existence.

The precisely controlled laboratory experiments of Gause in the study of populations of paramecium, their succession and survival, are explained by relatively complete mathematical theory expounded by Lotka (1925, 1932), Volterra (1926), and by Gause himself (1934 e, 1935 b).

In the competition for living, one species, because of advantages not possessed by its competitor, is able to displace it. We may take as an example clones of *P. caudatum* and *P. aurelia* living in the same environment. It is of course easily possible to cultivate these two species in the same culture where each will compete for the same bacterial food. As shown by Gause (1934 d, f, 1935 a), in such a mixed population, the growth curves for each population are very much alike for the first eight days, then the *P. aurelia* population continues to grow while that of *P. caudatum* declines reaching extinction in about 16 days. In the competition, *P. caudatum,* which requires 1.64 times as much food as *P. aurelia,* has the advantage of a greater coefficient of geometric increase. However, the greater growth rate is a liability in competition, for *P. aurelia* is less affected by excretion products since this species can live twice as long in a strong concentration of waste products of metabolism as can *P. caudatum.* Only *P. aurelia* is able to survive the competition of the mixed population with the amount of food available and the medium used.

In regard to a single species, the characteristics of competition in a homogeneous population of *P. caudatum* based upon a large amount of numerical data is shown graphically in Fig. 66.

The influence of biologically conditioned media on the growth of a mixed population of *P. caudatum* and *P. aurelia* was studied by Gause, Nastukova, and Alpatov (1934 a, b) in which they concluded that if the decisive factor of competition is a rapid utilization of the food sources, *P. caudatum* has an advantage over *P. aurelia* but if resistance to waste products is the essential point, then *P. aurelia* will take the place of *P. caudatum.*

To determine this feature, pure lines of *P. caudatum* and *P. aurelia* were cultured separately and when the population growth reached the saturation level (after 11 days at 26° C.) the filtered fluids, free of paramecia but with their metabolic waste products, served as the experimental media.

Gause (1934 a, f) studied individual

---

[18] As a source of food for paramecium in a saline medium, Gause used the yeast, *Torula utilis.*

and mixed populations of *P. caudatum* and *Stylonychia mytilus* cultivated in an infusion inoculated with the bacterium *B. subtilis*. Each species was found to grow better separately than in mixed populations; however, the influence of *Stylonychia* on *Paramecium* was about 40 times greater than the effect of the latter upon the former. When wild, mixed bacteria were added as food, paramecia grew to the same concentration in the mixed population as it did separately. When competing in the same environment with *Paramecium* as it would have alone, *Stylonychia* grew to about half the number and declined while *Paramecium* maintained its own concentration in spite of the competition.

When grown in competition with *P. aurelia*, *Glaucoma scintillans* survives while the paramecia perish.

In studying competition with food supplied by bacteria and yeast to *P. aurelia* or *P. caudatum* and *P. bursaria*, it was found that *P. bursaria* ate the yeast but the other species would not. It was possible to establish from the data varying equilibria of populations depending upon the initial concentrations of the organisms.

Gause (1939, 1942), Gause and Smaragdova (1939), and Smaragdova (1940) performed a great number of

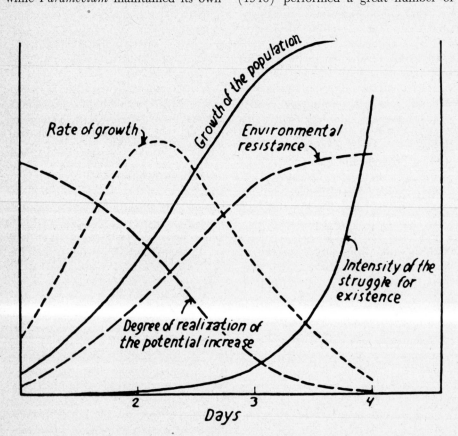

FIG. 66. The characteristics of competition in a homogeneous population of *Paramecium caudatum*. (Gause)

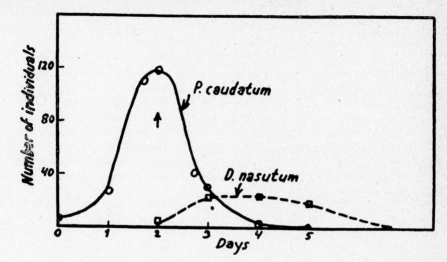

Fig. 67. The elementary interaction between *Didinium nasutum* and *Paramecium caudatum* (oat medium without sediment). Numbers of individuals per 0.5 cc. (Gause)

experiments upon various species of *Paramecium* in respect to adaptation to increased salinity and the relation of adaptability to adaptation.

Adaptability is considered as the capacity for adaptive modification in individual development which is due to a number of physiologic mechanisms acquired in the previous history of the species. When a population, say of *P. caudatum*, is subjected to the action of altered environmental conditions such as increased salt concentration of the medium, many complicated processes occur few of which are actually understood.

Gause (1939) and Smaragdova (1940) reported that *P. caudatum* and *P. aurelia* adapt themselves more readily to increased salinity than *P. bursaria*. *P. caudatum* and *P. aurelia* acclimatized themselves to 0.4–0.6 per cent of salinity while increased salinity over 0.21 per cent was usually fatal for *P. bursaria*. It is suggested that weak adaptability of *P. bursaria* to increased salinity is due to the fact that in the peat bogs where

this species is commonly found, the concentration of electrolytes is very low.

In *P. aurelia* the average resistance to salinity (Osterhout's salt medium) of 20 separately acclimatized clones was 1.129 (Table 20), while that of the acclimatized mixed population is seen to be greater: 1.230 and 1.245. Table 21 shows the same general effect in *P. caudatum* with the average salt resistance of 10 isolated clones being 0.557 but the resistance from a mixed population is 0.620.

Clones of paramecia which are genotypically weak at the start compensate this weakness by increased capability for adaptive modification. Those clones of paramecia which are genotypically strong in the initial resistance possess but weak adaptability. In the case of adaptation of *P. caudatum* and *P. aurelia* to increased salinity of the medium, the larger part of total adaptation is due to adaptability; clones that survived in mixed cultures are weak in the initial strength but powerfully adaptable.

Experiments were performed on individuals of opposite mating types of *P. bursaria* which were gradually acclimatized to salinity of 0.21 per cent for a period of 45 days. It is worthy of note that the disappearance from the mixed cultures of one of the two mating types showed that vigor or weakness of paramecia is closely bound with their mating type (Gause, Smaragdova, and Alpatov, 1942).

Gause found that the acclimatization of paramecia to quinine is essentially different from their acclimatization to salinity.

In his investigations dealing with the struggle for existence, Gause (1934 c, e, f) studied the destruction of one species of ciliate by another, namely, *Didinium nasutum,* the predator which devours *P. caudatum,* the victim. There exists a simple food chain as follows: bacteria→paramecia→didinia in which the last-named organism consumes on

Table 20

RESISTANCE TO SALINITY IN VARIOUS CLONES OF *P. aurelia* BEFORE AND AFTER ACCLIMATIZATION TO 0.4 PER CENT OF SALINITY*

Number of the Clone	Resistance before Acclimatization	Resistance after Acclimatization	Adaptive Modification
1a$_1$	0.470	1.140	+0.670
1a$_2$	0.422	1.080	+0.658
1b$_1$	0.500	1.145	+0.645
1b$_2$	0.477	1.090	+0.613
2a$_1$	0.525	1.165	+0.640
2a$_2$	0.458	1.230	+0.772
2b$_1$	0.480	1.240	+0.760
2b$_2$	0.485	1.100	+0.615
3a$_1$	0.480	1.195	+0.715
3a$_2$	0.465	1.120	+0.655
3b$_1$	0.520	1.015	+0.495
3b$_2$	0.520	1.230	+0.710
4a$_1$	0.520	0.900	+0.380
4a$_2$	0.540	1.210	+0.670
4b$_1$	0.507	1.130	+0.623
4b$_2$	0.510	1.105	+0.595
5a$_1$	0.550	1.160	+0.610
5a$_2$	0.430	0.990	+0.560
5b$_1$	0.455	1.125	+0.670
5b$_2$	0.430	1.200	+0.770
Average of 20 isolated clones	*0.487*	*1.129*	*+0.642*
Mixed population No. 1	*0.487*	*1.230*	*+0.743*
Mixed population No. 2	*0.487*	*1.245*	*+0.758*

* Resistance is expressed by the concentration of salt killing 50 per cent of the individuals per 24 hours. The test was made on the twentieth day from the beginning of the experiment. (Gause, 1942)

Table 21

RESISTANCE TO SALINITY IN VARIOUS CLONES OF *P. caudatum* BEFORE AND AFTER
RAPID ACCLIMATIZATION TO 0.36 PER CENT OF SALINITY* (GAUSE, 1942)

Number of the Clone	Resistance before Acclimatization	Resistance after Acclimatization	Adaptive Modification
41a$_1$	0.40	0.57	+0.17
41a$_2$	0.38	0.59	+0.21
41b$_1$	0.41	0.57	+0.16
41b$_2$	0.41	0.56	+0.15
42a$_1$	0.41	0.56	+0.15
42a$_2$	0.40	0.52	+0.12
42b$_1$	0.37	0.60	+0.23
42b$_2$	0.34	0.51	+0.17
44a$_1$	0.38	0.54	+0.16
44a$_2$	0.37	0.55	+0.18
Average of 10 isolated clones	—	0.557	—
Mixed populations	—	0.620	—

* The test was made on the thirtieth day from the beginning of the experiment.

the average one paramecium every three hours. When five specimens of *P. caudatum* are placed in a small test tube with 0.5 cm³ of oaten medium and after two days there are introduced three didinia, we have the picture as shown in Fig. 67.

It is thus seen that the didinia increase at the expense of the paramecia in such a homogeneous microcosm and after passing through a maximum, gradually decrease and disappear a few days later than paramecium. When the microcosm approached natural conditions by being made heterogeneous (which provided a place of shelter for the paramecia) then in eight out of 25 cases, the didinia died out, leaving the microcosm to the paramecia. By introducing more animals from time to time into the homogeneous microcosm, the fluctuations in the number of both species could be made periodic.

These experiments—treated in detail by Gause—should be of considerable interest to those concerned with the mathematical theory of the struggle for existence.

It is of interest to report that the first attempt at a quantitative study of the struggle for existence was made with Protozoa by Sir Ronald Ross in his theoretical investigations dealing with malaria. His conclusions are of great importance not only in epidemiology but also in the fact that Ross had expressed mathematical terms for the struggle for existence. This work is reviewed in detail by Gause (1934 f).

In an admirable book entitled "The Struggle for Existence" by Gause (1934 f), a large amount of valuable data based upon experimental and theoretical investigations are presented which deal with paramecium and other microörganisms. It is here that he presents detailed mathematical formulas and analyses found in many of his earlier papers. It is not within the scope of the present book to deal extensively with the work of Gause but since he and his associates have made such valuable contributions and present the latest thought in this type of research, the interested reader must consult it. Continu-ation of these carefully planned and executed studies should add much to our knowledge not only of the struggle for existence but growth, ecology, and the broader aspect of evolution as well.

Chapter 7

Metabolism: Respiration, Excretion and the Functions of the Contractile Vacuoles and Cell Surface

A. Respiratory Metabolism

1. Respiration.

To maintain the functions of life, paramecium, like every animal, must obtain energy from oxidative processes. By means of oxidation, there occurs a transformation of potential energy of complex chemical substances from the cytoplasm into active energy.

Paramecium takes up oxygen from its environment—the surrounding culture medium—and when a carbon compound is oxidized, one of the end products is carbon dioxide; the gaseous metabolism involved being called *respiration*. The problem concerned with the entrance of oxygen into the protoplasm of paramecium which is necessary for this physiologic combustion needs further investigation.

Spallanzani (1776) first gave the respiratory function to the contractile vacuoles and Frisch (1937) and many others share the same opinion, viz., that oxygen needed for respiration is obtained from water taken into the body through the ingestatory structures and carbon dioxide is removed by the contractile vacuoles. Frisch contended further that the pellicle of paramecium is impermeable to water and possibly gases. On the other hand, Ludwig (1928 b) maintained that the amount of water which enters paramecium through the oral passageway is too small to supply the oxygen for its respiratory needs and believed therefore that most of the oxygen enters through the pellicle.

From the work of many investigators, it seems likely that both contractile vacuoles and the pellicle play a part in the respiratory process.

Ordinarily, paramecia are seen to

gather at the surface of the media they inhabit where the oxygen concentration is at a maximum. They have been reported as inhabiting the profundal, oxygenless regions of lakes (Burge and Juday, 1911) and from low-oxygen environments such as sewage filters (Lackey, 1932; Lloyd, 1946 a, b) (see also section on ecology, p. 93). However, it can hardly be claimed that all traces of oxygen are absent at lake bottoms and certain other specified locations but the oxygen content of these environments is certainly much lower than that to which paramecium is accustomed. Pütter (1904 a, 1905) and others have claimed that paramecia and certain other ciliates can be cultivated in the absence of oxygen but Kitching (1939) noted that if care is taken to remove oxygen completely from the environment, paramecia die within 12 hours.

While it is a fact that paramecia require oxygen to carry on the functions of life, an examination of the large amount of literature dealing with respiration studies discloses results that are far from uniform even in the same species in respect to oxygen consumption (Table 22). Some of the factors which are perhaps responsible for this variation are as follows: (1) possible error in species identification, (2) bacterial contamination in respiration experiments, (3) racial characteristics of given clones, (4) physiologic condition of the paramecia such as age, nutritional state, conjugation, etc., (5) differences in size of specimens used, and finally (6) faulty technique in the handling of the paramecia and the precision apparatus.

In the main, studies upon respiration in paramecium have yielded facts which are comparable to those obtained from other microörganisms. Thus in paramecium, the rate of respiration is markedly accelerated with increased temperature. The rate of oxygen consumption in a freshly made culture containing young, growing, and actively dividing specimens is greater than in a culture containing older, non-dividing paramecia, hence the oxygen consumption decreases as the population becomes older. As would be expected, more active paramecia utilize more oxygen than others less active.

The ratio of the volume of carbon dioxide given off to the volume of oxygen consumed is called the *respiratory quotient* and is designated by the expression R.Q. Reducing processes occur in which oxygen can be obtained by the breakdown of oxygen-holding substances of the cytoplasm. Thus a carbohydrate such as glucose has an R.Q. of 1.0; fats give an average R.Q. of 0.707 and proteins have an R.Q. of 0.83 providing the nitrogen is eliminated as urea but an R.Q. of 0.93 if eliminated as ammonia. It is evident that the R.Q. as a definite value gives a useful indication of the nature of the substances which are being oxidized in the organism as well as being useful in comparative studies.

Respiration studies upon paramecium have been performed on either one to a few specimens or concentrated suspensions of them. The methods used have been by titration (Lund, 1918 a, b, c; Leichsenring, 1925), gas analysis (Amberson, 1928; Root, 1930), standard and capillary manometers (Kalmus, 1927, 1928 a; Howland and Bernstein, 1931), and the Cartesian diver ultramicrorespirometer (Boell and Woodruff, 1941; Claff and Tahmisian, 1948, 1949). One is referred to the original articles for descriptions of the methods but an excellent review of some of them is given by Jahn (1941) and Heilbrunn (1947).

Kalmus (1928 a), Howland and

Table 22

OXYGEN CONSUMPTION OF PARAMECIUM

Species	Temp. C.	mm^3 O_2 per Hour per Individual $(×10^{-4})$*	Average Rate of O_2 Consumption in mm^3 per Hour per mm^3 of Cell Substance	Method	Reference
P. aurelia	20°	.16	—	CO_2 production	Barratt (1905 b)
"	20°	.35	3.52	Barcroft and Kato (Warburg)	Pace and Kimura (1944)
"	25°	.62	6.13	"	"
"	30°	.83	8.27	"	"
"	35°	1.51	15.04	"	"
P. caudatum	26°	.14 (fed)	—	Winkler	Lund (1918 c)
"	26°	.04 (starved)	—	"	"
P. caudatum (as determined upon individuals during the conjugation cycle)	23°	.74	—	Thunberg-Winterstein	Zweibaum (1921, 1922)
	23°	3.48	—	"	"
	23°	.68	—	"	"
	23°	2.25	—	"	"
P. caudatum	18.5°–19°	3.85	—	Warburg microrespirometer	Necheles (1924)
"	25°	.15	—	Winterstein	Grobicka and Wasilewska (1925)
"	21°	5.20	—	Capillary respirometer	Kalmus (1927, 1928 a)
"	25°	.66	—	Warburg	Hayes (1930)
"	21.2°	.33	—	Capillary respirometer	Howland and Bernstein (1931)
"	23°	3.50	—	Capillary respirometer	Cunningham and Kirk (1942)
"	15°	1.78	2.78	Warburg	Pace and Kimura (1944)
"	20°	2.11	3.30	"	"
"	25°	3.86	6.03	"	"
"	30°	5.38	8.40	"	"
"	35°	9.70	15.16	"	"
P. multimicronucleatum	25°	1.00	—	Cartesian diver ultra-microrespirometer	Mast, Pace and Mast (1936)
"	25°	.60	—	"	Boell and Woodruff (1941)
P. calkinsi	25°	.25 (reactive for mating)	—	"	"
"	25°	.45 (non-reactive for mating)	—	"	"

* In some cases it has been necessary to compute oxygen consumption per individual from data of investigator.

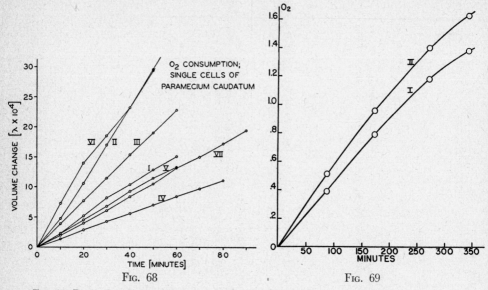

FIG. 68. Rate of oxygen consumption of single cells of *P. caudatum* in seven experiments. (Cunningham and Kirk)

FIG. 69. Graphic summary of the oxygen uptake of mating types of *P. calkinsi*. Each point represents the average of three determinations as described in the text. Ordinate denotes oxygen uptake in mμl per Paramecium. (Boell and Woodruff)

Bernstein (1931), and Cunningham and Kirk (1942) made direct measurement of the oxygen consumption of single specimens of *P. caudatum*. Howland and Bernstein expressed the opinion that the method of Kalmus failed to give an adequate temperature equilibrium. By modifying the Kalmus respirometer, they obtained values of oxygen consumption approximately one tenth of those given by Kalmus.

In what appears to be a carefully executed investigation, Cunningham and Kirk (1942) first constructed and calibrated capillary respirometers sensitive to $5 \times 10^{-5}\lambda$ and studied the oxygen consumption in single washed specimens of *P. caudatum*.[1] The results of their measurments are plotted in Fig. 68 and their mean value of oxygen consumption is given as $3.5 \times 10^{-3}\lambda$ per hour. It is in general agreement with

[1] $1\lambda = 1$ mm³ = 1 microliter

the findings of Kalmus and which Ludwig (1928 a) believes to be most reliable ($3.7 \times 10^{-3}\lambda$ per hour).

It is significant that Cunningham and Kirk have shown that variations of surface or volume of different specimens of paramecium are sufficient to account for differences in rates of oxygen consumption.

In their respiration studies upon Protozoa, Amberson (1928) reported that paramecium has an R.Q. of 0.69; Root (1930) an R.Q. of 0.62 and Mast, Pace, and Mast (1936) an R.Q. of 0.72 for *P. multimicronucleatum*. Root observed a definite trend in the direction of high R.Q. values in media of high CO_2 tension. As an example, the average R.Q. was found to be 1.43 at 238–423 mm. Hg.

According to Lund and Amberson, oxygen consumption of paramecium is fairly constant over a wide range of oxygen concentrations. The former, by the Winkler method, found that the rate of O_2 consumption was independent of oxygen tension between 0.04 cc. and 2.2 cc. of oxygen per 137 cc. By gas analysis, the latter reported a uniform rate of oxygen consumption with partial pressures of oxygen which varied from 50–220 mm. Hg. A slight decrease was observed at pressures as low as 11 mm. Hg. Kalmus (1928 b), however, reported a decrease of 35 per cent when partial pressure of oxygen is decreased 20 per cent below normal.

By means of chemical indicators, it is possible to determine colorimetrically the degree of oxidation and reduction. The relationship is called the oxidation-reduction or redox potential and is expressed by the symbol rH. It is apparent that oxidation-reduction phenomena are involved in respiration and growth of paramecium but little concrete information is available.

2. Effect of Temperature on Oxygen Consumption.

By means of titration, Barratt (1905 b) determined the CO_2 production of *P. caudatum* and observed that the rate at 27°–30° C. was more than twice that at 15° C. (2.7 per cent against 1.3 per cent). Using the Rideal-Stewart modification of the Winkler method, Leichsenring (1925) reported that a fall in temperature of the environment of paramecium produced a decrease in oxygen consumption and a rise in temperature brought about an increase in oxygen consumption. For *P. caudatum* respiration increased 35 per cent between 20° and 35° C. Paramecia at 0°, 5°, 10°, and 15° C. for a period of five hours showed a decrease of 58, 50, 34, and 30 per cent respectively as compared to the controls at 20° C. Fauré-Fremiet, Léon, Mayer, and Plantefol

(1929) noted that paramecium withstands lack of oxygen longer at 4° C. than at higher temperatures.

Pace and Kimura (1944) determined the rate of oxygen consumption and carbon dioxide elimination for *P. aurelia* and *P. caudatum* and their results are summarized in Table 23. Their results show a marked acceleration in respiration with increased temperature and their respiratory quotients indicate that paramecium has a greater carbohydrate metabolism at higher temperatures (30°–35° C.) than at lower temperatures (15°–20° C.).

They also determined the values of the temperature coefficient (Q_{10}) and the temperature characteristic (critical thermal increment or μ) for *P. aurelia* and *P. caudatum*. For *P. aurelia,* their Q_{10} values of 2.36 and 2.45 and the corresponding μ values of 15,144 and 16,457 were obtained between 20° and 30° C. and 25° and 35° C. respectively. The Q_{10} values for *P. caudatum* were found to be 2.17, 2.54 and 2.51 and the corresponding μ values 13,331; 16,608 and 16,888 between 15° and 25° C., 20° and 30° C., and 25° and 35° C. respectively. Between 15° and 40° C. the temperature coefficient was found to be greater than 2.0 and slightly greater for *P. caudatum* (2.5) than for *P. aurelia* (2.35–2.45). On the other hand Kalmus (1928 b) reported a Q_{10} of 1.5 for *P. caudatum* between 23° and 32° C.

3. Oxygen Consumption in Regard to the Physiologic Condition of Paramecium.

If paramecia are starved by simply removing them from a rich food source to water, there occurs a marked decrease in respiration concomitant with the utilization of reserve food materials in the cytoplasm. Upon feeding such

Table 23

COMPARISON OF OXYGEN CONSUMPTION AND CARBON DIOXIDE ELIMINATION IN *Paramecium aurelia* AND *Paramecium caudatum* AT DIFFERENT TEMPERATURES. EACH FIGURE REPRESENTS AN AVERAGE FOR FOUR TESTS (PACE AND KIMURA, 1944)

Temperature, °C.	Average Rate of Oxygen Consumption, mm^3 Per Hour Per mm^3 Cell Substance	Average Rate of CO_2 Elimination, mm^3 Per Hour per mm^3 Cell Substance	R. Q.
Paramecium aurelia			
20	3.16	2.30	0.73
25	6.60	5.30	0.80
30	8.50	7.70	0.90
35	15.50	13.70	0.88
Paramecium caudatum			
15	2.78	1.95	0.70
20	3.95	2.58	0.65
25	5.59	4.08	0.73
30	9.42	8.28	0.88
35	14.17	14.05	0.99

starved specimens with boiled yeast cells, the oxygen consumption is increased up to threefold even independent of cell division (Lund, 1918 c). When specimens of *P. caudatum* are starved for 24 hours, there occurs a 23 per cent decrease in oxygen consumption; after 72 hours, a 29 per cent decrease (Leichsenring, 1925).

The relationship of certain stages in the conjugation process to oxygen consumption was studied by Zweibaum (1921) who noted that the rate was approximately 0.73 mm³ of oxygen per 1000 specimens per hour immediately prior to conjugation. Throughout the conjugation process, the rate increased to 3.4 mm³ per 1000 specimens per hour, then decreased to 0.73 immediately after conjugation. The rate of oxygen consumption increased gradually to 2.0 mm³ per 1000 specimens per hour during the first eight or nine days following conjugation and the rate was maintained for approximately four and one-half months.

Boell and Woodruff (1941) studied the oxygen uptake at different intensities of the mating reaction in mating types of *P. calkinsi*. Fig. 69 shows the amount of oxygen consumed during a period which was capable of producing a mating tendency in the two mating types. Paramecia in the physiologic state in which they are reactive for clumping and pairing in conjugation respire at a lower rate than animals not reactive for mating. It will be noted that Zweibaum found essentially the same condition to exist earlier.

It was also reported by Boell and Woodruff that the oxygen consumption of mating type II specimens was greater than those of mating type I specimens and that regardless of their physiologic state, a statistically significant difference of about 12 per cent appeared in the rates of oxygen consumption for the two types. It occurs to the author that this may in reality be a racial or clonal difference rather than one of mating type. It is noteworthy that when metabolic activity of one mating type was high and the other low, the mating reaction occurred but it did not result in the formation of conjugating pairs.

In a newly developed Cartesian diver devised by Claff (Fig. 70) it is possible to make oxygen consumption determinations upon two opposite mating types of paramecium in the one diver before, upon, during, and after mixing and mating. Two clones of opposite mating type of *P. calkinsi* were supplied by the author to Claff and Tahmisian (1948, 1949) and their results are shown in the figure. One diver was used as a control and the contents were not mixed. After the combined respiration of mating types I and II was recorded for 50 minutes in each diver, the contents of the experimental diver were

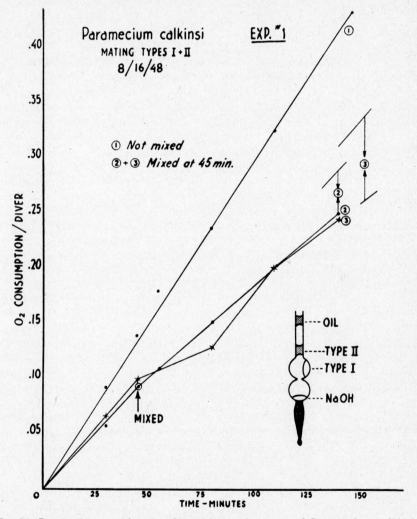

Fig. 70. Respiration rate changes of opposite mating types of *Paramecium calkinsi* before, during and after mixing for mating reaction. Figure at lower right shows diagram of new type of Cartesian diver. (×2.) (Claff)

mixed. For the next half hour, a reduction of respiration rate in the experimental diver was recorded, after which there was a partial return to the former rate of respiration. The initial clumping of paramecia soon after mixing was correlated with the decreased respiration. The breaking down of the large masses of mated paramecia occurred about 30 minutes after mixing and coincided with a recorded partial return to the original rate of respiration as shown in the figure.

4. Effects of Various Substances upon Respiration; Respiratory Enzymes.

It has been claimed by some that the respiratory mechanism of *P. caudatum* is insensitive to cyanide (Lund, 1918 b, 1921; Shoup and Boykin, 1931; Gerard and Hyman, 1931; Kitching, 1939 a; Jahn, 1941). If this were true, it would indicate that an oxidase or cytochrome-cytochrome oxidase enzyme system does not function in the respiration of paramecium.

Contrariwise, evidence has accumulated to indicate that paramecium is susceptible to cyanide and therefore possesses a cytochrome-cytochrome oxidase system with a respiratory mechanism comparable to that of typical plant and animal cells.

Child (1941) reported that Hyman (unpublished data) found a considerable decrease in oxygen consumption of paramecium in KCN. Pace (1945) wrote that he was informed by Dr. Hyman that she found an inhibition of oxygen consumption in *P. caudatum* when the organisms were exposed to KCN.

Boell (1942, 1946) noted that with appropriate dosages of cyanide and sodium azide, the oxygen consumption of *P. calkinsi* was reversibly depressed to 50 per cent of the normal. He and others expressed the opinion that failure of investigators to observe cyanide depression of respiration in paramecium may be due to the fact that certain technical details might have been faulty in the use of cyanide. One of these is possibly the failure to select properly balanced KOH-KCN mixtures.

With this fact in mind, the author suggested to Dr. Tahmisian that he test the cyanide sensitivity of a pure-line culture of *P. caudatum* (strain 352) which has been cultivated continuously in the laboratory for over 14 years. When tested in the Cartesian diver ultramicrorespirometer at 27.8° C. it was found that 1×10^{-5} M cyanide (cyanide and KOH mixtures prepared according to Robbie, 1946) will inhibit oxygen uptake 55 per cent when compared to the controls. After one hour when brilliant cresyl blue was added the recovery was 93 per cent of that of the controls. Therefore in the light of this and more recent experimental work, it must be concluded that paramecium is sensitive to cyanide and hence possesses a cytochrome-cytochrome oxidase system. Thiourea and thiouracil were found to have inhibitory effects on respiration in *P. caudatum* according to Nelson and Krueger (1950).

Also using a Cartesian diver respirometer, Humphrey and Humphrey (1948) observed that the endogenous oxygen consumption of *P. caudatum* had a value of 1.9 μl. per 10^4 animals per hour and was inhibited 60 per cent by 0.01 M cyanide and 40 per cent by 0.01 M azide. Methylene blue did not increase the endogenous oxygen uptake. Succinic acid doubled the oxygen consumption and this increase was inhibited by malonate. Oxygen consumption was increased by methylene blue in the presence of succinate still further and it also abolished the inhibition of this extra respiration by cyanide and azide. They concluded that *P. caudatum* resembles other animal tissue in possessing an active succinic dehydrogenase.

It has been reported by a number of investigators that one of the factors in sensitivity of the mechanism to cyanide is the degree of carbohydrate saturation within the cell. Pace (1945) noted that the rate of respiration in *P. caudatum* is increased with the addition of dextrose and that sensitivity depends in part upon the food content. Starved specimens of *P. aurelia* and *P. caudatum* were shown to be insensitive to cyanide while old specimens were not as sensitive to cyanide as young ones. He concluded that the effect of cyanide on respiration depended upon the degree of saturation of the respiratory mechanism with carbohydrates.

Leichsenring (1925) studied the effects of various substances upon the respiration of *P. caudatum*. It was found that oxygen consumption was increased with glycocoll, phenyl alanine, isoleucine, succinic acid, caprine, glutamic acid, peptone, and aminoids while tyrosine and cystine had little or no effect. An increase of 16 per cent was observed with lactose; other sugars and polysaccharides such as dextrose, galactose, sucrose, maltose, dextrin, and soluble starch increased the rate 3–10 per cent. Thyroxin increased the rate by 13 per cent. It is of interest that for the ciliate called by him *Glaucoma pyriformis*, Lwoff (1932) noted that if only sugar were present in the medium, the organism could live without oxygen for three days. According to Leichsenring, ethylene and nitrous oxide did not affect respiration of *P. caudatum* while ether and chloroform decreased the rate of oxygen consumption up to 25 per cent after two and one hours respectively and produced a deep anesthesia.

Pütter (1905) and others have shown that under anaerobic conditions, the glycogen content of paramecium decreased. He noted that when paramecia possessed little glycogen, they could live (possibly due to the utilization of albumen) for a relatively long time. Also observed was the fact that under anaerobic conditions, paramecia were able to live longer when the ratio of volume of medium to number of organisms is higher.

RESPIRATORY ENZYMES. Little work has been done with paramecium in regard to the possible role of the respiratory enzyme systems and pigments as related to the mechanism of respiration. The enzyme catalase, which is present in aerobic organisms, converts hydrogen peroxide to water and molecular oxygen. Its presence can be detected by adding hydrogen peroxide to a suspension of paramecia and measuring the oxygen evolved either chemically or manometrically. Burge (1924) found that the catalase action of *P. caudatum* was decreased by ether and chloroform but not by ethylene or nitrous oxide. Holter and Doyle (1938) showed the existence and some properties of amylase, peptidase, and catalase activities in their enzymatic studies upon Protozoa. The catalase activity of single specimens was determined for *Frontonia*, *P. caudatum*, and *Amoeba* to be 190:30:5 respectively but considerable variation was noted by them in individuals from the same and different cultures.

Sato and Tamiya (1937) reported the presence of the respiratory enzymes cytochrome *a* and *c* in *P. caudatum* by making a spectroscopic analysis of the absorption bands after being passed through thick suspensions of washed specimens. The examination under anaerobic conditions revealed three absorption bands with absorption maxima at 608, 551, and 523 Mμ respectively.

After aeration of the culture, these bands were replaced by ones with maxima at 581 and 545 Mμ. By alternate aeration and evacuation, the aerobic and anaerobic absorption bands, respectively, appeared and this reversible action indicated the respiratory nature of the substances causing the bands. Further treatment revealed that the respiratory pigments were a special type of cytochrome. Also revealed was the presence of a considerable amount of hemoglobin which had not been demonstrated previously in paramecium. As mentioned earlier, Humphrey and Humphrey (1947) have demonstrated the presence of a succinic dehydrogenase in homogenates of *P. caudatum* essentially similar to that found in the Metazoa.

A useful classification and description of respiratory enzymes dealing with Protozoa is given by Jahn (1941).

B. Excretion and the Functions of the Contractile Vacuoles and Cell Surface[2]

1. Excretion.

Excretion may be regarded as a type of activity in which waste products of cellular metabolism are eliminated from the protoplasm. An example of such a waste product is carbon dioxide which has been discussed earlier under the heading of respiration, but others would prefer to limit waste products of metabolism to nitrogenous substances only.

Excretion also functions in order to maintain osmotic equilibrium. It is clear that the protoplasm of paramecium has a much higher osmotic concentration than that of the surrounding medium. Water from the environment of the ciliate continuously enters the protoplasm especially while the animal is

[2] For information dealing with the structure, classification, origin, and operation of contractile vacuoles and the relation of Golgi apparatus and granules to them, see page 70.

feeding. Paramecia would increase abnormally in volume were it not for the regulative action of the contractile vacuoles and possibly the cell membrane in the elimination of excess fluid. Also, the contractile vacuole in paramecium as in other fresh-water ciliates, is a valuable indicator of general physiologic activities occurring within the cell-body. It is closely related to the water metabolism of the protozoön especially in regard to osmotic changes in the environment.

The structure and composition of crystals in paramecium are considered in detail on page 83. It is perhaps universally agreed that they represent catabolic products of metabolism. Schewiakoff (1893) observed them to disappear in paramecium in one to two days when the organisms were starved, only to reappear when food was again ingested. He noted that they are not defecated as are undigested food substances but are first dissolved and then disposed of presumably with the fluid of the contractile vacuoles.

Since the discovery of the contractile vacuoles by Joblot in 1718, many have ascribed to them peculiar and distinct functions which are mainly of historical interest. Spallanzani (1776) who studied their structure and rhythmic changes suggested that the contractile vacuoles are respiratory in function—a belief which was held by Dujardin (1841), Maupas (1863), Rossbach (1872), Bütschli (1876, 1887–89), Ehrmann (1895), and Haeckel (1873) as reported by Kent (1880–82). Ehrenberg (1838) held that the contractile vacuoles are organs which compose a part of the reproductive system.

Others, namely Lieberkühn (1856), Claparède and Lachmann (1854, 1858–60), Siebold and Stannius (1854), and Pritchard (1861) expressed the opinion that the contractile vacuole is a rudimentary beating heart.

That the contractile vacuoles are excretory in function was reported by Schmidt (1849), Stein (1859–83), Carter (1861), Griffiths (1888), Gruber (1889), Nowikoff (1908), Khainsky (1910), Woodruff (1911), Minchin (1912), Shumway (1917), Flather (1919 a, b), Marshall (1921), Riddle and Torrey (1923), Howland (1924), and others, some of whom gave additional functions beside excretory.

As first set forth by Hartog (1888), many including Degen (1905), Khainsky (1910), Züelzer (1910), Doflein (1911), Stempell (1914), Herfs (1922), Nassonov (1924), Day (1930), Kamada (1935), Kitching (1934, 1939), Gaw (1936), and Weatherby (1941) stated that the contractile vacuoles[3] are essentially organelles for maintaining the proper hydrostatic pressure within the body of paramecium. This interpretation is based mainly upon the fact that in the formation of food vacuoles, large amounts of water are taken into the body. Accordingly, the contractile vacuoles would then function as a device for the regulation of water balance within the organism.

The most widely accepted view held today is that the contractile vacuoles function primarily as organelles for the removal of excess water and in so doing remove excretory waste products of metabolism from the body. This interpretation is in reality a combination of two views mentioned earlier in which an abundance of water from the environment first passes into the cytoplasm then becomes laden with the nitrogenous waste products of metabolism only to be excreted through the contractile vacuoles. The contractile vacuoles would then function as osmo-regulators

[3] Lowering of the temperature of the culture medium of paramecium slackens the organisms and retards the rate of contraction of the vacuole; raising the temperature has the opposite effect (Day, 1930).

and excretory organelles at the same time.[4,5]

The origin of water excreted by the contractile vacuoles is said by Eisenberg-Hamburg (1925, 1926, 1929) and Frisch (1937) to enter the body of paramecium through the cytopharynx while Bozler (1924 a, b), Fortner (1926 b), and Müller (1932) contend that some of it enters through the pellicle. Indeed Kitching (1934, 1936, 1938) reports that in certain of his experiments, all water enters through the pellicle. On the other hand, Frisch is of the opinion that the pellicle of paramecium is impermeable to water, salts, and probably gases. Ludwig (1928 a) found that the amount of oxygen dis-

[4] Jennings (1904) observed that the contractile vacuoles of paramecium discharge their contents externally into the surrounding medium but earlier Stokes (1893) who studied *P. aurelia* and other ciliates reported (possibly referring to paramecium), "An Infusorian was encompassed by a cloud of bacteria and of similarly minute bodies or debris, through which, at every contraction of the vacuole a narrow path was swept with a quick puff, as a passage might be made through the duct by the sudden blast of a bellows." Lee (1941) described a technique for continuous or intermittent observation of the contractile vacuoles wherein rich cultures of paramecia dropped in agar may be studied for as long as 10 days.

[5] Although there are exceptions, generally marine and many parasitic Protozoa lack contractile vacuoles while the majority of fresh-water forms possess them. It would appear that the presence or absence of contractile vacuoles is related to the osmotic concentration of the environment. Thus in the parasitic amebas of man, contractile vacuoles are absent but the osmotic concentration of the environmental intestinal substance is high. A comprehensive investigation made upon a large number of free-living, fresh-water and marine ciliates and parasitic ones in respect to the osmotic concentration of their normal environments, presence or absence of contractile vacuoles and ingestatory structures would likely yield considerable information upon the function of contractile vacuoles. In studying pulsation frequency of the contractile vacuoles, Parnas (1926) concluded that they are principally excretory in function in marine Protozoa but both excretory and osmo-regulatory in fresh-water species.

solved in water that is taken with food in paramecium is insignificant when compared with the respiratory requirement of the animal and concluded that oxygen intake must also occur through the cell surface. He believed also that the contractile vacuole not only functions as an osmo-regulator but also in the excretion of carbon dioxide.

Experiments by certain investigators show that while water enters the body through the pellicle, a large amount also enters through the cytopharynx and may even pass directly into the cytoplasm without even entering the food vacuole. In support of the view that water does enter the pellicle, one need only observe a pair of conjugating paramecia in which the cytopharynx of each conjugant is completely dedifferentiated and lost from view. In such conjugants —each without a cytopharynx—the contractile vacuoles continue to function.

2. The Contractile Vacuoles in Respect to the Nitrogenous Waste Products of Metabolism.

In investigations upon paramecium and other Protozoa, Griffiths (1888), using the murexide test, claimed to be able to detect uric acid crystals in the fluid contents of some of the contractile vacuoles. He therefore believed the vacuoles to function in the manner of a kidney. Using the same method, Howland (1924) was unable to confirm the findings of Griffiths but when the Benedict blood-filtrate test for uric acid was used, positive results were obtained when culture fluid was tested.

Weatherby (1927, 1929, 1941) attempted to discover if the nitrogen in paramecium is excreted as ammonia, urea, uric acid, or a combination of these substances. He reported the presence of ammonia and urea in cultures of paramecium but held the opinion

that the ammonia was due to the hydrolysis of urea instead of being excreted as such. On the basis of his experiments, he held that urea is excreted by *Paramecium* and *Spirostomum* but not ammonia or uric acid and noted that many commonly used substances in culture media such as hay, wheat, barley, beef extract, etc. yielded positive tests for uric acid. It was his contention that a part of the urea which is an excretory product in these ciliates is excreted through the contractile vacuoles but that the greater part passes by dialysis directly to the exterior through the cell membrane. Like Weatherby, Darby (1929) concluded that urea was excreted by paramecium after which it was hydrolyzed by the bacteria present to produce ammonia. In studying *P. multimicronucleatum*, Jones (1933) found neither urea nor ammonia in any of his cultures but uric acid was present in greatest concentration in rich cultures 38–49 days old. In regard to nitrogenous excretion in *Colpidium campylum,* one is referred to the work of Nardone and Wilber (1950).

Believing that all of the urea excreted by paramecium could not be eliminated through the contractile vacuoles, Weatherby concluded that their function is not for the elimination of nitrogenous waste products of metabolism but rather the regulation of the hydrostatic pressure within the body.

There is considerable evidence to support this interpretation. It is well known that large amounts of fluid— primarily water—are discharged by the contractile vacuoles. This was observed quite early by Maupas (1883) who reported that the vacuoles of a single specimen of *P. aurelia* at 27° C. discharged a quantity of water equal to its volume in 46 minutes. After determining the volume of *P. caudatum*, Eisenberg-Hamburg (1926) stated that the two contractile vacuoles discharged a volume of liquid equal to that of the organism in 20 minutes, 51 seconds. It is easy to conceive that in the removal of such large amounts of water so quickly, an involved excretory mechanism is unessential for paramecium since extremely small traces of nitrogenous materials must be collected as the water passes from the cytoplasm to the contractile vacuoles.

Experiments based upon the effects of thyroid feeding and the reaction of paramecia to thyroxin were conducted by Nowikoff (1908), Shumway (1917), and Riddle and Torrey (1923) in which an excretory function of the contractile vacuoles was presented indirectly. A similar function was expressed by Flather (1919 b) after using adrenaline, posterior pituitary extract, and pineal gland extract.

Gelei (1925, 1928), with observations based principally upon morphologic studies in which he made homologies with the vertebrate kidney, is firm in his belief that the contractile vacuoles of paramecium function in an excretory manner although he suggests that they may remove excess water from the body.

Attempts to inject certain dyes into the cytoplasm of paramecium in an effort to determine whether or not they are excreted have thus far been unsuccessful.

The fact remains that there is really little substantial experimental proof of a purely excretory function of the contractile vacuoles in paramecium or for that matter in Protozoa generally but that they very likely play a role in the excretory process.

3. The Contractile Vacuoles in Respect to Their Functioning as Hydrostatic Organelles.

It is maintained by many that the contractile vacuoles function exclusively or nearly so in the regulation of hydrostatic pressure within the cell-body thereby preventing excess water from accumulating in the organism. Also, that the discharge rate of the contractile vacuole decreases in increasing concentrations of external medium. Pulsation frequency was found by Herfs (1922) to be decreased to approximately one-fourth the normal rate when paramecia were transferred from fresh water to 0.75 NaCl solution (Table 24).[6] Frisch (1935) was unable to adapt *P. caudatum* and *P. multimicronucleatum* to sea water, the ciliates dying when the concentration reached 40 per cent. Among other things, however, he noticed a marked decrease in pulsation rate of the contractile vacuoles. Similarly, in a study of the physiology of the contractile vacuoles of *P. caudatum, P. aurelia, P. multimicronucleatum,* and *P. polycaryum,* Gaw (1936) observed that the rate of contraction of the vacuole is decreased by increased osmotic pressure of the external solutions and the same phenomenon was reported by Yocom (1934). Gaw found considerable variation of the vacuole pulsations in paramecia from old and new cultures, hence used specimens from cultures five to seven days old. He concluded that all species investigated reacted in the same way to changes in osmotic pressure of the medium.

Kamada (1935) observed that paramecium flattens in concentrated solutions and was of the opinion that this may be an indication that the liquid cell contents are covered by a solid semipermeable envelope, the rigidity of

[6] Balbiani (1898) reported that paramecium has an osmotic pressure of 0.3 per cent sodium chloride solution (n/10). Spek (1919, 1920) in agreement with Estabrook (1910 a, b) gave a smaller value of n/70. Stempell (1914, a, 1924) investigated the behavior of the contractile vacuoles of paramecium in regard to the Hofmeister or lyotropic series.

Table 24

Concentration per cent NaCl	Temperature °C.	Vacuolar Frequency in Seconds	Time in which the Discharged Water-volume Would Be Equal to the Body-volume
0.00 %	22°–23°	6.2	14.7 minutes
0.25 %	20°–22°	9.3	21.3 minutes
0.50 %	19°–20°	18.4	42.7 minutes
0.75 %	19°–20°	24.8	56.9 minutes
1.00 %	19°–20°	163.0	6 hours, 19 minutes

which allows the cell only to take a smaller volume without changing its surface area. The result, therefore, would be an exosmotic flattening instead of the usual shrinkage of the cell-body. Like others, he noted that in the case of concentrated solutions, the discharge rate of the contractile vacuoles decreased gradually but after attaining a minimum, it increased again and approached a definite value in the long run. This final value, generally, was independent of the osmolar concentration of the applied solution. Accordingly, he maintained that some fraction of the amount of water discharged from the vacuoles under the normal condition must be of non-endosmotic origin. It was pointed out that this fraction—of some metabolic origin—masked the pure effect of osmosis of the new medium and the discharge rate took its course quite differently from what is expected from the theory of osmosis thereby suggesting that a special mechanism is involved.

Fortner (1926 b) expressed the opinion that the contractile vacuoles function for the protection of vital turgescence of the cell. He contended that in the interest of turgescence, the cell-body will accumulate water which is dammed up as long as possible, then discharged by a purely mechanical method.

4. Concluding Observations on the Functioning of the Contractile Vacuoles.

Frisch (1937), who has made an extensive study of the rate of pulsation and the function of the contractile vacuole in *P. multimicronucleatum,* observed that in flourishing cultures, the pulsation rate of the posterior vacuole is usually higher than that of the anterior vacuole. Unger (1926) found the same condition to exist in *P. aurelia* but in *P. caudatum* the anterior vacuole was found to pulsate more rapidly than the posterior one. On the other hand, he found that on the average, the anterior and posterior vacuoles contracted at the same rate in *P. calkinsi* (Table 25).

Child (1914), working with an unnamed species of paramecium (presumably *P. caudatum*) found that the anterior vacuole contracted more rapidly than the posterior one which is in agreement with Unger's findings for this species. On the basis of his experiments upon the susceptibility of paramecium to KCN, Child reported that the anterior end with its anterior vacuole was more susceptible than the posterior.

While Port (1927 a) reported that the average rate of pulsation is the same

Table 25

CONTRACTILE VACUOLE RATES (UNGER, 1926)

Approximately 24° C.	Contractions per Minute	
	Anterior	Posterior
P. aurelia	7.69	8.26
P. calkinsi	3.83	3.84
P. caudatum	8.27	7.44

in well-fed and in starving paramecia, Frisch maintained that the rate of pulsation of the contractile vacuoles varies directly with the rate of feeding; is lower in active than in resting animals with the magnitude of the difference depending upon the kind and the extent of the locomotion; and that variation in the rate of pulsation in one vacuole is accompanied by simultaneous and similar variation in the other.

He concluded that the passing of a constant stream of water through the ingestatory structures of paramecium serves for the intake of oxygen needed for respiration and the elimination of carbon dioxide and nitrogenous waste products, hence the contractile vacuoles serve not only to regulate the water content but have a respiratory and excretory function as well.

It is a fact that lowering of the temperature of the culture medium not only slows paramecium but retards the rate of the contraction of the vacuoles while raising of the temperature increases movement of the organism and rate of vacuolar contractions (Kanitz, 1907; Khainsky, 1910 a; Cole, 1925, and Gaw, 1936). Gaw repeated Cole's work and obtained similar results for the four species of paramecium under investigation in which the rate of pulsation of the anterior contractile vacuole was determined at approximately 16° C. (Fig. 71). The temperature characteristics calculated by means of the

van't Hoff-Arrhenius equation were nearly the same. He noted that below 16.1° C. the temperature characteristic was 25,000 calories for *P. caudatum, P. aurelia,* and *P. multimicronucleatum* and 23,300 calories for *P. polycaryum.* Between 16.1° and 25.1° C., 16,600 calories for *P. polycaryum,* 16,700 calories for *P. caudatum,* 16,500 calories for *P. multimicronucleatum* (between 17.7° and 26.8° C.), and 17,000 calories for *P. aurelia.* The figures above 22.5 ± 1° C. follow: 14,500 calories for *P. caudatum,* 12,400 calories for *P. polycaryum,* 10,900 calories for *P. aurelia,* and 9,500 calories for *P. multimicronucleatum* (above 26.8° C.). This and the work of others on paramecium would lead one to believe that the rate of pulsation within the temperature ranges is controlled by a catenary set of three reactions.

Gaw also noted that the rate of contraction of the vacuole is markedly affected by changes in the pH of the external medium as shown in Fig. 72 and that when studied over a pH range from 5.8 to 8.9, the optimum for contraction occurred around pH 7.0. It is of interest to note that the pH optima for the vacuole in each species are near the optimal pH for growth.

According to Taylor, Swingle, Eyring, and Frost (1933), the contractile vacuoles of *P. caudatum* cease pulsations in 92 per cent heavy water and greatly increase in size. For the same

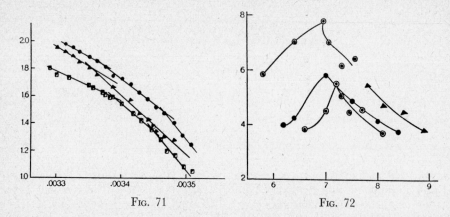

FIG. 71   FIG. 72

FIG. 71. Average rate of contraction of the anterior vacuole of *P. caudatum* at different temperatures (two series). The data of Cole (1925) are indicated by the points marked ◪. The log of the rate of contraction of the vacuole is plotted against the reciprocal of the absolute temperature as required by the Arrhenius equation. (Gaw)

FIG. 72. Average rate of contraction of the anterior vacuole of *P. caudatum* as effected by pH; two series of experiments (● first series and ▲ second series) and Eisenberg's data (1929) (☉). The rate of contraction of the vacuole per minute is plotted against the hydrogen-ion concentration (abscissae). (Gaw)

species, Gaw (1936) observed that the rate of pulsation of the vacuoles was reduced considerably in 30, 50, and 95 per cent heavy water and that the rate of reduction is proportional to the concentration of heavy water used.

Finally in the last analysis of the functions of the contractile vacuoles in paramecium or for that matter, freshwater ciliates as well, little concrete evidence is available. Existing evidence would indicate that their function is primarily one to regulate the water content within the cell-body hence serving as osmo-regulators. It is also plausible to believe that in so doing, certain products of metabolic activity such as carbon dioxide and nitrogenous wastes—however scant—are also eliminated through them as well as through the cell membrane or pellicle. Certainly it has never been demonstrated that the vacuoles function exclusively for the excretion of metabolic waste products.

Chapter 8

Movement and Motor Response (Stimulation; Irritability)

A. **Movement and Locomotion**
B. **Response to Stimuli and Types of Stimulation (Taxes)**
    1. **Reactions to Mechanical Stimuli (Barotaxis: Thigmotaxis, Rheotaxis, Geotaxis)**
    2. **Reactions to Chemical Stimuli (Chemotaxis)**
    3. **Reactions to Osmotic Stimuli (Osmotaxis)**
    4. **Reactions to Temperature (Thermotaxis)**
    5. **Reactions to Light; Radiation Stimuli (Phototaxis)**
    6. **Reactions to Electric Stimuli (Galvanotaxis; Electrotaxis)**
    7. **Reactions to Combined Stimuli**
C. **Trichocyst Extrusion**
D. **Conclusions on Behavior of Paramecium and the Question of Learning**

A. Movement and Locomotion

Paramecium swims in its environment by the beating of its cilia. They vary in length and distribution on the body and the reader is referred to page 60 for a discussion of their structure, organization, and distribution. When analyzed, the movement of a cilium is observed to consist of two parts; firstly an effective stroke and secondly, a recovery stroke. In an actively moving specimen, the cilium during its effective stroke moves as a fairly rigid rod though bending throughout its length and striking the water in such a manner that the organism tends to move in a direction opposite to that of the effective beat while the water moves in the direction of the beat (Fig. 73). However, during the faster recovery stroke, the cilium moves in a greatly flexed condition thereby exposing considerably less surface to the resistance of the water. Kraft (1890) reported that the velocity of the effective stroke of a cilium is about five times that of the recovery stroke.

The cilia are arranged in longitudinal or oblique rows on the body of paramecium and although these locomotor structures move at the same rate, they do not beat in the same phase. If we attempt to watch the behavior of a cilium in a single row, it will be noted that it is slightly in advance of the one behind it and slightly behind the one just in front of it so that taken collectively, the cilia, moving in a characteristic wave-like manner, are said to beat metachronously. This appearance is generally likened to the wave-like picture one sees while watching a field of grain or tall grass on a windy day. The crests of the ciliary waves moving in this metachronal manner are cilia at the peak of their effective stroke while the troughs of the ciliary waves are

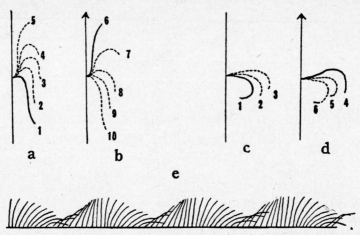

FIG. 73. Diagrams illustrating ciliary movements. (a–d) Movement of a single cilium in which (a) shows the recovery or preparatory stroke and (b) the beginning of the effective stroke for more rapid movement than shown in (c) and (d). (e) A longitudinal row of cilia showing metachronous movement. (Courtesy, Kudo: "Protozoology," Illinois, Charles C Thomas.)

cilia at or near the beginning of their recovery stroke. However, cilia in the same transverse row beat synchronously.

The effective strokes of all the cilia propel the animal forward or backward and since the stroke is generally oblique, paramecium, when swimming forward, is seen to rotate on its long axis in the manner of a left spiral (Figs. 7A and 74A) (except *P. calkinsi* which ordinarily spirals to the right as in Fig. 9C). The oral groove is not responsible for the direction of spiralling as can be determined by observing cut pieces of paramecium which, like the entire organism, spiral to the left (Jennings and Jamieson, 1902; Bullington, 1930).

From the very beginning of his work with the Protozoa, especially paramecium, Jennings made an exhaustive study on behavior and the effects of many different kinds of stimuli and his findings are reported in considerable detail in numerous publications (1897, 1899, 1899 a, b, c, 1900, 1901, 1902 with Jamieson [1902] and with Moore [1902]). The essential results of these

investigations are to be found in his book, "Behavior of the Lower Organisms" (1931).

Jennings contended that the spiralling of paramecium is due to the fact that while the cilia strike chiefly backward, they do so obliquely to the right thereby causing the animal to roll over to the left. Also this swerving of the body toward the aboral surface is due largely to the greater power of the effective stroke of the oral cilia which strike more directly backward. The result— the rotation of paramecium on its long axis—thereby enables this asymetric animal to follow a more or less straight course in forming large spirals. Bullington (1930) made a detailed study of spiralling in the eight species of paramecium and figures of the spiral paths of each are shown in Chapter 2. He reported that all except *P. calkinsi* are *characteristically* left spiralling but that *P. aurelia*, *P. bursaria*, *P. calkinsi*, *P. woodruffi*, and *P. polycaryum* are able to spiral in either direction, right or left. The characteristic direction of

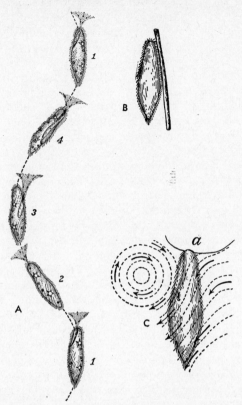

Fig. 74. (A) Spiral path of Paramecium. The figures 1, 2, 3, 4 show the successive positions occupied. The dotted areas with small arrows show the currents of water drawn from in front. (B) Paramecium at rest against a cotton fiber, showing the motionless cilia in contact with the fiber. (C) Paramecium at rest with anterior end against a mass of bacterial zoögloea (*a*), showing the currents produced by the cilia. (Jennings)

spiralling is the "automatic" one; the other direction occurs infrequently, not regularly. All eight species swim backward in right spirals. To observe spiralling, paramecia should not be placed on a slide and covered with a glass slip. Instead, they should be observed in a shallow watch glass under a low-power dissecting microscope. Additional references dealing with movement follow: Alverdes (1922 a, b), Glaser (1924), Fox (1925), Tsvetkov (1928), Ludwig (1929 d), Fauré-Fremiet, Léon, Mayer, and Plantefol (1929, 1929 a), Bullington (1930), Kalmus (1931), and Gelei (1937).

Hammond (1935) made stroboscopic observations of ciliary movement in *P. caudatum* and found three zones of cilia as follows: (1) the "body" cilia, (2) the longer cilia of the oral groove which were seen to begin at a point just aboral to the anterior tip of the animal, coming up around the tip and progressing posteriorly in a narrow triangular area with its base at the cytostome, (3) cilia of the gullet. The zones were differentiated by the differences in speed and nature of the beat of the cilia. Hammond was of the opinion that in respect to coördination, none of the so-called coördinating fibrillar systems explain the three different zones of ciliary activity in this organism.

In the classical book, "On Growth and Form," D'Arcy Thompson (1945) noted, concerning cilia, that one cannot reproduce or imitate them in any non-living drop or fluid surface. Also, that it is hopeless to understand them by comparison with a whip-lash or through any other analogy drawn from a different order of magnitude. It is his opinion that a ciliary surface is always electrically charged and that a point-charge is formed or induced in each cilium. It is not within the scope of this book to give a review of the numerous theories on ciliary movement but adequate reviews have been made by Pütter (1904, 1904 a), Erhard (1910), Prenant (1913), and Gray (1928, 1929, 1930).

It is to be remembered that all ciliary movement is essentially slow and that cilia move with low velocity. For an unnamed species of *Paramecium*, Chase and Glaser (1930) report that at 24° C. and at a pH of 7, the rate of movement is 833 microns per second. Thompson (1945) gives the rate at 1.3 mm. per second while Ludwig (1928 a) found that *P. caudatum* can swim with a velocity of almost 3 mm. per second or 15 body lengths per second for at least an hour without a pause. That a cilium is not a rapidly moving unit is shown by Gray (1928) who noted that if the length of a cilium be 10 microns and it oscillates through an amplitude of 180° 12 times every second, the total distance travelled in one second by the tip of the cilium is arrived at by the simple calculation as follows:

$$12 \left[\pi \left(10 \, \mu \right) \right] = 377 \text{ microns}$$

Attempts have been made to estimate the force exerted by the cilia of paramecium by adjusting these negatively geotropic organisms to a centrifugal force at which the ciliates are just unable to orientate themselves. In this manner, Jensen (1893) concluded that the ciliated surface of paramecium could exert a force of .00017 mg. or approximately nine times its own weight. According to him, each cilium exerted a force of approximately 4.5×10^{-7} mg. and each square centimeter of surface, a force of 21 mg. As reported by Thompson, an approximate estimate of the efficiency of the cilia alone is that they are able to exert a force 368 times their own weight.[1]

As noted by Jennings, food and other substances may be brought to paramecium by the currents created by the more strongly beating oral cilia. In this manner, paramecium is ever receiving samples or stimuli from the environmental medium some distance in front of it such as food, dyes, etc.

Not generally mentioned in a consideration of the movements and locomotion of paramecium—especially species of the "aurelia" group—is the great flexibility of the body. This feature can be seen readily by placing paramecia

[1] In an investigation to determine the effect of 56 different chemicals upon the direction of the effective stroke of the cilia in *P. caudatum*, Mast and Nadler (1926) reported that all the monovalent cation salts and hydrates tested (31) with the exception of $(NH_4)SO_4$ and $NH_4C_2H_3O_2$ induced reversal. None of the bivalent and trivalent cation salts tested (19) except $CaHPO_4$ and $MgHPO_4$ induced reversal. Reversal was induced with $Ba(OH)_2$, H_3PO_4, and $H_2C_2O_4$ but not with HCl and lactose. Oliphant (1938, 1941, 1942, 1943) also studied the effect of monovalent and bivalent cation salts on ciliary reversal in paramecium and his results with those of Mast and Nadler indicate that monovalent salts induce reversal but bivalent cation salts do not. In addition, they conclude that the duration of the reversed action varies with the kind and concentration of the salts and that the effect of the salts is due primarily to the action of the cations but that the anions have little if any effect. The aforementioned three investigators as well as Rees (1922 a) and others hold that the action of the cilia is controlled by the neuromotor apparatus.

among cotton fibers, dense debris, or upon semisolid agar.

B. Response to Stimuli and Types of Stimulation (Taxes)[2]

Response to stimuli is a fundamental characteristic of protoplasm and the characteristic is termed *irritability*. An organism like paramecium will respond to a given change in environment by some kind of activity. The amount of experimental work dealing with the reactions or responses of paramecium to various kinds of stimuli is enormous, hence no attempt will be made here to present an exhaustive treatment of the research done in the fields of the different types of stimuli.

In its daily life in a pond, paramecium, as mentioned earlier, is able to sample a region of its environment directly in front of it by the vortex of current created by the strong oral cilia. If the organism reacts favorably to the stimulus as in the case of bacterial food, paramecium may react positively by

swimming more closely into the food mass and begin feeding. On the other hand, if the stimulus be of an injurious or deleterious nature such as water of considerably varying temperatures, a strong chemical, or even when a piece of impassable vegetation blocks its path, paramecium simply swims backward in a right spiral by reversing its ciliary action thereby reacting negatively to the stimulus. This backward swimming is of course due to the reversal of the forward stroke of the cilia. In this manner paramecium will then attempt to avoid the unfavorable stimulus by backing out of the region, turning to one side and swimming forward again at a new angle and in a different direction. The animal may repeat this reaction until the stimulus is clearly avoided by the organism swimming free of it. Jennings has called this the *avoiding reaction* (Fig. 75A). Upon swimming backward for some distance in response to the stimulus, the swerving toward the aboral side is increased and this results in the anterior end swinging about in a large circle (Figs. 75B, 75C, and 75D). In this manner, the animal may set off in a new direction by swimming forward and mechanical obstacles in its path are avoided. Jennings reported that paramecia react negatively to the different classes of stimuli in the manner just described but in different intensities of the avoiding reaction to mechanical stimuli such as solid objects, water disturbances, chemicals, temperature changes, light, electric shocks, gravity disturbances, and centrifugal force. It is easy to understand how acclimatization is related to certain stimuli of low thresholds.

Finally, paramecia are seen to perform movements which subject it repeatedly and successively to a large number of different environmental conditions or stimuli. If the organisms react

[2] Some investigators prefer to use the term *tropism* as first used by de Candolle (1832) in regard to the turning of plants toward light. Its usual connotation is the orientation of an organism in relation to stimuli such as light, gravity, etc., in terms of its environment. Others prefer to use a word for each stimulus terminating in *taxis* or its adjectival form, *tactic;* this terminology has been used rather extensively in describing reactions to stimuli in the Protozoa. As an example, *chemotaxis* is a reaction to chemical stimuli in which the direction of locomotion of the organisms is determined by substances in their environment. If the direction be toward the stimulus, chemotaxis is positive (*positive taxis*); if away from the stimulus it is negative (*negative taxis*). If the direction of the movement is not clearly toward or away from the stimulus, chemotaxis is said to be absent. Other examples of *taxes* are *phototaxis*, or reactions to light; *thermotaxis* or reactions to heat or cold; and *barotaxis* or reactions to mechanical stimuli. Other mechanical stimuli generally included under the heading of barotaxis are *geotaxis* or reaction to gravity; *thigmotaxis* or reaction to the mechanical contacts of hard surfaces and *rheotaxis* or reaction to the pressure of currents in the surrounding environmental medium.

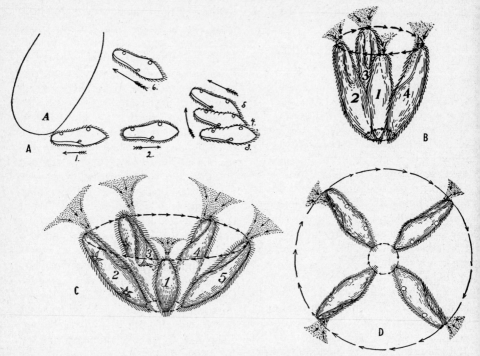

Fig. 75. (A) Diagram of the avoiding reaction of Paramecium. (*A*) A solid object or other source of stimulation; (1–6) successive positions occupied by the animal. (The rotation on the long axis is not shown.) (B) Paramecium swinging its anterior end about in a small circle, in a weak avoiding reaction; (1, 2, 3, 4) successive positions occupied. (C) More pronounced avoiding reaction. The anterior end swings about a larger circle; (1–5) successive positions occupied. (D) Avoiding reaction when revolution on the long axis ceases completely. The anterior end swings about a circle of which the body forms one of the radii. (Jennings)

positively, they will collect in certain regions near the point or source of stimulation. If they react *negatively,* they demonstrate the avoiding reaction (trial and error) which will cause the paramecia to avoid and escape from the place where the stimulus is acting.

1. Reactions to Mechanical Stimuli (Barotaxis: Thigmotaxis, Rheotaxis, Geotaxis) (see also Chapter 5).

Stimuli of this sort represent the type most frequently encountered by paramecium in its daily life in nature. An example of a mechanical stimulus (thig-

motaxis) is shown when a finely drawn glass tip is brought in contact with the body of paramecium and the response observed under a microscope. As reported by Jennings, after such an experiment the anterior tip is shown to be much more sensitive than the remainder of the body surface. In nature, the glass rod would be substituted by some object in the water and there would be a given response depending upon the nature of the stimulus.

Other examples of mechanical stimuli are currents of water (rheotaxis), agitation or jarring as in the case of sudden impact or pressure, and gravity (geotaxis). Various kinds of mechanical stimuli will now be discussed.

While the anterior part of paramecium is very sensitive to mechanical stimuli, a region around the mouth is similarly sensitive. When paramecium is touched with a finely tipped glass rod near the middle or posterior regions of the body, the animal moves forward. According to Bullington (1930), *P. aurelia* and *P. calkinsi* are very sensitive to touch but *P. multimicronucleatum* and *P. trichium,* less so. If the animal responds to stimuli, it will do so by turning toward the aboral side regardless of which side receives the stimulation.

Jennings observed that not all solid objects are necessarily avoided by paramecium. Under certain conditions, especially if it is swimming slowly, the animal may respond with weak avoiding reactions—say, to a piece of torn filter paper—then come to rest with its side adjacent to the object in such a manner that those cilia against the solid become stiff and set, seemingly anchoring the organism (Fig. 74B). While at rest in this position, oral cilia beat normally but body cilia beat only weakly with those behind the point of contact usually at rest. Figure 74C shows ciliary action and currents created when the organism is at rest with the anterior end attached to material.

Paramecium reacts to water currents, gravity, and centrifugal force in such a manner as to result in a definite orientation with the long axis of the body in line with the external force (Jennings).

Dale (1901) and Statkewitsch (1903 a) have reported that the animals swim with the water current but these investigators used irregular currents as in stirring. Jennings, on the other hand, noted that with a certain velocity of water current, the paramecia placed themselves with the anterior end upstream in line with the current. He believed that this reaction is essentially a response to a mechanical disturbance comparable to the one caused by the touch of a solid body.

In a study of hydrostatic pressure upon paramecia, organisms were placed in a pressure bomb by Regnard (1891) and subjected to pressures up to 300 atmospheres with no effect. He reported that at 400–600 atmospheres, the ciliates became immobile. Somewhat similar results were obtained by Ebbecke (1935, 1936) who reported that under high pressure, paramecia decreased their speed of locomotion or stopped entirely. Even after exposure to 800 atmospheres, Ebbecke was able to recover living specimens.

Dognon, Biancani, and Biancani (1932) and Biancani and Dognon (1935) described an apparatus for producing sound waves above the limit of audibility (20,000 per second) with a vibration frequency of 250,000 per second. When the temperature of 2 cc. of medium containing paramecia was raised 2°–10° C., in 10 seconds, movements of the organisms stopped, then the animals disintegrated.

Paramecia are negatively geotactic animals. They place themselves with anterior end directed upward and swim toward the top of a container, in which region they are found in greatest numbers. This may be readily demonstrated in a stoppered test tube containing medium with a large number of paramecia but little or no food. After the paramecia have collected at the top of the tube and it be inverted, the organisms will be seen to swim and collect in a region below the meniscus of the medium. The author, after centrifuging numerous cultures of paramecia representing many different races and species, is impressed by the fact that almost immediately after centrifugation and when the centrifuge tubes with concentrated paramecia are lifted from the machine, the ciliates will swim toward the top of the tube with considerable speed.

It was believed by Jensen (1893) that this reaction is due to the difference in pressure between the upper and lower parts of the organism but this appears to be an inadequate interpretation. Davenport (1897) was of the opinion that the reaction to gravity is due to the fact that the resistance in moving upward is greater than that moving downward since the animal is heavier than water. While Jennings (1904 a) held this view earlier, he later was inclined to follow Radl (1903) and Lyon (1905, 1918). Radl believed that under the uniform action of gravity, it is not apparent how any such difference of resistance could be detected by the organism. As Lyon (1905) maintained—and it has been amply demonstrated since— paramecium contains substances of differing specific gravity and these substances change with changing positions of the animal. This reaction to gravity is seen clearly only in the absence of other stimuli, hence if the medium contains many solids in suspension such as clumps of bacteria, frequently no reaction to gravity occurs. In this regard, it should be mentioned that Sosnowski (1899) and Moore (1903) have shown that many different stimuli modify the reaction to gravity. Both of these investigators were able to demonstrate a positive geotaxis or swimming down of paramecia by the use of low temperatures and after shaking the cultures. Jarring of paramecia which have collected at the upper surface of a tube (negative geotaxis) will generally cause them to swim downward showing a change in this reaction. King and Beams (1941) observed that mechanical agitation in *P. caudatum* caused a marked decrease in locomotion and ingestion,

due probably to effects upon ciliary activity.

In a series of papers, Dembowski (1923, 1923 a, 1928–29, 1929, 1929 a, b, 1931, 1931 a) reported upon the vertical movements of *P. caudatum* which play such an important part in geotaxis and noted that these movements are influenced by various factors. Dembowski is of the opinion that the posterior end of paramecium is heavier than the anterior end and this is the basis for his explanation of geotaxis. His theory is rejected by Koehler (1922, 1922 a, 1928, 1930) who presents a theory of his own (statocyst theory).

At the present time, there does not appear to be a satisfactory explanation of the mechanisms involved in the negative geotaxis of paramecium. Other references dealing with experimental work upon geotaxis are as follows: Verworn (1889, 1915), Jensen (1893 a), Ludloff (1895), Harper (1901, 1912, 1912 a), Statkewitsch (1905), Bozler (1926), and Merton (1935).

Jensen (1893) observed that if a tube containing paramecia is placed in a horizontal position in a centrifuge and then the centrifuge turned at a given rate, the ciliates tended to swim toward that end of the tube next to the center. It appears that paramecium, as one would expect, reacts to centrifugal force in the same way as it does to gravity.

2. Reactions to Chemical Stimuli (Chemotaxis) (see also Chapters 5 and 6).

Generally, paramecia respond to chemical stimuli by means of the avoiding reaction providing the chemical be of a certain concentration. They collect in weakly acid solutions, i.e. sulfuric, hydrochloric, nitric, and many other inorganic acids and acetic, formic, carbonic, propionic, and other organic acids and under certain circumstances, oxygen. Paramecia in pronounced alkaline water collect more readily in acids than even a neutral fluid (Jennings). This may be effectively demonstrated by placing a drop of $\frac{1}{100}$ to $\frac{1}{50}$ per cent of hydrochloric, acetic, or sulfuric acid on a slide containing a rich concentration of paramecia. Shortly thereafter, the organisms are seen to be collected in the drop of weak acid. According to Nagai (1907) calcium produces an increase of locomotion in paramecium.

Jennings (1931) and Barratt (1904 a, b, 1905 a, b, c) in an extensive series of experiments with chemicals have shown that certain ones are more injurious than others, i.e. chromic acid is 150 times as injurious as potassium bichromate. Jennings tested the effect of a large number of chemicals upon paramecium in respect to injurious effects, repellent power and the avoiding reaction and his work should be consulted by those who seek more information. On the basis of his work, Jennings concluded that the relative repellent power of different substances bears a relation to their chemical composition. Alkalies and compounds of the alkali and earth alkali metals (except the alums) have a relatively strong repellent effect while most other compounds have not. Paramecia even collect (i.e. react positively) in solutions of poisonous acid salts as corrosive sublimate and copper sulfate in which they are quickly killed. In such instances, the organisms swim into the solution without an avoiding reaction but show this reaction upon leaving it. However, strong acid solutions cause the avoiding reaction as shown when a drop of the acid is introduced into a preparation of paramecia.

Jennings observed that under usual conditions, paramecia do not collect in oxygen but under certain conditions the reverse is true. He performed a simple experiment to demonstrate this as follows: a bubble of oxygen (or air) was introduced into a slide preparation of paramecia and they failed to collect around it immediately. When the outer air was excluded by a petroleum jelly seal and the preparation allowed to stand, paramecia collected around the oxygen (or air) due to the fact that oxygen in the preparation became exhausted or nearly so.

Paramecia in the course of their respiration produce carbon dioxide which, when dissolved in water, produces an acid solution, the acidity being due to carbonic acid. This was demonstrated by Jennings in which a bubble of carbon dioxide was introduced into a slide preparation of paramecia. The organisms were seen to collect quickly and become concentrated around the bubble (Fig. 76, *left*, A). After a few minutes, the paramecia migrated with the diffusion of carbon dioxide (B) and still later the animals were found chiefly around the margin of the area containing carbon dioxide (C). As reported by Jennings (1931), this tendency of paramecia to gather in regions containing carbon dioxide plays a large part in their life under natural conditions. The fact that they themselves produce carbon dioxide explains many peculiar phenomena in their behavior.[3] Additional information dealing with the effects of various chemicals, hydrogen-ion con-

_____

[3] As emphasized by Jennings, all experimental work on the reactions of paramecia to stimuli must take these facts into account. Paramecia must not be taken with a pipette directly from a dense collection in a culture jar, and at once mounted on a slide since such collections contain carbon dioxide. Because paramecia quickly gather in the region containing most carbon dioxide, their reactions to other substances are inconstant and irregular. For experimental work it is always necessary before each experiment to place a few drops of the water containing paramecia in the bottom of a shallow dish and to aërate it thoroughly with air by means of a pipette.

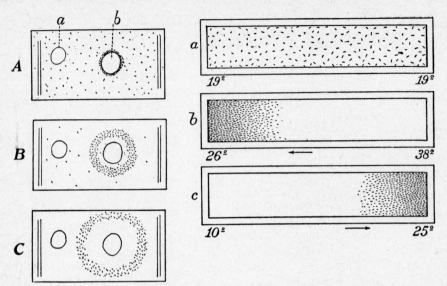

Fig. 76. (*Left*) Collection of paramecia about a bubble of CO_2. (*a*) A bubble of air, (*b*) of CO_2. (*A*) Shows the preparation two minutes after the introduction of the CO_2. (*B*) Two minutes later. (*C*) Eighteen minutes later. (*Right*) Reactions of paramecia to heat and cold, after Mendelssohn (1902). At *a* the infusoria are placed in a trough, both ends of which have a temperature of 19 degrees. They are equally scattered. At *b* the temperature of one end is raised to 38 degrees while the other is only 26 degrees. The infusoria collect at the end having the lower temperature. At *c* one end has a temperature of 25 degrees, while the other is lowered to 10 degrees. The animals now collect at the end having the higher temperature. (Jennings)

centration, hormones, vitamins, organic extracts, etc. is given in Chapters 5 and 6.

3. Reactions to Osmotic Stimuli (Osmotaxis) (see also Chapters 5 and 6).

It is at times difficult to determine whether or not a response to a given stimulus, i.e. a chemical reagent, is due to the chemical *per se* or to the osmotic concentration produced by the chemical.

Jennings has shown that paramecia enter without reaction, solutions of sugar (20 per cent) and glycerin (10 per cent), each possessing osmotic pressure many times greater than that of a solution of sodium chloride, yet in the salt solution the organisms react negatively with the avoiding reaction. The sugar and glycerin solutions each cause plasmolysis to such an extent that the paramecia resemble flattened plates. However, as the shrinkage occurs, paramecia react in the usual negative manner even though the specimens die in the solutions. It is thus seen that the organisms respond to osmotic pressure or stimuli with the avoiding reaction as with other stimuli but the onset of this reaction comes later than with other stimuli.

Large numbers of chemicals were examined by Jennings (1899 b) and Barratt (1904 a, b, 1905 a, b, c) in an effort to determine the effects of chemical and osmotic stimuli upon paramecium as well as the relative toxicity of the chemicals. Their papers should be consulted for additional information. (See also Chemotaxis.)

4. Reactions to Temperature (Thermotaxis).

While it is a fact that paramecium shows an optimum range around 24°–28° C., the organism can withstand temperatures which are considerably lower or higher than its optimum range. Khainsky (1910 a) reported that an increase in temperature to 24° C. or above resulted in a pronounced increase in the rate of the digestive processes. However, at 30° C. and above, paramecia ingested little, if any, food and the food vacuoles thus formed contained mainly water.

When a temperature change occurs markedly above or below the optimal range, paramecia show the avoiding reaction. In a gradual change, paramecia tend to collect near the optimal temperature and fail to show the avoiding reaction. The stimulus of heat is seen to increase the rate of movement while cold reduces movement, each stimulus producing the avoiding reaction.

The reactions to heat and cold were strikingly demonstrated by Mendelssohn (1895, 1902, 1902 a, b) who placed paramecia in long glass tubes or troughs. One end of the device was heated to a desired temperature and the other end maintained at normal temperature or a cooled surface. Paramecia were seen to collect in a region nearest the optimum (Fig. 76, *right*). They seek this optimum—reported to be between 24° and 28° C.—by the trial and error pattern of the avoiding reaction ultimately finding one route which does not cause thermal stimulation.

For additional information on the effect of temperature on paramecium, refer to the sections on acclimatization (p. 203), Chapter 5, p. 127 and Chapter 6, p. 191.

5. Reactions to Light; Radiation Stimuli (Phototaxis).

With the exception of the green paramecium, *P. bursaria,* which is positively phototactic, other species are indifferent to ordinary light. However, when the light intensity is suddenly and sharply increased, a negative reaction generally follows.

Joseph and Prowazek (1902) observed that paramecia exhibited a negative taxis to roentgen rays and responded 10–15 minutes after exposure by a concentration of organisms in a region not exposed to the rays. The effect of x-rays has been studied recently by the author (Wichterman, 1947 b, 1948 b, e, Wichterman and Figge, 1952, and Figge and Wichterman, 1952) and this work with others is treated in more detail in Chapter 5, p. 139.

The effects of radium rays upon various Protozoa including paramecium were studied by Züelzer (1905) who noted that responses shortly after exposure to the rays resulted in accelerated movement. Upon longer exposure, movement is considerably decreased below normal and there follows an injurious action resulting in death. The present author has observed that paramecia respond in the same manner to x-rays.

Urbanowicz (1927) found no positive effect of Gurwitsch's "mitogenetic rays." In this regard, refer also to Baron (1926).

Paramecia exhibit an immediate negative response to ultraviolet light. Upon longer exposure, a number of phenomena follow which are considered in detail in Chapter 5, p. 135, and Chapter 6, p. 193.

6. Reactions to Electric Stimuli (Galvanotaxis; Electrotaxis).

Paramecia respond to electric stimuli but the underlying causes remain complex and obscure. When two electrodes are placed opposite each other in a shallow dish containing paramecia and a constant current applied, all the organisms swim in the same direction toward the *cathode* or *negative* electrode where they concentrate in large numbers (Fig. 77A, *right*). If the direction of the electric current is reversed while the paramecia are swimming toward the cathode, the organisms reverse their direction and swim toward the new cathode.

Verworn (1889, 1889 a, b, 1892, 1896, 1896 a, 1915) observed and reported that when the stimulating current is weak, paramecia fail to respond sharply with comparatively few swimming toward the cathode. With an increased strength of current, more of the animals respond with increased speed until at a given current all are rapidly seen swimming toward the cathode. If after this, the strength of the current is still increased, the movement of organisms toward the cathode becomes slower and a still higher rate of current strength results in the organisms not swimming in any direction but with their anterior ends pointed toward the cathode and their posterior ends toward the anode. With a still greater increase in current, the animals now become shorter and thicker than before and they swim backward toward the anode. Finally a last increase in strength of current causes the paramecia to burst at one end and distintegrate.

Statkewitsch (1903, 1904) examined galvanotaxis by means of single induction shocks and found that while move-

ments of paramecia are considerably reduced, they react to that part of their bodies which are exposed next to the anode. The cilia are suddenly reversed and strike forward instead of backward with a contraction of the ectosarc and extrusion of the trichocysts, all on the anodal side (Fig. 77B). With a weak current only, ciliary reversal occurs but with a very strong current, contraction of the body and extrusion of the trichocysts occur even at the cathodal side. Still stronger currents result in deformation and disintegration of the body.[4]

Ludloff (1895) made the important observation that when the circuit of an electric current is closed, the direction of the stroke of the cilia on that surface of paramecium which is directed toward the cathode, reverses. However, if the long axis of the body is placed in an oblique position to the direction of the current, reversal occurs on all sides of the end of the cell nearest the cathode. Reversal extends to a line encircling the body in a plane at right angles to the direction of the current. The effect of this reversal depends upon the strength of the stimulus in which a stronger current affects a larger portion of the body (Fig. 77C, *right*). Both Ludloff and Statkewitsch noted that in galvanotaxis, the fluid in the body of paramecium is carried endosmotically toward the surface upon which the stroke of the cilia reverses. Ludloff (1895), Verworn

[4] Kamada (1928), in studying the time intensity factors in the electro-destruction of the membrane of paramecium, observed the "surface envelope" to be very firm and elastic. In its rupture by a galvanic current of considerable strength, he noted polar differences in its destruction. In anodal destruction of the body (especially the anodal half), contraction results, then expansion before rupture. In cathodal destruction, no such change in shape occurs but a transparent region forms on the cathodal side where rupture occurs. In anodal destruction, he reported that the macronuclear membrane bursts on the anodal side outside the paramecium but never within the body.

(1896 a), and Koehler (1925) claimed functional division of the organisms into anterior and posterior halves in which one half responds in one manner and the other half in a different manner.

Roesle (1902) noted that paramecium reacts more readily when the oral surface is toward the anode suggesting a more sensitive region around the mouth.

When frequent induction shocks are passed in a given direction through the medium containing paramecia, all the organisms become pointed forward and swim in the direction of the cathode, even when the current is so weak that one induction shock shows no reaction. Paramecia react more effectively to the stronger "break" currents of the inductorium than to the weaker "make" shocks (Birukoff, 1899, 1904, 1906; Statkewitsch, 1903, 1903 a, 1904, 1905, 1907).

As reported by Statkewitsch, when paramecia are placed in the field of a weak electric current and then observed closely, the organisms, while swimming in the direction of the cathode, show the cilia reversed at the anterior tip, with their effective stroke forward (Fig. 46 [1]). When the current is increased, this effect becomes more pronounced; the cilia becoming reversed farther and farther back. Later, with a stronger stimulus, the cilia on the anterior half of the body strike forward while those on the posterior half strike backward (Fig. 46 [3]). Fig. 77E [b] shows the water currents produced in such an animal. It will be seen that the cilia from each half of the body oppose each other with the organism seemingly trying to swim in both directions at the same time with the result that the animal remains in one position but revolves on its long axis or occasionally moves forward or backward for short distances. Finally with an increased stimulus of electricity, the effect is such

that the reversed cilia overwhelm the animal and it swims backward in the direction of the anode (Fig. 46 [4] and [5]). Ultimately after all the cilia have become reversed, the animal becomes deformed and disintegrates (Fig. 46 [6]).

When paramecium is transverse or oblique to the direction of an electric current of medium strength and the circuit closed so as to cause reversal in half the body cilia, the result is strikingly pictured in Fig. 77D. It will be seen that cilia on the anodal side strike backward, as usual, while those on the cathodal side strike forward. When in a transverse position, the animals react by turning directly toward the cathode side, even when the oral side is in the direction of the cathode. When the anterior end is pointing in the direction of the anode at the time the circuit is closed, paramecium turns toward its aboral side until its anterior end is directed toward the cathode.

Statkewitsch applied electric stimulation after the paramecia had been stained with neutral red and found that with weak currents, stained structures within the body became violet or acid but that in strong currents the structures became yellow or alkaline. Using Nile blue sulfate in addition to neutral red, Kinosita (1936, 1938, 1939), among other things, found a color difference at opposite ends after electric stimulation and concluded that paramecium becomes acid at the cathodal end but alkaline at the anodal end. Also noted was the fact that the alkaline region extended rapidly forward to encompass the entire body.

Jennings noted that if the direction of the electric current is frequently reversed, certain peculiar effects are produced. If reversal occurs at the moment when the anterior end has become directed toward the cathode, then the

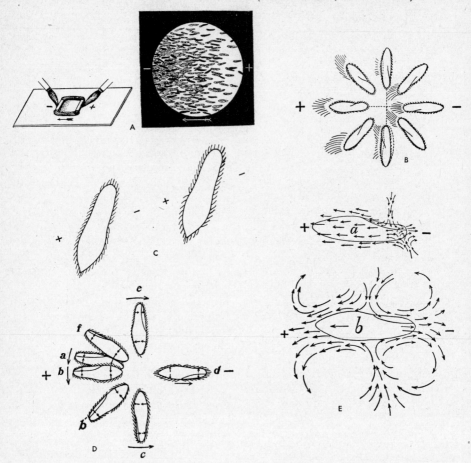

FIG. 77A. (*Left*) General appearance of paramecia reacting to the electric current. After Verworn (1899). The current is passed by means of unpolarizable brush electrodes through a cell with porous walls. The organisms have gathered at the cathodic side. (*Right*) Magnified view of a portion of the swarm as it moves toward the cathode. (B) Effect of induction shocks on paramecia in different positions. After Statkewitsch (1903). Trichocysts discharged, cilia reversed, and contraction of the ectosarc, at the anodic side or end, in a moderate current. (C) Paramecium showing reversal in the direction of the stroke of the cilia in a galvanic current. (*Left*) Weak current. (*Right*) Strong current. (+) Anode. (−) Cathode. (D) Effects of the electric current on the cilia of paramecia, and direction of turning in different positions. The oral side is marked by an oblique line. The large arrows show the direction toward which the animal turns. The small internal arrows indicate the direction in which the cilia of the corresponding quarter of the body tend to turn the animal. In all positions save *c* and *e* the cilia of different regions oppose each other. From *a* to *d* the turning is toward the aboral side; from *d* to *f*, toward the oral side. At *f* the impulse to turn is equal in both directions, and there is no result till by revolution on the long axis the animal comes into a position with aboral side to the cathode. (E) Water currents produced by the cilia in the electric current. (*a*) Electric currents weak; water currents reversed only at cathodic tip. (*b*) Electric currents stronger; water currents reversed over cathodic half as far back as the middle.

animal continues to turn toward the aboral side until the anterior end is pointed toward the new cathode. By repeated properly timed reversals, the animals can be caused to spin around and around, always toward the aboral side. If the intervals between the reversals of the current become shorter so that the animals have not yet become pointed toward the cathode, the animals may be made to swing back and forth like animated galvanometers with the anterior ends pointing out the direction of the current. Birukoff (1899) and Statkewitsch (1903, 1903 a) have described other interesting phenomena in regard to geotaxis which have been reviewed briefly by Jennings (1931).

It is seen that galvanotactic stimuli result in responses differing from other stimuli. Generally, other stimuli considered here result in coördinated movements but with electric stimulation, different parts of the body can be made to oppose each other. Galvanotactic stimuli are more localized and direct, wherein opposed reactions occur upon different surfaces of the body but in the main, all these reactions appear to be due to the cathodic reversal of the cilia.

Loeb and Budgett (1897) hold that the reaction to electric stimuli is due to electrolytic effects or changes in the chemical constitution of the environmental medium produced by the electric current. Studying galvanotaxis from the standpoint of physical chemistry, Carlgren (1899–1900, 1905) investigated localized changes in water content within the ciliates and believed that endosmotic streaming played an important role in the response. It is maintained by Coehn and Barratt (1905) that the movements of paramecia in response to galvanotaxis are purely cataphoretic (electrophoretic) but their theory ignores cathodic reversal of the cilia. Bancroft (1906 a, b) expressed the view that the responses caused by electric stimulation are due to local changes in the calcium content of the protoplasm in relation to other ions. He noted that if potassium, sodium, barium, and certain other salts are added to the environmental medium, paramecia swim forward toward the anode.

Kamada (1928, 1928 a, 1929, 1934, 1940), who studied ciliary reversal, galvanotaxis in respect to different salts, and potential differences across the ectoplasmic membrane of paramecium, demonstrated that some salts which induced forward swimming toward the anode, did not induce backward swimming. On the basis of his experiments he held that, contrary to the Loeb-Bancroft theory of galvanotaxis, the mere difference in the relative migration velocity of the cations or the anions does not seem to explain the reversal phenomenon in paramecium although he was unable to give a satisfactory explanation.

Obviously, many theories have been presented to explain the behavior of cilia and paramecia as a result of galvanotactic stimuli. It is the opinion of Statkewitsch, Jennings, and the writer that the phenomena of cataphoretic action or electric convection does not adequately explain the behavior of paramecium under electric stimulation.

In conclusion, it may be stated that in spite of the easily repeatable experiments concerning galvanotaxis, we still lack a satisfactory explanation of the underlying causes of the various reactions of paramecium to electric stimuli.

Additional references dealing with galvanotaxis and electricity upon paramecium follow: Pearl (1900), Massart (1902), Wallengren (1902, 1902–03, 1903), Greeley (1903, 1904), Herter (1927), Gelfan (1927), Kopaczewski

(1928), Metalnikow (1933), and Kinosita (1936 b, c, d, e, f, g, 1938 a). Also refer to Chapter 5.

7. Reactions to Combined Stimuli.

It has been shown that if paramecium is subjected to a second stimulus while the first is in effect, there results an altered response of what would be expected for either stimulus if each were allowed to act independently. On the other hand, paramecium may react to the first stimulus without responding to the second or to the second stimulus without responding to the first, whichever one is more effective.

Jennings observed that if paramecium is at rest against material under the action of the contact stimulus, and then struck with a finely pointed glass rod, the organism may not react to the second stimulus. However, a stronger blow on the anterior end caused the organism to move during which it gave the avoiding reaction. Pütter (1900) observed that a contact stimulus in paramecium is sufficient to prevent the organism from reacting normally to thermal or electric stimuli.

Both heat and cold stimuli also interfere with the contact reaction as well as immersion of the organisms in strong chemicals. Reactions to chemicals often interfere with the reaction to the electric current. As mentioned earlier, the reaction to gravity may be prevented by the presence of small pieces of solid matter in the culture medium and Sosnowski (1899) and Moore (1903) have shown that many other different stimuli may modify this reaction.

Jennings (1931) noted that when reacting to the contact stimulus, paramecium is less easily affected by other stimuli; when reacting to the other stimuli, it is less easily affected by the contact stimulus.

C. Trichocyst Extrusion

Paramecia may also react to stimuli by local contractions of the ectosarc and extrusion of the trichocysts. Their structure and operation have been discussed on page 64. Massart (1901) and Statkewitsch (1903, 1903 a) have presented valuable information upon their behavior after the application of many different stimuli.

Trichocyst extrusion can be produced by many different agents and stimuli such as crushing the organisms, quick application of heat, and many different chemicals including acids. On the other hand, weak mechanical stimuli and cold do not evoke trichocyst extrusion. Some reagents kill the animals without trichocyst extrusion while others may evoke extrusion at a localized area and still others may result in complete extrusion of all trichocysts. Statkewitsch noted that weak induction shocks result in trichocyst extrusion at the anodal side only while a greater shock causes discharge at both anodal and cathodal directions. A still greater current results in trichocyst extrusion over the entire body surface.

As mentioned by Jennings, although their function is uncertain, there is demonstrated in their reaction, a phenomenon comparable to a reflex action of higher animals. Their extrusion would appear to be more of an expression of injury, hence a secondary or pathologic phenomenon like blistering of the pellicle under certain stimuli. Commonly thought organelles for defense, they provide little if any safety against their chief enemy, *Didinium* (Fig. 135).

D. Conclusions on Behavior of Paramecium and the Question of Learning

From the preceding account, it is seen that paramecium reacts by varied movements of the entire body or its cilia to responses caused by different stimuli. Most of these movements result in the organism seeking an environmental stimulus to its liking or if not, avoiding the harmful stimulus. Thus, it may be said that in the main, paramecium rejects injurious environmental stimuli and accepts those which are beneficial. Many theories have been presented to explain this behavior in part or in its entirety but the fact remains that there exists today no suitable explanation to account for paramecium rejecting certain stimuli or conditions and retaining others.

Some are led to believe that in the so-called simplest animals, we may find simple explanations of life phenomena; but it is clearly seen that no such simple explanation is forthcoming in paramecium. When paramecium is compared to man, we note some stimuli which bear a striking similarity of response. For instance, in man the response to heat and cold beyond the optimum is by drawing back, just as in *Paramecium*.

Other examples of behavior and responses in paramecium which are fundamentally similar to those in metazoan animals follow: spontaneous action; existence of more sensitive regions of the body; the conduction of an impulse through the body; the summation of stimuli (Statkewitsch's experiments with induction shocks) ; changes in reactions to certain stimuli; the existence of positive and negative responses to stimuli and optimal levels.

Many protozoölogists have sought special structures in paramecium which, if not likened in morphology to the nervous system of metazoan animals, are claimed to function in the manner of a nervous system. These structures, called the "neuromotor mechanisms" are said by some to be responsible for the reception, conduction, and response of stimuli. It should be remembered that there are many unicellular organisms, especially amebas, which react like other organisms but do not possess so-called "neuromotor mechanisms." (See p. 48 for a detailed account of these structures and a discussion of the subject.)

Also, many writers have ascribed phenomena to paramecium and other Protozoa as comparable to learning which involves memory, as in higher animals. They maintain that paramecium is not entirely a "reflex machine" but that it can modify its behavior as a result of previous activity and experience.

If drops of a heavily concentrated culture of suitable bacteria are placed in a dish containing previously starved paramecia and observed under the microscope, it will be seen that the ciliates soon gather in groups of 100 or more about the masses of bacterial food. The paramecia cease active swimming and lie quite motionless except for the action of cilia in the oral groove and cytopharynx which produce the food-bearing current of water (Fig. 52).

The question of food selection in paramecium was investigated by Metalnikow (1907, 1912, 1914, 1917) who claimed that the organism can learn to recognize indigestible substances. He observed the number of food vacuoles formed and their contents under different conditions. Unassimilable substances such as chalk, carmine, and aluminum were rejected by starved paramecia in 20–30 minutes whereas food vacuoles containing assimilable food such as albumin and bacteria were retained for two to four hours. It was his contention that since paramecium ceased to ingest the unassimilable substances (until after the next fission stage) it had learned to do so. In a number of other experiments, he claimed phenomena analagous to conditioned reflexes but the data suporting his claim are not conclusive. Metalnikow observed that after paramecia were fed on unassimilable grains of carmine, the organisms ceased to ingest them, brushing the grains aside with their cilia although other digestible substances were taken up as before. This last effect only persisted until the next division of the body into two daughters in which each daughter ingested the grains of carmine in the usual manner.

Paramecia were fed on carmine in red light until they no longer ingested this substance. When next transferred to a yeast suspension, they ingested less yeast in red light than in daylight. Similar experiments were performed using alcohol as the conditioned stimulus instead of red light. In these as in other experiments, Metalnikow considered them to be conclusive but explained that the many cases in which they were unsuccessful were due to technical difficulties. Criticism has been directed both at the experimental methods employed by him and at the errors of his conclusions and deductions by Wladimirsky (1916), Bozler (1924 a, b), and others.

Wladimirsky noted that when paramecia are transferred to fresh culture medium every day, they continue to ingest carmine for as long as three and one-half months. He maintained that the refusal to ingest it in Metalnikow's experiments was due not to learning by experience but to a condition of depression as a result of the accumulation of harmful metabolic substances in the culture medium. Contrariwise, Lozina-Lozinsky (1929) a, e) claimed that in cultures containing carmine, the number of food vacuoles containing carmine diminished while the reproductive rate and capacity for feeding on other assimilable substances remained unimpaired, thereby indicating no general state of depression in the organisms. It

is to be regretted that these briefly reported experiments are difficult to evaluate.

Alverdes (1922 a, b, c, d, e, 1937) also studied food selection and reported that when drops of rich suspensions of bacteria are placed some distance away from previously starved specimens of *P. caudatum,* the ciliates swim directly toward the diffusing bacterial solution even before the bacteria reach the paramecia. According to Alverdes, since the paramecia were able to detect the bacteria at a distance, presumably as a result of a chemical stimulus, there occurred a direct orientation toward the food source and not a "trial and error" activity as set forth by Jennings.

Bozler (1924 a) repeated many of Metalnikow's experiments on *P. caudatum* and demonstrated in them a number of mistaken conclusions drawn by Metalnikow. For example, the distinction between carmine grains and India ink disappears if the carmine is ground to the same degree of fineness as other substances. Bozler observed that if sufficiently small particles (carmine, sulfur, chalk) of about the same size as bacteria came within the range of action of cilia in the ingestatory structures, the particles were driven into the food vacuole. Larger granules were swept away in the ciliary current but granules of intermediate size often were moved about for a considerable time in the oral structures before being either ingested or rejected although it is admitted that size is not the only factor. When paramecium is removed from a carmine suspension and placed into a yeast or starch suspension, the organism is observed to contain yeast cells up to 6 microns in length and starch grains up to 11 microns but only very small carmine grains measuring 1–2 microns.

Bozler maintained that ciliary activity of the ingestatory apparatus of paramecium is responsible for ingestion or rejection of particles and that selection seems to depend on the physical rather than the chemical characters of the objects taken into the body. Readily ingested yeasts, algae, and starch grains possess smooth, rounded surfaces while carmine, sulfur, and chalk particles which are more frequently rejected are more irregular in shape.

It was also noted that different paramecia varied greatly in their reactions to the larger particles and that this reaction was found to depend largely on previous experience. For instance, when paramecium was placed in a mixed suspension of bacteria and carmine particles, the formed food vacuoles contained carmine grains of various sizes. However, when paramecium was kept for the previous day in a carmine suspension, the organism took into the body only a few small grains of carmine with the bacteria.

Concerning other factors involved in food selection and food vacuole formation, see page 163 dealing with ingestion.

Smith (1908) and Day and Bentley (1911) attempted to demonstrate that paramecia can learn to double on themselves in order to turn around in capillary tubes so narrow in diameter as to restrict ordinary free movement. They reported that the average time and number of unsuccessful attempts to turn was greatly reduced with practice and that this learning was not due to the effect of an accumulation of carbon dioxide. It was claimed that the results of practice were retained by the paramecia even after the animals had been allowed to swim for 20 minutes in an open dish.

This explanation was criticized by Buytendijk (1919) who believed that instead of demonstrating true learning, the behavior showed a change in the physiologic state of paramecium which simply resulted in its greater flexibility. Organisms first treated with chloroform became flexible enough to permit turning easily in the capillary tube.

According to Bramstedt (1935), *P. caudatum* can learn to associate the avoidance reaction to heat with the stimuli of mechanical shock or light and his results on the association of light with heat were confirmed later by Alverdes (1937). Bramstedt also claimed that paramecia adapt themselves to the shape of a container since upon their removal, the ciliates continue to swim in a course similar to the outline of the container. Grabowski (1939) and others, however, disagreed with Bramstedt's conclusions that learning is involved and presented evidence to show that the paramecia were merely reacting to a chemical change in the liquid arising from heat action. Grabowski also held that the observations of Bramstedt, upon adaptation of the shape of paramecium to the shape of the container, are inconclusive to warrant that learning had occurred. Instead, the results merely indicate an increased sensitivity to the factors involved, according to this critic.

Soest (1937) was unsuccessful in forming an association between dark or light with cold. Although electric shocks could not be associated with dark, they did produce an avoidance of light in the ciliates. As reported by Tschakhotine (1938), *P. caudatum* will avoid a beam of ultraviolet light, even continuing to avoid the same region for a half-hour after the beam has been removed. This experiment was also criticized by Grabowski who suggested that a chemical change in the liquid as a result of the ultraviolet light may have been responsible for the behavior of paramecia.

Alverdes (1939) attempted to answer the criticisms raised by Grabowski and the former maintained that Protozoa can remember and form associations much like higher animals. However, there follow claims and counterclaims and replies to criticisms by Koehler (1939), Alverdes (1939), Bramstedt (1939), Diebschlag (1940), and others.

French (1940) considered trial and error learning in paramecium by enclosing a specimen in a glass tube (0.6 mm. internal diameter; 4.6 mm. in length) which was large enough to permit free swimming and turning. The object of the experiment was to observe any change with practice in the time taken to escape from the lower end of the tube. This investigator reported that the average time for escape decreased on successive trials. He concluded that the change in behavior was not related to general activity and seemed best understood in terms of learning.

A number of investigators have reported upon individual differences in regard to behavior patterns in paramecium. Jennings (1899) noted individual variation in sensitivity to chemicals and to electric stimulation. He observed that while paramecia generally swim toward the cathode, there are always a few specimens which either react in the opposite manner or not at all. Day and Bentley (1911) found that certain paramecia were able to turn around in a capillary tube more quickly than others and French (1940 a) reported individual variations in the rate of learning by trial and error. In regard to the selection of food by paramecium, Bozler (1924 a) observed that certain specimens of *P. caudatum* varied greatly in their reactions to larger food particles and claimed that this reaction was due largely to previous experience. In studying individual differences, French reported that the tendencies for some paramecia to form groups in the presence of food and for others to continue to swim freely, as well as for some to enter a chemical region and others not, persisted for many hours under different experimental conditions. Accordingly, he favored the interpretation that the individual reactions were due to persistent individual characteristics.

In the last analysis and taking into account the claims and counterclaims, it is difficult to draw definite conclusions in regard to convincing demonstrations of conditioned reflexes, associations, and true learning in paramecium. Until more convincing data are published, it seems unwise to draw conclusions at the present time. Some of the work in attempting to demonstrate learning in paramecium is due to faulty techniques and failure to know the general physiologic activities of the organism under different conditions. Chapters 5, 6, 7, 11, and the present one in this book contain much information that should be taken into account by the experimenter attempting to demonstrate true learning in paramecium.

Additional references in regard to learning are Binet (1889) and Jennings (1899, 1931). The behavior of paramecia while undergoing fission, mating, conjugation, and various nuclear phenomena should also be consulted under their respective headings.

Concerning social life and behavior, the transition from the individual to the social level, and the beginnings of social behavior in paramecium and other unicellular organisms, refer to Jennings (1940 a, 1941, 1945 a).

Chapter 9

Reproduction, Nuclear Processes, and Sexuality

A. Asexual and Other Similar Processes Involving Nuclear Reorganization

Paramecium, like the majority of ciliates, is unique in its nuclear dimorphism. Each typical specimen possesses one large macronucleus and (depending upon the species) one or more small micronuclei (p. 87). The macronucleus is a massive structure commonly found in the center of the cell with the micronucleus close to it. In some species, the dimorphic nuclei can be seen clearly in the living animal provided slight pressure is applied; they are readily stained with nuclear dyes so that their behavior may be followed.

The macronucleus and micronucleus have a common origin as will be shown in sexual processes described later. A nucleus of micronuclear origin undergoes a number of divisions to produce nuclei which develop into macronuclei and micronuclei. After a macronucleus degenerates in paramecium, a new one must originate from a micronucleus. On the other hand, the macronucleus, while it may give rise to daughter macronuclei as in binary fission, cannot produce micronuclei. Although the dimorphic nuclei have a common origin, when fully developed the macronucleus and micronucleus bear no close resemblance to each other in regard to structure, behavior, and function. Their structures are considered in greater detail on page 86.

A large number of observations and experiments disclose that the macronucleus of paramecium is closely related to—indeed possibly the center of—the entire series of metabolic activities; in its absence, the animal soon dies. On the other hand, the micronucleus is not immediately indispensable to the well-being of the cell since many amicronucleate races of paramecium and other ciliates have been described. Indeed such races not only reproduce by binary fission but also conjugate with other races.

In the accounts of nuclear processes that follow, it will be seen that paramecium in some manner ultimately obtains a reconstituted macronucleus which is necessary for continued existence of the organism. Sometimes this is simply performed as in binary fission; other times it is involved and complicated as in conjugation. Aside from fission, the replacement of a macronucleus must, of course, first involve the disintegration and ultimate disappearance of the old macronucleus from the cell. Before its dissolution, the macronucleus may pass through a stage resembling a thick, irregularly coiled, thread-like structure called a *skein*, as in autogamy and conjugation (Figs. 90 and 101), after which it becomes fragmented. On the other hand, the macronucleus may, instead, extrude small pieces of chromatin which are "sausage-shaped," spherical, or subspherical as in hemixis. These pieces of macronuclear chromatin ultimately disappear from the cell. In some instances, the macronucleus may hypertrophy (Fig. 84).

In normal binary fission, amitosis of the macronucleus is a normal, regularly occurring process, chromosomes never being evident. On the other hand, the micronucleus, like that of a metazoan cell, divides mitotically. In its division, one can readily identify the chromosomes as well as characteristic stages in mitosis. The chromosomes are so small, numerous, and compactly arranged that it becomes a difficult task to determine their exact number with certainty. Unlike mitosis or meiosis in metazoan cells, in paramecium as in other Protozoa, the nuclear membrane persists and surrounds all the mitotic stages.

It appears that the nuclei and cytoplasm in paramecium are always in delicate equilibrium. One wonders what stimuli underlie the causes for initiating division and the varied behavior of that labile organelle, the macronucleus. Interesting problems of cellular dynamics are involved in the many nuclear processes of paramecium.

1. Normal Binary Fission.

Binary fission is the commonest type of reproduction in paramecium. It is a distinctly unique asexual process in which one fully grown specimen divides into two daughters without leaving a parental corpse (Figs. 78 and 79). The plane of division is through the center of the cell and in a plane at right angles to the long axis of the body. Division of the cell-body as a whole is always preceded by division of the nuclei; indeed it appears that reproduction is initiated by nuclear activity and division. Early indications that division will occur are the doubling of cilia on the ventral side of paramecium (Downing, 1951) and the appearance of two new contractile vacuole pores (King, 1951). The details are treated more fully in the sections dealing with cilia and contractile vacuoles.

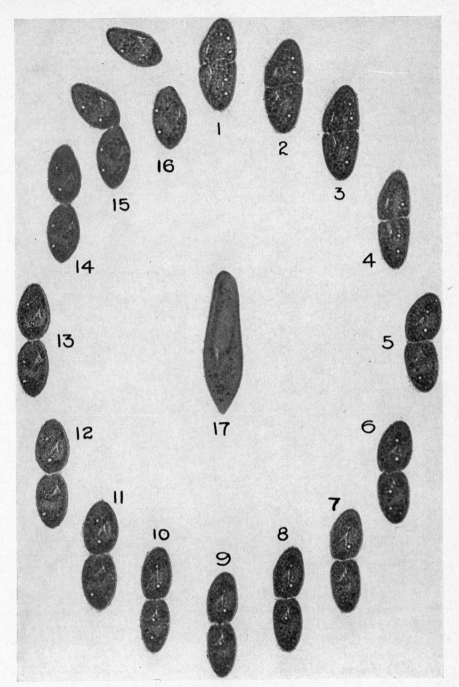

FIG. 78. Division of a single isolated specimen of *P. caudatum* as photographed in the living condition. The time taken to complete fission (stages 1–16) consumed 22 minutes; (17) is a ventral view of a vegetative specimen showing oral groove and cytopharynx. (Leica, unretouched photographs.) (Stages 1–16; × 100.) (Wichterman)

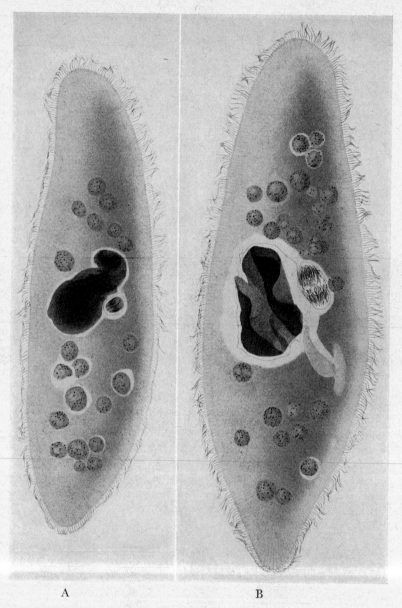

A   B

FIG. 79. Binary fission in *Paramecium caudatum*. (A) Small micronucleus to right of large macronucleus. (B) The micronucleus shows two groups of chromosomes at the poles in the anaphase stage. (C) The two daughter micronuclei are widely separated in the telophase stage. The macronucleus has lengthened prior to constriction as the animal begins to constrict in the middle. (D) Late telophase stage. The constriction of the animal becomes more pronounced. Note recently divided macronucleus into two daughter marcronuclei. (Approx. × 500.) (Chen)

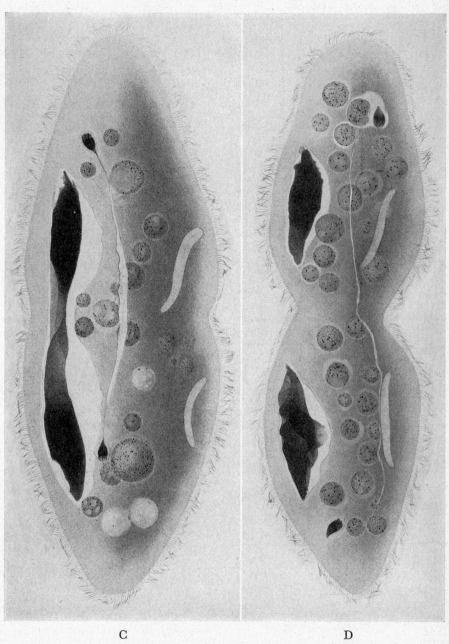

C D

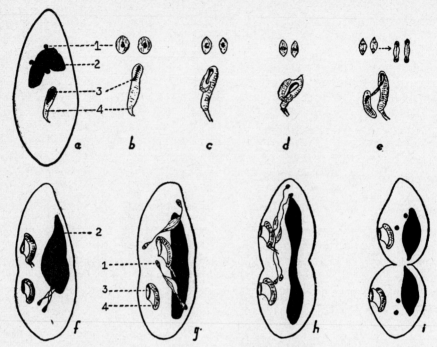

FIG. 80. Binary Fission in *P. aurelia*. Whole animals shown in a, f, g, h, i; only micro-nuclei, mouth and gullet shown in figures b, c, d, e. After Hertwig and Sonneborn, except a and i. Approximate magnifications: 300 diameters for figures a, f, g, h, i; 600 diameters for figures b, c, d, e. (1) Micronucleus; (2) macronucleus; (3) mouth; (4) gullet. (a) Resting animal. (b) Early micronuclear spindles, mouth almost closed. (c) Later pro-phase, mouth and gullet beginning to divide. (d) Metaphase, new gullet and mouth en-larging. (e) Early and late anaphase, new mouth and gullet about to separate from old ones. (f) Early telophase, mouths and gullets separated, macronucleus beginning to elongate. (g) Middle telophase, macronucleus more elongated. (h) Late telophase, body and macronucleus beginning to constrict. (i) Division of micronuclei and macronucleus complete, body nearly divided.

Under carefully controlled conditions such as temperature (27° C.) and with ample bacterial food, it is possible to obtain as many as five divisions in a 24-hour period for *P. aurelia* and two to three divisions for *P. caudatum*. The time taken to complete division in *P. caudatum* from the moment a living specimen can be recognized as begin-ning the process is approximately 30 minutes but occasionally it requires up to two hours before the daughters part from the thin strand of protoplasm which holds them together at the end of fission.[1]

Since *P. aurelia* has been so exten-sively studied by protozoölogists, binary fission will be discussed in this species although the process is fundamentally alike in the other species.

[1] It is the writer's experience that dividing specimens are less active than vegetative, non-dividing ones. In a fresh mass culture containing rapidly dividing specimens, more fission stages can be found and obtained from the bottom of the container than elsewhere in the culture. Also, the body of a dividing paramecium tends to be sticky.

Prior to division, it should be remembered that the two micronuclei lie close to the macronucleus in this species. In the interphase stage, the small, vesicular micronuclei measure approximately 3 microns in diameter and each consists of a deeply staining chromatic center surrounded by a clear achromatic space which is bounded by a thin nuclear membrane. At the beginning of fission, the micronuclei increase slightly in size, become ellipsoidal and form spindles (Fig. 80). In each spindle a group of chromatic rod- or thread-like elements, the *chromosomes*, appear which because of their extremely small size, large number, and close association, make counting their exact number a difficult task.

Diller (1936) estimates their number in *P. aurelia* as 20–30.[2] The chromosomes arrange themselves on the equatorial plate in the short, spindle-shaped metaphase stage which is then followed by a rapid and great increase in length to produce the anaphase and telophase stages. During this time, the achromatic region shows a longitudinal fibrillation or striation. The rapidly dividing micronucleus becomes dumbbell-shaped during late anaphase and telophase stages as a connecting portion or strand continues to elongate. The telophase stage, which extends to nearly the entire length of the cell-body, is characterized by knob-like ends which contain the chromosomes. Still joining the knobbed ends is the long, thin, connecting strand which becomes spindle-shaped to form the *separation spindle* located in the center of the mitotic figure. The separation spindle is passed into the cytoplasm and is resorbed while the knobbed ends produce the daughter micronuclei which soon become characteristic of interphase nuclei. A unique feature of the division is that the micronuclear membrane does not break down at any stage of mitosis. The mitotic divisions in *P. aurelia* are ordinarily synchronous so

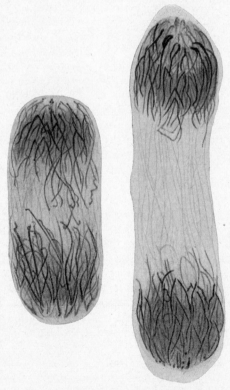

FIG. 81. Chromosomes in a race of *P. caudatum*. Anaphases of ordinary vegetative reproduction. (× 2516.) (Chen)

[2] It appears that the chromosome number varies considerably in the different races of species of *Paramecium*. Penn (1937) reported that the diploid number of chromosomes in his race of *P. caudatum* to be 36 while Calkins and Cull (1907) reported that more than 150 appeared in the race under investigation. Figure 81 shows the numerous chromosomes during the anaphase stage of division in *P. caudatum*. In his study of polyploid (heteroploid) races of *P. bursaria,* Chen (1940 f) estimated the diploid number to be about 80 in one race and "running up to several hundred" in other races. It appears that polyploidy is of common occurrence in paramecium (p. 290). Chen has shown that the micronuclei in different races of *P. bursaria* differ in size and quantity of chromatin and that these differences are constant and correlated with the difference in chromosome number, i.e. races with small micronuclei possess fewer chromosomes than races with larger micronuclei (Fig. 98).

that at their completion, two micronuclei are present at each end of the dividing paramecium before the macronuclear division is completed.

The macronucleus divides by amitosis. At the beginning of division, the structure becomes slightly enlarged, moves away from the mouth region, then increases in length while it decreases in thickness. It then becomes constricted through the middle while the animal is constricting into two daughter cells until there results two daughter macronuclei. A thin macronuclear strand of chromatin finally breaks to mark completion of its division into two.

Changes in macronuclear constitution during fission have been reported and described for other ciliates but not for paramecium. Calkins (1926) is of the opinion that each granule of the macronucleus elongates and divides into two, thus doubling their number. In view of the importance of the macronucleus, a thorough investigation of its detailed structure and behavior during binary fission is needed.

Before the beginning of fission, paramecium ceases feeding and in early stages of the process there occurs a change in shape of the body. Best seen in living specimens, paramecia become somewhat spindle-shaped at the beginning of the process. The oral groove disappears and a new one is formed in each daughter. Concerning the behavior and origin of other cytoplasmic structures during fission, Hertwig (1889) reported that a new mouth and gullet arise in the posterior daughter by being budded off from the original ingestatory structures which are retained by the anterior daughter (Fig. 80). Very likely, there is a partial dedifferentiation of original ingestatory structures, then redifferentiation while fission is progressing.

In binary fission, the contractile vacuoles separate, one going to each daughter. The anterior parental vacuole is retained by the anterior daughter as its posterior vacuole and a new one originates anteriorly. The posterior daughter retains the posterior vacuole and a new one forms anteriorly. DeGaris (1927) reported on their behavior and accession during fission of *P. caudatum* and this is treated more fully on page 79.

The ciliary granules (basal granules; kinetosomes) are self-multiplying. After each divides, a new cilium is produced. Cilia are always present during the division process.

Each of a pair from a recently divided paramecium does not resemble the typical vegetative parent and for a while each daughter can be recognized as being an anterior one or posterior one because of shape differences. Extensive shape changes therefore occur in each daughter shortly after the animals have separated after fission. Gelei (1934 b) gave additional details of fission, especially in regard to the silverline system, and Köster (1939) described the process in *P. multimicronucleatum* containing variable numbers of micronuclei.

2. Endomixis (Parthenogenesis).

As a result of their cytological investigations dealing with rhythms in long-maintained, pedigreed, isolation cultures of *P. aurelia,* Woodruff and Erdmann (1914) made the pronouncement of a new nuclear reorganization process which they termed *endomixis* in single animals. The process was described as occurring periodically in which a new macronuclear apparatus is produced without synkaryon formation (Fig. 82A).

According to Woodruff and Erd-

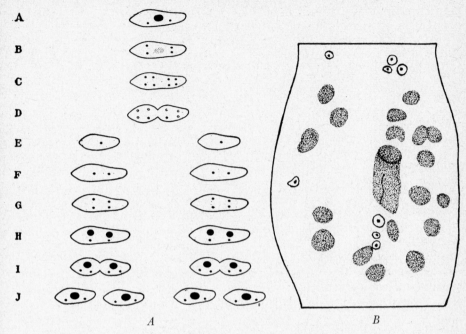

FIG. 82. (*A*) General plan of the usual nuclear changes during endomixis in *Paramecium aurelia*. (A) Typical nuclear condition; (B) degeneration of macronucleus (chromatin bodies not shown) and first division of micronuclei; (C) "climax": second division of micronuclei; (D) degeneration of six of the eight micronuclei; (E) division of the cell; (F) first reconstruction micronuclear division; (G) second reconstruction micronuclear division; (H) transformation of two micronuclei into macronuclei; (I) micronuclear and cell division; (J) typical nuclear condition restored. (Constructed from the description and figures of Woodruff and Erdmann, 1914.) (*B*) Climax of endomixis in *Paramecium aurelia*. The old macronucleus is merely in the form of a membrane from which the numerous chromatin bodies have been ejected and are free in the cytoplasm. Eight so-called reduction micronuclei. (Redrawn from Woodruff and Erdmann, 1914.)

mann, the old macronucleus degenerates into the cytoplasm (B) and is lost from view while each of the two micronuclei divides twice to produce eight micronuclei (C) of which six or seven degenerate (D) (Figs. 82A and B). With only one or two micronuclei remaining, the animal divides by fission (D) so that each daughter receives one micronucleus (E). The solitary micronucleus in each daughter then divides twice to produce four nuclei (F and G), two of which develop into macronuclei (H). As the organism proceeds to

divide by fission, the two micronuclei divide once (I) so that each daughter paramecium contains the typical nuclear complement of one macronucleus and two micronuclei (J). Alternative variations of the process are shown in Fig. 83.

Later, Erdmann and Woodruff (1916) reported the process as occurring in *P. caudatum;* Woodruff and Spencer (1923) in *P. polycaryum;* Erdmann (1925) in *P. bursaria;* Chejfec (1928, 1930) in *P. caudatum;* Stranghöner (1932) in *P. multimicronuclea-*

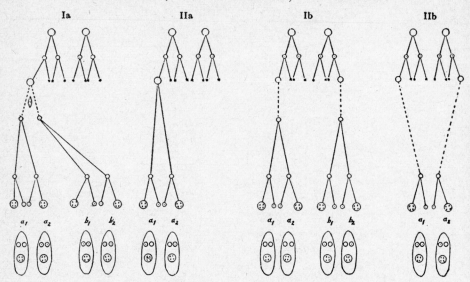

FIG. 83. Possible methods of micronuclear and cell division at the climax of endomixis in *Paramecium aurelia*. Ib is typical. (From Woodruff and Erdmann, 1914.)

tum, and Gelei (1938) in *"P. nephridiatum."* Stranghöner found that in *P. multimicronucleatum,* the so-called rhythms were due to external factors and that no causal relationship existed between a depression of fission-rate and nuclear reorganization. He reported that endomixis required two to three weeks in mass cultures but that a longer time was required in isolation cultures with the entire process extending over nine to 10 generations.

The process of endomixis has also been reported as occurring in many other ciliates. Additional references dealing with endomixis in paramecium follow: Erdmann and Woodruff (1914), Erdmann (1919, 1920), Diller (1936), Sonneborn (1937 a, b, 1938), and Woodruff (1941).

Soon after the announcement of endomixis[3] in *P. aurelia* by Woodruff and

Erdmann, a somewhat similar process was described in the same year in *P. aurelia* by Hertwig (1914) who preferred to call the nuclear reorganization by the term *parthenogenesis.* Hertwig's parthenogenesis has been generally considered to be the same as endomixis or at least equivalent to it.

3. Hemixis.

Diller (1936), besides describing autogamy (p. 298), reported the existence of still another nuclear process involving macronuclear behavior. He named this process *hemixis,* to include "a series of autonomous changes which the macronucleus undergoes in vegetative life, exclusive of those of normal binary fission" (Fig. 84). Hemixis is primarily a process of macronuclear fragmentation and division without any unusual micronuclear activity.

Diller classified hemixis into four types, namely A, B, C, and D, as shown in the figure for *P. aurelia,* but he also

TYPICAL INDIVIDUAL

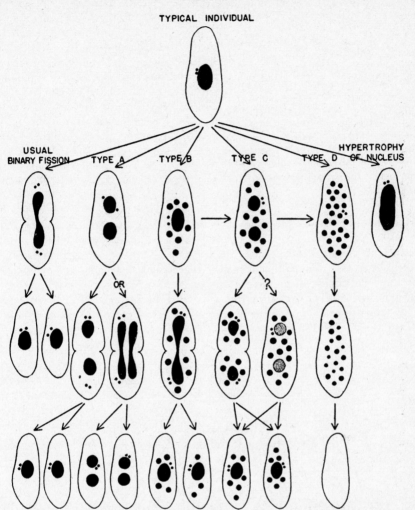

USUAL
BINARY FISSION   TYPE A   TYPE B   TYPE C   TYPE D   HYPERTROPHY
OF NUCLEUS

Fig. 84. Diagram of the macronuclear behavior (exclusive of conjugation and autogamy) of *Paramecium aurelia*. The macronuclei are represented by large solid ovals; macronuclear fragments by smaller circles; micronuclei by small round dots; "anlagenlike" macronuclei by stippled circles. The interrelationships of the various forms are indicated by arrows. Type A, Type B, Type C and Type D are types of hemixis. The similarities in appearance of some of these animals and of certain stages in autogamy and post-conjugation are to be noted. (Diller)

encountered all types in mass cultures of *P. caudatum* and *P. multimicronucleatum*.

Type A is the simplest form of hemixis, characterized by a precocious splitting or division of the macronucleus into two or more parts. This splitting is not synchronized with micronuclear division.

Type B, considered to be common, is characterized by the extrusion of one to 20 or more chromatin balls from the macronucleus into the cytoplasm. During the process, chromatin balls occasionally may be found connected to the fragmenting macronucleus.

Type C, also considered to be common, is characterized by the simultaneous splitting of the macronucleus into two or more major portions and the extrusion of macronuclear balls into the cytoplasm.

Type D is considered to represent pathologic conditions in which the macronucleus undergoes complete fragmentation into chromatin balls that eventually disappear from the cell. Micronuclei generally disappear before the dissolution of the macronucleus.

4. Macronuclear Regeneration.

Sonneborn (1940, 1941 c, 1947) described a process of macronuclear regeneration in stocks of *P. aurelia* in which the organisms reorganize their nuclear apparatus both after conjugation and autogamy. Disintegrated prod-

ucts of the old macronucleus—numbering 20–40 or more—develop into new macronuclei, possibly by multiple amitosis, instead of from products of the synkaryon (Fig. 85). These fragmented pieces of the old macronucleus segregate into approximately equal numbers during each fission until there is only one fragment per organism. During these segregation stages, the macronuclear pieces grow in size but do not divide. Ultimately, when there is only one piece per animal, the piece has regenerated to the full size of the typical, vegetative macronucleus, after which it divides amitotically at each fission. The method for induction of macronuclear regeneration is given on page 375.

It would appear that the macronucleus is a compound structure with its fragment-nuclei (subnuclei) each con-

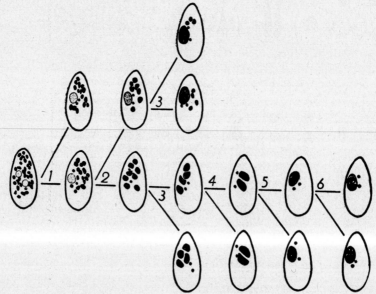

Fig. 85. Macronuclear regeneration in *P. aurelia.* The animal on the left shows the nuclear condition before the first fission following fertilization (conjugation, cytogamy, or autogamy): the solid black spheres represent the 30 or more pieces of the old disintegrated macronucleus; the two stippled circles represent the developing new macronuclei; the small circles containing a single dot represent the two micronuclei. The numbers represent successive fissions. (Sonneborn)

taining all that is needed for development into a complete macronucleus. Sonneborn reported that paramecia with regenerate macronuclei commonly gave rise to cultures in which micronuclei disappeared and similar regenerations recurred at short intervals. He concluded that since each fragment contains all that is required for macronuclear functioning, even without micronuclei, the normal macronucleus must contain at least 40 complete and distinct genomes.

Sonneborn also reported that after macronuclear regeneration, many animals die but living specimens produce clones of diverse characteristics but all of the same mating type as before fragmentation.[4]

5. General Considerations and Status of the Described Nuclear Processes.

One immediate effect of the announcement of endomixis in *P. aurelia* was to stimulate investigators to examine other ciliates for nuclear phenomena. This resulted in a long list of publications dealing primarily with the occurrence in ciliates of endomixis or endomictic-like processes, some of which might be questioned because of their incompleteness.

It remained for Diller (1936) to make a detailed cytological investigation of the complex nuclear behavior in the commonly studied species, *P. aurelia*, which resulted in his descriptions of autogamy and hemixis. In his examination of a number of races of this species, including the famous Wood-

ruffian race from which endomixis was described, Diller was unable to confirm the micronuclear behavior as reported by Woodruff and Erdmann in their account of endomixis and was inclined to deny the existence of endomixis as a valid reorganization process. He believed that Woodruff and Erdmann combined stages of hemixis and autogamy into one process, endomixis, and missed the maturation and synkaryon stages in autogamy. In like manner, Diller is of the opinion that Hertwig's parthenogenesis (generally considered to be endomixis), represents isolated stages of autogamy.

Concerning its role in the life-cycle, Woodruff and Erdmann reported that the processes of endomixis and conjugation occurred simultaneously in mass cultures and Diller found that both autogamy and conjugation occurred at the same time in his cultures. It would appear therefore that the conditions which initiate conjugation also initiate autogamy. The sequence of nuclear events in conjugation and autogamy is exceedingly close as will be shown later. Both processes involve the three pregamic divisions and synkaryon formation. Also, no appreciable differences appear to exist in exconjugants of *P. aurelia* and comparable stages of autogamous animals.

Concerning endomixis and autogamy, the former is distinguished by the absence of synkaryon formation which, of course, occurs in the latter process. In autogamy, the macronucleus transforms into a skein before it fragments into the characteristic "sausage-shaped" pieces; the same is true for Hertwig's parthenogenesis although it should be remembered that Hertwig did not rule out the possibility of the occurrence of autogamy in his process. In respect to macronuclear fragmentation in endomixis, Woodruff and Erdmann speak of ex-

[4] It is of interest to note that as early as 1900 and 1901, Hertwig and his associates reported that during depression periods, the macronucleus became enlarged, fragmented, and partly dissolved after which the macronucleus was replaced by one of the fragments thereby affecting a cure of the depression (as reported by Jennings, 1929, p. 192).

trusion of chromatin balls from the macronucleus although, as pointed out by Diller, many of their figures show macronuclear fragments such as occur in autogamy. Strictly speaking, therefore, one cannot differentiate between the two processes on the basis of macronuclear fragmentation.

As reported by Diller, all the nuclear events up to and including the formation of macronuclear anlagen in autogamy occur in one cell-generation. Although there are variations depending upon cultural conditions, less than a day is required for all nuclear stages up to and including the formation of the two small anlagen. The first fission occurs about two days after the beginning of autogamy and about five cell-generations, extending from seven to 10 days, are required to complete resorption of the old macronuclear fragments according to Diller. In endomixis, Woodruff and Erdmann report that the number of generations was variable.

Woodruff (1941) replied in part to some of the criticisms of endomixis made by Diller (1936). The former stated that the possibility of combining stages of hemixis and autogamy in one scheme as suggested by Diller was precluded since he and Erdmann studied pedigreed series of daily isolation cultures. According to Woodruff, he and Erdmann did not observe maturation spindles and crescents, synkaryon formation, and the paroral cone because they did not occur in their material. Woodruff also maintained that he and his co-worker observed only one endomictic animal out of many hundreds which showed even a slight similarity to macronuclear ribbon formation which is so characteristic of conjugating and autogamous animals.

While there is cytological and genetic foundation to support the existence of autogamy, one cannot disregard completely the extensive endomictic studies made upon *P. aurelia* and other ciliates by other investigators. As suggested by Woodruff, cultural conditions may play a role in nuclear phenomena as evidenced by the work of DeLamater (1939) who found that certain strains of deleterious bacteria in culture media produced unusual macronuclear changes. Also, one should not overlook the possibility of ultramicroscopic parasites or even viruses as inducing consistent nuclear changes and aberrations as reported by Wichterman (1946 c).

What, then, is the status of Woodruff's endomixis in present-day biology? Investigators who have used newer and more exacting cultural conditions and staining techniques, not only with Woodruff's original race—now over 40 years old—but with many other races of *P. aurelia,* have been unable to confirm endomixis. Sonneborn worked in collaboration with Woodruff and attempted to utilize the latter's old methods of cultivation with the endomictic race but endomixis could not be demonstrated. Instead, autogamy regularly occurred. However, it is unlikely that Sonneborn could exactly duplicate Woodruff's original and very likely mixed bacterial infusions. We have very little information upon the bacteria used by Woodruff and paramecia ordinarily depend upon suitable bacteria as food.

If Woodruff did not mistake autogamy for endomixis, it occurs to the author that unsuitable, possibly toxic, bacteria in the cultures may have induced the nuclear behavior known as endomixis. If endomixis occurs, it is a rarity while autogamy is common. It is therefore exasperating to find in many recent textbooks of general biology, a full account of endomixis—a phenomenon of doubtful status—in the absence of autogamy which is widespread.

Regarding hemixis, Diller offers no information as to its place and function in the life-cycle, including its relation to autogamy and conjugation. Morphologically, he suggests that hemixis occupies a position intermediate between binary fission and autogamy.

In a typical flourishing mass culture of paramecium as growing in the laboratory, the writer has often noticed from time to time specimens in different physiologic vigor. When paramecia are examined from the bottom of the container, one will find not only many dividing specimens but also less vigorous ones in various stages of slow degeneration and abnormality. One wonders if stages reported as hemixis—said to be closely allied to binary fission—are not abnormal forms. Also there is the question of the fate of hemictic specimens. Are such specimens part of a normal life-cycle and will they reproduce or are such hemictic organisms on their way to slow degeneration and death like some of those on the bottom of the container?

More than ever before, there is a need for a reinvestigation of the so-called endomictic phenomena in the cases where the process has been described. In the future, cytologists will have to exercise great care and caution in interpreting their results. At least for the moment, all work—especially in the field of genetics and serology—will have to take into account all the described nuclear processes.

It occurs to the writer that part of the difficulty may lie in antiquated and faulty techniques on the part of the investigator. A common procedure, for instance, is to fix and stain large numbers of paramecia from mass cultures and interpret the material. This becomes a difficult task since several nuclear processes can occur at the same time; hence in fitting together certain stages in a given process—perhaps critical stages—the chance for error is considerable. Newer methods of study, perhaps similar to those reported by the author in studying living specimens (Wichterman, 1940), are needed.

B. Sexual Processes and Related Phenomena

1. Conjugation: Historical Review and Modern Concept.

Ordinarily paramecium multiplies by binary fission for long periods of time, but at intervals this may be interrupted by the joining of two animals along their oral surfaces for the sexual processes of conjugation and cytogamy.

King (1693) appears to have been the first to assign a sexual process to ciliates. He observed *Euplotes* in transverse fission and erroneously interpreted this as a sexual process which he believed to be analagous to that of higher animals. The sexual union of two individuals appears to have been first observed by Leeuwenhoek who reported in 1695 ". . . I looked at the animalcules again, and saw, to my wonder, that many were coupled; nay some even coupled before my eye; and at the beginning of their copulation they had a wobbling motion, but after coupling swam forward together." O. F. Müller (1786) described and showed figures of conjugation in paramecium and correctly interpreted the union as a sexual process. Yet prior to 1786 and for many years after, a large number of observers held the view that such coupled animals were dividing longitudinally by fission. It remained for Balbiani (1858) to demonstrate that the so-called cases of "longitudinal fission" (as figured and described by Ehrenberg in 1838 for *P. bursaria*) were in reality the stages of sexual union of two animals undergoing conjugation. Indeed, Ehrenberg presented the groundless hypothesis that the contractile vacuole was a reservoir for sperm whose function was to fertilize the ova within the body of the protozoön.

The interpretation of conjugation by Balbiani (1858, 1858 a, 1861) for paramecium, and Stein (1854, 1859–78) for other ciliates was mainly responsible for the general acceptance of an erroneous concept of the sexual process up to 1870. As set forth by Balbiani, conjugation in paramecium was regarded as a sexual process in the hermaphroditic organisms in which the structure now called the macronucleus was considered to be the ovary and the micronucleus, the testis. So-called "sperm" from the "testis" were in reality the striations of the early mitotic figure of the micronucleus. The so-called ovary during conjugation was said to produce "ova" which we now know to have been fragments of the degenerating macronucleus, a feature which normally occurs during the process. Balbiani believed that since the animals were hermaphroditic, during conjugation the sperm of one of the two individuals fertilized the eggs of the other in the manner of reciprocal fertilization. Stein, on the other hand, believed that self-fertilization occurred in the ciliates.

Although subsequent details differed depending upon the observer, it was generally held that after the so-called "ova" (macronuclear fragments) were fertilized, they developed into "embryos" within the body of paramecium and that the viviparous parent then gave birth to the living embryos. Actually, ciliated stages of members of the class *Suctoria* were seen inside, which were interpreted as the living embryos of paramecium. So accepted was the theory that Engelmann (1862) emphatically declared (p. 347) that the observations of Balbiani and Stein in regard to conjugation placed on firm foundation (*sic*) the concept of sexual reproduction with eggs and sperm although later (1876) he was to change his views. A remarkable feature about this erroneous concept of the sexual process is that the observations of structures described by the early workers were, in the main, faultless but that their interpretations were awry.

For correctly interpreting the sexual process of conjugation, we are indebted largely to Bütschli (1873, 1873 a, 1876), Engelmann (1876), and Jickeli (1884). These investigators finally recognized the dimorphic nature of the nuclei; a macronucleus instead of an ovary and a micronucleus instead of a testis. The macronucleus was correctly described as fragmenting and disappearing during conjugation only to be replaced by a new macronucleus. Bütschli and Engelmann surmised that there occurred exchange of micronuclear products but this had not as yet been observed. It remained for Jickeli to observe the important feature of exchange of micronuclear products surmised but not demonstrated earlier. However, it was not until Maupas (1889) and Hertwig (1889) working independently, described in two classical papers, the details of the pregamic divisions of the micronucleus and reciprocal fertilization in the conjugation process of paramecium. Their excellent accounts are the basis for modern descriptions of the phenomenon. Since that time, numerous investigators have added further detailed cytological observations on conjugation in paramecium as well as in other ciliates.

The salient features of conjugation are similar although variations in details exist depending upon the species. In conjugation, two individuals join temporarily and there typically follows three micronuclear (pregamic) divisions in each conjugant. Generally, all micronuclei or micronuclear products degenerate, save one, before the third division. This remaining micronuclear product enters into the third division which results in the formation of two pronuclei or gametic nuclei. Simultaneously one pronucleus, generally referred to as the migratory pronucleus, crosses over to the other conjugant and unites with the so-called stationary pronucleus of its co-conjugant to form a synkaryon in each member of the pair. Before or after synkaryon formation, there occurs a gradual degeneration of the macronucleus, depending upon the species of *Paramecium* but soon after synkaryon formation, the conjugants separate. Generally speaking, conjugants of *P. aurelia* and *P. caudatum* remain joined together in the sexual process for 12–15 hours; *P. bursaria,* for 20–48 hours. However, the reorganization stages of the exconjugants take considerably longer (Wichterman, 1948 a). Finally, there follow a series of nuclear and cytosomal divisions to restore the original number of macronuclei and micronuclei to successive filial generations, the details of which will be discussed later.

In brief, this is the modern concept of conjugation as it has come down to us today with nuclear exchange a *sine qua non* of the process if we are to interpret correctly the writings of Maupas, Hertwig, Calkins, Jennings, and contemporary workers.

2. Cytological Details of Conjugation in the Species of Paramecium.

Conjugation is here defined as the temporary union of two ciliates involving exchange of micronuclear elements. It is a unique type of a sexual process in which the two organisms separate soon after fertilization but in which the two organisms fail to yield offspring in the manner common to higher animals. Instead, as a result of conjugation there arise new genetic characters and combinations which ultimately are passed on to daughter progeny by asexual reproduction through simple binary fission. Hartmann used the term *automixis* to designate fusion between two nuclei which originate in a single nucleus of the ciliate in contrast to *amphimixis* as used by Weismann to designate complete fusion of two nuclei originating in two organisms.

The cytological details of the process of conjugation in the species of *Paramecium* have been described by a large number of investigators including the following for the species listed: *P. aurelia:* Gruber (1886 a), Hertwig (1889), Maupas (1889), and Diller (1936); *P. caudatum:* Hertwig (1889), Maupas (1889), Prandtl (1906), Calkins and Cull (1907), Dehorne (1911, 1920), Klitzke (1914), Fermor-Adrianowa (1925), Bêlâr (1926), Ilowaisky (1926), Chejfec (1928, 1930), Müller (1932), Penn (1937), Wichterman (1937 b, c, 1938), and Diller (1940, 1942, 1948 b, 1949, 1950, 1950 a); *P. bursaria:* Hamburger (1904), Chen (1940, 1940 a, b, c, d, e, f, 1946, 1946 b, 1949, 1951, 1951 a), and Wichterman (1943, 1946 a, b, e, f, 1948 a); *P. multimicronucleatum:* Landis (1925), Stranghöner (1932), Müller (1932), Köster (1933), and Diller (1940); *P. trichium:* Diller (1934, 1947, 1948, 1949 a) and Wichterman (1937 a).

Conjugation has been reported in *P. calkinsi* by Wenrich and Wang (1928), Boell and Woodruff (1941), and Wichterman (1948 e, 1950, 1951 b) but details of the process have not as yet been described. As far as the author is aware, conjugation has not been observed in the remaining two species, *P. polycaryum* and *P. woodruffi*. Bütschli (1876) and Doflein (1916) described conjugation in a species of *Paramecium* which they called *P. putrinum* but which appears to have been *P. trichium*. The stages of conjugation described by Bütschli (1873) in a species which he called *P. aurelia* appear to be stages in the conjugation of *P. trichium*.

EARLY PHENOMENA OF CONJUGATION.[5] Conjugants of paramecium are usually smaller in size than typical vegetative forms. Individuals entering into conjugation are generally of the same size; conspicuous and persistent size differences such as are found in the *Vorticellidae*-conjugants do not ordinarily occur in paramecium. Preconjugants which enter into conjugation as well as conjugants are usually free of food vacuoles indicating that active feeding has at least temporarily been halted. As a matter of fact, recently well-fed specimens will not immediately undergo conjugation. Movement in preconjugants is slower than in well-fed individuals and this in part may be responsible for their tendency to aggregate in the mating reaction leading to conjugation.

While the preconjugants appear to be somewhat adhesive or sticky, a region from the anterior end extending down to a point anterior to the mouth is most adhesive. At the time of conjugation, two individuals place themselves with their oral surfaces together and swim forward. The union at first is very insecure and they may be separated mechanically soon after joining, such as with ejection from a pipette. Preconjugants at the beginning of the process may break apart after swimming together in an erratic manner. If two animals come together and remain united after attempts have been made to break apart, they swim forward together in a more or less coördinated manner, spiraling with the oral grooves directly opposite each other. The fusion then becomes more secure as the oral surfaces anterior to their mouths become cemented together. All feeding soon stops after the onset of conjugation and there follows a slow dedifferentiation of ingestatory structures.

NUCLEAR PHENOMENA. It appears that the union of two individuals in conjugation is the signal for the beginning of nuclear events.[6] In the process, several different phenomena occur simultaneously but for convenience each will be discussed singly.

MACRONUCLEAR BEHAVIOR (Fragmentation). The macronucleus in typical vegetative individuals is concerned mainly with metabolic activities, hence is referred to as the trophic or vegetative nucleus. Soon after the animals have entered into the conjugation process, the macronucleus takes a passive, indeed, degenerative role. In each conjugant, the macronucleus is seen to lose its typical shape and move from its usual position. Later it breaks into pieces and, depending upon the species, may first form a thick, coarse, threadlike structure called a *skein,* prior to its fragmentation (Fig. 90 A). Finally, the scattered fragments in the protoplasm now take on the appearance of spherical, subspherical or sausage-shaped pieces of chromatin which are ultimately resorbed by the cytoplasm (Fig. 90 B). The macronucleus of paramecium contains large amounts of desoxyribonucleic acid (DNA). It is suggested by Seshachar (1947, 1950, 1950 a) that the breaking up and absorption of the macronucleus in the cytoplasm of

---

[5] Details of the mating reaction leading to conjugation in the various species are described in Chapter 10.

[6] Klein (1927 b) reported that once conjugation is under way, the silverline system of the pellicle of both conjugants comes together as a single, unified network.

the conjugating ciliate is essentially dispersal of DNA into the cytoplasm. It may take several generations after separation of the conjugants before all fragments of the macronucleus are finally resorbed and lost from view. Generally, the most rapid degeneration of the old macronucleus occurs during the development and differentiation of the new macronuclei from the synkaryon. A newly developing macronucleus in the exconjugant is called an *anlage* or *placenta*.

MICRONUCLEAR BEHAVIOR. A close study of the process in all species where conjugation has been described discloses that with few minor differences, the process is strikingly uniform. It will be seen that the micronucleus assumes conspicuous activity and with this in mind, Maupas (1889) characterized *eight stages* of the process which is here modified as follows:

A. Union of conjugating pairs followed by the first pregamic (meiotic, maturation) division involving growth in size of the micronucleus and, depending upon the species, the formation of the characteristic prophase *crescent* stage. This crescent stage is present in *P. aurelia, P. caudatum, P. bursaria, P. multimicronucleatum,* and *P. calkinsi,* but not in *P. trichium.*

B. Second pregamic (meiotic, maturation) division.

C. Third pregamic division in which the pronuclei or gametic nuclei are produced.

D. Simultaneous and mutual exchange of the pronuclei and their union to form the synkaryon (amphinucleus).

E. First postgamic (metagamic, amphinuclear) division.

F. Second postgamic (metagamic, amphinuclear) division.

G. Third postgamic (metagamic, amphinuclear) division.

H. Subsequent reorganization stages of the exconjugant.

A. THE CONJUGATION PROCESS OF *Paramecium aurelia.* The details of nuclear events in the conjugation of *P. aurelia* have been described by Maupas (1889), Hertwig (1889), and Diller (1936).

After joining together, the two members undergo the same nuclear events at approximately the same time but for simplicity these are shown in only one animal in Fig. 86. The macronucleus loses its characteristic shape and develops into a long complex skein (Fig. 86 d, e). Constrictions develop (Fig. 86 f, g, h) resulting in the formation of 20–40 separate pieces of macronuclear chromatin (Fig. 86 i, j, k) which finally disappear as they become resorbed in the cytoplasm. Hertwig maintained that resorption ordinarily occurred prior to the first postgamic division; contrariwise, Diller maintains that macronuclear resorption occurs considerably later and only after several stages of fission of the exconjugant. Fig. 86 l shows the segregation of macronuclear fragments in a dividing specimen and their persistence in recently produced daughters (Fig. 86 m, n, o). It would appear that the latter account is of more general occurrence although as stated by Sonneborn (1947), the resorption may be correlated with differences in nutritive conditions. He reports that under poor nutritive conditions, macronuclear fragments are commonly resorbed before exconjugant fission but under better nutritive conditions, resorption occurs later.

Stage A

Soon after the animals have become joined in conjugation and before macronuclear fragmentation, micronu-

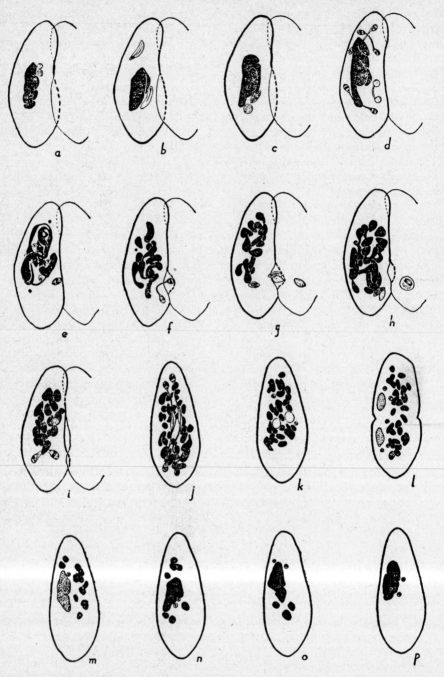

FIG. 86. Conjugation of *P. aurelia*. Based on Hertwig. (a–j) Show mating pairs. (j–p) Show exconjugants and early progeny. Nuclei shown in left conjugant only, except in two figures (g and h) in which the micronuclei are shown in both mates. (Sonneborn)

clear activity begins. The two micronuclei (Fig. 86 a) increase in size from
3 microns to approximately 20 microns
in which each takes on the characteristic shape of a crescent or sickle (Fig.
86 b). The crescents change into
broadly ellipsoidal metaphase spindles
(Fig. 86 c) and chromosomes, in the
form of extremely small granules or
short rods, become apparent (Fig. 87).
Diller and Sonneborn agree in approximating the diploid number of chromosomes at this stage to be around 30–40.
Anaphase and telophase stages quickly
follow to produce the four micronuclear
products as a result of this first pregamic (prezygotic) division. This is
considered to be the first meiotic division and it occupies at least two-thirds
the length of time that the conjugants
are joined together in the process.

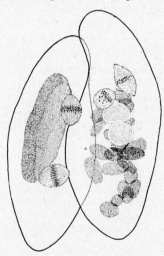

FIG. 87. Conjugation in *P. aurelia.*
(Woodruff race) showing small chromosomes of micronuclei in first prezygotic
division. (Diller)

Stage B

The four micronuclear products of
the first pregamic division enter quickly
upon the second pregamic division without a resting stage between them (Fig.

86 d) to produce eight micronuclear
products. This is considered to be the
second meiotic division.

Stage C

The account as originally given by
Maupas and Hertwig and accepted by
most investigators states that seven of
the eight micronuclear products in each
conjugant degenerate (Fig. 86 e) leaving one to enter into the third pregamic
division (Fig. 86 e, f) to produce the
two pronuclei or gametic nuclei (Fig.
86 g).[7] It appears that the micronuclear
product nearest the old mouth is the one
destined to undergo this third pregamic
division. One does not find examples of
this unique third division in the
Metazoa.

Stage D

One of the two pronuclei, the so-
called migratory gametic nucleus, is
located in a recently formed protrusion
near the old mouth and named the
paroral cone by Diller (Fig. 86 f, g).
The migratory pronucleus of each conjugant passes into the other through
these cones (Fig. 86 g) to unite with
the stationary gametic nucleus of its
co-conjugant. There is thus formed the
fertilization nucleus or synkaryon (amphinucleus) as a result of this reciprocal
fertilization (Fig. 86 h).

The gametic nuclei are considered to
possess the haploid number of chromosomes and the diploid number is restored upon formation of the synkaryon.

Stages E and F

Generally the synkaryon in each conjugant of *P. aurelia* divides twice (Fig.
86 i, j, k) to form four nuclear prod-

[7] Diller maintains that two to five of the
eight micronuclear products resulting from
the first two pregamic divisions begin to enter
a third division with two or more proceeding
to complete it. According to him, if four or
more potential gametic nuclei are formed as
a result of this division, all degenerate save
two that become the functional ones.

ucts. It is usually soon after the first postgamic division that the conjugants separate. Of the four nuclear products formed as a result of the second postgamic division, two become micronuclei and two increase in size to become the macronuclear anlagen (Fig. 86 k) which later develop into two macronuclei.[8]

Stage G, characterized by a third postgamic division such as is found in *P. caudatum* is absent in *P. aurelia*.

Stage H

An exconjugant containing two micronuclei and two macronuclear anlagen divides once, segregating to each daughter a macronuclear anlage, while the two micronuclei divide as in ordinary fission to restore the typical bimicronucleate condition (Fig. 86 l–p).[9] Fragments of the old macronucleus are more or less evenly distributed to each daughter (Fig. 86 l).

b. THE CONJUGATION PROCESS OF *Paramecium caudatum*. After joining together, the two conjugants undergo the same nuclear events at approximately the same time as shown schematically in Fig. 88.

Stage A

Soon after the two mated individuals place themselves with their oral surfaces together and become more permanently fused, the micronucleus leaves its position adjacent to the macronucleus where it is found frequently in a cleft of that structure, and increases in size

[8] Sonneborn (1947) reported that in many of the fertilized paramecia of certain stocks, the number of nuclei formed from the synkaryon and the number that become macronuclei vary with up to 10 macronuclei forming in some exconjugants.

[9] Hertwig maintained that the macronuclear anlagen fused prior to the first exconjugant division of paramecium. Balbiani (1881–82) and more recently Sonneborn (1947) asserted that this fusion may occur under conditions of starvation.

(Fig. 89). First the micronucleus (a) becomes long and slender with the clear achromatic "polar cap" at one end (b) and chromatin arranged in fine threads forming a reticulum or branching network throughout the remainder of the structure. This is followed by a characteristic bending in which the structure assumes the form of a crescent—found only in the first pregamic division—which is widest in the center with two sharply curved and pointed ends (c). The polar cap, which becomes the "division center" for this stage, leaves its terminal position to become located in the approximate center of the crescent but on its convex side (c). The division center of the prophase crescent stage divides into two as the figure loses its sharply crescentic appearance (d, e). According to Calkins and Cull (1907), the long axis of the first maturation spindle which is derived from the crescent is always at right angles to the longitudinal axis of the crescent. The late prophase stage (e) now passes into the broadly ellipsoidal metaphase stage (f) where spindle fibers and numerous, thread-like chromosomes are visible. Immediately following is the anaphase stage (g), then the telophase stage which results in the formation of two micronuclear products of the first pregamic division (h). Calkins and Cull report that the chromosomes divide longitudinally in the crescent stage and reduction in chromosome number takes place during synapsis of this first maturation division. Penn (1937) also claimed reduction division to occur in this stage, but it is held by many that reduction division occurs in the second meiotic division.

Stage B

The two micronuclear products resulting from the first pregamic division pass at once without a resting phase

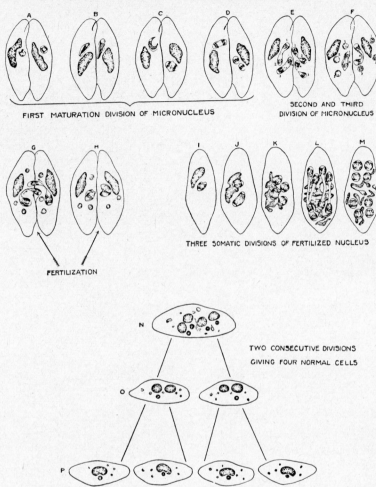

FIRST MATURATION DIVISION OF MICRONUCLEUS

SECOND AND THIRD
DIVISION OF MICRONUCLEUS

FERTILIZATION

THREE SOMATIC DIVISIONS OF FERTILIZED NUCLEUS

TWO CONSECUTIVE DIVISIONS

GIVING FOUR NORMAL CELLS

FIG. 88. *Paramecium caudatum*. Diagram of the fertilization processes. (Modified after Calkins.)

into the second maturation spindle and four micronuclear products are formed. Soon after their formation, three of the four micronuclear products degenerate and disappear (Fig. 88).

Stage C

The remaining fourth micronuclear product which tends to localize in the old mouth (paroral) region divides a third time to produce the two pronuclei or gametic nuclei in each conjugant. It is claimed by Prandtl (1906) and Cal-

kins and Cull (1907) that this third division is heteropolar, i.e. one pronucleus being measurably smaller than the other. The smaller pronucleus of each conjugant occasionally has been designated as the "wandering" or male pronucleus while the remaining one, said to be larger, is named the "stationary" or female pronucleus.[10]

[10] In spite of the claimed size differences of the pronuclei, the author seriously questions the existence of any such morphological differences in the formed pronuclei of a given conjugant.

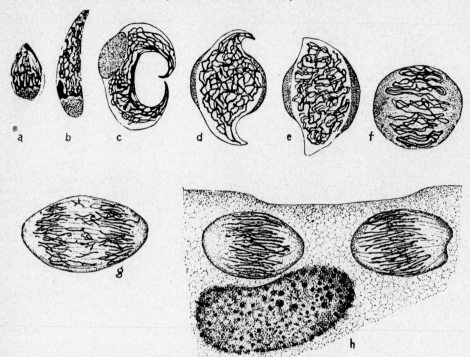

FIG. 89. Stages through which the micronucleus passes during the first pregamic division in the conjugation of *Paramecium caudatum*.

Stage D

Next there follow the simultaneous movement and exchange of the migratory pronuclei in the paroral region where the conjugants are tightly fused (Fig. 88). Hence, each migratory pronucleus passes directly into the other member of the conjugating pair in which it unites with the remaining so-called stationary pronucleus to form the synkaryon or amphinucleus. In this manner, the haploid pronuclei produce a diploid synkaryon in each conjugant in which the two conjugants after union of the pronuclei, possess the same chromosome combinations in each synkaryon. Soon after pronuclear exchange, the conjugants separate and each is referred to as an exconjugant.

Stages E, F, and G

The synkaryon or amphinucleus in each exconjugant divides three times, thus producing eight amphinuclear products in a single individual.[11] Four of the eight increase in size and are known as the macronuclear anlagen or placenta—nuclear bodies destined to develop into functional macronuclei. Accounts vary concerning the fate of the four smaller amphinuclear products. Calkins and Cull state briefly and inconclusively that all of them develop into micronuclei. According to these workers such an exconjugant with four macronuclear anlagen and four micro-

[11] In the race of *P. caudatum* under investigation, Diller (1948 b, 1949, 1950) reported the occurrence of an extra postzygotic division.

nuclei undergo binary fission once and then once again to establish the normal nuclear condition wherein the amphinuclear products are equally distributed to the dividing individuals. Although this method may possibly occur, the generally accepted method as reported by Maupas (1889), Klitzke (1916), Jennings (1920), and Doflein and Reichenow (1928) states that three of the four amphinuclear products which resemble micronuclei in each exconjugant degenerate with the remaining one being the functional micronucleus in company with the four macronuclear anlagen. Such an exconjugant, through two successive fission stages, distributes evenly the macronuclear anlagen as shown in Fig. 88 while the micronucleus divides mitotically to result in individuals with one micronucleus and one macronuclear anlage. Thus, individuals of the usual vegetative type are produced in which a macronuclear anlage grows to become the typical vegetative macronucleus.

Stage H

It appears that from the onset of conjugation, the macronucleus begins the slow process of degeneration which is not conspicuous until at later stages. During later periods while the conjugants are still joined—especially prior to their separation—the macronucleus of each individual frequently leaves its usual position and presents a more irregular contour. The first striking appearance of the disintegration of the old macronucleus in each conjugant occurs toward the end of the metagamic divisions (Fig. 90A). Here the material of the old macronucleus arranges itself into the coarse, irregular thread or *skein*. Later the strands of chromatin show fragmentation at their thinnest places which results in the production of more or less ellipsoidal pieces (Fig.

90B). These later become more or less spherical, lose their staining capacity, and ultimately disappear. Evidences of macronuclear fragments may persist even beyond the first few normal divisions of the individual.

c. THE CONJUGATION PROCESS OF *Paramecium bursaria.* For the cytological study of conjugation, *P. bursaria* is extremely favorable since opposite mating types of this species can be cultivated readily in the laboratory (Wichterman, 1944, 1949); the ciliates readily enter into the mating reaction and conjugate; the micronuclei and their subsequent divisions in the conjugation process are very large and conspicuous and the chromosomes can be seen with great clarity. Also, a great deal of research pertaining to the mating reaction, cytology of conjugation, genetics and serology has been done and is easily accessible in the literature.

The first detailed account of conjugation in *P. bursaria* was made by Hamburger (1904) although earlier, others made scattered observations upon the process. More recently Chen (1940, 1940 a, b, c, d, e, f, 1946 b, 1949, 1951 a) and Wichterman (1943, 1944, 1945, 1946 a, b, e, f, 1947 b, 1948 a, d) have added to our knowledge of conjugation in this species. Fig. 91 shows the behavior of the nuclei during the entire conjugation process.

Stage A

Soon after a pair of paramecia have come together for conjugation, the micronucleus of each increases in size as it prepares for the first pregamic division (Fig. 91 [1, 2, 3]). This is followed by the characteristic crescent stage (prophase) (Fig. 91 [4, 5, 6]) after which metaphase, anaphase, and telophase stages result in the formation

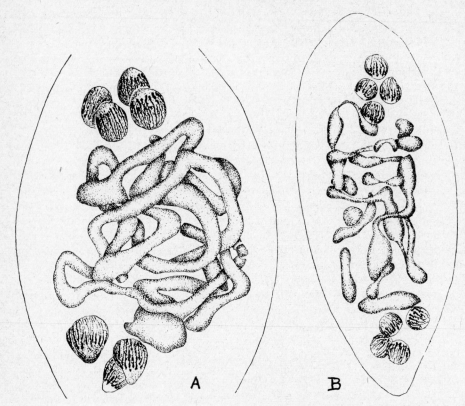

FIG. 90. Exconjugants of *Paramecium caudatum*. (A) Note "loosely wound skein stage" of degenerating macronucleus and eight micronuclear products which have not as yet become differentiated. (B) Slightly later stage. Note fragmentation of macronuclear *skein* at thin regions. Micronuclear structure is the same. (Calkins and Cull)

of two micronuclear products (Fig. 91 [7, 8, 9]) one of which is destined to degenerate in each conjugant.

Stage B

The second pregamic division is the most rapid one of the process and is characteristic of this particular stage in the conjugation of ciliates. Even while the degenerating micronuclear product is still present, the remaining functional one enters directly into the second pregamic division (Fig. 91 [10, 11]) to yield two micronuclear products (Fig. 91 [12, 13]) one of which again degenerates in each conjugant.

Stage C

The remaining functional micronuclear product is invariably located in the mouth region and enters into the third pregamic division (Fig. 91 [13]). It arranges itself at nearly right angles against the membrane at the point where nuclear exchange is to occur (Fig. 91 [14, 15]).

Stage D

It requires about 20 minutes for the migratory pole of the third pregamic division spindle to pass from one conjugant to the other (Fig. 91 [15, 15a,

16]) to produce the pronuclei (Figs. 91 [17] and 92). During and after simultaneous and mutual exchange of the pronuclei, a careful examination shows the sister pronuclei to be of the same size. It occurs to the writer that size differences of pronuclei as given in the literature may be erroneous due to faulty interpretation. Observers may have seen the pronuclei in different positions which, when measured, gave slight size differences.

The pronuclei thus produced as a result of the third pregamic division (Fig. 91 [17]) fuse to form the fertilization nucleus or synkaryon (Fig. 91 [18]).

Stages E, F, and G

After syngamy, three postgamic (amphinuclear, metagamic) divisions occur before the conjugants separate as first observed by Chen (1940). The author observed the first to occur approximately 20 hours after mating; the second, 22 hours; and the third, 24 hours after mating[12] (Fig. 93).

The first postgamic division of the synkaryon (Fig. 91 [19, 20]) results in two nuclear products (Fig. 91 [21]) one of which degenerates (Fig. 91 [22]). The remaining nuclear product in each conjugant enters into the second postgamic division (Fig. 91 [23, 24]) and yields two nuclear products (Fig. 91 [25]). Each of these nuclear products divides to complete the third postgamic division which results in four being

[12] Chen (1949) reported that the first postgamic division occurred 29–31 hours after the onset of conjugation and that two further nuclear divisions occurred at 33–36 hours and by the thirty-sixth hour, exconjugants may be found. While the time taken to complete nuclear divisions in conjugation may be a racial character, temperature invariably is another factor since in the same races, lower temperatures slacken nuclear divisions and increase the length of time the conjugants remain joined while the reverse is true with higher temperatures.

present in each conjugant (Fig. 91 [26, 27, 28, 28a]). Conjugants are frequently seen completing the third postgamic division at the time of their separation (Fig. 91 [28]).

Stage H

The recent exconjugant is shorter and wider than it was before entering into the sexual process (Fig. 91 [29]). As early as 30 hours after mating, one can recognize in the exconjugant, two of the four nuclear products as macronuclear anlagen (Fig. 91 [31]). These two macronuclear anlagen gradually increase in size until about 80 hours after mating; each anlage is approximately the same size as the old, degenerating macronucleus (Fig. 91 [34]). One cell division of the exconjugant follows which generally distributes equally the new nuclear components (Fig. 91 [35, 36]). Most exconjugant stages appear to divide approximately four days after mating but some divide before and after that time. On the other hand, some exconjugants are unable to survive the conjugation process and die before division.

The old degenerating macronucleus, which does not undergo skein formation, persists beyond the stage of fission so that one daughter receives this element with a new macronucleus and micronucleus while the other daughter receives only a new macronucleus and micronucleus. Variations may also occur in respect to segregation of nuclear components during the division of the exconjugant which appears to have been first observed by Hamburger (1904). In this regard, Woodruff (1931, 1931 a) reported on amicronucleate and bimicronucleate races and Chen (1940 c) on amicronucleate races of P. bursaria. It is easy to understand how this segregation of nuclear components may occur during division of the excon-

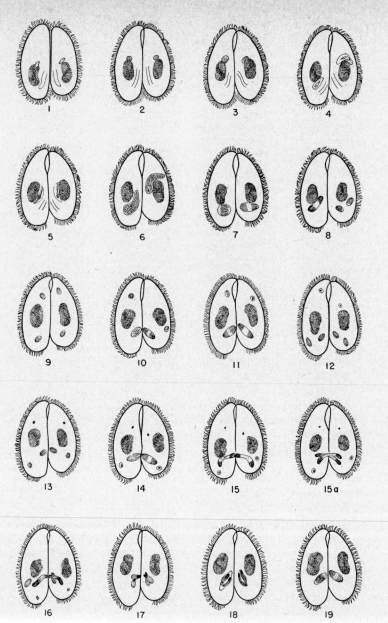

Fig. 91. Characteristic stages in the conjugation of *Paramecium bursaria* based on a study of living and stained (Feulgen-reaction) preparations at precisely known time intervals after mating. (1–9) First pregamic division. Note the characteristically large prophase stage in 6 and the gradual disappearance of cytopharynx and the cilia at posterior ends of conjugants. (10–12) Second pregamic division. Degenerating micronuclear product of first division in anterior part of each conjugant. (13–17) Third pregamic division and pronuclear exchange. Degenerating micronuclear product of first division still visible in anterior end of each conjugant in 13–15a. Larger degenerating micronuclear product of second division visible in posterior end of each conjugant in 13–16. Exchange pronuclei in 17 showing concentration of chromatin at pointed ends of pronuclei. (18) Syngamy (synkaryon formation). Fusion of exchanged pronuclei. (19–21) First postgamic

280

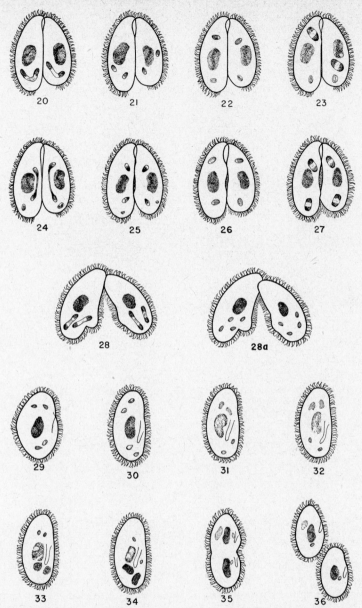

division. Production of two nuclear products. (22) Degeneration of one nuclear product of first postgamic division seen in posterior end of each conjugant. (23–26) Second postgamic division. Production of two nuclear products. Note redifferentiation of cilia between conjugants which are beginning to separate and the reappearance of cilia at posterior ends. (27–28a) Third pregamic division and separation of conjugants. Production of four nuclear products in each conjugant. (29–34) Post-conjugant (ex-conjugant) stages. Note two of four micronuclear products increasing in size to develop into macronuclear anlagen. Shown also is gradual disorganization of the old macronucleus. (35–36) Post-conjugant division. Micronuclei and new macronuclei in characteristic positions while old macronucleus is seen degenerating in anterior daughter. (Wichterman)

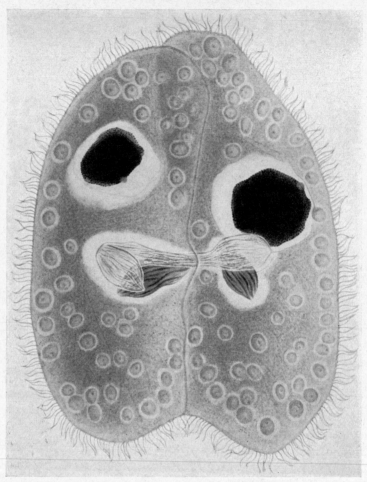

FIG. 92. Exchange of pronuclei in the conjugation of two different races of *Paramecium bursaria*. (Chen)

jugants. One daughter may receive one new macronuclear element only and no micronuclei at the time of fission while the other daughter would receive not only a new macronuclear element but two micronuclei as well. The first daughter would therefore be amicronucleate and the second, bimicronucleate. Woodruff reported that conjugation did not occur in his amicronucleate races and believed therefore the micronucleus to be essential for the conjugation process. After removing the micro-

nucleus, Schwartz (1939) reported the conjugation of amicronucleate *P. bursaria* with normal micronucleate animals as well as between two amicronucleate animals. Chen also reported the successful conjugation in *P. bursaria* of amicronucleate races with micronucleate ones and this kind of conjugation has been reported by Diller (1936) for *P. aurelia*.

Once conjugation has proceeded, there occurs a dedifferentiation of cilia as well as of trichocysts along the fused

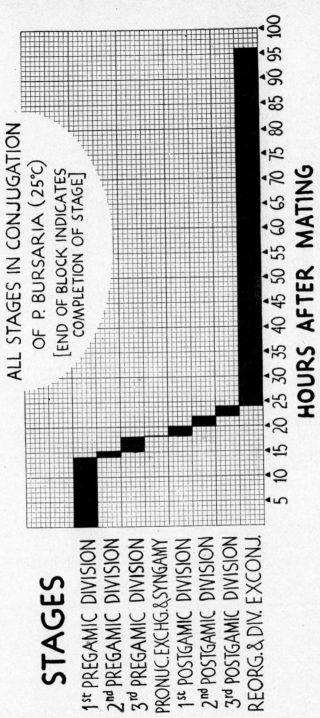

STAGES

1st PREGAMIC DIVISION
2nd PREGAMIC DIVISION
3rd PREGAMIC DIVISION
PRONUC.EXCHG.&SYNGAMY
1st POSTGAMIC DIVISION
2nd POSTGAMIC DIVISION
3rd POSTGAMIC DIVISION
REORG.&DIV.EXCONJ.

TIME

REQUIRED TO COMPLETE

ALL STAGES IN CONJUGATION

OF P. BURSARIA (25°C)

[END OF BLOCK INDICATES
COMPLETION OF STAGE]

HOURS AFTER MATING

5 10 15 20 25 30 35 40 45 50 55 60 65 70 75 80 85 90 95 100

FIG. 93. Time relationships of nuclear processes in the conjugation of *Paramecium bursaria* based on a study of living and stained (Feulgen-reaction) preparations at precisely known time intervals after mating. (Wichterman.)

oral surfaces of the conjugants. The de-differentiation of cilia proceeds below the region where the animals are joined, showing frequently no cilia even at the extreme posterior ends of the conjugants on their oral surfaces (Fig. 91 [3–26]). The cytopharynx in each conjugant is conspicuous as late as eight hours after mating, then begins to de-differentiate until it disappears about 12–14 hours after mating only to become redifferentiated near the end of the process. Joined, living conjugants when irritated can extrude trichocysts during the sexual union.

d. THE CONJUGATION PROCESS OF *Paramecium trichium*. It appears that Bütschli (1873) was the first to describe and show figures of stages in conjugation in *P. trichium* although he called the species *P. aurelia*. Later (1876) he described stages in conjugation for "*P. putrinum*" which again may possibly have been *P. trichium*. Doflein (1916) showed, with descriptions and figures, the details of conjugation in a species which he called "*P. putrinum*" but which appears to have been *P. trichium*.

More recently, Diller (1934, 1947, 1948, 1949 a) and Wichterman (1937 a) have reported on the conjugation of *P. trichium*. In the material studied by the author, it was found that the preconjugants, which were smaller than the vegetative individuals, fused along their oral grooves. The centrally located, ellipsoidal macronucleus, during conjugation, became completely fragmented after first undergoing skein formation. The macronuclear skein resembled a twisted ribbon which became thinner and longer, finally resulting in small, irregular rod-like elements. These then became spherical or subspherical, then disappeared in the cytoplasm after exconjugant reorganization.

In conjugation, the ellipsoidal micronucleus divided three times. The first pregamic division resulted in two micronuclear products; the second division, four products, three of which degenerated. The remaining micronuclear product entered into the third pregamic division to produce the pronuclei. After exchange of pronuclei, a synkaryon was formed which divided three times to produce eight products. Two of these postgamic divisions occurred while the conjugants were joined; the third postgamic division occurred in the exconjugant. Exconjugants with four macronuclear anlagen and a single micronucleus were commonly encountered which divided with two anlagen passing to each daughter while the micronucleus divided mitotically. In exconjugants with two macronuclear anlagen and one micronucleus, a final cell division distributed a single macronuclear anlage to each daughter as the micronucleus again divided mitotically, thus restoring the original nuclear condition to each animal (Wichterman, 1937 a).

While this description confirmed in most respects the observations of Diller (1934), he has extended considerably our knowledge of conjugation in *P. trichium* in later publications (1947, 1948, 1949 a). His investigations have disclosed that this species is extremely variable and labile in regard to macronuclear and micronuclear activity. He has shown that conjugation in *P. trichium* may exhibit a number of alternative procedures with the so-called "standard" process being a sequence of three pregamic divisions to produce the pronuclei and the synkaryon. There follow three postzygotic divisions to produce eight nuclear products, four of which become macronuclear anlagen, one a micronucleus, and three degenerate. Two cell divisions with the micronucleus dividing at each one, segregate the nuclei to their normal condition

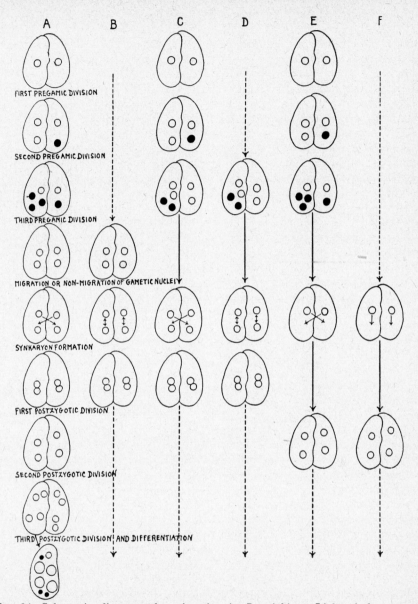

Fig. 94. Schematic diagram of conjugation in *P. trichium*. Light circles represent functional nuclei; dark circles, degenerating nuclei which are not indicated as persisting beyond the generation in which they arose. Macronuclei are omitted, except for the lowest figure. In the left partner of a pair, degeneration is represented as occurring after the second pregamic division; in the right partner, after the second and third divisions. Broken lines indicate a history similar to that of the column to the left of it; solid lines, that a stage has been omitted. (A) Reciprocal fertilization following 3 pregamic divisions. (B) Autogamy after 3 pregamic divisions. (C–F) Omission of third pregamic division; (C) reciprocal fertilization, (D) autogamy, (E) parthenogenetic development after gametic interchange, (F) parthenogenesis with development of the gamete in the partner in which it arose. (Diller)

285

(Fig. 94). Omission of the third pregamic division as shown in schemes C, D, E, and F of the figure are followed either by fertilization or parthenogenesis.[13]

As noted earlier by Wenrich (1926), Diller reported that the number of micronuclei is variable both in vegetative stages and at the time of conjugation. Normally one is present but two or three may be found and specimens may be aneuploid, hyperploid, or normally euploid.

Crescent stages so characteristic of the first pregamic prophases of *P. caudatum, P. aurelia, P. multimicronucleatum, P. calkinsi,* and *P. bursaria* are absent in *P. trichium.*

Diller has shown macronuclear skein formation to be rather extensive in *P. trichium.* In addition, he described the occurrence in some conjugants of macronuclear interchange during the process. In some instances, cytosomal continuity of joined pairs did not appear to exist but occasionally cytosomal continuity was so pronounced that strands of the macronuclear skein extended conspicuously from one conjugant to the other across the paroral cone bridge (Fig. 95A).

While macronuclear interchange has been noted in a number of other ciliates, it has been described in no other species of *Paramecium* except *P. trichium.*

Thus in this species, joined pairs may enter into conjugation involving reciprocal cross-fertilization with synkaryon formation, cytogamy, or self-fertilization with synkaryon formation or parthenogenesis.

More recently, Diller (1949 a) has reported still greater versatility of nu-

clear behavior in which a process of "abbreviated" conjugation occurs in one race of *P. trichium.* In his account, the number of micronuclear divisions was reduced to three (or possibly four) from the so-called "standard" pattern of five or six. According to Diller, exchange of micronuclei may take place at the conclusion of the first division. He stated that the products of the first division proceed, directly, to reconstitute the new nuclear apparatus by synkaryon formation, parthenogenetic development, or a combination of the two, usually dividing twice; no degeneration of nuclei occurs between divisions.

e. THE CONJUGATION PROCESS OF *Paramecium multimicronucleatum.* The details of nuclear events in the conjugation of *P. multimicronucleatum* have been described by Landis (1925) and Müller (1932). The gross features of conjugation are similar to those described for *P. caudatum* (Figs. 96 and 97).

As reported by Landis, conjugating animals are smaller and more transparent than vegetative ones. The manner of union, as in other species, begins at the anterior end of the oral surface and gradually extends backward until the animals are attached over a maximal area on the oral side. The race described by Landis possessed four vesicular micronuclei (Fig. 96 [A]) all of which take part in the first two pregamic divisions (B, C). In the first pregamic division, the four micronuclei form the characteristic crescent prophase stages, then produce eight micronuclear products (B). Each then enters into the second pregamic division to yield 16 micronuclear products of which 12 degenerate in each conjugant (C). The remaining four presumably possess the ability of producing the functional pronuclei (D) but the micronuclear product nearest the point of nuclear

---

[13] Cases of parthenogenesis have been reported in joined pairs of *P. bursaria* by Chen and similar instances appear to be represented in the same species by Hamburger (1904).

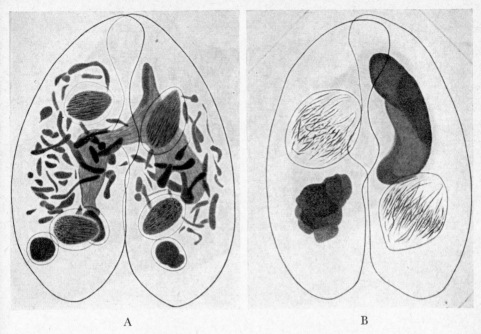

A   B

Fig. 95. (A) Two products of the first postzygotic division in each conjugant of *P. trichium*. Nuclei about the same size in both members. Delayed skein formation and passage of a heavy macronuclear lobe from the right conjugant to the left. Two degenerating nuclei in the left; one in the right. (B) First pregamic metaphase in conjugation of *P. trichium*. Both nuclei have approximately the same number and type of chromosomes which are very numerous. (Diller)

exchange undergoes the third pregamic division to produce the two functional pronuclei while the remaining three micronuclear products degenerate before, during, or after their division (E).

According to Landis, the two functional pronuclei are dissimilar in size and shape as well as in their behavior and chromatin arrangement (E, F). One, called the "stationary pronucleus" is said to be larger and commonly found in the middle of the cell body. The other, called the "migratory pronucleus" is smaller and found near the place of formation of the cytoplasmic bridge. After the dissolution of the contiguous membranes to form the cytoplasmic bridge, each migratory pronucleus moves over to the co-conjugant

thus bringing about nuclear exchange and the resulting synkaryon formation of a migratory pronucleus with a stationary one (G). From the synkaryon, three postgamic divisions result to yield eight nuclear products of which seven degenerate (H, I, J). The remaining nuclear product divides twice to form four products (K, L), two of which become macronuclear and two micronuclear anlagen six to seven hours after the conjugants separate (M). Ten hours after separation of the conjugants, all of the anlagen divide once to yield four of each type (N). Two divisions of each exconjugant are required, each preceded by a division of the micronuclei to bring the nuclear constituents back to the typical vegetative

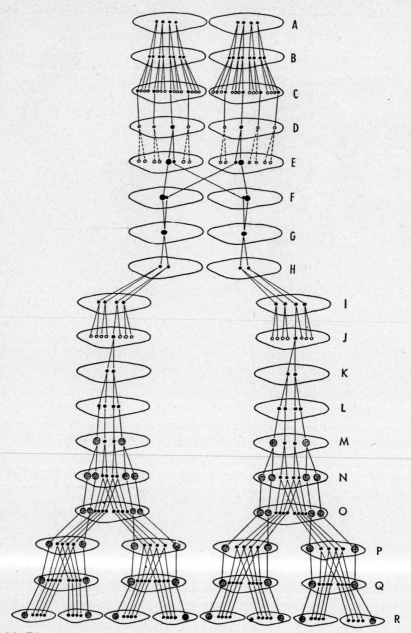

FIG. 96. Diagram summarizing conjugation of *Paramecium multimicronucleatum*. Micronuclear behavior shown only. (Landis)

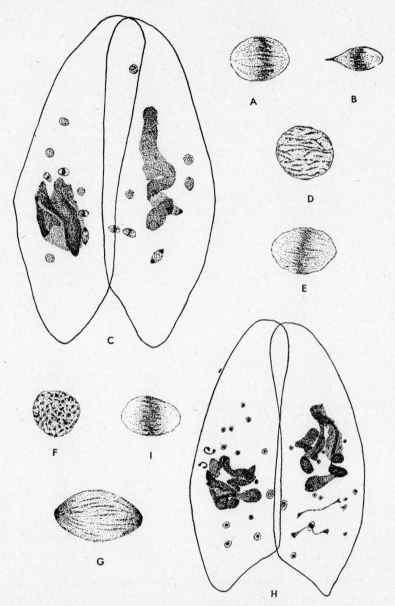

FIG. 97. (A) Metaphase of first pregamic division. (B) Telophase of first pregamic division. (C) Macronucleus forming short ribbon, micronuclei in various stages of the second pregamic division. (× 380.) (D–G) Stages in the second pregamic division. (F) An optical section, showing polar view of equatorial plate. (H) Macronucleus is stretched out into a longer ribbon or skein. Pronuclei are in process of formation. There is one normal metaphase in each animal, lying near the common cell wall. The other micronuclei are degenerating. (× 380.) (I) Metaphase of the third pregamic division producing pronuclei. (Landis)

condition (O, P, Q, R). On the other hand, Müller (1932) reported considerable variation in the reorganization of the nuclear apparatus in exconjugants of *P. multimicronucleatum.*

While the paramecia are undergoing conjugation, the old macronucleus forms a long skein prior to its fragmentation (Fig. 97).

3. The Chromosomes of Paramecium and Polyploidy (Heteroploidy).

An organism or a cell contains a basic number of chromosomes usually designated as *n* number and referred to as *haploid* (*haploos*, single). Ordinarily, sexually reproducing organisms possess a double set of chromosomes and are known as *diploids* (*2n*). A doubling of a diploid set results in a *tetraploid* (*tetra*, four). In organisms where there occurs a series of chromosome numbers that are multiples of the basic number, the condition is known as *polyploidy* (*poly*, many; *ploid*, fold).

In the species of *Paramecium*, the exact chromosome number has not been determined with certainty but estimates of their number have been made. The term *heteroploidy* has been used by Diller (1940, 1948) to distinguish differences in number of chromosomes and the amount of chromatin between individual micronuclei within the species.

Polyploidy or heteroploidy appears to be of common occurrence within the species of *Paramecium*. Evidence has been presented by Diller and others that at least in *P. caudatum, P. trichium, P. multimicronucleatum, P. aurelia,* and *P. bursaria,* the range of heteroploidy appears to be less than the diploid condition (hypoploid) as well as greater than the diploid condition (hyperploid).

In higher organisms it has been possible to obtain polyploids by colchicine, high and low temperatures, x-rays and centrifugation, but little or nothing has been done in this regard with paramecium.

SIZE, SHAPE, AND NUMBER. As mentioned earlier, the chromosomes in *P. aurelia* appear as numerous granules or very short rods measuring approximately several microns or less. Diller (1936) and Sonneborn (1947) estimate the diploid number as being 30–40. Hertwig's (1889) estimate of eight to 10 chromosomes for the diploid number appears to be in error since he counted not chromosomes but groups of spindle fibers.

For *P. caudatum,* Calkins and Cull (1907) and more recently Chen (1940 e) describe the numerous chromosomes as fine, thread-like elements (Fig. 81). Penn (1937) estimated the diploid number of chromosomes in his race of *P. caudatum* to be 36 while Calkins and Cull reported the presence of more than 150 chromosomes in their race. In the race of *P. caudatum* studied by Chen (1940 e) in which specimens contained a large micronucleus and numerous chromosomes, he concluded that it was like the one investigated by Calkins and Cull in being polyploid. The evidence suggests that polyploidy is of common occurrence in *P. caudatum*. The work of Diller (1948) indicates that for *P. trichium,* chromosomes vary in size, number, and stainability and that heteroploidy or polyploidy is pronounced and common even in the same stock.

Diller depicts large numbers of clearly defined, thread-like chromosomes in the conjugation of *P. trichium* but gives no actual count of their number (Fig. 95B). Landis (1925) reported that in the conjugation of *P. multimicronucleatum,* the chromosomes appear as dark strands but that little can be discovered of their individuality because of their small size (Fig. 97). In an examination of the chromosomes during conjugation of *P. calkinsi*, the present author finds them to be so extremely small that they are barely recognizable.

The most favorable material for the study of chromosomes and polyploidy in paramecium appears to exist in races of *P. bursaria* and *P. trichium* in which the chromosomes, while quite numerous, exist as fairly conspicuous rods or threads of varying length and thickness (Fig. 98). For the determination of the chromosome number in paramecium, the most favorable stage for counting is during the late prophase of the first pregamic division at conjugation.

Chen (1940 a, b, c, d, e, f, 1944, 1946 b, c, 1949) has made many fundamental discoveries and contributions in regard to the chromosomes of *P. bursaria* and his work should be consulted by the student who is interested in more details. With patience, extreme care and exacting techniques and procedures, this investigator has studied the micronuclei and chromosomes in many different races of *P. bursaria* collected from the United States and foreign countries. He concluded that the size of the micronucleus is correlated with the number of chromosomes in the different races while the macronucleus of all races is approximately of the same size.[14] The race with the smallest micronucleus (Fd of Fig. 98) possessed the smallest number of chromosomes, about 80, while in each of the other races, the chromosome number was found to be much greater, running up to several hundred.

As demonstrated by the work of Chen in *P. bursaria*, polyploidy generally results from the fusion of more than two pronuclei during the conjugation process. This investigator found conjugants in which three or four pronuclei fused, resulting in an increase in the number of chromosomes in the fusion nucleus. Additional suggestions to account for polyploidy by Chen are (1) the failure of one of the two products of the first or second pregamic divisions to degenerate and (2) conjugation between a normal unimicronucleate animal and one that is bimicronucleate.

In a cytological study of conjugation between normal green and colorless races of *P. bursaria*, the author (Wichterman, 1946 e) noted that approximately 2 per cent of the conjugants produced the polyploid condition. The critical stage where polyploidy occurred was found during the period of pronuclear transfer approximately 16–18 hours after the animals had been mated. This confirms the suggestion made by Chen in accounting for polyploidy, namely the failure of a migratory pronucleus in one conjugant to migrate to the co-conjugant. The result is an individual with one small pronucleus (the "hemicaryon"[15] or stationary pronucleus) which is haploid and the conjugant with three pronuclei (two migratory pronuclei and one stationary pronucleus) which fuse and form a larger triploid synkaryon.

In each conjugant, what is the fate of each nuclear body that is now comparable to the normal synkaryon? The subsequent micronuclear stages showed a conspicuous and persisting size difference in all later stages. In the haploid conjugant, late anaphase stages (com-

---

[14] In conjugation, the macronucleus originates from the synkaryon as a simple diploid nucleus after which it grows enormously. The investigations of Sonneborn (1945) have led Subramaniam (1947) to suggest that the macronucleus of paramecium and other ciliates is endopolyploid.

[15] The term "hemicaryon" was coined by Chen (1940 c) and is used when the single pronucleus in a conjugant starts to divide. This is in contradistinction to synkaryon which is made up of two or more pronuclei.

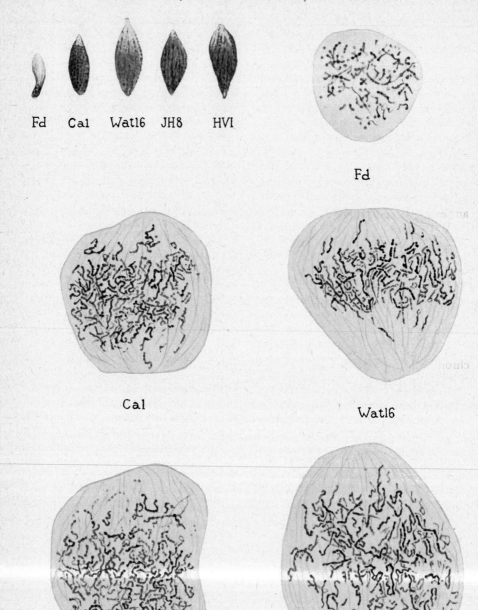

Fig. 98. (*For legend see facing page.*)

parable to postgamic ones) measured 10.8 microns in length and were very thin; similar stages in the triploid co-conjugant measured 27 microns in length and were proportionately wider than the haploid stages. Their division products measured 8 microns in the haploid forms and 15.5 microns in the triploid individuals respectively. While polyploidy occurred in only 2 per cent of the cases in this material, it nevertheless created variation in micronuclear composition and is therefore of evolutionary significance (Wichterman, 1946 e).

Chen mated races of *P. bursaria* each with very different chromosome numbers and found that all nuclear phenomena behaved normally, with the result that the exconjugants possessed the same number of chromosomes. In observing conjugation between animals with and without micronuclei, Chen found that the micronucleus of the former gave rise to the normal two pronuclei, one of which migrated into the amicronucleate conjugant. The amicronucleate conjugant of course produced no pronuclei. This resulted in a condition in which each conjugant possessed one pronucleus or "hemicaryon" which behaved like a normal synkaryon since it divided three times. As a result of this conjugation, the nuclear apparatus became alike.

From his study, Chen (1940 c) concluded that animals with many chromosomes can conjugate with animals with relatively few chromosomes and that animals with micronuclei can conjugate with animals without micronuclei. He also noted that one of the results of conjugation in *Paramecium* is the elimination of the great diversities in the nuclear apparatus and great difference in chromosome number.

4. Some Unusual Aspects and Variations of Normal Conjugation.[16]

a. RECONJUGATION, INTERSPECIES, AND INTERVARIETAL CONJUGATION. A reconjugant is an animal which has separated from its mate and paired with another before its reorganization has been completed (Diller, 1942) (Fig. 99). This phenomenon long has been known to European workers who called the reconjugants *Wiederconjugante,* a name coined by Enriques (1908).

Recently, reconjugation in paramecium has been studied extensively by Diller (1942, 1948) who reported that Bütschli (1876) and Doflein (1907) apparently saw this phenomenon in species of *Paramecium*. Klitzke (1914, 1915) noted reconjugation in *P. caudatum* which he regarded as a pathologic process. Müller (1932) described stages of reconjugation in *P. multimicronucleatum* and reported conjugating pairs consisting of so-called "normal" conjugants joined to either exconjugants or those undergoing endomixis. According to this investigator, the mated exconjugant and the mated endomictic animal did not take an active part in conjugation while the normal partner completed typical micronuclear divisions to produce pronuclei in which further development was probably autogamous, possibly parthenogenetic.

[16] See Chapter 10 for variations in respect to the mating phenomenon.

Fig. 98. The vegetative resting micronuclei of five races of *Paramecium bursaria* are shown at the upper left. The other five drawings show the micronuclei during the first pregamic division during conjugation. The chromosome number varies in different races belonging to the same mating type. There appears to be no correlation between chromosome number and mating type. ($\times$ 1640.) (Chen)

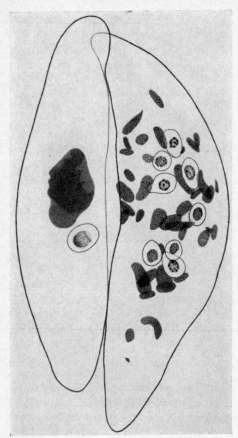

FIG. 99. Specimen of *P. caudatum* conjugating with a reconjugant. Micronucleus of specimen on left slightly swollen in first pregamic prophase stage. Reconjugant on right shows eight products of postzygotic divisions, four of which are prospective macronuclear anlagen while the remaining four are prospective micronuclear anlagen. Sausage-shaped fragments of the old macronucleus appear in reconjugant. (Diller)

Diller (1936) figured a stage which he believed represented either a reconjugation or conjugation-autogamy in *P. aurelia* and for the same species, Sonneborn (1936) reported as many as eight successive reconjugation unions in a period of 17 days. Diller (1948) found that reconjugations are not uncommon in his races of *P. trichium*.

According to Diller, the reconjugant is almost always associated with a normal partner which is freshly involved in the conjugation process. He reported that union of two reconjugants is very infrequent in his races of *P. caudatum*. The cytological details of reconjugation are fully described by Diller and his work should be consulted for additional information.

Generally following the normal conjugation process, there occur periods of asexual reproduction by fission before the beginning of another period of conjugation. The work of Diller, however, shows that this is not always the case since exconjugants occasionally reconjugate before they have reorganized completely from a previous mating. His studies show clearly that the closer the ex-conjugant to the indifferent (8-nuclei) stage, the more pronounced will be the changes induced by a second conjugation.

Finally, it must be kept in mind that stages in reconjugation be interpreted correctly. The author has observed the union of a living specimen with dead and dying ones resembling conjugating pairs (Wichterman, 1940). Hence, it is possible that one member of a pair be not only dead but dying at a given stage while the other normal member undergoes the usual pregamic micronuclear divisions (resulting in self-fertilization) and be incorrectly interpreted as a case of reconjugation when seen in killed and stained animals.

Concerning conjugation between different species of *Paramecium*, Müller (1932) reported three cases between *P. caudatum* and *P. multimicronucleatum*, but this has not been confirmed by others. The author has attempted interspecies matings with a number of different species of *Paramecium* known to be reactive for mating but with no success.

Animals belonging to any one variety

do not, as a rule, mate with other members of other varieties. However, Chen (1946 b) found an exception to this rule in the mating and subsequent conjugation of specimens belonging to certain Russian and American clones belonging to different varieties of *P. bursaria*. In considerable detail, this investigator demonstrated that nuclear changes in such conjugations are decidedly abnormal, possibly due to physiologic incompatibility between the varieties. As noted by Chen,

Abnormality in nuclear changes during conjugation between varieties includes: (1) the arresting of nuclear changes so that they do not proceed beyond the end of the first pregamic division; (2) the slowing down of the rate of nuclear changes; (3) the deformation of chromosomes; (4) abnormal elongation of nuclei; (5) the formation of tripartite nuclei; and (6) the initiation of the process of nuclear exchange even though no pronuclei are formed.

He reported that most of the conjugating pairs failed to separate even as late as 70 hours after the onset of conjugation in *P. bursaria*.

Gilman (Table 30) has also reported cases of intervarietal mating in *P. caudatum* but cytological studies of the conjugants have not been made.

b. CONJUGATION INVOLVING THREE OR MORE ANIMALS, MONSTERS, AND DEAD ANIMALS. In addition to the usual coupling of two ciliates for conjugation, instances have been reported in which three or more specimens were involved. Stein (1859–78) apparently was the first to notice this phenomenon, reported as occurring in *P. caudatum* and *Amphileptus*. Later, Jickeli (1884) and Maupas (1889) reported the occurrence of triples in sexual union in *P. caudatum*, while Jennings (1931, p. 103) shows a figure of triples and quadruples in what appears to be *P. caudatum*. Concerning them, Jennings

reported that often one paramecium succeeds in freeing itself, and then swims away while others remain caught in such groups indefinitely. The occurrence of triples and quadruples engaged in sexual union as observed in other ciliates by various investigators has been reviewed by Chen (1946) who was the first to make a detailed cytological study of this phenomenon.

For his study, Chen used known mating types of *P. bursaria*. To obtain large numbers of triples engaged in the sexual process, the following requirements are necessary, according to Chen: (1) large (rich) cultures of paramecia; (2) strongly reactive animals for mating; (3) concentrations of animals prior to mixing the opposite mating types. The ratio of triples to pairs after certain matings may be as great as one to 10 respectively although generally it is considerably less.

In the conjugation of three specimens of *P. bursaria*, two form an ordinary conjugating pair while the third member is attached anteriorly to a small part of the posterior end of one of the pairs (Fig. 100). As reported by Chen (1940, 1946), the usual micronuclear (pregamic) divisions occur to produce pronuclei in all three members of the union at about the same rate as those in ordinary conjugating pairs but often the nuclear events lag behind in the third, posterior member. Pronuclear exchange, however, occurs only between the two conjugants in the normal conjugating position. In the third posteriorly located member, pronuclear exchange never occurs; instead, the so-called migratory pronucleus moves to the vicinity of the mouth region where it remains to fuse with the so-called stationary pronucleus in the same individual to form the synkaryon. Thus, true conjugation involving pronuclear exchange occurs in the pair joined in

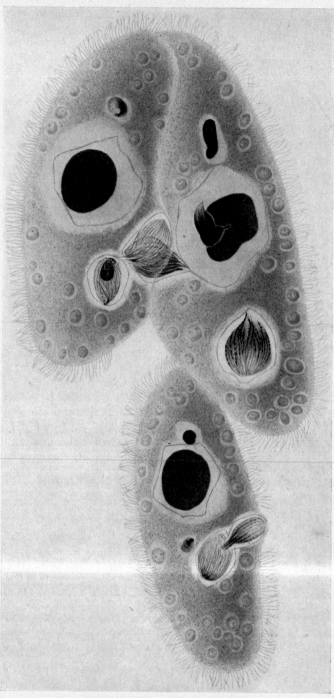

FIG. 100. (*For legend see facing page.*)

the normal position of conjugants while self-fertilization occurs in the posterior third member. Chen also described the nuclear events in quadruples and his paper should be consulted for details.

Chen's observations are of considerable significance in respect to the cytodynamics of the conjugation process. He has shown that the path of the migratory pronucleus appears to be predetermined and that only a small region of body contact as evidenced in the third posterior member is all that is required to initiate nuclear activity.

DeGaris (1935, 1935 a, b) reported the conjugation of normal specimens of *P. aurelia* with L-shaped monster forms of *P. caudatum*. He produced double-monsters of *P. caudatum* by using cyanide vapor and placed such specimens into a conjugating population of *P. aurelia*. According to him, such monster *P. caudatum* specimens readily mated with *P. aurelia* and within one or two hours, "conjugation" was well under way with such unions lasting from five to 12 hours. It is regrettable that he presented no cytological descriptions of nuclear changes. He reported that such crosses were invariably lethal to both members as a result of this type of conjugation.

On the other hand, DeGaris reported the successful conjugation of double-monsters of *P. caudatum* with normal specimens of the same or of different pure lines of the same species. He claimed that such crosses gave rise to durable lines just as if the conjugation were between normal members.

The conjugation between double-monsters and single animals of *P. bursaria* has been reported by Chen (1949, 1951) who also described the nuclear events of each in such a union. The double-monsters—chains of two animals—resulted when specimens during fission failed to separate but were held together by a slender protoplasmic bridge. Each of the two components of the monster (the anterior half and the posterior half) possessed a micronucleus and a macronucleus. A double-monster may conjugate with one, two, or occasionally more single animals but, typically, one single animal is attached to the anterior component of the double-monster at the adoral surface.

According to Chen, micronuclear (pregamic) divisions occur simultaneously in all three micronuclei at about the same rate as in ordinary conjugating pairs. In conjugation, the posterior component of the double-monster fails to come in contact with the single animal but the micronucleus in this lower half undergoes the usual nuclear changes. As pointed out by Chen, this would indicate that the influence of the contact between the anterior component and the single animal must have passed to the posterior component of the monster through the slender protoplasmic bridge.

Pronuclear exchange takes place only between the anterior component of the double-monster and the single animal. In the posterior component of the monster, the sister pronuclei, produced as a result of the third pregamic division, fuse and form a synkaryon resulting in autogamy.

FIG. 100. Conjugation of three animals in *P. bursaria*. Beginning of exchange of pronuclei. The larger anterior conjugant is clone McD₃, with two relatively large and darkly staining pronuclei. The smaller anterior conjugant and the third conjugant belong to clone Fd, each having two relatively small and lightly staining pronuclei. In the third conjugant (Fd) the migratory pronucleus has moved to the vicinity of the mouth region. About 30 hours after onset of conjugation. (Chen)

The author (Wichterman, 1940, p. 437) reported the union resembling conjugating pairs of a living *P. caudatum* with a dead specimen and raised the question of nuclear behavior occurring in the living member. Boell and Woodruff (1941) reported a specific mating reaction between living *P. calkinsi* of one mating type and a single dead animal of opposite type. In these two instances, the authors made no attempt to study nuclear behavior in the living animals of such a union.

Later Metz (1947) reported that when specimens of strongly reactive living *P. aurelia* of one mating type were mixed with equally reactive but *Formalin-killed* paramecia of complementary type, a large proportion of the living and dead animals clumped strongly, similar to that of the normal mating reaction. Such clumps sank to the bottom of the container and remained relatively quiescent for 60–90 minutes, after which time the clumps broke down to release single living animals and pairs of living animals joined in conjugation called "pseudo-selfing" pairs. These pairs remained united for about five hours in which there occurred in each member, macronuclear breakdown and micronuclear divisions (presumably meiosis, according to Metz). His figures show micronuclear activity extending to the second pregamic divisions.

Metz reported that nuclear behavior in "selfing" pairs is similar to that found in ordinary conjugants formed when opposite mating types of living specimens are crossed. It is suggested that the initiating mechanism for inducing micronuclear activity begins at or before the time that free-swimming "pseudo-selfing" pairs escape from the clumps of living and dead animals.

X-ray-killed specimens of one mating type of *P. bursaria* will not conjugate with living members of the opposite mating type (Wichterman, 1947 b, 1948 b) nor will such matings induce "selfing."

5. Autogamy.

Fermor (1913) appears to have been the first to describe autogamy in a ciliate; synkaryon formation as a result of micronuclear divisions and fusion in the encystment of *Stylonchia pustulata*. A year later, Hertwig (1914), in describing parthenogenesis in *P. aurelia*, admitted that the process might in reality be autogamy. Indeed, as late as 1928, Wilson suggested that possibly, a process of autogamous syngamy had been overlooked in endomixis.

In a detailed study of the nuclear behavior of *P. aurelia*, Diller (1934 a, 1936) described a process of self-fertilization or *autogamy* occurring in single animals (Figs. 101 and 102). He reported that in autogamy (Fig. 101), three micronuclear (pregamic) divisions, involving maturation, produce the gametic nuclei (pronuclei) (A–F). The first division of the two vegetative micronuclei (A) is characterized by their enlargement and the formation of long, thread-like crescents (B) which results in four nuclei (C); the second division results in the production of eight nuclei (D). Considerable variation appears to occur after the production of the eight nuclei prior to the formation of the pronuclei (gametic nuclei). Some of the eight nuclei degenerate while others prepare to divide (E) but which may not complete division. Eventually, however, the third nuclear division results in the formation of the gametic nuclei (F) which lie in a newly formed bulge near the mouth of the animal which Diller named the "paroral cone" (E–H).

Concerning the potential gametic

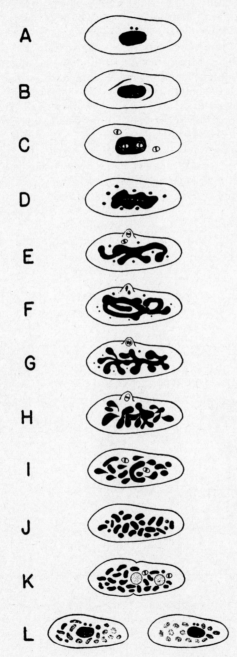

FIG. 101. Diagram of the nuclear changes during autogamy in *Paramecium aurelia* (Diller)

nuclei formed as a result of the third division, Diller stated that their number varies with the number of nuclei which have completed the third division. It would appear that their position in the cell determines their fate since the two gametic nuclei nearest the mouth region (paroral cone, F) appear to be the ones which will fuse to form the synkaryon (G) while the others degenerate rather quickly. The synkaryon (G) divides twice (H, I) to produce four nuclei, two of which remain as micronuclei while two become macronuclear anlagen (J, K). At the time of the first fission, a macronuclear anlage is passed to each daughter (K) while the two micronuclei divide so that ultimately

each cell possesses the normal nuclear apparatus of one macronucleus and two micronuclei.

An examination of Fig. 102 will show the changes undergone by the old macronucleus which involves skein formation prior to its fragmentation and final dissolution. Accompanying autogamy is a change in position of the ingestatory structures (possibly partial dedifferentiation) with the mouth opening partly occluded.

In his investigation dealing with the mixing of living and dead *P. aurelia* of opposite mating type, Metz (1947) reported the induction of nuclear reorganization, possibly autogamy, in living single isolated animals, and Diller (1948

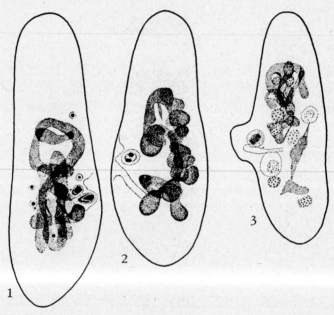

FIG. 102. (1) Autogamy, Woodruff race (gamete nuclei in contact in the paroral cone at the right; five or six degenerating nuclei are visible; macronucleus in skein). (2) Autogamy, Woodruff race (synkaryon formation; gamete nuclei enclosed within a common membrane; paroral cone; no degenerating micronuclei visible; macronucleus in skein). (3) Autogamy, isolation, Philadelphia race (synkaryon, in paroral cone, in metaphase of first division; no degenerating micronuclei seen; macronuclear skein fragmenting; a number of macronuclear bodies of various stages of degeneration present in the cell). Animals 1 and 2 from mass cultures. (Diller)

a) observed the induction of autogamy in single specimens of *P. calkinsi* following mixture of two mating types. The Burbancks (1951, 1951 a) reported induction of autogamy in *P. aurelia* when grown in non-living media.

See page 265 for discussion of the process of autogamy in relation to endomixis and page 374 for the method of inducing autogamy.

6. Cytogamy.[17]

For a long time the author (Wichterman, 1937 a, b, 1938, 1939, 1940, 1940 a, 1944 a, 1946 a, b, f, 1948 a) has been interested in studying living joined pairs of paramecia engaged in sexual processes in a special device known as a microcompression chamber (Fig. 106 s). Its use has led to the description of a new sexual process called *cytogamy* and has added new information relating to conjugation not disclosed by the use of killed and stained specimens.

A description of the microcompression chamber and method for studying living single specimens or isolated pairs of joined paramecia will be found on page 412.

As described earlier, two basic sexual processes have been reported for paramecium in which three micronuclear or pregamic divisions lead to synkaryon formation. The first process, called conjugation (cross-fertilization) may be defined as the temporary union of two individuals involving micronuclear divisions, nuclear exchange, and the establishment of a new synkaryon in each

conjugant. Thus a *sine qua non* of conjugation is nuclear exchange.

The second process, called autogamy (self-fertilization), was described by Diller (1936) for *P. aurelia* as occurring in single animals only and without the coöperation of another individual. Here three micronuclear divisions, as in conjugation, lead to the formation of gametic nuclei which fuse and form a synkaryon in the single animal.

The author has observed a third sexual process in joined *Paramecium caudatum* which appears to be intermediate to true conjugation on one hand and autogamy on the other (Figs. 103, 105, and 106). The process resembles conjugation since two ciliates join together along their oral surfaces and three micronuclear or pregamic divisions occur as in conjugation but there is no nuclear exchange between the members of the pair. Instead, the three micronuclear divisions lead to the formation of the gametic nuclei, but a synkaryon is established in the same individual, as in autogamy. The author proposed the term cytogamy (Wichterman, 1939) for this phenomenon to distinguish it from autogamy on the one hand (occurring in single individuals without the coöperation of another member), and from conjugation on the other hand, which involves a nuclear transfer in the joined specimens.

EARLY PHENOMENA OF CYTOGAMY. Cytogamous specimens of *P. caudatum* are considerably smaller than typical vegetative ones, indicating that there are divisions immediately prior to the sexual process that result in this smaller size. The onset of cytogamy appears to be correlated with a shortage of available food after the animals have been well fed. These smaller individuals (precytogamonts) are not as active as the typical, larger, well-fed ciliates and they possess protoplasm which is very

[17] After nearly 10 years of fairly wide usage, it has been brought to the writer's attention that the term *cytogamy* was first used by Rhumbler (1898) to designate one of the early steps in foraminiferal development wherein cell fusion occurs without karyogamy. Although the author attempted to coin a new term, it is regretted that he failed to encounter the term in Rhumbler's paper of 50 years ago. In its original connotation, the term is scarcely used.

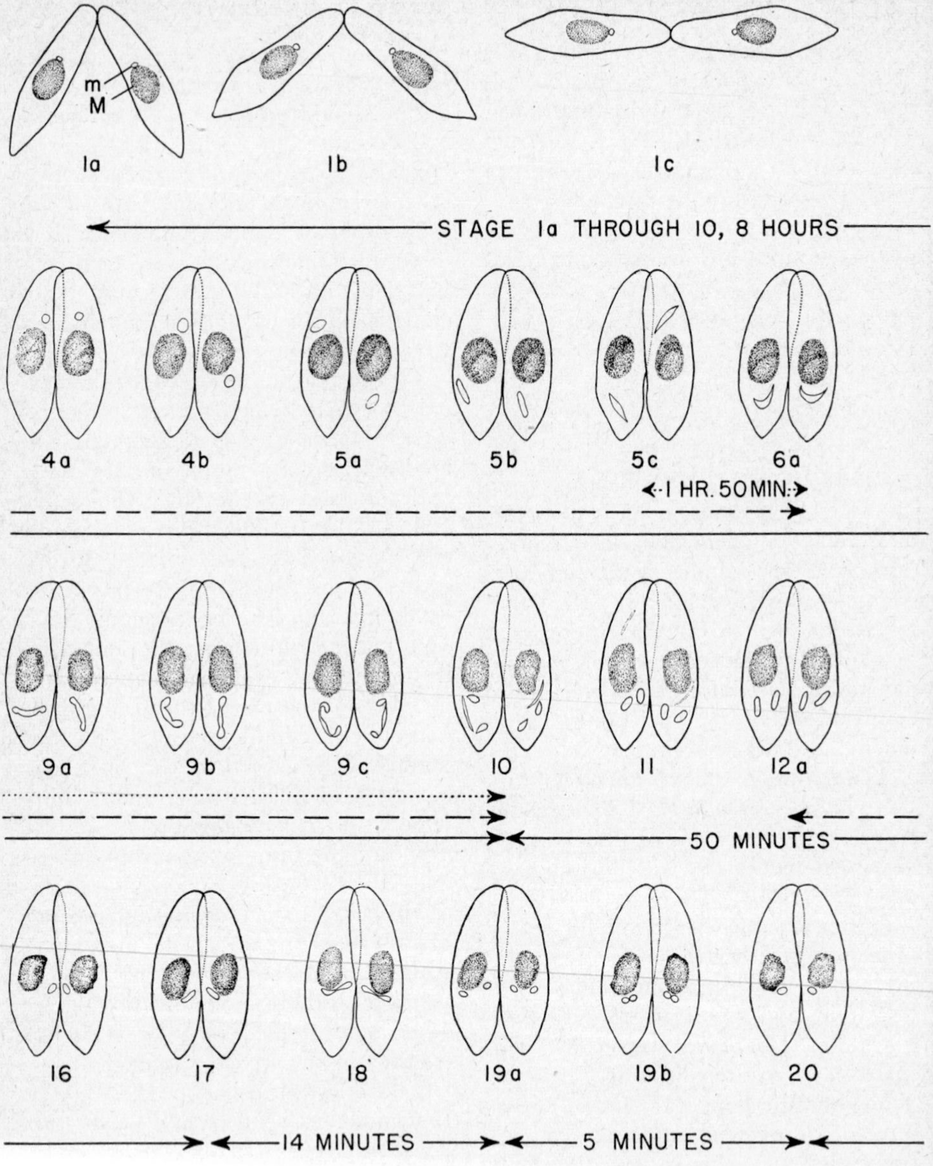
m
M
1a
1b
1c
STAGE 1a THROUGH 10, 8 HOURS
4a
4b
5a
5b
5c
6a
1 HR. 50 MIN.
9a
9b
9c
10
11
12a
50 MINUTES
16
17
18
19a
19b
20
14 MINUTES
5 MINUTES

Fig. 103. Characteristic stages in cytogamy of *Paramecium caudatum.* (M. macronucleus, m. micronucleus). Cytogamonts remain together for approximately 13 hours at 26° C. (1a–1d) First place of union at adhesive anterior ends (flexing reaction). (2) Second place of union in paroral region resulting in oral surfaces being opposite each other. Note space between cytogamonts. (3) Increase in size of micronuclei which lie against macronuclei. (4a–4b) Separation of micronuclei from macronuclei with continued enlargement of micronuclei. (5a) Lengthening out of micronuclei which become ellipsoidal in shape. (5b) Increased lengthening of micronuclei resulting in long, thin spindle with rounded ends. (5c) Sharply pointed spindle which precedes "crescent" formation. (6a) Characteristic crescent stage (prophase). (6b) Stage in transformation of crescent to

(*Continued on facing page.*)

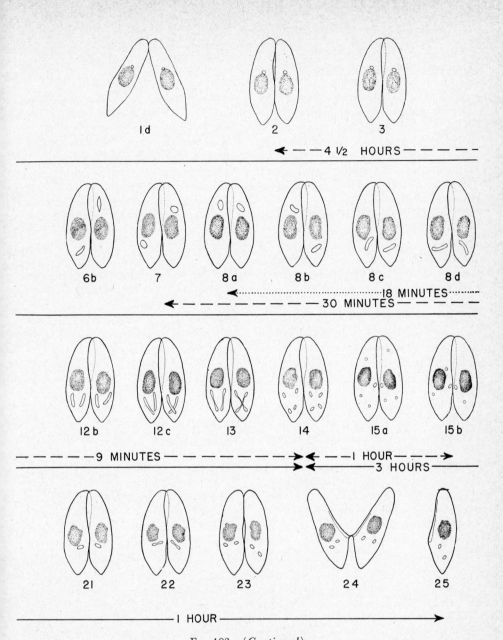

FIG. 103—(*Continued*)

ellipsoidal stage. (7) Broadly ellipsoidal stage (metaphase). (8a–8d) Increase in length of ellipsoidal stage (anaphase). (9a–9c) Telophase stages of first pregamic division. Separation spindle (connecting strand) clearly visible in 9c. (10) Completion of first pregamic division showing two pyriform micronuclear products and separation spindle in each cytogamont. (11) Degeneration of separation spindles and transition of pyriform micronuclear products into oval ones (metaphase). (12a–12c) Increase in length of two micronuclear products (anaphase). (13) Telophase stage of second division. (14) Completion of second pregamic division showing four micronuclear products in each cytoga-

(*Continued on following page.*)

303

clear and homogeneous but with crystals and few or no food vacuoles.

For the convenience of narration and identification, the sexual process has been divided into fairly clear-cut descriptive stages from one to 25 (Fig. 103). Many, if not most of the stage numbers can be applied with equal accuracy to comparable stages in autogamy and conjugation.

Two precytogamonts usually come together in direct contact at their anterior ends which appear to be very adhesive or sticky. The union at first is very insecure and they may separate; in the first hour or two of union they can be easily disjoined with a pipette. These preliminary steps to cytogamy are very similar to those described for conjugation by Jennings (1911). Joined cytogamous paramecia are observed to flex back and forth, at one time lying side by side along their oral surfaces, then later, apart, with one apparently pulling the other (stages 1a, 1b, 1c, and 1d). While the animals always are seen to join first at their sticky adhesive anterior ends, another region near the cytostome is also adhesive, and if in the beginning of the process the paramecia are not separated the anterior union becomes more secure. Then when the joined members of a pair swim in the same direction so that they lie with their oral regions opposite each other, a second point of fusion takes place here (stage 2). The paramecia are joined at these two extreme ends of the union in this early stage and the more secure union follows with the fusion between these ends. In all pedigree cultures, individuals when joining in cytogamy were never strongly attracted to each other as is so strikingly demonstrated in Sonneborn's (1937 c) and Jennings' (1938, 1939) mating reaction of paramecia. The "cluster formation" followed by pairing as reported by these investigators is conspicuously absent in cytogamy.

MICRONUCLEAR DIVISIONS AND TIME RELATIONSHIPS AS SEEN IN LIVING ANIMALS. *First pregamic division*. In animals joined along their oral surfaces in this early stage, the micronucleus of each may be easily seen lying against the macronucleus (stages 2 and 3). Micronuclei (as in stage 2) are nearly spherical in shape and measure about 8.5 microns in diameter. A gradual increase in size of the micronuclei (stage 3) up to 16 microns follows. In stage 4a the larger spherical micronuclei leave their place adjacent to the macronuclei and may be moved about in the cytoplasm quite actively, anteriorly and

---

FIG. 103—(*Continued*)

mont. (15a–15b) Stages in degeneration of three of the four micronuclear products in each cytogamont. Note persistence of one micronuclear body near the paroral region. (16) Remaining micronuclear product after degeneration of all others in each cytogamont and prior to division (metaphase). (17) Increase in length of micronuclear product (anaphase). (18) Telophase stage of third pregamic division. (19a) Completion of third pregamic division to show resulting micronuclear products, the pronuclei. (19b) Pronuclei of each cytogamont prior to fusion. (20) Synkaryon formation resulting from fusion of two pronuclei from same individual. (21) Increase in length of synkaryon (anaphase). (22) Telophase stage of synkaryon division. (23) Completion of synkaryon division resulting in formation of two products. Note space between cytogamonts. (24) Separation of cytogamonts along oral surfaces to paroral region. Paramecia now held by fused and stretched apices of paroral cones. Two division products of synkaryon present. (25) Excytogamont immediately after separation showing two division products of synkaryon and wrinkled macronucleus. (Wichterman)

posteriorly (stage 4b) by the streaming protoplasm as shown in Fig. 104.

This is followed by a lengthening out of the micronuclei (Fig. 103, stage 5a) which become broadly ellipsoidal or egg-shaped measuring 20 microns. Further lengthening occurs (stage 5b) at the expense of a decrease in diameter of the spindle which may attain a length of approximately 45 microns with the ends of the spindle sharply tapering to

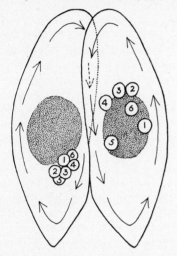

Fig. 104. Pair of living cytogamonts ($\times$ 309) 2½ hours after union to show cyclosis and resulting movement of micronuclei. Positions of micronuclei drawn at 1 minute intervals. (Wichterman)

a point (stage 5c). This stage precedes the characteristic "crescent" but in fact is an early stage in its formation. The time consumed up to the crescent formation is approximately four and a half hours after the cytogamonts have joined together along their oral surfaces. The average temperature at which all time relationships were determined was 26° C. While temperature is only one element which will hasten or lengthen the process of cytogamy, the security of the fusion at the initial union is also a factor in determining time relationships

especially in the beginning of the process.

The long sharply pointed spindle in each joined paramecium may be seen to transform gradually into the "crescent" (stage 6a). The transition into the crescent stage and its persistence so characteristic of conjugation, last for approximately one hour and 50 minutes. The typical mature crescent is a very evanescent and lucent stage but the largest in a volumetric sense. It is significant that the smallest size of the micronucleus with its condensed chromatin offers the greatest contrast and visibility. The chromatin of course becomes more diffuse as the micronucleus increases in volume up to the crescent (prophase) stage. This is readily understandable if one compares these observations to those of stained slides of similar stages of dividing micronuclei.

The "crescent" then loses its characteristic shape (stage 6b) and transforms into an ellipsoidal body of the metaphase stage, measuring about 32 microns in length (stage 7). While the times given above have been averaged, the slow and gradual process in the beginning is subject individually to a certain amount of variation; but the completion of division of the micronuclear body into two from metaphase through the anaphase (stages 8a, 8b, 8c, and 8d) and telophase (stages 9a, 9b, 9c) is comparatively rapid and constant, being consummated in 27–30 minutes. The metaphase stage may be seen to pass into the anaphase stage by the obvious increase in length. Of approximately the 30 minutes required for completion of this division, the anaphase and telophase stages require only about 18 minutes. As the dividing micronucleus becomes rapidly longer, the very thin but long "separation spindle" (connecting strand) is clearly visible (stage 9c). The completion of the first pregamic division is the separation of the daughter nuclei from the separation spindle (stage 10). Both products of the first micronuclear division, and the separation spindle are moved about in the streaming cytoplasm, but the separation spindle soon degenerates in the cytoplasm (stage 11). The average length of time from the fusion of two cytogamonts to the first pregamic division is eight hours. During the entire division, the dividing micronuclei of each cytogamont are constantly moved about in the streaming cytoplasm as well as crystals and other cytoplasmic inclusions.

In such joined cytogamonts, generally both micronuclear divisions are in the same phase although exceptions have been noted; i.e. one stage was 13 minutes ahead of its co-cytogamont. An interesting observation was that of a pair of joined paramecia in which one member in the first three hours of union died and was being pulled about by the living member whose micronucleus was in the crescent stage preparing for the first division.

After cytogamous paramecia have joined firmly, there is a slow progressive dedifferentiation of ingestatory structures which ultimately appear to pass into a sol state. In the crescent stage, however, movement of the penniculus may still be observed in the cytopharynx. Joined individuals in early stages of cytogamy were seen to egest through the cytopyge.

Second pregamic division. Each product of the first micronuclear division enters into the second where again two long spindles in each individual are visible. This division requires approximately 50 minutes for completion from the time the first division products are formed (stages 12a, 12b, 12c, 13, and 14). Immediately after the first division the two products are slenderly pyriform (stage 10) but soon become ellipsoidal

(stage 11) and finally lengthen into the anaphase (stages 12a, 12b, and 12c) and telophase (stage 13). Both stages are rapid, requiring only nine of the 50 minutes. Divisions of the two micronuclear products in each cytogamont into four (stage 14) were generally simultaneous, although occasional discrepancies were noted; e.g. one pair was observed in which the division in one cytogamont was 38 minutes ahead of the other. Four micronuclear products become clearly visible in each cytogamont and these even prior to division and while dividing were moved about in the streaming protoplasm.

Third pregamic division. Perhaps the most critical stage in the process of cytogamy is the fate of one of the four micronuclear products formed after the second division. While there is marked active cyclosis of the cytoplasm in the first and second divisions, joined pairs immediately after the second division become sluggish and physiologically appear to be in poor condition. This is not due to confinement in the chamber because pairs that had joined earlier in the cytogamous process and then placed in the chamber for the first time at this stage, exhibited this same condition. However, the four micronuclear products are moved about slowly in the cytoplasm and persist for about one hour, when degeneration of three of the four follows (stages 15a and 15b). An exception was noted however when all four of the products instead of degenerating passed through the third division, producing eight bodies. Degenerating products generally become spherical and small, simulating minute clear vacuoles before disappearing. The cytopharynx is visible for about two hours after the second division. Before degeneration of three of the four micronuclear products, one of them is found to remain in the region of the "paroral cone" (stages 15a and 15b). This usually is the one which enters into the third pregamic division to produce the two pronuclei.

The pronuclei and syngamy. Division of the remaining micronuclear product (stage 16) ordinarily occurs after all three in each cytogamont degenerate, approximately three hours after the second division products are formed.

The long, thin anaphasic-telophasic stages however require only about 14 minutes for completion (stages 17 and 18). The pronuclei in each cytogamont are generally found in the center or posterior part of each individual (stages 19a and 19b). It is significant that they appear to be of equal size and move to the same extent in the cytoplasm of each joined individual. Very soon after the third division, fusion of the pronuclei occurs in the same individual to form the synkaryon (stage 20). Almost immediately, the synkaryon passes through the anaphase and telophase stages (stages 21 and 22 respectively). Stage 23 shows the two division products of the synkaryon. The animals now prepare to separate. First, a space is noticed between the anterior ends of the joined animals and the fused apices of the paroral cones (stage 23). Next the anterior ends become free and the cytogamonts are then held only by the apices of the paroral cones (stage 24). Some few animals were seen to break apart in three minutes after the anterior ends became separated but generally the animals are held together for about an hour by these stretched fused apices, which takes on the appearance of a very thin elastic thread attaining a length of approximately 25 microns before breaking. Separation of cytogamonts takes place after the animals have remained joined together for approximately 13 hours at 26° C. Excytogamonts still show the old macronucleus

and the two products of the synkaryon division, but no trace of ingestatory structures is evident (stage 25).

At the present time, it appears that excytogamont reorganization stages of *P. caudatum* are similar to exconjugant stages.

Discussion

A criticism of this new method of studying living joined paramecia may be that an interference in the normal process is incurred by retaining them for study in the precision chamber. In answering, extraordinary precautions have been taken to reduce undue compression in the joined pairs. To simulate normal conditions, the cytogamonts were allowed to spiral freely except during those stages when critical observations were made. In fact, over long periods of time, the joined paramecia were removed from the small drop of culture fluid in the chamber and placed in a larger volume of fluid until a further observation was to be made. Since practically all the other stages of micronuclear behavior can be verified with stained slides of joined paramecia, it seems very unlikely that the stage of nuclear transfer would be the only one affected by the relatively slight amount of compression to which they are subjected at certain times. In all of the author's pedigree cultures, individuals when joining in cytogamy are never strongly attracted to each other as is so strikingly demonstrated in Sonneborn's (1937 c) sex reaction of paramecia. The characteristic agglutination or clumping reaction is conspicuously absent in cytogamy. In cytogamous union, they become rather feebly and insecurely joined in the beginning of the process in mass pedigree cultures left standing in the laboratory for approximately seven days after inoculation.

In respect to the behavior of micro-

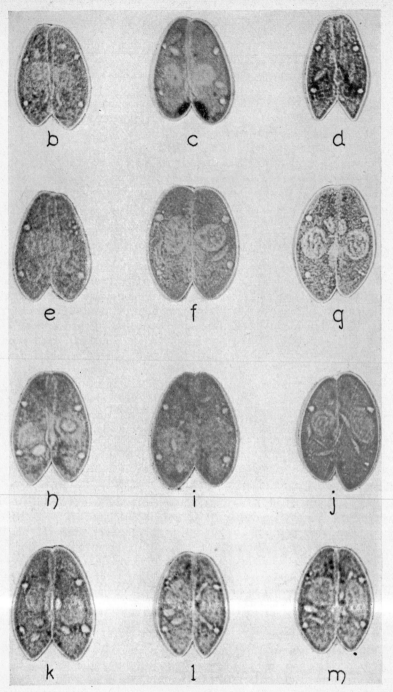

Fig. 105. Photographs of living cytogamonts printed from motion picture film. All approximately × 145. Contractile vacuoles, when photographed appear as small circles close to edge of each individual. (b) Early stage in cytogamy (stage 4a) showing the

(*Continued on facing page.*)

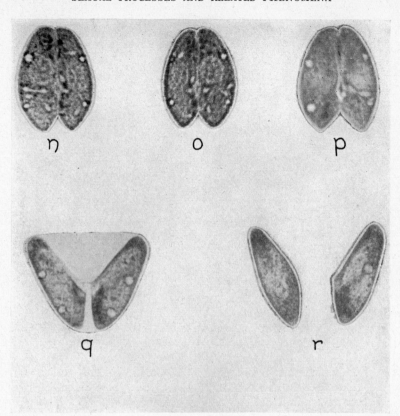

FIG. 105—(*Continued*)

freed micronucleus in anterior part of each Paramecium. Note prominence of ingestatory structures. (c) Similar stage but with micronuclei in different positions. Large masses of crystals (black in photograph) found posteriad in each cytogamont. (d) Long pointed spindle in each individual (stage 5c) which precedes crescent formation. (e) Crescents of stage 6a, each in posterior part of cytogamont. (f) Stage 6b which immediately follows crescent formation showing its disappearance. (g) End of crescent (prophase) stage and beginning of metaphase. (h) Large, metaphase stage (stage 7). (i) Telophase stage 9c showing the formation of "separation spindle" in center of dividing micronucleus of left cytogamont. (j) Completion of first pregamic division (stage 10). Note "separation spindle" between the two micronuclear products in left cytogamont. (k) Stage 11 showing two products in each cytogamont, the result of the first pregamic division. (l) Dividing micronuclear products of second pregamic division. Note nonsynchronous nuclear activity of cytogamonts with one on right ahead of the left. (m) Same pair as (l) but 1½ minutes later. (n) Same pair as in figure m but 2 minutes later. (o) Same pair as n but showing result of second pregamic division with four micronuclear products in each individual (stage 14). (p) Third pregamic division shown clearly in right individual (stage 18). (q) Cytogamonts about to separate (stage 24). Note thin strand holding them together. (r) Same pair as q but after separation (stage 25). (Wichterman)

FIG. 106. Photograph of micro-compression chamber (about × 1). (Wichterman)

nuclear phenomena in conjugation, cytogamy and autogamy, there is a remarkably close parallel with regard to the three pregamic divisions leading up to pronuclear formation. Autogamy and cytogamy are fundamentally alike in nuclear behavior since in cytogamy all nuclear activity is restricted to a single member of the joined pair. In conjugation, of course, the distinctive characteristic is transfer of pronuclei.

Concerning duration of time of the conjugation process, Maupas (1889) stated that it required 12 to 14 hours at 24° to 25° C. while cytogamy requires about the same length of time: i.e. 13 hours at 26° C. The extent of time for autogamy, according to Diller, was less than a day for all stages up to and including the formation of the two small anlagen with the first cell division occurring about two days after the beginning of the process.

Calkins and Cull (1907) show in *P. caudatum,* and there is shown again in text figures of the former's book (1933, p. 273), a cytolysis of the membranes between the two joined individuals in very early stages of conjugation. In addition to a nuclear exchange, this would infer transfer of cytoplasm. In direct contrast, cytogamy is characterized by the persistence of these membranes.

At present, the author is not in a position to give quantitative data concerning the place or function of cytogamy in the life of *P. caudatum*. This much appears clear, however, that precytogamonts are not as active as well-fed individuals; not only is their movement slower but the volume of crystals in cytogamonts appears higher than in vegetative individuals. It seems probable that the completion of cytogamy is a physiologic crisis through which joined paramecia pass with considerable difficulty. Animals when first joined in cytogamy show active cyclosis of the protoplasm but this is followed by a gradually diminishing rate of activity, especially shortly after the second pregamic division.

The occasional non-synchronous character of the nuclear activity in the individual members of a pair was noteworthy; that is, one member may show a nuclear stage definitely more advanced than that of its co-cytogamont (Fig. 105 l and m). This phenomenon was observed in *P. caudatum* by Diller, and more recently Penn (1937) using stained slides also observed this occurrence in a reinvestigation of the cytology of conjugation in *P. caudatum*. Although Penn believed he was studying conjugation, it appears very likely from his account and photographs that cytogamous individuals were seen instead. In such joined individuals, where the micronuclear divisions are out of phase with each other, the conclusion that there is a simultaneous transfer of pronuclei which characterizes conjugation seems irrational. In opposition to nuclear transfer are the number of cases noted where there is extreme inequality in size of joined individuals, which was also observed by Jennings (1911, p. 53). It seems unlikely that in such a union, nuclear exchange would occur. Also, there is the difficulty of reconciling the observation of a living *P. caudatum* joined to a dead one where nuclear exchange was impossible.

The fact that a dividing third pregamic or maturation spindle presses against the contiguous membranes of joined paramecia is not evidence enough to enable one to infer pronuclear exchange, even though the membranes may seem to be stretched. It does appear, however, that in joined cytogamonts, the membrane in the region of the paroral cone is very elastic. Diller suspected that exchange of gametic nuclei may not always occur in conjugation of *P. aurelia*. According to him, if nuclear transfer does not occur the animals would be undergoing a special type of autogamy. The present work justifies his belief, but in another species (*P. caudatum*), for the process of cytogamy is in a sense a distinctive type of autogamy but in joined pairs.

In spite of the notable studies conducted in the past on *P. caudatum*, there is a marked inadequacy in the description of the conjugation process and there exists now more than ever before, the urgent need for detailed accurate cytological investigation on the

exchange of pronuclei that characterizes conjugation.

More recently, a rather lively interest has developed in the problems of exchange of pronuclei and cytoplasm during the conjugation process. On the general problem, papers have appeared giving cytological, genetic, and serologic evidence of the transfer of pronuclei or cytoplasm or both during conjugation. Chen (1940 a), using fixed and stained preparations of *P. bursaria*, presents cytological evidence of pronuclear transfer. The author (Wichterman, 1946 f), for the same species, also has observed transfer of pronuclei in living specimens. However, Chen reported some cases in which the two pronuclei remain in the same conjugant and fuse to form a synkaryon (cytogamy). Recently Diller (1950 a) has reported on cytological evidence for pronuclear interchange in *P. caudatum*.

Sonneborn (1944) presents genetic evidence to show that there is nuclear and cytoplasmic transfer in *P. aurelia* during conjugation and he also shows genetic evidence for cytogamy. He reports that the transfer of cytoplasm is crudely measured by the extent of the time interval between separation of conjugants at their anterior ends and separation at their paroral cones. If the interval is less than three and a half minutes, exchange of cytoplasm is not detected, regardless of what races were crossed, but when the interval is 20 minutes or more, "cytoplasmic factor" is invariably transferred. Further, he reports that exchange of cytoplasm at conjugation never occurs in crosses in certain races (although nuclei are exchanged in these crosses) but does occur in others. Concerning nuclear exchange, Sonneborn (1941) reported that there is a definite temperature factor involved in cross-fertilization and some indication that calcium increases the frequency of it and that sodium decreases it.

It is of interest to report that Harrison and Fowler (1945 b) present serologic evidence of cytoplasmic interchange during conjugation of their races of *P. bursaria*.

Since cytogamy has been shown to occur, any conclusions genetic or otherwise based on the assumption that there is always a transfer of pronuclei in joined paramecia are open to serious question. The two processes, autogamy and cytogamy, may be identical except for the union of another member in cytogamy.

Finally, observations have led the author to question the work of others where a transfer of pronuclei is presumed to occur in "conjugation" of *P. caudatum* and other species especially when the contiguous membranes of the two joined individuals are not cytolyzed.

Speculation is incited as to how often cytogamy in other ciliates has been mistaken for conjugation by investigators. Cytogamy occurs with such frequency and regularity in cultures that one is inclined to believe it of general common occurrence in cultures left standing after inoculation where joined pairs are observed. Perhaps true conjugation is a rarity in such cultures.

Chapter 10

Genetics: The Mating Reaction, Mating Types, System of Breeding Relations and Inheritance in the Various Species of Paramecium

A. Introduction

Valuable reviews of the different sexual conditions among Protozoa bearing upon inheritance, genetics, and variation have been written by Dobell (1914), Calkins (1933), Jennings (1929, 1941 b), Kimball (1943), Sonneborn (1941 b, 1947, 1950 a, 1951, 1951 a) and Catcheside (1951). The recent literature on the subject in paramecium is vast and evergrowing and clearly reveals that this phase in the life of paramecium is receiving more attention by more investigators than any other. The study has and will continue to unfold the answer to many fundamental problems in heredity and genetics.[1]

In the ciliates belonging to the order *Peritricha*, a well-defined differentiation into two sex types is shown in which one type, called the macroconjugant, is sessile and large while the other type, called the microconjugant, is small and free-swimming. Sexual union and conjugation occur only between these different types.

In certain other ciliates belonging to different orders, less conspicuous sexual differences have been noted between the members of a pair.

While it is true for paramecium that conjugating members are generally smaller than non-conjugating specimens, the members of a pair entering into sexual union are approximately of the same size. However, morphologic differences in the sex types have been

[1] Extensive studies concerned with sexuality in unicellular organisms have been made upon the flagellate *Chlamydomonas* by Moewus (1933–40). A bibliography, review, and criticism of Moewus' work is given by Sonneborn (1941 b) and should be consulted by the student interested in this field. Investigations upon the genetics of the ciliates have been confined essentially to *Paramecium* and *Euplotes*. An extensive review of the sexuality and genetics in these two ciliate genera has been presented by Sonneborn (1947). Detailed information on methods in the genetics of paramecium is given by Sonneborn (1950 b).

noted. Size differences between the two conjugants have been observed by Doflein (1907) in a species which he called "*P. putrinum.*" According to Zweibaum (1922) approximately 70 per cent of the conjugating pairs of *P. caudatum* showed a difference in the amount of glycogen that was detected in the two members. In observations extending over many years, the present author (unpublished) has observed occasional differences between two members of a pair in regard to size, the number of contained crystals, and the amount of contained zoöchlorellae as in *P. bursaria.* Obvious differences are to be found when normally green specimens of *P. bursaria* mate with colorless ones; amicronucleate specimens (races) mate with micronucleate ones. In the vast majority of cases, however, one cannot detect a morphologic sex difference between the conjugant individuals in the species of *Paramecium.*

On the other hand, instead of the few instances of easily recognizable morphologic sex differences, inconspicuous *physiologic* sex differences have been reported between conjugant individuals. Boell and Woodruff (1941) reported a consistent difference in respiratory metabolism between the two members of pairs entering into conjugation in *P. calkinsi* and the author (Wichterman, 1948 b) observed a consistent difference in x-ray susceptibility between the members of pairs in the conjugation of *P. bursaria.*

Finally, it must be admitted that these differences may be only racial ones and not in reality sexual differences in the true sense of the word.

A thorough statistical treatment by Jennings (1911) and Jennings and Lashley (1913, 1913 a) disclosed that while the two members of a pair occasionally differed in their characters, generally one member mated with a like member even in regard to vitality and viability. Frequent viability differences were noted by Calkins and Cull (1907) between the two conjugating members in *P. caudatum.*

With the spectacular discovery of the existence of mating types in *P. aurelia* by Sonneborn in 1937, the genetics of this and other ciliates entered into a new phase. With this important fact in the hands of biologists, it now became possible to actually cross-breed stocks of diverse genetic ancestry just as in higher organisms. To date only certain species of *Paramecium* and *Euplotes* have been investigated; the great wealth of material in other ciliates is as yet untouched. The existence of mating types in other ciliates may demonstrate more effectively the answer to some of the problems in genetics, sexuality, and cytogenetics.

Perhaps the first clue to the existence of a sexual or mating type difference may be found in the extensive writings of Maupas (1889). In cultures of ciliates from a single natural source, conjugation never occurred. However, when ciliates from different sources were maintained in a single culture, conjugants were obtained. This led Maupas to believe that diversity of ancestry was necessary for conjugation.

More recently, we are indebted to Sonneborn not only for his discovery of the existence of mating types but also for his valuable, long-continued and extensive contributions into the genetics of paramecium.[2]

---

[2] These publications and those of his students, associates, and others dealing with sexuality, mating types, and genetics follow: Sonneborn (1937 c, 1938 a, b, 1939, 1939 a, b, 1940, 1941, 1941 a, b, c, 1942, 1942 a, b, c, d, e, f, 1943, 1943 a, b, c, 1944, 1945 a, b, 1946, 1946 a, 1947, 1947 a, b, 1948, 1948 a, b, 1949, 1950, 1950 a, all dealing with *P. aurelia*), Kimball (1937, 1939, 1939 a, 1943, 1947, 1947 a, b, 1948, all dealing with *P. aurelia;* 1939 b, 1941, 1942, dealing with *Euplotes*), Jennings (1938, 1940, 1941 a, b,

Sonneborn (1937 c) discovered the existence of mating types while working with different clones of *P. aurelia*. Upon mixing together animals from certain clones, he observed that the paramecia clumped or agglutinated into large masses. From these masses emerged pairs of individuals which entered into and completed conjugation. However, when he mixed together certain other clones no clumping or conjugation occurred. It soon became evident that the clones could be divided into two classes; those which demonstrated clumping and conjugation and those which did not. Sonneborn then proceeded to mix together, two at a time, all possible combinations of clones and listed the instances when clumping and conjugation did and did not occur. The different classes were first referred to as sexes but are now designated as *mating types*.

After Sonneborn's original discovery, he and his associates attack the problem of sexuality and genetics in paramecium in a most intensive manner and from many different aspects. In a book of this kind, it will be possible to cover only the major genetic contributions in this rapidly developing field wherein publications are appearing with frequent regularity. However, references will be given to all related publications

concerned with inheritance, sexuality, mating phenomena, etc. for the interested reader.

At the beginning, it may be useful to define certain terms used in the genetic study of paramecium. A *population* is considered as an aggregation of individual organisms, i.e. paramecia of the same species living together in a limited or defined region. A series of successive generations descended in culture from a member of a population that may be obtained from a pond, stream, or culture is commonly called a *line, series,* or *stock*.

The individuals of a population may be of one or many biotypes. A *biotype* is a group of individuals having the same hereditary constitution while a *genotype* indicates the constitution of an organism from the genic standpoint and represents the underlying constitution of a biotype. A type determined by the visible characters common to a group, as distinguished from their hereditary characters, is called a *phenotype*.

A *clone* exists when all the individuals in a culture are descended by uniparental reproduction from a single individual—that is by binary fission alone and in the absence of all fertilization processes—in which animals are of identical genotype and phenotype. When macronuclei are descended only by divisions from a preëxisting original macronucleus, the group of individuals of such a progeny is called a *caryonide*, a term useful for certain kinds of genetic work. A given caryonide will end when the original unit macronucleus degenerates, as in nuclear reorganization stages, and new caryonides are produced by the micronuclei as in conjugation.

As reported by Sonneborn (1947), if cultures of different caryonides of the same stock are mixed together, two at a

1942 a, 1944, 1944 a, b, c, 1945, Jennings and Opitz, 1944, all dealing with *P. bursaria*), Gilman (1939, 1941, 1946–47, 1946–47 a, 1949, all dealing with *P. caudatum*), Giese (1938, 1939, 1941, all dealing with *P. multimicronucleatum*), Giese and Arkoosh (1939, dealing with *P. caudatum* and *P. multimicronucleatum*), Hiwatashi (1949, 1949 a, 1950, dealing with *P. caudatum*). The following publications deal with *P. aurelia*: Sonneborn and Cohen (1936), Sonneborn and Dippell (1943, 1946, 1946 a), Sonneborn, Dippell, and Jacobson (1946), Sonneborn, Jacobson, and Dippell (1946), Sonneborn and Lesuer (1948), Dippell (1948), Ewer (1948), Preer (1941, 1946, 1948, 1948 a, 1950), Chao (1952) and Nanney (1952).

time, in all possible combinations and under appropriate conditions, most stocks show immediately that they consist of two classes of caryonides with respect to mating. When the mixture consists of one caryonide of each of the two classes, the paramecia agglutinate into clumps. Later, the animals emerge as conjugated pairs, each pair consisting of one member of each of the two caryonides. These pairs undergo reciprocal fertilization, then separate. When two caryonides of the same class are mixed together, neither agglutination nor conjugation occurs.[3] With rare exceptions, every caryonide of the stock may be classified to one class or the other. These two classes of caryonides within a stock are designated *mating types:* as a rule, all individuals of any one caryonide are of the same mating type and no stock contains more than two mating types.

B. Paramecium aurelia

When individuals of a number of different caryonides are cultivated singly in separate dishes and then mixed with samples of each of the others, the mating reaction may or may not occur. If mating does not occur after a mixture of two caryonides, it will be seen that individuals swim about independently. In other mixtures, however, the animals quickly unite in large clusters indicating that the two caryonides or classes of paramecia are of opposite mating type. It must be remembered that animals to be tested for mating should be ripe for

conjugation. Recently fed or over-fed specimens, which commonly appear plump, are not ripe for mating and conjugation. Starvation, however, is not required.

In the formation of a cluster, paramecia are seen to collide in their random swimming movements, beginning first with two individuals of opposite mating type and building up into larger clumps by the addition of other specimens. According to Sonneborn (1941 b), animals not in contact do not attract each other, nor are they in a specially sticky condition as has been often stated. Caryonides do not show the least trace of stickiness until the animals are mixed and then only when paramecia of different caryonides meet. Within the first minute after mixture, clusters of three or four individuals are formed; within five minutes, clusters may consist of approximately 40 or more specimens. Large clusters may persist for an hour or less, then break down leaving only pairs in conjugation and possibly some single specimens that were unable to obtain mates.

It is important to state that the two different classes or mating types do not differ in size, shape, or other visible structures; they appear exactly alike. However, it can be demonstrated that members of one sex type mate and conjugate only with members of the opposite sex type and not with themselves.[4]

Sonneborn (1938 b, 1939 a, b, 1947)

---

[3] It seems pertinent to apply these facts to Maupas' observations and to those of his critics. As noted by Sonneborn (1941 b), the fact that individuals of the same caryonide do not conjugate with each other agrees with Maupas' observation (who found conjugation within a clone) that closely related individuals do not interbreed. However, when several caryonides are present in the same culture, even though all come from a single original individual, conjugation may occur.

[4] As first reported by Sonneborn (1937 c) in his discovery of the mating reaction this may be demonstrated by introducing a single individual of one sex type into a drop containing many individuals of the opposite sex type. From the cluster of several individuals thus formed, there finally emerges only a single pair of conjugants. That this remaining pair includes the single individual of the introduced sex type may be demonstrated by selecting an organism that is larger or smaller than usual, so that it can be recognized. The single final pair is always found to contain this selected individual.

analyzed over 50 stocks of *P. aurelia* from many different and widely scattered sources in nature between Canada and Florida and from the Atlantic to the Pacific Coast. By mixing together all possible combinations of two clones or caryonides, under appropriate conditions, the breeding relations and number of different mating types in the species were discovered. It became apparent that the stocks could be classified into eight groups or, as they are now designated, *varieties*, even though the eight sexually isolated groups of stocks are effectively genetic species. When clones of different varieties are mixed, clumping and conjugation ordinarily do not occur. Within each variety, two mating types exist. Paramecia of the same variety but of different mating types are characteristically the only ones which will clump and conjugate when they are mixed. The mating types of the eight varieties were designated by the Roman numerals I and II, III and IV, V and VI, etc. (Table 26). It will be seen from the table that only one stock has been found in variety 7 containing only the one mating type XIII. Stocks containing two mating types always contain the two that occur in one variety. As reported by Sonneborn (1947) and observed in the table, under suitable mating conditions, the two mating types of the same variety, when mixed, show nearly 100 per cent immediate clumping and conjugation. This holds regardless of whether or not the two mixed mating types are derived from the same stock or from different stocks of the same variety.

It is, of course, a simple matter to ascertain the mating type of a culture by mixing members of it with standard, known cultures of other mating types. Since members of the unknown culture will ordinarily mate and conjugate with one of the known mating types, its mating type would then belong to the variety with which it mates. As an example, if the unknown type mates with type V, the type in question is of mating type VI, belonging to variety 3 as shown in the table.

According to Sonneborn (personal communication) and shown in Table 26,

The system of breeding relations in *P. aurelia* (Table 26) is best described with reference to the two classes of varieties, known as groups A and B, the grouping being based partly on genetic features, partly on breeding relations, and partly on other features. Group A includes varieties 1, 3, 5 and 7; group B includes varieties 2, 4, 6 and 8.

Among the four varieties of group A, intervarietal mating reactions occur most intensely between varieties 1 and 5; in mixtures of mating type I with mating type X and in mixtures of mating type II with mating type IX, as many as 40 per cent of the animals may conjugate. Lesser percentages of intervarietal mating take place in mixtures of type II (variety 1) with type V (variety 3) and of type II with type XIII (variety 7). Extremely feeble incomplete mating reactions (temporary adhesions between animals never leading to conjugation) also occur between mating type XIII and mating type VI (variety 3) and between mating type XIII and mating type X (variety 5). Aside from providing intervarietal hybrids, intervarietal reactions are important in showing that the mating types of group A varieties are homologous from variety to variety. Types II, VI and X are similar in that they react with type XIII; types I, V, IX and XIII are similar in that they all react with type II. In general, these intervarietal reactions are limited to combinations between an even and an odd type. Thus in the varieties of group A, the mating types are divisible into two classes that differ in the same way that the two mating types in any one variety differ: types that belong to the same class cannot interact sexually or mate with each other; mating reactions and conjugation occur

Table 26

THE SYSTEM OF BREEDING RELATIONS IN *P. aurelia** (After Sonneborn)

			A							B							
			1		3		5		7	2		4		6		8		
Group	Variety	Mating Type	I	II	V	VI	IX	X	XIII	III	IV	VII	VIII	XI	XII	XV	XVI	General Type
A	1	I	0	95	0	0	0	40	0	0	0	0	0	0	0	0	0	−
A	1	II	95	0	0	1	40	0	10	0	0	0	0	0	0	0	0	+
A	3	V			0	95	0	0	0	0	0	0	0	0	0	1	0	−
A	3	VI			95	0	0	0	3 Inc.	0	0	0	0	0	0	0	0	+
A	5	IX					0	95	0	0	0	0	0	0	0	0	0	−
A	5	X					95	0	1 Inc.	0	0	0	0	0	0	0	0	+
A	7	XIII							0	0	0	0	0	0	0	0	0	−
B	2	III								0	95	0	0	0	0	0	0	−
B	2	IV								95	0	0	0	0	0	0	0	+
B	4	VII										0	95	0	0	<50	0	−
B	4	VIII										95	0	0	0	0	0	+
B	6	XI												0	95	0	0	−
B	6	XII												95	0	0	0	+
B	8	XV														0	95	−
B	8	XVI														95	0	+

* The numbers in the body of the table give the maximum percentage of conjugant pairs formed in mixtures of mating types in the corresponding row and file. The abbreviation "Inc." stands for the "incomplete mating reaction" that never leads to conjugation; the number preceding "Inc." gives the maximum percentage of animals that give this reaction at any one time in a mixture of the two types.

only between types belonging to different classes. This seems to mean that each class of mating types is itself a mating type of a more general sort. The even types may therefore be designated as plus types and the odd types as minus types. Each variety of group A thus has one plus and one minus type.

Among the varieties of group B, mating reactions are confined to reactions between varieties 4 and 8. Type VIII of variety 4 reacts with type XV of variety 8 as strongly as do the two mating types of either of these varieties with each other. The initial reaction between type VII of variety 4 and type XVI of variety 8 is also as strong or nearly so; but the reaction does not proceed to conjugation in most of the agglutinated cells. These reactions show that types VII and XV are homologous and that types VIII and XVI are homologous.

Finally, one reaction occurs between groups A and B (Sonneborn and Dippell, 1946): type V of variety 3 reacts with type XVI of variety 8. Although usually less than one per cent of the animals in such a mixture form tightly united conjugant pairs, the initial agglutination reaction may involve a considerably larger proportion of the animals. This connection between groups A and B permits the extension of the plus and minus designations to varieties 4 and 8, again the odd types being minus types and the even types plus. At present, therefore, the homologies among the mating types of *P. aurelia* include the mating types of six of the eight varieties, only varieties 2 and 6 failing to show any mating reactions outside their own variety.

Collections of *P. aurelia* have now been obtained not only in the United States from coast to coast, but also from Puerto Rico (1), Peru (2), Chile (8), Scotland (1), France (2), Japan (10), and India (4). All of these foreign collections have been identified as belonging to varieties already described from the United States, namely varieties 1, 2, 4 and 6. The collections from Puerto Rico and India all belong to variety 6, previously found only once in the United States. The collections from Scotland and France all belong to variety 1; those from Japan belong to varieties 1

and 4; and those from Peru and Chile belong to varieties 1, 2 and 4. It now appears that, if varieties exist other than the eight already described, they could hardly be common or widespread.

While it is true that the eight varieties of *P. aurelia* conform to the species characters and superficially appear the same, careful examination by Sonneborn shows differences, especially physiologic, in addition to mating type. Varietal differences disclose the following characters in combination: size, fission-rate, temperature and light conditions upon which the mating reactions may occur, minimum lethal temperature, periodicity between successive self-fertilizations, antigenic constitution, and certain basic rules of inheritance (Sonneborn, 1942 e, 1945, 1947, 1950 a).[5]

A valuable, comprehensive review with complete bibliography of earlier work on the genetics of paramecium is to be found in Jennings (1929) classical study, *Genetics of the Protozoa*. His later review (Jennings, 1941 b) and especially the reviews of Sonneborn (1941 b, 1947, 1950 a) set forth the most recent knowledge of the subject.

It is not possible, within the scope of this book, to present any but the fundamental and most recent contributions beginning primarily with Sonneborn's discovery of the mating reaction in 1937. With this important discovery as a powerful genetic tool, Sonneborn and his associates have added greatly to our understanding of inheritance in para-

[5] As noted by Sonneborn, Jennings (1908 b, 1910) with Hargitt (1910) first observed that when races of *P. aurelia* differing markedly in size were grown together in the same culture, interbreeding never occurred since large specimens conjugated only with large ones and small specimens with like members. Also, the conditions for conjugation were observed by Jennings to be different for the different races, suggesting that the races in his cultures belonged to different varieties.

mecium and the ciliate Protozoa in general.

Before considering the genetics of paramecium, it will be of value to understand certain of the finer details of conjugation and nuclear behavior as described in the preceding chapter. Having a bearing on certain genetic problems are the following questions: the place where reductional division occurs in the three pregamic nuclear divisions in conjugation; the occurrence and constancy of conjugation (reciprocal fertilization) and/or cytogamy (double self-fertilization) in given crosses; the occurrence and constancy of cytoplasmic exchange in joined pairs of paramecia; the genetic constitution of the macronucleus (or caryonide of a clone).

Paramecium, like other Protozoa, possesses chromosomes which are similar to those in other multicellular organisms. The science of genetics shows that great numbers of diverse genetic factors, called the *genes,* have specific effects on certain characteristics and that these are transferred from parent to offspring. Also, that this genetic material reproduces itself, gene for gene, in development and reproduction. Recently, investigations with paramecium have disclosed that the cytoplasm may play a greater role in inheritance than has been suspected. This concept will be considered later in the chapter.

Inheritance in Uniparental Reproduction.

In this type of vegetative reproduction, a parent gives rise to two daughters by binary fission, and a typical clonal culture would represent many individuals of this type. Here the genetic material of the parent is divided, then duplicated to each of the offspring. Thus, all members of a clone are identical in genetic constitution with certain important exceptional cases discussed later.

A species consists of many diverse *biotypes;* that is, races differing in certain inherited characteristics such as size, shape, structure, and physiologic functions. Jennings has shown that, commonly, members of a single clone belong to the same biotype, i.e. a large specimen of paramecium will produce a clone of large specimens, a small specimen will yield consistently small specimens, etc. In addition, inheritable biotypes have been demonstrated showing the following characteristics: fission-rate differences, respiration differences, general resistance differences, and differences in constancy of mating type. Jennings also considered classes of non-heritable differences of a biotype, such as age (sexual immaturity and maturity in mating), nutritional differences, and environmental diversities as a result of differences in temperature and chemical conditions.

Many accounts in the literature show that when paramecium is cultivated in isolation culture for long periods of time, the organisms decline in vigor and die with structural changes characteristic of degenerating specimens. The many graphs showing decline and degeneracy may be due to continued, progressively slow environmental conditions that are unfavorable.

In paramecium, continued exposure to unfavorable conditions may result in acclimatization to the condition or immunity. Removal of the organism from the unfavorable environment has shown the inheritance of acquired acclimatization or immunity. It has been shown that the inheritance may continue for long periods of time through hundreds of cell divisions. However, when cultivated again under favorable conditions, i.e. with the particular acclimatization factor removed, the acquired immunity slowly decreases and is finally lost. (See p. 203 for additional information on acclimatization.)

Jollos (1913 a, 1920, 1921, 1934) studied acclimatization of *P. aurelia* and *P. caudatum* to high temperatures and to chemicals including arsenic (arsenious acid), and the possible occurrence of mutations as a result of these environmental factors. While he found that *P. aurelia* could not be acclimatized to arsenic, certain biotypes of *P. caudatum* could be acclimatized successfully.

Essentially, his method consisted in the exposure of paramecia to gradually increased concentrations of the acid followed by selection of surviving specimens. As an example, specimens at first were left in very weak solutions of acid for days or months which ordinarily resulted in no loss of life. Gradually, the concentration of acid was increased until most died. Survivors were selected, then placed again in a weak solution and allowed to multiply after which the process of selection was repeated to the extent that final survivors were able to withstand greater and greater concentrations of the acid. Such highly resistant, selected survivors thus were found to possess an acquired resistance as shown when the specimens were placed in arsenic-free media, allowed to reproduce and tested again. This inherited acquired resistance lasted in some cases for approximately eight months after 250 generations and to these long-lasting modifications Jollos gave the name *Dauermodifikationen*.

Jollos gave paramecia the same treatment with high temperatures and in some instances specimens were acclimatized for two and a half years. As with arsenic, the inherited acquired tolerance to high temperatures finally disappeared. (See also Chapter 6.)

On the basis of his discoveries, Jollos concluded that certain of his experiments showed that mutations had occurred in his specimens which were expressible in terms of resistance to temperature and arsenic as well as the fission-rate. Later, Raffel (1932, 1932 a) proceeded to show that mutations occurred with high frequency in *P. aurelia*, but it appears that his conclusions are unconvincing in the light of more recent knowledge and techniques. Dawson (1924, 1926, 1928), Moore (1927), and Sonneborn (1942 e) have reported observations on what appear to be inherited permanent morphologic alterations, considered to be mutations, in members of stocks of paramecium. A number of antigenic variations which persisted for one and a half years and arose in various clones of *P. aurelia* were reported by Harrison and Fowler (1945 a).

As noted by Sonneborn (1947), the genetic nature of these variants or mutations is not as yet understood, but they should be analyzed on the basis of present knowledge and with the newer techniques. His discovery of macronuclear regeneration (p. 264) and analysis of specimens have demonstrated that variants may arise as a result of the regenerations and that such organisms should not be considered as mutations.

Inheritance in Biparental Reproduction.

In the study of inheritance in conjugation as a result of crossing specimens of one mating type with those of another, it will be of value to keep in mind the details of the sexual process (p. 269). In conjugation, two diploid specimens join together and exchange pronuclei each of which contains a haploid set of chromosomes. Upon separation, each specimen contains two haploid sets—one from each of the conjugating specimens—which then fuse to form a diploid synkaryon. Each member of the conjugating pair now possesses identical genic constitution insofar as the synkaryon is concerned (Figs. 107 and 114). In addition, these figures show the genetic consequences of alternative chromosomal behaviors at the third maturation division of conjugation as depicted by Sonneborn (1947). Jollos indicated that not only do the members of a pair after conjugation possess identical nuclei (synkaryons) but that re-

duction division occurs in the first two nuclear divisions with the equational division occurring in the third nuclear division.

In reference to the aforementioned figure, Sonneborn reasoned as follows:

For a single pair of original heterozygotic genes, if the third division is equational, with reduction occurring at preceding divisions, one obtains identical heterozygotes in both mates of half the pairs, identical dominant homozygotes in both mates of ¼ of the pairs and identical recessive homozygotes in both mates of ¼ the pairs. If the

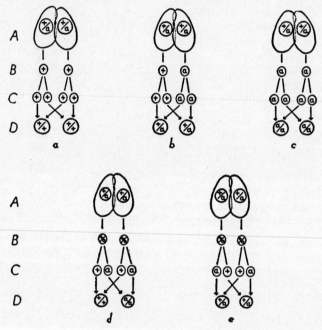

Fig. 107. Genetic consequences of alternative chromosomal behaviors at the third maturation division of conjugation. Cross of $+/a \times +/a$. The upper series of diagrams shows the three possibilities if the third division is equational, with reduction at the earlier maturation divisions. The lower series of diagrams shows the two possible alternatives if the third division is reductional, the earlier divisions equational. (On both alternatives, the gamete nuclei are assumed to arise from a single product of the second maturation division.) In both series of diagrams, row A shows the conjugants and their micronuclear genotypes at the start of conjugation; the remaining rows show the functional micronuclei only. Row B represents the single functional product of the second maturation division; row C represents the gamete nuclei; row D represents the synkaryons. (Sonneborn)

third division is, on the other hand, reductional, one obtains identical heterozygotes in both mates of ½ the pairs, but diverse homozygotes in the two mates in the other ½ of the pairs. Thus, reduction at the third division yields diverse clones from the two members of ½ the pairs, but equational third division requires that the two members of a pair of conjugants always yield genetically identical clones. The latter is moreover true, regardless of how many heterozygotic loci are involved in the original mating and regardless of whether the mates were genetically alike or different before the mating. This analysis tacitly assumes further that the two functional gamete nuclei are sister nuclei arising from a single haploid nucleus, for without this assumption one cannot regularly obtain genic identity between mates.

Therefore, members of a conjugating pair in which there occurred reciprocal fertilization emerge with the same genotype. As mentioned by Sonneborn, this is only possible if the third prezygotic division is equational with reduction division occurring during the two earlier divisions. Therefore, the plan shown in Fig. 107 a, b, c is the correct one.

The method of genic transmission is shown in Fig. 108. Figure 109 shows not only inheritance at conjugation but also the result when joined pairs do not undergo reciprocal fertilization such as at cytogamy.

It is, therefore, apparent that true conjugation in paramecium is like fertilization in the higher animals in which each specimen is endowed with a new set of diploid chromosomes and repeated conjugations in diverse clones give new recombinations of chromosomes. The diploid synkaryon after conjugation then divides with certain of its products eventually increasing in size to become macronuclei while others are destined to become micronuclei for distribution at the time of fission of the exconjugant.

As summarized in the earlier work of Jennings (1929) and before the dis-

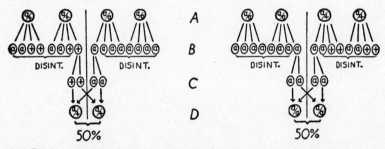

FIG. 108. Genic recombinations at conjugation in *P. aurelia*. Cross of $a/+ \times a/a$. The two possibilities, which occur with equal frequency, are shown in two diagrams, one to the left of the column A, B, C, D and one to the right. In each diagram, the vertical line separates the nuclei of one conjugant from those of the other conjugant of the pair. Row A represents the two nuclei in each conjugant, with the genic content at the start of conjugation. Row B represents the four reduced nuclei descended from each original nucleus, at the end of the two meiotic divisions; nuclei that disintegrate are marked "Disint." Row C represents the two gamete nuclei produced by the third division of one of the reduced nuclei in each conjugant. Row D shows the synkaryons formed by fusion of the gamete nuclei after exchange of the migratory gamete nuclei between the two conjugants of a pair. The synkaryons in both mates of half the conjugant pairs acquire the genotype $a/+$; the synkaryons in both mates of the remaining conjugant pairs acquire the genotype a/a. (Sonneborn)

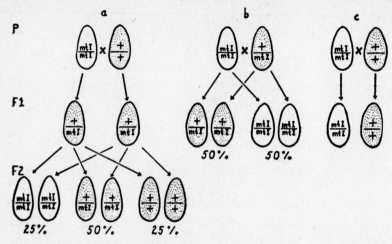

FIG. 109. Inheritance at conjugation and cytogamy in variety 1 of *P. aurelia*. Two-type condition: phenotype represented by stippling; gene, by $+$. Pure type I condition: phenotype represented by absence of stippling: gene, by *mtI*. Cross of $mtI/mtI \times +/+$. In column *a* are shown the results of conjugation in F_1 and F_2; the F_2 consists of 3 kinds of pairs of exconjugant clones in a $1:2:1$ ratio. In column *b* are shown the results of backcrossing the F_1 to the pure type I parent; two kinds of pairs are produced in a $1:1$ ratio. In column *c* are shown the results of the original mating if cytogamy occurs instead of conjugation: each mate produces a clone like itself. (Sonneborn)

covery of mating types, qualitative relations on the inheritance in conjugation were clearly demonstrated. He showed that conjugation in paramecium resulted in the production of many hereditarily diverse biotypes from the two involved in the conjugation process in regard to the following types of characteristics: fission rate, mortality rate and presence of abnormalities. Research on the production of inherited differences at conjugation was extended by the work of Raffel (1930) and by Jennings, Raffel, Lynch, and Sonneborn (1932) to include various other characteristics such as size and form, vigor, resistance, and degeneration. It was also shown that conjugation caused the descendants of the two members of a pair of paramecia to become similar in fission and mortality rate, and in the occurrence of abnormalities.

Geckler (1950) studied the inher-

itance of changes induced by nitrogen mustard in *P. aurelia* in which reduction in viability of the animals after fertilization was used as a measure of the effect. He presented data to show that some of the changes induced in *P. aurelia* are inherited through nuclear mechanisms and that in some of the changes, the cytoplasm is involved.

Inheritance of Mating Type in P. aurelia.

In a very striking manner, Sonneborn (1937 c) demonstrated the inheritance of mating type in *P. aurelia*. He crossed mating types I and II of variety 1; then after conjugation, each exconjugant was allowed to divide once before a new clone was begun from each specimen. Thus, with two individuals from each exconjugant, the resulting four specimens were isolated, and each allowed to reproduce by fission to yield clones. For about a week after conjugation, the animals were not reactive to mating.

Sonneborn discovered that all members of any one of the four clones belonged to the same mating type which is inherited within the clone until autogamy or until the next conjugation. Exconjugants, which were originally of a given mating type produced descendants of both mating types and no tendency for the two clones descended from a single exconjugant to be alike in mating type. It is seen that the first fission of the recent exconjugant produces two individuals each yielding clones which may be of different or of the same mating type. He reported that this segregation of the diverse mating types occurring at the first fission after conjugation is the result of separation of the two macronuclei in which one tends to produce mating type I and the other, mating type II.

Sonneborn found that the temperature during conjugation has an effect upon the proportion of the mating of the different mating types in the exconjugant progeny. In variety 1, higher temperatures resulted in the occurrence of a greater proportion of mating type II individuals, while in variety 3, higher temperatures produced more of mating type VI specimens than those of mating type V.

C. Paramecium bursaria

Soon after Sonneborn's discovery of mating types in *P. aurelia*, Jennings (1938) reported the existence of mating types in the green paramecium, *P. bursaria*. In many respects, *P. bursaria* is more suitable for studies of this kind and in the field of cytogenetics, extensive studies have been made by Jennings (1939, 1939 a, 1941 a, b, 1942 a, 1944, 1944 a, b, c, 1945), Jennings and Opitz (1944), Chen (1940, 1940 a, b, c, d, e,

f, 1945, 1946, 1946 a, b, d, 1949), Wichterman (1943, 1944, 1944 a, 1945, 1946 a, b, f, 1947 b, 1948 a, b), and Harrison and Fowler (1945 b, 1946, 1946 a).

Unlike the situation in *P. aurelia* where there exist only two opposite mating types in any one variety or group, a system of multiple interbreeding mating types is found to occur in *P. bursaria*. Also, the different groups of mating types are, with one exception, sexually isolated (Table 27).[6] In *P. bursaria*, the varieties have been designated by Roman numerals I–VI and the mating types within the varieties by the capital letters A–X. As will be seen from the table, there are at present six known varieties. Variety II contains eight mating types, each of which mates with every other mating type of the variety but of course not with itself. The three varieties called I, III, and VI each contain four mating types in which any mating type in a variety mates with all others of the given variety but not with itself. One variety, called IV, has two mating types that mate only with each other, and one variety from Russia, called V, contains only one mating type which has not been found to mate with any known mating type or with itself.

It is of interest to note that varieties I, II, and III represent clones from the United States and varieties IV, V and VI represent clones from Europe. Of the European varieties, IV and V are from Russia and variety VI from Eng-

[6] An exception occurs to complete sexual isolation of the varieties in which type R of variety IV conjugates with the four mating types E, K, L, and M of variety II but not with the other four mating types F, G, H, and J of the same variety. Some interpret this to indicate homologies of mating types between varieties, but in such matings, the hybrids die without completing the conjugation process (Jennings and Opitz, 1944; Chen, 1946 b; Sonneborn and Dippell, 1946).

Table 27

THE SYSTEM OF BREEDING RELATIONS IN *P. bursaria**

Variety	Mating Type	I				II								III				IV		V	VI			
		A	B	C	D	E	F	G	H	J	K	L	M	N	O	P	Q	R	S	T	U	V	W	X
I	A	−	+	+	+	−	−	−	−	−	−	−	−	−	−	−	−	−	−	−	−	−	−	−
	B	+	−	+	+	−	−	−	−	−	−	−	−	−	−	−	−	−	−	−	−	−	−	−
	C	+	+	−	+	−	−	−	−	−	−	−	−	−	−	−	−	−	−	−	−	−	−	−
	D	+	+	+	−	−	−	−	−	−	−	−	−	−	−	−	−	−	−	−	−	−	−	−
II	E	−	−	−	−	−	+	+	+	+	+	+	+	−	−	−	−	−	−	−	−	−	−	−
	F	−	−	−	−	+	−	+	+	+	+	+	+	−	−	−	−	−	−	−	−	−	−	−
	G	−	−	−	−	+	+	−	+	+	+	+	+	−	−	−	−	−	−	−	−	−	−	−
	H	−	−	−	−	+	+	+	−	+	+	+	+	−	−	−	−	−	−	−	−	−	−	−
	J	−	−	−	−	+	+	+	+	−	+	+	+	−	−	−	−	−	−	−	−	−	−	−
	K	−	−	−	−	+	+	+	+	+	−	+	+	−	−	−	−	−	−	−	−	−	−	−
	L	−	−	−	−	+	+	+	+	+	+	−	+	−	−	−	−	−	−	−	−	−	−	−
	M	−	−	−	−	+	+	+	+	+	+	+	−	−	−	−	−	−	−	−	−	−	−	−
III	N	−	−	−	−	−	−	−	−	−	−	−	−	−	+	+	+	−	−	−	−	−	−	−
	O	−	−	−	−	−	−	−	−	−	−	−	−	+	−	+	+	−	−	−	−	−	−	−
	P	−	−	−	−	−	−	−	−	−	−	−	−	+	+	−	+	−	−	−	−	−	−	−
	Q	−	−	−	−	−	−	−	−	−	−	−	−	+	+	+	−	−	−	−	−	−	−	−
IV	R	−	−	−	−	−	−	−	−	−	−	−	−	−	−	−	−	−	+	−	−	−	−	−
	S	−	−	−	−	−	−	−	−	−	−	−	−	−	−	−	−	+	−	−	−	−	−	−
V	T	−	−	−	−	−	−	−	−	−	−	−	−	−	−	−	−	−	−	−	−	−	−	−
VI	U	−	−	−	−	−	−	−	−	−	−	−	−	−	−	−	−	−	−	−	−	+	+	+
	V	−	−	−	−	−	−	−	−	−	−	−	−	−	−	−	−	−	−	−	+	−	+	+
	W	−	−	−	−	−	−	−	−	−	−	−	−	−	−	−	−	−	−	−	+	+	−	+
	X	−	−	−	−	−	−	−	−	−	−	−	−	−	−	−	−	−	−	−	+	+	+	−

* + indicates the occurrence of conjugation, − indicates that no conjugation occurs, in mixtures of the two mating types represented on the corresponding row and file. (Sonneborn)

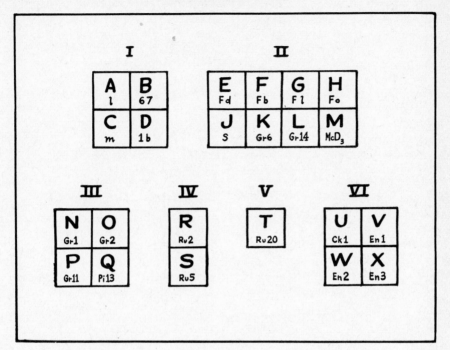

FIG. 110. Diagram of the six known non-interbreeding varieties of *Paramecium bursaria*. The capital letters are the designations of mating types in each variety: four in Variety I, eight in Variety II, four in Variety III, two in Variety IV, four in Variety VI. In Variety V apparently only one mating type is known. The figures or numbers in the lower half of the squares are the designations of clones that first exemplified each mating type. (Chen)

land, Ireland, and Czechoslovakia (Fig. 110).

From many aspects, *P. bursaria* is the most favorable species for demonstrating the mating reaction. Specimens can be collected in nature and maintained easily in continuous culture, apparently indefinitely (Chapter 4); many races have been in continuous cultivation by the author for over 11 years. In all that time selfing (conjugation) never has been observed in a given clone of one mating type and the mating types have remained stable throughout the period. It is therefore a simple matter after a single specimen has been isolated from nature, to obtain progeny and test for mating type.

The mating reaction between two clones belonging to opposite mating types occurs when the clones are mixed, provided they are in the proper physiologic condition. There is a period of immaturity in a clone, especially in the start of a culture, during which the mating reaction either does not occur or does so weakly. Clotting, mating, and conjugation do not occur when individuals to be tested are taken from a rich medium containing plump, rapidly dividing specimens. While plumpness of body tends to inhibit the mating reaction, starvation is not necessary. The reaction readily takes place in well-fed, slender individuals.

In newly made transfer cultures containing animals undergoing binary fission, peak reactions are obtained a week to 10 days after inoculation. Jennings noted that the shortest period observed from conjugation to the beginning of the mating reaction was 12 days, but it is usually considerably longer.

Time of day when the clones are mixed is an important factor in regard to the occurrence of the mating reaction. An examination of Table 28 shows when the reaction does not occur, when it occurs feebly, and when the optimum reaction occurs as based on an analysis of many clones by the author (Wichter-

man, 1948 a). With a few exceptions,[7] mating does not occur in the early morning hours or after 5 or 6 p.m. while the reaction is greatest at 12 o'clock noon.

When opposite mating types of *P. bursaria* are in the proper physiologic condition and mixed, specimens immediately will enter into the mating reaction. Upon being mixed, a few paramecia are first seen to stick together; then the clumps become larger and larger until 10–15 minutes later im-

[7] According to Jennings, in group (Variety) III, mating and conjugation may occur at any hour of the day or night.

Table 28

TIME-TABLE OF MATING REACTION IN *Paramecium bursaria**

Time	Result of Mixing
4:00 a.m.	No reaction; paramecia slow moving
4:15 a.m.	No reaction; paramecia slow moving
4:30 a.m.	No reaction; paramecia slow moving
4:45 a.m. (Dawn)	Faint reaction. Only a few instances of pairs seen and these poorly joined in various positions
5:00 a.m.	As above but more pairs; some small clumps of three and four
5:15 a.m.	As above but singles are more active
5:30 a.m.	As above
5:45 a.m.	As above but with more groups of three, four, and five individuals in the groups
6:00 a.m.	Same as above but clumps numbering up to eight individuals
6:15 a.m.	Same as above with clumps numbering ten or more individuals
7:15 a.m.	Almost immediate clumps of 20 and more individuals
8:00 a.m.	Large clumps soon forming of 50 and more individuals
9:00 a.m to 1:30 p.m.	Extremely large clumps of 100 and more individuals
2:00 p.m.	Large clumps but smaller than above
2:30 p.m.	Smaller clumps than above
3:00 p.m.	Clumps of approximately twelve individuals
3:30 p.m.	As above but clumps smaller
4:00 p.m.	Very slight mating reaction with clusters of two, three, four, and five individuals
4:30 p.m.	As above but only pairs and clusters of three individuals loosely joined
5:00 p.m.	Few pairs loosely joined
6:00 p.m.	No reaction

* Rich cultures of opposite mating types of *P. bursaria* and highly reactive for the mating reaction were mixed at times indicated; temperature approximately 25° C. The examinations were made immediately after the mixture. (Wichterman)

mense clumps of a hundred or more individuals are formed (Fig. 111). These larger masses then become smaller and smaller until several hours later only small groups of individuals and some singles and joined pairs remain. Finally, after approximately five hours, only pairs of conjugants (and single ones that were unable to obtain mates) are left. The behavior of agglutination, mating, and pairing in *P. bursaria* has been described in detail by Jennings (1939). Two specimens, after touching any part of the body, stick together exactly as if their surfaces were covered with some strong adhesive material. Once they adhere, they begin to move in an irregular way resultant upon the divergent action of their free cilia. Figures 112 and 113 show a number of observed cases of such irregular adherence; frequently one individual drags another backward or sidewise through the water. At first it is the cilia that thus adhere together; and one may see a definite open space between the two bodies spanned only by the cilia, which are immobilized where in contact. After swimming irregularly, a third individual adheres to the two, then a fourth, and this continues until large masses are formed, containing 20 to 100 individuals or more.

Jennings (1941 a, b, 1942 a) observed two major types of inheritance of mating types in certain of his clones of *P. bursaria*. The first, called by him self-differentiation, is a type in which the mating type remains unchanged over a period of many cell divisions in the clone but with the occasional appearance of pairs. Testing of the pairs disclosed that the animals consisted of different mating types even though they were descended from a single specimen. Although he has no experimental proof, Jennings suggested that this change of mating type was due possibly to endomixis or autogamy. However, Erdmann (1925) reported that nuclear reorganization in single specimens (called by her endomixis) is rare in this species. In a cytological examination of many specimens from different races, the author has not encountered any nuclear changes resembling autogamy, but it is conceivable that such changes may occur in a few specimens within a clone and escape unnoticed. On the other hand, self-differentiation through binary fission is a possibility that should be investigated.

Table 29

MATING TYPES OF DESCENDANT CLONES FROM 131 PAIRS (INCLUDING 279 DESCENDANT CLONES), IN THE SIX POSSIBLE CROSSES OF THE FOUR MATING TYPES OF VARIETY 1, *Paramecium bursaria* (JENNINGS, 1939)

Cross	Number of Pairs	Mating Types of the Descendant Clones: Numbers of Pairs Yielding Each Type					
		A	B	C	D	Like Parents	Unlike Parents
A × B	25	13	12	0	0	25	0
A × C	6	3	1	1	1	4	2
A × D	61	30	4	4	23	53	8
B × C	10	0	10	0	0	10	0
B × D	15	5	8	0	2	10	5
C × D	14	0	0	6	8	14	0
Totals	131	51	35	11	34	116	15

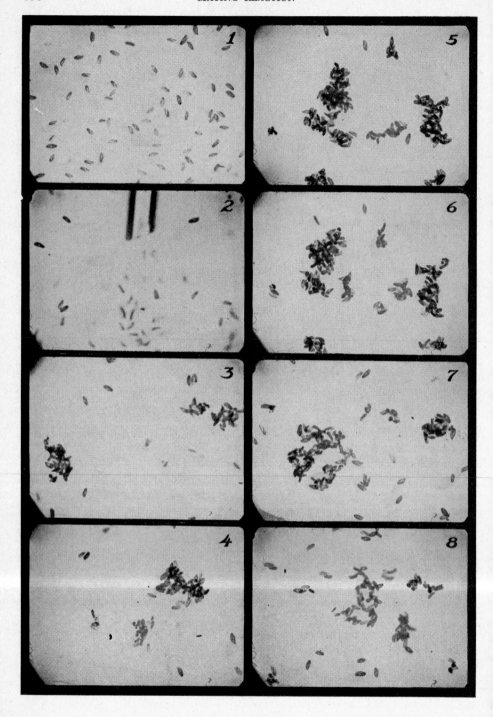

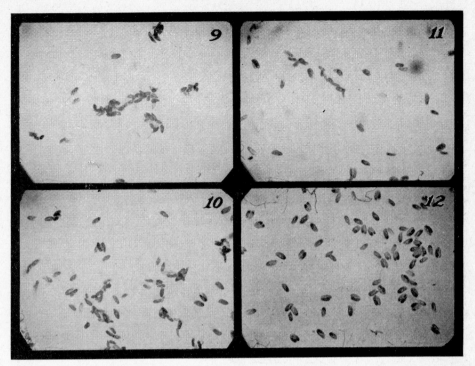

Fig. 111. Photographs of the mating reaction and conjugation in living *Paramecium bursaria*. All photographs are unretouched and have been printed directly from the 16 mm. negative of the motion picture. (1) Living individuals of one mating type swimming normally. (2) Individuals from culture of opposite mating type being introduced by means of a capillary pipette to individuals shown in fig. 1. (3) The mixture of individuals of the two opposite mating types about one minute later showing formation of clumps of paramecia. (4–7) Mixture of opposite mating types from two to seven minutes after mixing showing formation of large clumps. (8–9) Breaking apart of larger clumps into smaller ones about one hour after mixing. (10) About one and a half hours after mixing showing only smaller groups, paired and single individuals. (11) A chain of seven paramecia showing individuals of the two mating types alternating. About one and a half hours after mixing. (12) About twenty-four hours after mixing showing predominantly pairs undergoing conjugation and some single individuals that were unable to obtain mates. (Wichterman)

When a new mating type has been produced by self-differentiation and the progeny cultured, a second self-differentiation may occur. When this happens, the new type so produced is always the same as the original type prior to the first self-differentiation.

In the second type, in which there is inheritance following conjugation, Jennings noted that when crosses were made between clones of known mating types, the descended clones from the exconjugants of any one pair were of the same mating type in about 97.4 per cent of the pairs (Table 29). Thus the original two conjugants of different mating types became of the identical mating type after pronuclear exchange. This would suggest that mating type is determined genetically by the pronuclei

FIG. 112. Characteristic small groups formed by the adhesion of individuals in the clumping that precedes conjugation in *P. bursaria*. The arrows present in certain figures show the direction in which the individual's cilia tend to carry it. (Jennings)

and that both members would, after nuclear exchange, possess identical synkaryons as shown in Figs. 107 and 114.

However, it is more difficult to explain the exceptional cases on genetic grounds. Concerning these, Jennings observed that in a few cases there are two mating types represented among the clones descended from a single pair.

It is therefore concluded that in all crosses, the greatest majority of descendant clones of the exconjugants are of one or the other of the two original parental mating types, but a few exconjugant clones are of other mating types.

Jennings (1939 a, 1941 a, 1944 a, b,

c) has studied intensively the changes that occur within clones started with exconjugant lines. He observed also that repeated inbreeding rapidly increased the mortality rate. This, with other effects of conjugation, is discussed on page 365.

Chen's comprehensive studies on chromosomes in *P. bursaria* have shown the existence of many polyploid (heteroploid) races, a situation which may tend to make analysis of genetic studies difficult (p. 290). Sonneborn's (1947) suggestion of using selected mating types of known homozygous diploid material after the methods of Chen for studying polyploidy is worthy of experimentation and likely to yield fruitful results.

Also, the role of cytoplasm in inheritance—if and when it is exchanged during the conjugation of *P. bursaria*—should be considered and investigated.

D. Paramecium caudatum

Mating types in *P. caudatum* have been investigated by Gilman (1939, 1941, 1946–47, 1946–47 a, 1949, 1950), Giese and Arkoosh (1939), Y. T. Chen (1944), and Hiwatashi (1949). Chen found the existence of four varieties in China, and Hiwatashi reported the same number in Japan. Gilman (1950) obtained from Hiwatashi, the four noninterbreeding Japanese varieties col-

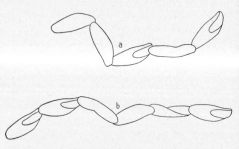

FIG. 113. Chains of individuals adhering together at their tips, in a late stage of the agglutination that leads to conjugation in *P. bursaria*. (Jennings)

lected from the Island of Honshu. Each variety consisted of two interbreeding mating types. He then mixed samples of these mating types with varieties from the United States and found that two varieties (1 and 3) occur in both the United States and Japan. Nine varieties (2, 4, 5, 6, 7, 8, 9, 10, and 11) occur in the United States only and two

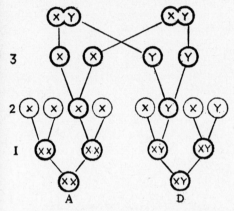

Fig. 114. The three maturation divisions (1, 2, 3) and the exchange of pronuclei in the micronuclei during conjugation. (A and D) The micronuclei of the two conjugants respectively. Reduction at the second division, represented by the separation of XX and XY. The diagram illustrates the fact that the two diploid nuclei produced (above) are alike in constitution. (Jennings)

varieties (12 and 13) occur only in Japan (Table 30).

It appears that the breeding system of this species is similar to that of *P. aurelia* in which there exist a number of sexually isolated varieties each containing two interbreeding mating types. At present there exist 13 varieties containing 25 mating types designated by the Roman numerals I–XXVI. Instances of intervarietal mating have been reported by Gilman (1949) as noted in and below the interbreeding table. Each variety contains the two mating types except variety 10 in which

mating type XX is sexually isolated. It would appear that a presumed mating type, namely XIX has not as yet been found.

In studying the effect of diverse nutritive conditions in *P. caudatum* in regard to the mating reaction and conjugation in lettuce-infusion cultures (0.75 Gm. dried lettuce per liter distilled water), Gilman (1939) observed the following:

1. When animals are very well fed and plump, no mating reaction or conjugation occurs.

2. When animals are well fed but not plump, a weak mating reaction occurs in which a few pairs cling together but break apart without conjugating.

3. When animals are of moderate size and not well fed, a strong mating reaction occurs and many clumps are formed. These disintegrate into pairs which fuse and complete conjugation.

4. When animals are small and thin, a strong mating reaction occurs. Many clumps are formed but these disintegrate. Few or none of the animals proceed to conjugate.

5. When animals are very small and starved, no mating reaction and no conjugation occur.

To demonstrate the mating reaction and conjugation, one may culture clones of opposite mating types in lettuce infusion in test tubes plugged with cotton. On the day before mating, add approximately 20 drops of a bacterized lettuce infusion containing suitable bacteria. After about 24 hours, these clones of fed paramecia are ready for mating and conjugation when mixed at room temperature. It will be seen that recently fed paramecia are fat and will not mate, but after 24 hours, specimens become slender and should be reactive. To obtain cultures in optimal conditions for mating, specimens may be fed daily with a definite number of drops of bacterized lettuce infusion.

When nutritive conditions are met,

Table 30

The System of Breeding Relations in *Paramecium caudatum**† (Gilman).

Variety		1	1	2	2	3	3	4	4	5	5	6	6	7	7	8	8	9	9	10	10	11	11	12	12	13	13
Variety	Mating Type	I	II	III	IV	V	VI	VII	VIII	IX	X	XI	XII	XIII	XIV	XV	XVI	XVII	XVIII	−	XX	XXI	XXII	XXIII	XXIV	XXV	XXVI
1	I	−	+	−	−	−	−	−	−	−	−	−	−	−	−	−	−	−	−	−	−	−	−	−	−	−	−
	II	+	−	−	−	−	−	−	−	−	−	−	−	−	−	−	−	−	−	−	−	−	−	−	−	−	−
2	III	−	−	−	+	−	−	−	−	−	−	−	−	−	−	−	−	−	−	−	−	−	−	−	−	−	−
	IV	−	−	+	−	−	−	−	−	−	−	−	−	−	−	+	−	−	−	−	−	−	−	−	−	−	−
3	V	−	−	−	−	−	+	−	−	−	−	−	−	−	−	−	−	−	−	−	−	−	−	−	−	−	−
	VI	−	−	−	−	+	−	−	−	−	−	−	−	−	−	−	−	−	−	−	−	−	−	−	−	−	−
4	VII	−	−	−	−	−	−	−	+	−	−	−	−	−	−	−	−	−	−	−	−	−	−	−	−	−	−
	VIII	−	−	−	−	−	−	+	−	−	−	−	−	−	−	−	−	−	−	−	−	−	−	−	−	−	−
5	IX	−	−	−	−	−	−	−	−	−	+	−	−	−	−	−	−	−	−	−	−	−	−	−	−	−	−
	X	−	−	−	−	−	−	−	−	+	−	−	−	−	−	−	−	−	−	−	−	−	−	−	−	−	−
6	XI	−	−	−	−	−	−	−	−	−	−	−	+	−	−	−	−	−	−	−	−	−	−	−	−	−	−
	XII	−	−	−	−	−	−	−	−	−	−	+	−	−	−	−	−	−	−	−	−	−	−	−	−	−	−
7	XIII	−	−	−	−	−	−	−	−	−	−	−	−	−	+	−	−	−	−	−	−	−	−	−	−	−	−
	XIV	−	−	−	−	−	−	−	−	−	−	−	−	+	−	−	−	−	−	−	−	−	−	−	−	−	−
8	XV	−	−	−	+	−	−	−	−	−	−	−	−	−	−	−	+	−	−	−	+	−	−	−	−	−	−
	XVI	−	−	−	−	−	−	−	−	−	−	−	−	−	−	+	−	−	−	−	−	−	−	−	−	−	−
9	XVII	−	−	−	−	−	−	−	−	−	−	−	−	−	−	−	−	−	+	−	+	−	−	−	−	−	−
	XVIII	−	−	−	−	−	−	−	−	−	−	−	−	−	−	−	−	+	−	−	−	−	−	−	−	−	−
10	−	−	−	−	−	−	−	−	−	−	−	−	−	−	−	−	−	−	−	−	+	−	−	−	−	−	−
	XX	−	−	−	−	−	−	−	−	−	−	−	−	−	−	+	−	+	−	+	−	−	−	−	−	−	−
11	XXI	−	−	−	−	−	−	−	−	−	−	−	−	−	−	−	−	−	−	−	−	−	+	−	−	−	−
	XXII	−	−	−	−	−	−	−	−	−	−	−	−	−	−	−	−	−	−	−	−	+	−	−	−	−	−
12	XXIII	−	−	−	−	−	−	−	−	−	−	−	−	−	−	−	−	−	−	−	−	−	−	−	+	−	−
	XXIV	−	−	−	−	−	−	−	−	−	−	−	−	−	−	−	−	−	−	−	−	−	−	+	−	−	−
13	XXV	−	−	−	−	−	−	−	−	−	−	−	−	−	−	−	−	−	−	−	−	−	−	−	−	−	+
	XXVI	−	−	−	−	−	−	−	−	−	−	−	−	−	−	−	−	−	−	−	−	−	−	−	−	+	−

*A plus sign (+) indicates the occurrence of mating and conjugation, a minus sign (−) indicates that no mating and conjugation occur in mixtures of the two mating types represented on the corresponding row and file.

†According to Gilman (1949), weak intervarietal reactions involving mating phenomena but no conjugation occur as follows: type VI variety 3 with type XI variety 6; type III variety 2 with type XVI variety 8; type XVIII variety 9 and type XX variety 10; type XVII variety 2 and type XVI variety 8.
Strong intervarietal mating reactions followed by conjugation occur as follows: type XV variety 8 with type IV variety 2 and type XX variety 10; type XVII variety 9 with type XX variety 10.

334

the animals react strongly between 18° and 24° C. The mating reaction is very similar to that shown when specimens of opposite mating types of *P. aurelia* are mixed together. The effects of different temperatures on the reaction have been studied by Gilman (1939).

Maupas (1889) reported that *P. caudatum* conjugates at about 4 A.M. but Gilman observed no such diurnal periodicity in his clones.

Comparatively little modern genetics work has been done with this species. The fact that considerable conjugation (selfing) occurs within a clone in most of the mating types makes this material more difficult to readily demonstrate mating and conjugation.

As noted by Gilman, all clones reproduce true to mating type for periods of at least two weeks; in some cases, somewhat longer. Some stocks give rise to both mating types characteristic of their group while others remain of one type. Giese and Arkoosh (1939) have shown that in one of the mating types of *P. caudatum,* conjugation within a clone occurs only after the process called by them endomixis but which is apparently autogamy. It would appear that this is the basis for change of mating type in certain clones of this species, but Gilman maintains that change of mating type can occur in *P. caudatum* without previous autogamy.

E. Paramecium multimicronucleatum

The mating reaction and mating types of *P. multimicronucleatum* have been studied by Giese and Arkoosh (1939) and Giese (1939 a, 1941; also unpublished manuscript) in which there is a system of multiple mating types. Giese reported the existence of one set of four interbreeding mating types as shown in Table 31, after a study of mating reaction in 16 stocks from 12 states in the United States.

Giese observed that many stocks of *P. multimicronucleatum* would not demonstrate the mating reaction and conjugation soon after isolation from nature but only after a prolonged subculturing in the laboratory. The same was true in regard to selfing within a clone. After this prolonged period of residence in the laboratory, the mating reaction and conjugation could then be demonstrated by mixing the proper

Table 31

THE SYSTEM OF BREEDING RELATIONS IN *P. multimicronucleatum* (Giese)

Mating Types	I	II	III	IV
I	−	+	+	+
II	+	−	+	+
III	+	+	−	+
IV	+	+	+	−

* A plus sign (+) indicates the occurrence of mating and conjugation; a minus sign (−) indicates that no mating and conjugation occur in mixtures of the two mating types represented in the corresponding row and file.

mating types. In such mixtures of clones, the mating reactions at first were weak but later, after a period of maturity, large clumps were formed which resulted in conjugating pairs. Clones were found to vary in regard to the length of time needed for them to reach sexual maturity.

As with other species, the mating reaction is greatest immediately following a decline in nutritive conditions. According to Giese, selfing occurs so frequently that it greatly complicates the study of mating reactions between stocks. This factor would also make genetic studies more difficult.

Giese studied a number of environmental factors in regard to the occurrence of the mating reaction and conjugation in *P. multimicronucleatum.* Food appears to be the most important single factor, i.e. a decline in available food after a period of plenty (as noted earlier by Maupas and Jennings). Paramecia grown at 15°, 20°, and 26° C. will conjugate when the food is depleted with the conjugants appearing earliest at the highest temperature. Animals cultivated at 30° C. seldom were found to demonstrate conjugation according to Giese but do so when placed at lower temperatures. Although Enriques (1909) and Zweibaum (1912) claimed that salts of a specific type and concentration were necessary for the induction of conjugation, Giese found that this is apparently not the case with his races of *P. multimicronucleatum.* Concerning pH in respect to conjugation, this investigator found little difference in the ranges studied, i.e. pH 6.0–8.0.

F. Paramecium trichium and Paramecium calkinsi

Sonneborn (1938 b, 1939 b) reported the existence of mating types in *P. trichium* in which three interbreeding types indicated a system of multiple types such as Jennings found in *P. bursaria.* (Table 32 A.)

Diller (1948), who has studied extensively cytological details in conjugation of this species, noted that it is notorious for its instability of mating type even in the absence of conjugation. In his races, which appear to be unusually labile ones, he observed conjugation in progeny within a few days after isolation of single animals. On the other hand Wichterman (unpublished) has had a clone of this species in continuous cultivation for 12 years and has never observed conjugation (selfing) in the cultures. He has, however, observed races in which conjugation occurred readily (Wichterman, 1937 a).

Little, if any, genetic work has as yet been done with *P. trichium.*

The existence of mating types in *P. calkinsi* was first reported by Sonneborn (1938 b), and they have been studied by Boell and Woodruff (1941), Wichterman (1948 e, 1951 d), and Metz and Fusco (1949).

The author has maintained in continuous culture for several years opposite mating types I and II of brackish

Tables 32 A, B

THE SYSTEM OF BREEDING RELATIONS IN *P. trichium* AND *P. calkinsi**
(Table 32 A, Sonneborn; B, Wichterman)

A				B					
P. trichium				*P. calkinsi*					
Mating Type	I	II	III	Variety		1		2	
					Mating Type	I	II	III	IV
I	−	+	+	1	I	−	+	−	−
II	+	−	+		II	+	−	−	−
III	+	+	−	2	III	−	−	−	+
					IV	−	−	+	−

* A plus sign (+) indicates the occurrence of mating and conjugation; a minus sign (−) indicates that no mating and conjugation occur in mixtures of the two mating types represented in the corresponding row and file.

water *P. calkinsi* from Yale University and two additional brackish water clones, one of which is bimicronucleate. Like mating type II, the two additional clones mate and conjugate readily with mating type I. The author (Wichterman, 1950, 1951) discovered another variety of *P. calkinsi* containing two opposite mating types from sea water of high salinity content. At the present time, there exist two non-interbreeding varieties each containing two opposite mating types (Table 32 B). Nutritive conditions for mating and conjugation are much like those for the other species. Well-fed, plump, actively dividing animals do not mate and conjugate but after becoming thinner do so soon after feeding. Opposite mating types react any time of day at room temperature and produce a clumping pattern similar to that of *P. bursaria*. Animals remain joined in the conjugation process for approximately 24 hours after being brought together at 22° C.

Up to the present, mating types have not been reported for the remaining two species, *P. woodruffi* and *P. polycaryum*. This is very likely due to the fact that enough clones have not been made available in testing for the existence of mating types.

G. Cytoplasmic Inheritance, Dauermodifikationen, Cytoplasmic Lag, and Plasmagenes

It may be said that in general, genetic diversities in paramecium are inherited in a manner comparable to those in higher animals. On the other hand, some diversities show conditions which appear to differ from the usual concept of inheritance, some of which may be imperfectly understood.

According to Sonneborn (1947), the inheritance of stock differences in certain varieties appears to be due entirely to genic differences, but in other varie-

ties, such differences appear to be due to cytoplasmic factors. Mating-type characteristics, varietal differences, and antigenic differences have been reported as being inherited in typical Mendelian fashion. Some of the hereditary differences within a stock are permanent even through fertilization processes as well as through binary fission. These appear to be due to recombinations, mutations, and macronuclear regeneration (Sonneborn). Others (the *Dauermodifikationen* of Jollos) are long lasting but impermanent, while still others increase with time to become cumulative as in ageing or under the influence of substandard cultural conditions.

In many of his papers, Sonneborn has shown how caryonidal inheritance occurs regularly in the inheritance of mating type in certain of the two-type stocks of some varieties of *P. aurelia*. Although the mechanism of this type of heredity is not fully understood, Sonneborn has presented a number of possible explanations for this phenomenon.

As set forth by Jollos (1913 a, b, 1920, 1921, 1934) and discussed earlier in this chapter, *Dauermodifikationen*—a term now universally used by protozoölogists—means long-lasting but temporarily inherited alterations produced by environmental conditions. Depending upon the environmental stimulus over a given period of time, the effects may persist for many hundreds of fissions, ultimately disappearing because of fertilization or environmental changes. These inherited environmental modifications or *Dauermodifikationen* may be an intrinsic character of the cytoplasm, the macronucleus or both. (See also Chapter 6 and Chapter 12 on serology.) These modifications are not to be considered as a part of the permanently and genuinely inherited characters such as are demonstrated by genetic factors of the chromosomes.

The name *cytoplasmic lag* ("maternal effect") means a relatively short-time effect which has no real genetic basis, while *Dauermodifikationen* refers to long-lasting effects which must have some temporary genetic basis. This distinction is not merely one of relative length of persistence but becomes one of a qualitative difference when one considers the possible mechanisms which are involved. As noted by Sonneborn (personal communication, 1950),

Cytoplasmic lag is probably due to materials formed under the action of the genes of the ancestral cell, these materials continuing to have effects until they are diluted out during subsequent multiplication of the cells or until alternative materials formed under the action of the new genotype come into action. On the other hand, *Dauermodifikationen*, which persist in some cases for hundreds of cell generations, cannot possibly be accounted for in that way. In my opinion, the phenomena which have been referred to by this term are probably of at least two different kinds. Some are based on self-duplicating particles such as kappa. Others are manifestations of different alternative and mutually exclusive capacities for development. A shift under environmental action, from one to another alternative characteristic, such as antigenic type, is an example of this type of Dauermodifikationen. Once the shift is made, cytoplasmic mechanisms of an as yet unknown sort can perpetuate the new trait.

The cytoplasm of paramecium, at fission, is reduced to half its volume but, upon growth, is restored to its original volume. Thus at each fission, the protoplasm is diluted to one-half with a resultant gradually diminishing effect. As noted by Jennings (1940), "The original cytoplasm seemingly must therefore have to some extent the power of reproducing itself in its distinctive nature, at the time that growth occurs. In this respect it partakes of the character of a gene or genetic material, in that it affects the characteristics of the individuals and reproduces itself in some degree true to type. But in time it is made over by the nucleus . . . (which) asserts itself, impressing its own constitution on the cytoplasm."

This type of temporary, cytoplasmic inheritance or cytoplasmic lag was strikingly demonstrated by DeGaris (1935 b) who crossed conspicuously large and small clones of *P. caudatum*. Differing greatly in hereditary size, the specimens conjugated, with each exchanging halves of their nuclei (pronuclei) but with each conjugant retaining its own cytoplasm. Upon separation, each exconjugant was now alike as to nuclear (synkaryon) constitution, but diverse as to cytoplasmic constitution. Subsequent divisions of the exconjugants of each tended to result in eventual equalization of size: the small exconjugant fissions resulted in specimens gradually larger, while the large exconjugant produced smaller specimens. After about 22 generations requiring as many days, the descendants of both large and small parents reached approximately the same size, with all the descendant progeny intermediate in size between the two original parents. Thus it is seen that this type of cytoplasmic inheritance is transitory, but the nuclear one permanently hereditary.

In much the same manner, Sonneborn and Lynch (1934) observed a cytoplasmic effect as shown when clones differing in fission-rate were crossed. The difference in progeny was seen to disappear after about 10 generations.

The procedure and some examples used by Jollos to obtain *Dauermodifikationen* are set forth earlier in this chapter and elsewhere in the book. Sonneborn (1943 b) gives a striking example of Jollos' *Dauermodifikationen* in his demonstration of acquired immunity to a specific antibody and its

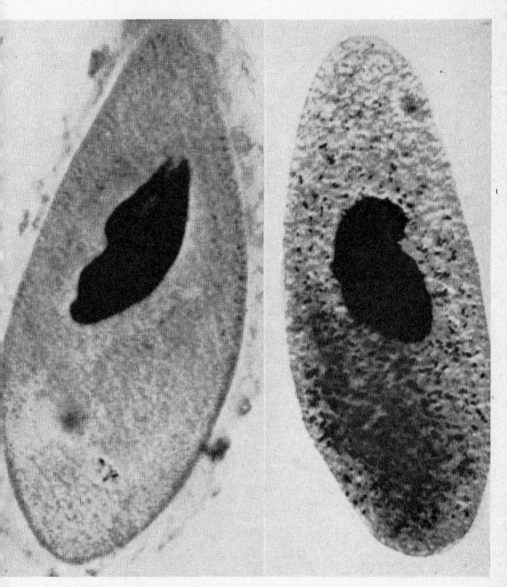

FIG. 114A. (*Left*) A sensitive *Paramecium aurelia*, Giemsa stain, showing macronucleus and a
od vacuole containing bacteria, but no kappa particles in the cytoplasm. × 1287. (*Right*) A killer
ramecium aurelia, Giemsa stain, showing the macronucleus and the larger number of kappa par-
les in the cytoplasm. × 1287. (Sonneborn)

inheritance in *Paramecium aurelia* (see also Chapter 12). His results indicated that this acquired immunity was not an ordinary persistent mutation but was only temporarily inherited for long periods during vegetative reproduction.

Another example of this phenomenon in *P. aurelia* is given by Kimball (1947 a, 1948) who studied the induction of inheritable modification in reaction to antiserum. He reported that the transmission of resistance for many fissions could be accounted for most readily by self-duplication or by the formation of some kind of semiautonomously formed cytoplasmic substances.

The work of Sonneborn (1939 a, 1943 c, 1947 a, 1948 a, b, 1949, 1950), Preer (1946, 1948 a), and others has shown the possible existence of *plasmagenes:* gene-like determiners controlling hereditary traits not found in the nucleus but in the cytoplasm of paramecium.[8]

In a series of experiments, Sonneborn obtained descendants from one paramecium showing six hereditarily distinct antigenic differences (types) designated A, B, C, D, E, and G. He demonstrated that this difference in antigenic type is due not to a difference in genes or gene mutation but to a difference in cytoplasm (Figs. 116 and 117). An examination of these two figures shows the following: when antigenic type A is mated with antigenic type B, conjugation (reciprocal fertilization) occurs but not cytoplasmic exchange (Fig. 116). The resulting progeny from type A parent remains type A and the clones derived from type B remain type B. Thus, it is seen in such matings that after completion of conjugation, the

exconjugants are alike in genes but not in cytoplasmic character; hence difference in antigenic type is due to a difference in cytoplasm. As shown in Fig. 117, when a cytoplasmic bridge is formed between the mating members thereby permitting a cytoplasmic exchange, the members after separation both give rise to pure cultures of type A. Progeny therefore become alike in antigenic character when they share a common cytoplasm although they remain diverse when their cytoplasms remain separate. Antigenic types thus differ in cytoplasm but not in genes (Sonneborn).

Serologic studies in relation to the genetics of paramecium are currently active and the newer discoveries may place the subject in a different light. Additional information bearing upon genetic studies and serology is found in Chapter 12.

Sonneborn reported a plasmagene in *P. aurelia* not initiated by gene action but which is not entirely independent of the genes. According to him, not only is a particular gene needed for its maintenance, but in its absence the plasmagene is soon lost. The pair of alternative characters concerned has been designated by Sonneborn as *killer* and *sensitive*. This plasmagene or killer factor, which has been shown to reside in the cytoplasm, is known as *kappa* (Fig. 114A). Killers are paramecia of certain stocks which liberate into their environmental fluid a substance called *paramecin* which makes the fluid in which they live poisonous for other stocks. Paramecia of stocks that are killed by such fluid (paramecin) are called sensitives. Stocks possessing the killer character have been reported from varieties 2 and 4 of *P. aurelia*, in which stock 51 of variety 4 has been studied most extensively by Sonneborn (1939 a, 1943 c, 1945, 1945 a, 1946, 1947 a). Additional

---

[8] Also known as cytogenes (cytoplasmic genes), cytoplasmic factors, genoids, neurogenes, pangenes, plastogenes, plasmatic genes, etc.

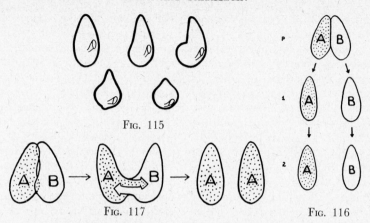

FIG. 115

FIG. 117

FIG. 116

FIG. 115. The sequence of morphological changes preceding death when sensitive animals of *P. aurelia* are exposed to the action of killer stock 51 of variety 4.

FIG. 116. Inheritance of antigenic types A and B in stock 51 of Paramecium. (P) Mating of parents. (1) First generation offspring. (2) Second generation, obtained by self-fertilization of animals of first generation. Each animal in generations 1 and 2 symbolizes an animal and its progeny produced by repeated divisions.

FIG. 117. Mating between antigenic types A and B of Paramecium, with flow of cytoplasm (symbolized by arrows) between mates. (Figs. 115, 116, 117 all after Sonneborn)

references dealing with kappa and paramecin follow: Chen (1945), Austin (1946, 1948), Preer (1946, 1948, 1948 a, 1950), Sonneborn, Dippell, and Jacobson (1946), Sonneborn, Jacobson, and Dippell (1946), Dippell (1948), Sonneborn and Lesuer, (1948), van Wagtendonk and Zill (1947), van Wagtendonk (1948), Geckler (1949), Sonneborn (1950 a, 1951, 1951 a) and Chao (1952).

When sensitive stocks are exposed to the poisonous fluid of killers, striking morphologic changes occur in the sensitives preceding death (Fig. 115). As described by Sonneborn, a slight hump appears after several hours on the aboral surface near the posterior end of the body which enlarges while the anterior part gradually wastes away and the posterior part is seemingly pushed into the humped region. Such animals then become smaller and spherical and finally die. Sensitives can be mated to

killers without any evidence of injury, (1) if mating begins soon after the two kinds of paramecia are brought together, (2) if the conjugant pairs are removed to fresh culture fluid soon after union, and (3) if the two members of each conjugant pair are put into separate culture dishes soon after conjugation has been completed.

Two killer substances have been found in *P. bursaria* by Chen. The first killer substance, paramecin 22 (Chen, 1945) is produced by a Russian killer strain (Ru 22). When sensitive animals are mixed with the killers or with animal-free culture fluid of the killers, sensitive animals become sluggish in movement, dark in color, and distorted in shape. The paramecia become sticky and clump together in pairs or small groups after which nuclear changes occur. Some of the single animals undergo nuclear changes which may be autogamy. However, some of the sensitive animals die as a result of their exposure to the killer substance.

The second killer substance, an antibiotic called paramecin 34 (Chen, unpublished) is produced by an American killer strain of *P. bursaria* (Mi 34) which was collected in Minnesota (Fig. 118). When used on the same strain of sensitive animals, the paramecia become swollen, develop blisters, and finally burst. Death follows in some animals 17 hours after the sensitive paramecia have been mixed with animal-free culture fluid of the killers. By the twenty-fourth hour, approximately half of the sensi-

tive animals die and 40 hours after the paramecia have been placed in the culture fluid of the killer paramecia, nearly all the animals have died. In such fluid, sensitive animals become abnormally active and darker in color. According to Chen, the killer substance affects only the cytoplasm and not the nuclei of sensitive animals.

The production of cytoplasmic killer factor, paramecin, and sensitivity to it are inherited in *P. aurelia*. This is demonstrated by mating killers to sensitives (Fig. 119) with the result that the conjugant which received cytoplasm from

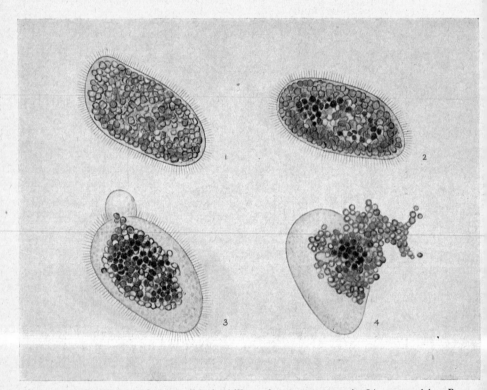

Fig. 118. The effect of the antibiotic, killer substance paramecin 34, on sensitive *Paramecium bursaria*. (1) Shows a normal sensitive animal before it is placed in the culture fluid containing the killer substance. (2) Approximately 12 hours after the sensitive animal has been placed in the culture fluid containing the killer substance. Note that it has become darker and swollen. (3–4) Approximately 16–24 hours after the sensitive animal has been placed in the culture fluid containing the killer substance. Note the large blister at one end of the animal in 3. (4) Shows animal in process of bursting. (Chen)

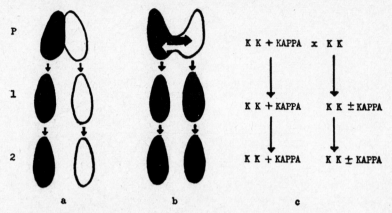

FIG. 119. Inheritance of killer and sensitive traits in Paramecium. Black symbolizes killers, white symbolizes sensitives. (P) Parents. (1) First generation. (2) Second generation, obtained by self-fertilization in animals of first generation. Column *a* gives results when no cytoplasm is exchanged between parents, column *b* when the parents exchange cytoplasm. Column *c* represents the genes and cytoplasmic condition (presence or absence of kappa) in the various animals and cultures. (Sonneborn)

a killer produces a killer clone; the one that receives cytoplasm from a sensitive specimen produces a sensitive clone even though both killer and sensitive clones contain specimens with identical genes.

Sonneborn has shown that the amount of cytoplasm exchanged at conjugation is roughly proportional to the length of time during which a visible cytoplasmic bridge—actually a fusion of the paroral cones—connects the conjugants after they have separated elsewhere along the oral surfaces. Most pairs of conjugating paramecia separate at the bridge within 3.5 minutes from the time such a bridge is visible and the only connection between the two; occasionally it persists for longer periods of time, such as 30 minutes or longer. When a killer is mated with a sensitive, Fig. 120 shows the result in regard to transfers of different amounts of cytoplasm and killer character (Sonneborn, 1945 a, 1947 a).

The unique properties of kappa according to Sonneborn are as follows:

(1) kappa mutates; (2) the rates at which kappa multiplies have been ascertained with certain kinds of kappa multiplying more rapidly than others; (3) when a killer paramecium divides, the contained kappa is distributed approximately at random but usually not with absolute equality to the two daughters; (4) kappa is readily destroyed by exposing the kappa-paramecia to high temperatures or x-rays; (5) a killer paramecium contains between 200 and 1000 particles of kappa; (6) the characteristics of the paramecia depend not only on whether they possess kappa but also on the amount possessed; they are killers only when the concentration of kappa is high. With less kappa, they are resistant to paramecin but produce none, and with still less kappa, they are sensitive to paramecin; (7) an animal needs to possess only a single particle of kappa in order to form more and become a killer; (8) the particles of kappa are believed to be from 0.2 to 0.8 of a micron in length, hence can be seen with the ordinary light microscope; (9)

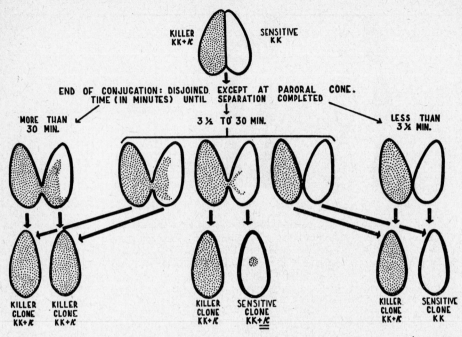

FIG. 120. Effects of transfers of different amounts of cytoplasm between mates in crosses of *KK*-plus-kappa killers by *KK* sensitives. The amount of cytoplasm exchanged is roughly proportional to the time the paroral bridge persists after the conjugants have separated elsewhere. (Sonneborn)

kappa contains desoxyribonucleic acid (Feulgen-positive); (10) sensitive animals "containing the K gene" may acquire kappa and transform into hereditary killers as a result of exposure to the broken-up bodies of killers, but this happens only when small numbers of sensitives are exposed to high concentrations of kappa; (11) the killing substance, paramecin, does not dialyze through cellophane. It is destroyed almost immediately at 55° C.; paramecin is lost in a day or less, but it can be preserved at 7° C. for several days.

It will be seen that kappa resembles a gene in its "determination of a hereditary trait, its self-duplication, its mutability, its chemical composition, and its dosage effect," according to Sonneborn (1949). He reports further, "It differs from a gene in the usual number per

cell, the mode of distribution at cell division, the occasional lack of synchronism between its rate of duplication and the rate of cell duplication, its high sensitivity to environmental conditions, its size, its cytoplasmic localization, and its capacity to enter a cell from the milieu and be maintained thereafter."

It has been suggested by Altenburg (1946, 1946 a) and others (see discussion by Lindegren et al. in Sonneborn, 1947 a) that kappa resembles a parasitic microörganism or symbiont, i.e. a large virus, a rickettsia, or a bacterium. Refer to Chapter 13 which deals with microörganisms living in paramecium. During the past century, many observers, including the author, have reported upon the different kinds of microörganisms of uncertain taxonomic status in paramecium nuclei and cytoplasm. Nearly all these parasitic micro-organisms of paramecia, described earlier, resulted in the eventual death of the host-paramecium. On the other hand, if kappa be a kind of commensalistic microörganism of paramecium, its residence in the cytoplasm gives the host-paramecium immunity providing kappa be in sufficient concentration. Also, the killer effects of kappa are not exerted by a kappa-contained paramecium if there is insufficient kappa in the cytoplasm.

If we dismiss the plasmagene theory in regard to kappa, we note that in a large number of the described cases, the word "microörganism" or virus can be substituted equally well for kappa and be understandable. Sonneborn (1949) reported that when kappa is present, it will not only fail to be formed, but will disappear in the absence of gene K. He was of the opinion that only when this gene is present with kappa, can kappa be maintained and multiplied. In extensive investigation of kappa, Preer (1946, 1948, 1948 a) concluded that his studies were inconsistent with the theory of the combination of kappa with gene K. Making use of mathematical techniques and numerous formulas, he has calculated the number and rate of reproduction of kappa-particles in cellular divisions.

Dippell (1948) has reported on mutations of kappa; Austin (1946, 1948) on the killing activity of paramecin, and van Wagtendonk (1948) upon the chemical nature of paramecin.

In the minds of many, the thought must occur as to whether or not "kappa" is in reality a plasmagene or a foreign microörganism similar to a virus which assumes some of the characteristics of a plasmagene. Viruses are common in multicellular plants and animals, and in view of the larger number of different foreign microörganisms in paramecium (Chapter 13), it is not unreasonable to look for viruses in this organism. Only continued research will provide the answer.

H. Genetic Evidence of Autogamy and Lack of It for Endomixis

It should be remembered that in endomixis as described by Woodruff and Erdmann (p. 260), a third pregamic nuclear division does not occur as in autogamy and conjugation. Therefore, since gametic nuclei are not formed, fertilization (synkaryon formation) is absent. On the other hand, autogamy (p. 298) is the counterpart of conjugation but in single animals; the cytological processes are the same, but in autogamy the synkaryon is formed by the union of the two gametic nuclei (pronuclei) which arise in the one individual instead of from two diverse individuals as in conjugation.

The status of endomixis is discussed on page 265, but it should be mentioned here that in his description of autogamy, Diller (1936) was unable to find any cytological evidence for the existence of endomixis. Sonneborn (1947) has made extensive cytological studies of many stocks of *P. aurelia,* and while he has never found critical evidence of endomixis, he has observed autogamy in every race examined. Also, only autogamy was found by these investigators in the main stock of *P. aurelia* which was studied by Woodruff and Erdmann to describe endomixis.

In addition to cytological evidence for autogamy and lack of it for endomixis, Sonneborn offers genetic proof for the former process (Fig. 121).

The commanding place of autogamy in the genetics of paramecium is clearly set forth by Sonneborn as follows:

The process of autogamy, in the form in which it is now known to occur, is of such surpassing importance in the genetics of *P. aurelia* that this aspect of it must here be emphasized. . . . The fact that, at autogamy, fertilization regularly is accomplished by the fusion of two sister haploid nuclei, necessarily brings about homozygosis for all genes. The heterozygotic condition can be brought about only by crosses of diverse stocks or by mutation; it can persist, in the absence of further crosses and mutations, only until the first autogamy, which normally occurs within a few weeks or even sooner. Thus, although *P. aurelia* is a diploid organism, it is as favorable for genetic work as haploids, for recessive genes cannot long remain hidden. Finally, the regular oc-

single stock can be recombinations only when the conjugation is one of the first to occur after the stock has been brought in from nature.

Mating Type Segregation at Autogamy.

As in conjugation, segregation of the diverse mating types may occur at the first fission after autogamy. Sonneborn (1937 c) and Kimball (1937, 1943) found that in certain stocks of variety 1 of *P. aurelia*, autogamy produced clones of both the mating types of that particular variety although later Sonne-

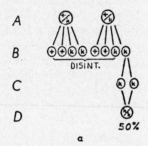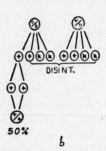

FIG. 121. Genic recombinations in a heterozygote $(+/k)$ at autogamy in *P. aurelia*. The two possibilities, which occur with equal frequency are shown in diagrams *a* and *b*. Row A represents the two original nuclei (in each animal) with their genes at the start of autogamy. Row B shows the 4 reduced nuclei produced by the two meiotic divisions from each original nucleus; nuclei that disintegrate are marked "Disint." Row C shows the two gamete nuclei produced at the third division from the surviving reduced nucleus. Row D shows the synkaryon formed by the fusion of the gamete nuclei. Diagrams *a* and *b* differ in the genic content of the surviving reduced nucleus (row B); as half the nuclei contain *k* and half contain +, each of these genes has a 50–50 chance of being in the surviving nucleus. (Sonneborn)

currence of autogamy enormously simplifies genetic analysis. The simplest way to discover the genotype of a clone is to induce autogamy (p. 374); each heterozygotic locus then segregates into the two homozygous classes in a 1:1 ratio. In modern genetic work on *P. aurelia,* therefore, autogamy rather than conjugation is the method of choice for genetic analysis following hybridization of diverse stocks. Finally, it has been pointed out that the knowledge of autogamy explains and confirms the view of Jollos (1913 b) that the hereditary variations appearing after conjugation within a

born (1939) found stocks in the same variety in which no change of mating type occurred. Clones which show no segregation into different mating types at autogamy but instead produce specimens of the same type are known as *single-type* clones. When both mating types of a variety are produced from a single clone at autogamy, the clones are designated as *double-type* ones. The latter appear to be of more common occurrence than the former.

When single-type clones are crossed

with double-type ones, it is found that the single-type condition is recessive to the double-type one. This factor for the difference resides in one chromosome and is inherited in typical Mendelian fashion. Ordinary crosses and back-crosses clearly demonstrated diploid Mendelian inheritance with double-type being dominant to the single-type condition (Sonneborn, 1939, 1939 a).

Kimball (1937, 1939, 1943, 1947, 1947 a, b, 1948), who has made extensive studies upon the genetics of *P. aurelia,* reported that after a change of mating type from type II to type I (variety 1) as a result of autogamy, the change in all the specimens may not be immediate. Instead, a lag occurs in the change from type II to type I, but no such lag occurs for the reverse change.

I. Some Unusual Aspects of the Mating Phenomenon and Characteristics of the Mating Substance

Chen (1945) observed that *animal-free* fluid (which now appears to be a *killer* fluid) from a Russian clone of *P. bursaria* induced clotting and conjugation among animals of another mating type even though the latter belonged to a different variety. It was noted that the induced conjugation did not occur until approximately four to six hours after the animals were introduced into the fluid of the other mating type. Many of the conjugating pairs were found to be atypical, and some did not appear to be as firmly joined in the conjugation process as in normal matings.

Also, animal-free fluid of one clone induced nuclear changes (possibly autogamy) in single animals of another clone.

Chen (1946 a) also observed temporary pair formation in *P. bursaria* after mating members of one clone with those of diverse mating type. In such mix-

tures, the typical agglutination of paramecia occurred and within an hour, many pairs were formed. However, within a few hours after mixture, the pairs broke apart into single animals. With the exception of a slight swelling of the micronucleus, no conspicuous nuclear changes occurred in the pairs or in the animals after separation.

It is of interest to report that Tartar and Chen (1940, 1941) observed the mating reaction after crossing cut fragments of *P. bursaria* of opposite mating type. Provided the animals were of opposite mating type, they found that the reaction occurred in the following cases: (a) between nucleate fragments and whole animals, (b) between enucleate fragments and whole animals, (c) between enucleate fragments. Thus, even in the absence of all nuclei, mating reactivity and diversity of mating type was exhibited. Concerning fragments, Metz (1948) noted that large fragments of *dead P. aurelia* clumped with living animals of opposite mating type.

The atypical intervarietal lethal crosses of *P. bursaria* (Jennings and Opitz, 1944; Chen, 1946b) and the matings resulting in the conjugation of three animals (Chen, 1946) are discussed in Chapter 9.

We are indebted to Metz (1946, 1946 a, 1947, 1948); with Butterfield (1950, 1951); with Foley (1949); and with Fusco (1948, 1949) for valuable contributions toward an understanding of the nature and mode of action of the mating type substances of paramecium. It was found that when *P. aurelia* were killed by appropriate treatment with Formalin, the dead specimens clumped strongly and specifically with living animals of the opposite mating type.[9] In

[9] For the preparation of Formalin-killed animals to demonstrate the "pseudo-selfing" reaction, it is necessary to obtain strongly reactive paramecia from cultures which must be freed of foreign material. Turbid cultures

60–90 minutes, the clumps of living and Formalin-killed paramecia broke down to release living single specimens and living joined pairs, the latter being called *pseudo-selfing* animals and of course, belonging to the same mating type. The union of the pseudo-selfing pair members occurred at the anterior or *holdfast* region only (Fig. 122). Although paroral cones were formed in such joined specimens, fusion of these structures did not occur. Pairs, which

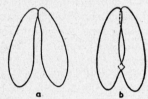

Fig. 122. Types of union in conjugating, and pseudo-selfing paramecia. (a) Holdfast union in early conjugants and pseudo-selfing animals. (b) Holdfast and paroral cone union in more advanced conjugants. (Metz)

remain united for about five hours, undergo meiosis and macronuclear breakdown. The two selfing pair members originate from a single clone of a given mating type, and each gives rise to clones of the original mating type.

Metz also observed that autogamy occurred in the single, isolated specimens of the one mating type of *P. aurelia* after they were mixed with Formalin-killed paramecia of opposite type.

In another species, namely *P. calkinsi*, living specimens of one mating type exhibited the mating reaction with Formalin-killed paramecia of opposite type, yielding pseudo-selfing pairs in which meiosis and macronuclear breakdown occurred (Metz, 1948).

In addition to Formalin-killed paramecia, Metz (1946) used other killing agents and reported as follows:

Animals killed by heating to 46°C. for 5 minutes clumped strongly and specifically with living but not with dead animals of opposite type. The animals failed to react after 5 minutes exposure at 52°C. Animals killed by brief (15–30 minutes) treatment with iodine-potassium iodide solution, ½ saturated picric acid, 1.1M urea, M/2 sodium chloride, 25% acetone or alcohol, or water saturated with chloroform, methylal or toluene clumped with living type VIII animals only. Animals retained their specific reactivity after treatment with acid to pH 3.6 or alkali to pH 9.8. No fraction containing mating type substance activity (specific action on animals of opposite type; inhibition of mating reaction) has yet been obtained by direct extraction of reactive animals by heating, grinding, freeze-thawing, or by treatment with organic solvents, salt solutions, urea (6.6M), acid or alkali. In fact all activity is lost when the animals are thoroughly broken up by mechanical or chemical means.

Metz and Fusco (1949) obtained highly reactive Formalin-killed *P. aurelia*, which were then quickly frozen in a solid CO_2-acetone bath. Specimens were next dried from the frozen state (lyophilized). When such dried *P. aurelia* of a given mating type were suspended in saline with living paramecia of opposite mating type, a strong clumping reaction occurred. However, when the same mixtures were made but with specimens first frozen in the living condition, only a weak mating reaction occurred. On the other hand, Formalin-killed lyophilized *P. calkinsi* gave only

---

with suspended particulate matter will not react. The optimum Formalin concentration used usually falls between ⅟₃₂ and ⅟₆₄ of commercial Formalin (40 per cent formaldehyde) or 1 per cent formaldehyde. After remaining in the Formalin solution for 60–90 minutes, the supernatant is aspirated from the settled dead animals which are then washed twice by centrifugation in tap water or dilute salt solution (0.0025N NaCl). Killed specimens are then suspended in 1–5 cm³ of the dilute salt solution after which they are ready for use.

weak mating reactions but strong ones when frozen first in the living condition.

These investigators reported that properly dried *P. calkinsi* (but not *P. aurelia*) retained their mating reactivity at room temperature for several months, possibly indefinitely.

The lyophilized *P. aurelia*, besides exhibiting the mating reaction, also induced macronuclear breakdown and pseudo-selfing pair formation in living paramecia of opposite mating type, hence the activation initiating mechanism is not destroyed by lyophilization. The mating reactivity of dried *P. aurelia* and *P. calkinsi* is not destroyed by extraction at room temperature for 30 minutes with the following agents: absolute ether, acetone, benzene, or chloroform. However, absolute alcohol inactivates the paramecia. As a result, Metz and Fusco concluded that loosely bound fat-soluble substances do not appear to be the essential constituents of the mating type substance or substances.

As suggested by them, their discovery and technique now makes possible: (1) the study of the effect of anhydrous solvents on the mating substances, (2) electron microscope studies of paramecia with mating-reactive surfaces, (3) the storage of mating-reactive animals for many different types of experimentation.

Using the method of Metz, Hiwatashi (1949 a) also induced pseudo-selfing pairs by Formalin-killed *P. caudatum*. Only Formalin concentrations of about 1.5–6.0 per cent were successful in demonstrating the mating reaction with dead and living specimens. Clumps were formed even using half-cytolized dead animals provided a large portion of cell surface remained. The mating reaction of living *P. caudatum* of one mating type with Formalin-killed specimens of opposite mating type occurred in all four groups (varieties) of Japanese clones. Hiwatashi noted that the living specimens of Japanese mating types 2, 4, 5, and 8 clumped with Formalin-killed animals of types 1, 3, 6, and 7 respectively. However, in the reciprocal combinations, such as crossing mating type 1 with 2, 3 with 4, 6 with 5, and 7 with 8, the mating reaction was negative.

Hiwatashi (1950) found that ammonium sulfate and glycerin were good killing agents for preserving the mating reactivity. This investigator treated the bodies of Formalin-killed *P. caudatum* with heat and different concentrations of ethanol, methanol, acetone, acetic acid, and urea in an effort to determine some of the characteristics of the mating substance, and it was observed that the mating reactivity of the Formalin-killed paramecia was inhibited by all of these agents.

In summation, it has been reported that the mating reactivity is inactivated by heat, extreme pH, complete disruption of bodies (Metz, 1946), specific antiserum (Metz and Fusco, 1948), treatment of Formalin-killed animals with antiserum (Metz, 1948), and treatment of lyophilized animals with absolute alcohol (Metz and Fusco, 1949) in *P. aurelia;* roentgen rays in *P. bursaria* and *P. calkinsi* (Wichterman, 1947 b, 1948 b, e) ; and treatment of Formalin-killed animals with heat, extremes of pH, ethanol, methanol, acetone, acetic acid, and high concentrations of urea (as well as the complete disruption of Formalin-killed bodies) in *P. caudatum* (Hiwatashi, 1950).

Several enzymes were tested for ability to destroy the reactivity of formalin or picric acid-killed paramecia in an attempt to characterize the mating type substances of *P. calkinsi* by Metz and Butterfield (1951). The non-proteolytic enzymes lecithinase (bee venom), hyaluronidase, lysozyme, ptyalin, and ribonuclease had no detectable effect upon the mating reactivity and it was concluded that the natural substrates for these enzymes were not essential constituents of the mating-type substance of this species. However, crystalline preparations of the proteolytic enzymes trypsin and chymotrypsin destroyed the mating reactivity.

In his discussion of the mechanics of fertilization in paramecium, Metz (1948) concluded that in conjugation, natural autogamy, pseudo-selfing, and probably nuclear reorganization as induced in single, isolated animals, the

sessed slight regenerative ability when compared with other Protozoa, ample work since has proved that this species and others of the genus show typical powers of regeneration. Recent important reviews of regeneration and morphogenesis in paramecium and other ciliates follow: Tartar (1939, 1941), Balamuth (1940, 1940 a), Summers (1941), Fauré-Fremiet (1948), and Lwoff (1950).

The method for observing regeneration or morphogenesis consists essentially of disturbing the normal body of the organism generally by means of

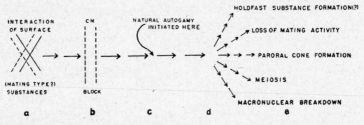

Fig. 123. Scheme for activation in Paramecium. (a) Initiating reaction (mating type substance interaction?) in sexually induced activation. (b) CM block, here assumed to lie "internal" to the initiating reaction, (a). (c) Position where main chain is activated in natural autogamy. (d) Break-up of main activation chain into side reactions leading to (c) the various end effects of activation. (Metz)

following series of events occur: (1) loss of mating activity, (2) paroral cone formation, (3) meiosis, and (4) macronuclear breakdown. It is tenable to believe that the given chain of reactions follows a common initiating mechanism (Fig. 123). With the exception of natural autogamy, this initiating mechanism may be an interaction of mating type substances which "activate" the paramecia in the same manner that the spermatozoa activates the metazoan egg (Metz).

J. Regeneration and Morphogenesis

Although Balbiani (1888, 1891, 1892, 1893) was of the opinion that *P. caudatum* (which he called *P. aurelia*) pos-

microdissection or micrurgic procedures. The process of cutting a protozoön into fragments is called *merotomy* and the fragments so produced are called *merozoa*.

Fragments move in the same general manner as the entire animal (Balbiani, 1888; Verworn, 1889; Jennings, 1901; Horton, 1935). In his study of the reactions of isolated parts of *P. caudatum*, Horton found that the spiral movements of the pieces are not necessarily related to the structure or relationship of the oral groove. Fragments without oral cilia spiralled similar to the entire animal. Jennings and Jamieson (1902) and Horton (1935) studied the reactions of paramecium-fragments to mechanical and chemical stimuli and concluded that reaction abilities compared favorably with those of the entire animal. Although Alverdes (1922 c) claimed that the anterior fragments of paramecium were more responsive to stimuli, Horton (1935) reported that the posterior ones were more sensitive (to weak acid). It is of interest to note that Peebles (1912) found the anterior cuts to cause greater physiologic disturbances in paramecium than the posterior cuts.

For the mating reaction of paramecium-fragments, see page 347.

Hosoi (1937) examined fragments of *P. caudatum* in his study of the forces involved in cyclosis. Cyclosis occurred in all the fragments within a few minutes after operation. However, the protoplasmic streaming concerned with the formation and release of the food vacuoles occurred only in those fragments possessing a large portion of the gullet. He believed that the nuclei play no direct role in cyclosis.

A large number of investigators have reported for many different Protozoa that both nucleus and a certain amount of cytoplasm are essential in merozoa for regeneration and the continuation of the vegetative and reproductive activities of the organism. Enucleate (non-nucleate) fragments may exhibit fairly normal behavior when compared to the entire organism but eventually die in a relatively short period of time.

A more complicated nuclear situation exists in paramecium and other ciliates because of nuclear dimorphism in which the nuclear material is organized into large macronuclei and small micronuclei. At the present time, considerable differences of opinion have been expressed not only in regard to the function and essential nature of dimorphic nuclei but also to their respective roles in regeneration. It should be remembered that dimorphic nuclei have a common origin from the synkaryon in conjugation. Nevertheless, once the two kinds of nuclei are fully developed in typical vegetative specimens, a micronucleus cannot give rise directly to a macronucleus nor is the reverse true.

As far back as 1886, Gruber considered micronuclei of *Stentor* as of secondary importance in regeneration since it occurred only when macronuclear material was present even in the absence of micronuclei. In Stentor and many other ciliates, a small fragment of the macronucleus can usually reconstitute the entire macronucleus. In some instances, cytological evidence indicates that the macronucleus undergoes complex morphologic changes including macronuclear extrusion in the regenerating fragments usually before the start of cytoplasmic differentiation.

Lewin (1910) and Schwartz (1934) found that the micronucleus of paramecium was dispensable for the actual regenerative process. From the investigations in merotomy upon paramecium and other ciliates, it may be stated that the macronuclear material is not only essential in regeneration but may be a controlling factor. Because of this fact, one is led to the conclusion that the macronucleus of ciliates exercises direct control of vegetative processes with the micronucleus assuming an unimportant secondary role or none at all. Amicronucleate races of paramecium and other ciliates are capable of continued existence, but it is significant that an amacronucleate race of paramecium never has been described.

Enucleate fragments of paramecium which contain neither the macronucleus nor the micronucleus fail to regenerate, but the pieces may live for days and behave in a manner comparable to the entire animal. The enucleate fragments, therefore, are capable of some energy metabolism as shown by their continued activity, but the anabolic processes cease and differentiation never begins.

Tartar (1938) observed that completely enucleated fragments of *P. bursaria* may survive for one to six days. In observations upon 100 enucleated fragments of this species, he and Chen (1941) found that all remained alive for one day and 50 per cent for two days. Tartar's observations upon the fragments in respect to cyclosis, positive thigmotactic reaction to bacterial masses, and coördination of the cilia led him to believe that these factors are not under the direct control of the nuclei.

It is a fact that a nucleated fragment of paramecium is frequently capable of regenerating to produce a complete organism indistinguishable from the normal animal. It is therefore apparent that a fundamental organizational pattern is inherent in the fragment.

Calkins (1911) reported that four giant races of *P. caudatum* (in reality, possibly *P. multimicronucleatum*) possessed different potentials of regeneration varying from 1–100 per cent. Similarly, Peebles (1912) noted that her races of *P. caudatum* varied in regenerative power from 23–67 per cent after cutting off anterior ends of the organisms. Upon removal of the posterior ends, regenerative power varied from 25–100 per cent. She concluded that regeneration will occur in about 90 per cent of the cases provided the paramecia are well fed and in a "viscid" state. It was her opinion that regenerative power is not so much a racial character as it is a physiologic one of the particular individual. This is in general agreement with the findings of Balbiani (1892, 1893) and Calkins (1911), who observed a correlation between regenerative ability and physiologic condition of *P. caudatum*, since fragments cut during periods of depression failed to regenerate or divide but did so under

favorable conditions. Differing in opinion, Chejfec (1932) reported that when paramecia were subjected to conditions of starvation or to acidified media, regeneration occurred more readily than those in hay-infusion controls.

Calkins (1911) noted that if *P. caudatum* is cut at the anterior or the posterior end, the body divides as in a normal animal. The division plane is not in the geometric center of the cut body but in the geometric center of the body before the operation (Fig. 124).

According to Peebles (1912) and Tartar (1941), in *P. caudatum*, posterior ends are regenerated much more rapidly than anterior ends. The latter investigator noted that when the specimen is cut in such a manner that the oral structures are removed, the fragment never regenerates a new mouth and eventually starves. The fact that no replacement of feeding apparatus occurs once the oral structures are excised even when the remainder of the body is left intact shows the rigid differentiation of this ciliate.

In carefully controlled experiments, Tartar (1939, 1940, 1941) has shown conclusively that regenerative capacity in paramecium is much greater than one would believe from earlier reports. He investigated the problem of racial and species regenerative differences in 25 races of seven species of *Paramecium*. Of 865 anterior transections, 509 specimens survived with morphologic and complete physiologic regeneration occurring in 98 per cent of the survivors. He contended that any paramecium able to recover from the cutting injury is able to regenerate completely and that little difference is noted in regenerative capacity in the different parts of the body. According to Tartar, there is no racial or species variation in regeneration of paramecium. In more extensive operations performed upon the

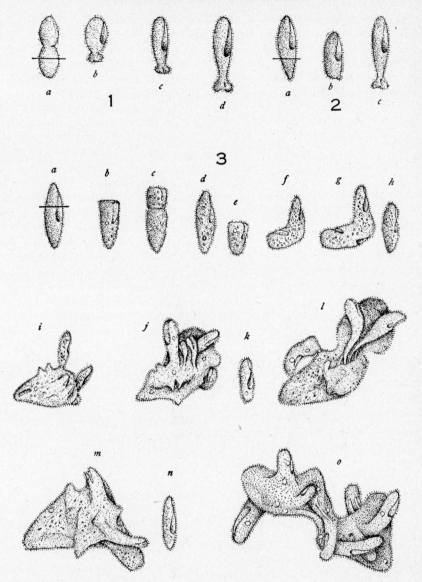

FIG. 124. Merotomy in *Paramecium caudatum*. Different experiments are designated by numbers 1, 2, 3 with the straight line indicating the plane of cutting. The history of a monster is shown in 3 in which an original cell, 3*a*, was cut. The posterior fragment *b* divided *c* into *d* and *e* in which *e* formed a monster with 16 mouths (3 *f–o*). Enucleated individuals (*h, k, n*) occasionally separated from the parental mass. (Calkins)

larger posterior ends of five races of *P. caudatum,* Tartar found that complete regeneration occurred in 93 per cent of the survivors.

An injured (cut) *P. caudatum* need not regenerate to divide; it may or may

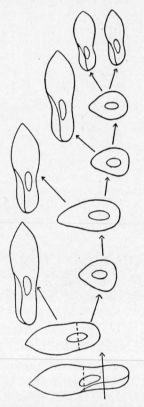

FIG. 125. Diagram showing delayed regeneration in *Paramecium caudatum*. The nucleated posterior part of a Paramecium transected anterior to the division plane may divide several times before form restoration occurs in all of its descendants. (Peebles)

not regenerate before dividing. Peebles observed that a truncated paramecium frequently produced a truncated and a fully formed daughter in which the former occasionally divided again before regenerating (Fig. 125). Tartar (1940) found that fragments of para-

mecium may divide before they have attained the definitive maximal volume of a given race. This would indicate that the attainment of a certain volume is not the factor which determines the time of division.

According to Peebles (1912), when *P. caudatum* is cut at two different stages of fission, no marked differences in regenerative ability are observed. If the transection is made in the division plane while the body is constricted for fission and the macronucleus elongated, both halves of the cell survive and produce clones of normal progeny. It was reported that the regenerative power was present in paramecia obtained two to five hours after separation, but that approximately 90 per cent of the operated paramecia died as a result of the dissection. Peebles attributed this to a lowering of cytoplasmic viscosity of growing paramecia.

While Calkins believed that the power of regeneration varied in different stages of the life of paramecium, Tartar (1938, 1939) concluded that the stage in the division cycle has no such influence on regeneration. In 50 operations performed during and after division in a race of *P. caudatum,* 24 specimens regenerated and subsequently divided, according to Tartar.

Little work has been done in merotomy of conjugants of paramecium. Based upon comparatively few experiments, Calkins (1911) asserted that the incidence of regeneration was greater in operations performed upon exconjugants or in joined conjugants than in typical vegetative specimens. On the other hand, in dissections made at the anterior and posterior ends of joined conjugants of *P. caudatum,* Peebles (1912) reported that survivors regenerated slowly and eventually died.

It is a fact that most of the investigations on regeneration lack the precision of technique such as is found in more recent work in the field of nutrition, respiration and genetics of paramecium. With newer and more refined methods in which carefully controlled, genetically distinct clones of paramecia are used, the field of regeneration and morphogenesis should yield important discoveries.

K. Monsters and Monster Formation

As mentioned earlier in regeneration, merotomy experiments upon paramecium frequently produce abnormal or monstrous specimens. Certain types of dissections or mechanical injuries may induce deep-seated changes in the organizational pattern of paramecium in which the precision of division in the operated cell is destroyed and a monster is produced (Fig. 124). In investigations upon regeneration after cutting paramecium, monsters were observed by Balbiani (1893), Calkins (1911), Peebles (1912), and Alverdes (1922 a). Unusual, ameba-like monsters were produced from *P. caudatum* by Merton (1928). Simpson (1901 a) discovered a double-monster which consisted of two fully grown paramecia with organic union between the parts. The monster lived for a week after which the anterior and posterior components produced average-sized daughter cells.

Jennings (1908 a), who was concerned with the genetic aspect of monstrosity, studied a great number of distorted forms in paramecium in order to observe the fate of recently acquired structures in subsequent generations. He noted that various types of abnormalities such as spines, appendages, and truncations were transmitted, as a general rule, to only one of the progeny. These non-heritable abnormalities tended to disappear in a relatively short number of generations (Fig. 126). Jennings, Simpson, Wichterman, and others

have observed another type of abnormal growth in paramecium called *chains* or *chain formation*. Chain formation is brought about by the fact that cytosomal division is incomplete, hence two or more animals adhere together where they would normally have separated in fission.

By exposure of specimens to hard, chlorinated tap water, Herzfeld (1926) was able to produce double-monsters of *P. caudatum* many of which possessed but one macronucleus. According to this investigator, the monsters remained motionless on the bottom of the container. Double-monsters and chains consisting of three and four specimens were also produced by Hinrichs (1928), who irradiated *P. caudatum* with ultraviolet light. In the same species, giant C-shaped monsters (Fig. 129) can be produced with x-rays (Wichterman, 1951).

DeGaris (1927 a, b) developed a method for producing double-monsters of *P. caudatum* by exposing specimens to low temperature and by exposing dividing specimens to the vapors of potassium cyanide, ethyl alcohol, Formalin, ether, and chloroform.[10] Such induced monsters were removed to hay-infusion media for cultivation and examination. His genetic results of the monsters of a pedigreed race of *P. caudatum* revealed that before and after monster formation, no significant change occurred in fission-rate or in mean cell length. Since these two factors are genotypic measures, he concluded that the genotype remains unchanged by the occurrence of monster formation.

Jennings (1913) described heritable

[10] Large numbers of monsters were produced by placing a depression slide of well-fed paramecia into a moist finger bowl after which the bowl was covered and surrounded with cracked ice. Temperature within the bowl dropped to 3° C. within an hour and returned to room temperature as the preparation remained overnight.

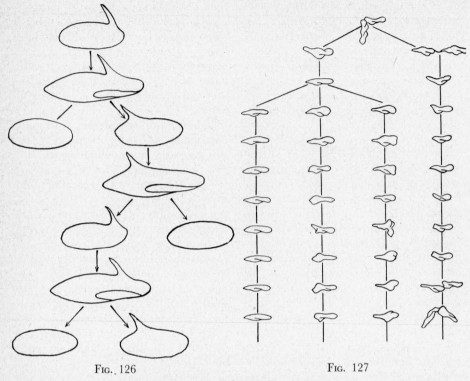

FIG. 126   FIG. 127

FIG. 126. Non-heritable abnormality in Paramecium. Transmission of an abnormal projection at reproduction. The projection is passed to but a single individual at each fission; so that at a given time only one individual bears it. (Jennings)

FIG. 127. Heritable abnormalities in *Paramecium caudatum,* after Stocking. Diverse lines of descent, some more abnormal, some less, derived from the same parent. (Stocking)

abnormalities after conjugation in *P. caudatum,* which appeared time and again in certain descendant lines of the exconjugants. These abnormalities after conjugation were studied more extensively by Stocking (1915) who reported that in the abnormal clones of *P. caudatum,* the tendency to produce abnormality is inherited. Certain types of abnormality were particularly common in some lines and other types of abnormality in other lines. Also, diverse types, which varied from normal-appearing individuals to complex monsters, appeared in the different individuals of the same clone (Fig. 127). The normal-appear-

ing individuals in an abnormal clone reproduced the abnormality to their progeny in later generations. When abnormal specimens were regularly discarded, leaving for reproduction only normal-appearing paramecia, abnormalities or monsters continued to reappear. By careful and gradual selection, Stocking was able to remove partly or completely, the tendency to produce abnormalities or to intensify the tendency for abnormality. In 25 diverse clones, she was successful in isolating one normal-appearing strain and one that continued to produce monstrosity.

Dawson (1924, 1926, 1928) described

a peculiar inherited abnormality characterized by a notched and truncated anterior end in *P. aurelia*. This condition, which was regarded by Dawson as a mutation, was bred in pedigreed culture for 412 generations. After fission, each daughter possessed the abnormality which even persisted after conjugation between members of the biotype.

In a clone of *P. aurelia,* Moore (1927) observed specimens possessing a "lump" on the oral side near the posterior end of the body. With some size variations, this abnormality was regularly inherited by both daughters at fission.

The behavior of monsters and fragments in regard to conjugation (De-Garis 1935, 1935 a, b; Tartar and Chen 1940, 1941) is discussed elsewhere in this and the preceding chapter. Monstrosity and the mechanism of morphogenesis in paramecium and other ciliates has been studied in detail by Fauré-Fremiet (1945, 1948).

Mottram (1940, 1941, 1942, 1944)

obtained monstrosity in clones of *Colpidium* sp. (which he earlier reported as *Paramecium aurelia*), by exposing the ciliates to heat, cold, solutions of different hypertonicity, high pH concentrations, gamma and ultraviolet radiations and the carcinogenic hydrocarbons, 3,4-benzpyrene, methylcholanthrene, and 1,2,5,6-dibenzanthracene. From abnormal clones, normal-appearing specimens continued to arise; careful selection of abnormals never resulted in a pure clone of abnormal specimens.

Concerning his abnormal specimens and those reported by others, Mottram concluded that such forms must follow laws of inheritance which are different from genetic inheritance, i.e. no guiding genes appear to control the great variation of abnormal specimens. Mottram did not discover until the end of his study that he was dealing with an amicronucleate race of *Colpidium*. He likened the abnormality of the ciliates as "resembling in many respects, the assemblage of cells of which tumors are

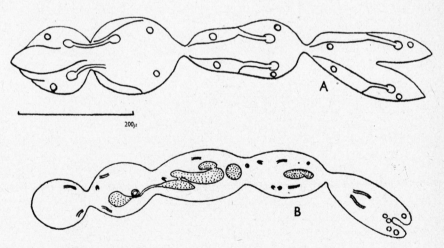

FIG. 128. Monsters (Siamese pair type) of *Paramecium caudatum*. (A) Chain of 4 cells from life, showing doubling of organelles and bifid last cell. The two front elements are in division. (B) Similar chain showing disorganization of nuclei and intercalary division. Macronuclei stippled, healthy micronuclei black, decadent micronuclei circles in B. Gullets are represented by slits and contractile vacuoles by circles in A. (Lloyd)

built." It is of interest to report that using paramecium, Tittler and Kobrin (1941, 1942) were unable to obtain similar results.

Lloyd (1947, 1948, 1949) studied the effects of benzene hexachloride or "gammexane" (the γ-isomer of hexachlorocyclohexane) upon two clones of *P. caudatum*. Gammexane is a compound widely used as an insecticide. It is toxic to paramecia; when introduced into cultures in concentrations of one to

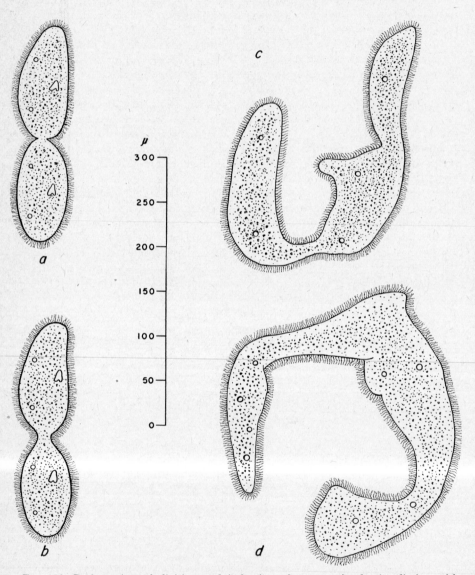

FIG. 129. Prolongation of division and induction of monstrosity by irradiation with x-rays of late divider (a) of *P. caudatum* with 300,000 r.; (b, c, and d) represent changes in form 24, 49, and 67 hours respectively after irradiation. (Wichterman)

10 parts per million, the action of the poison is slow and cumulative. Lloyd reported that the greatest development of abnormality occurred eight to 10 weeks after exposure to the compound. Near the threshold of toxicity, fission may be arrested before separation to produce fully grown specimens but joined anterior to posterior. This investigator obtained regularly and in abundance, many different kinds of monsters by transferring gammexane-conditioned paramecia to normal media. Occasionally, the monsters divided into large amorphous pieces and some monsters in turn gave off normal specimens. According to Lloyd, the continuance of abnormality, with one exception, was 38 days.

In another instance, a typical Siamese pair was found to arise spontaneously in a stock culture and the abnormality persisted in subcultures through 14 months. In such monsters, the posterior end is often deeply notched or bifid with a tendency toward bilateral symmetry in which many cells have two mouths and four contractile vacuoles (Fig. 128).

It is suggested by Lloyd that the persistence of abnormality is due not to genetic inheritance but to a breakdown in the timing between the end of nuclear division and parting of the daughter cells at cytosomal division. The result is abnormal cell growth and failure, later, in synchronization of the kinetics of division.

Irradiation with x-rays of late dividing stages of *P. caudatum* not only greatly prolongs the division process but also may produce giant C-shaped monsters from which normal appearing specimens may originate (Fig. 129). When late dividing stages are irradiated with 300,000 roentgen units, such stages commonly remain suspended in division, yet swim actively for 6–48 hours, occasionally longer. Generally the x-rayed dividing cells do not live to produce daughters but if daughters are formed, they usually persist for a day or more then die. In certain other instances an irradiated divider, after gradual recovery from x-ray effects, may produce viable daughter cells which yield survivors that later approach the normal fission rate (Wichterman, 1951, 1951 c).

Chapter 11

Vitality Studies and the Life-Cycle

A. **Isolation Cultures and the Fission-Rate as an Index to Vitality**
B. **Causes and Effects of Conjugation and Its Relation to Vitality**
C. **Relation of Endomixis, Autogamy, and Other Nuclear Phenomena to Vitality**
D. **The Question of Cyst Formation in Paramecium**

A question which intrigued the early protozoölogists was whether or not a ciliate and its descendants when provided with adequate food were capable of living and multiplying indefinitely. Over a century ago, Ehrenberg (1838) came to the conclusion that the Protozoa are potentially immortal. Later Weismann (1884, 1890) stated that the protoplasm or cells of a metazoan animal could be classified into two groups, namely the body or somatic protoplasm and the germinal protoplasm. He proposed the idea that the former is mortal but that the latter, the germinal protoplasm, is potentially immortal. Weismann contended that old age and natural death are penalties demanded of the Metazoa because of their specialization and differentiation into somatic and germinal protoplasm whereas the Protozoa, without this protoplasmic specialization, are potentially immortal like the germ cells.

On the other hand, Maupas (1888, 1889), Joukowsky (1898), Simpson (1901 a, b, 1902), Calkins (1902 c,

1904, 1906 a), and many others maintained that ciliates in isolation cultures passed through several hundred or more generations over a period of three or more months in which the fission-rate ultimately decreased until the animals died. It was therefore concluded that the ciliates completed their life-cycle, then died a natural death similar to metazoan animals. Accordingly, it was claimed that old age, a natural condition of protoplasm, was applicable to ciliate protoplasm as well as metazoan, both following the same physiologic laws of ageing. With improved methods and by changing the type of medium in his isolation cultures of paramecium, Calkins (1902 c) was able to prolong the series for 24 months.

Dujardin (1841), Calkins (1902 b, c, 1904, 1915), Calkins and Lieb (1902), and others observed that ciliates in culture passed through more or less regular periods of vigor and depression which alternated with each other in fairly regular succession. It was believed that the depression periods indicated evidence that the organisms were worn out physiologically. To these periodic fluctuations or variations in vitality, Woodruff (1905) gave the name "rhythms," which is a term commonly employed by protozoölogists. Woodruff defined a rhythm as a minor periodic rise and fall of the fission-rate, due to some unknown factor in cell metabolism, from which recovery is auton-

360

omous. Rhythms may be due to food, chemical changes in the culture medium as a result of growth of bacteria and/or paramecia, nuclear changes, and possibly temperature changes.

Like every animal, paramecium is able to respond to stimuli which may alter the behavior or structure of the organism. If the stimuli are slight, paramecium demonstrates the ability of adaptation or acclimatization (p. 203). If the stimuli or changes in the environment are of such a nature that visible changes in the organism become apparent and measurable, we may use such changes as a yardstick to measure its reaction or vitality. This feature of vitality involves intensity, with an increased fission-rate as its expression, and endurance, as observed in prolongation, or survival through a period of time.

Calkins (1933) reported that vitality, which is variable, is the sum-total of all the protoplasmic activities set up in response to internal and external stimuli. For instance, increased temperature increases oxidation which leads to more rapid movements including food-taking activities, more active digestion, assimilation, growth and reproduction. Conversely, decreased temperature slows up the entire series of physiologic activities and vitality is reduced. Any environmental condition which tends to quicken, weaken, or nullify any one link in the chain of vital activities will have its effect on the general vitality (Calkins, 1933).

In a study of vitality, it is therefore necessary to observe and measure one or more functions of the animal since any one is intimately related to the whole physiologic series. Some functions, such as respiration, nutrition, irritability, or excretion, are difficult to measure in long-continued experiments. The systolic rate of the contractile vac-uoles has been used as an index of vitality, but, as pointed out by Fortner (1924), there are variations in the contractions of the two vacuoles, emphasizing that an anterior or posterior vacuole or both should be observed consistently. For long-continued experiments on vitality, this method is admittedly unsuitable, but the retardation of vacuolar activity due to environmental action can be a useful indication in shorter experiments. Fortner also noted that the presence of numerous food vacuoles in the endoplasm of paramecium may prevent, in a purely mechanical way, the union of radial canals with the reservoir of the vacuole, resulting in uneven emptying of the contractile vacuoles.

A. Isolation Cultures and the Fission-Rate as an Index to Vitality

Since the work of Maupas, and based upon his isolation culture method, it has been shown by him, and many others since, that the reproductive or division rate is a reliable measure of vitality. With proper environment such as culture medium, food, and temperature, the fission-rate represents the summation of all the other physiologic activities. This study of the fission-rate in relation to vitality attracted the attention of a great number of protozoölogists and resulted in many publications after considerable labor. From this vast sea of literature on the subject, it is difficult to distill truth and fact from the cauldron of data, claims, and counterclaims. Most of the early work was done with out-moded techniques, and for that reason not much reliability can be placed upon the results or conclusions. The subject is reviewed by Calkins (1933) and Calkins and Summers (1941) to which the interested reader is referred.

The basic method for determining

the rate of fission of paramecium has changed little from the time of Maupas. What has been changed, however, is the greater regard for and attention to the technical aspects of the method.

Essentially, the method consists in placing a single individual (sometimes an exconjugant) in a depression slide or small glass dish containing a few drops of culture medium. The preparation is then placed in a moist chamber to prevent evaporation and contamination and the chamber placed in a constant temperature incubator. Since it is difficult, in fact unnecessary, to observe the animals without interruption, one may establish the division products after certain periods of time; mainly daily but occasionally every 12 or 48 hours. As an example, a single specimen so isolated and with appropriate food in the medium may show after 24 hours, two, four, or eight specimens.[1] This would represent one, two, and three divisions respectively. It is customary to isolate the progeny after these first division stages and establish five or so independent lines or series all representing protoplasm which had been a part of the original isolated individual. A parallel line is also begun from a sister clone as a control. After the 24-hour period, a single specimen is isolated from each of the lines and a daily record of generations kept for all lines. After the total numbers of divisions are recorded for all lines, they are next averaged for five- or 10-day periods (or longer) and the results placed on a graph to show the division rate which is then taken as a measure of vitality.

[1] Isolated specimens in such cultures maintained under optimal conditions at 27° C. and with appropriate bacterial food may yield as many as five to seven divisions per day (32–128 specimens) for *P. aurelia* but fewer per day for *P. caudatum*. The effect of supercentrifugation on the fission-rate is discussed by Yancey (1931).

The graph, known as a histogram-plot, shows the ordinate represented by the average fission-rate and the abscissa, the consecutive periods of time (Fig. 130). This figure is a long-term plot; five- or 10-day intervals are more commonly used.

In a just criticism of the histogram-plot method of analyzing data, Richards and Dawson (1927) have shown that the graph is not only theoretically incorrect but may be misleading in the interpretation of data. They report that the bars of the plot denote average division rates for 10-day periods thereby showing unit events which have no intermediate values. Information is presented by them to show how data on the division rate may be obscured with such plots. These investigators and Phelps (1934) present formulas and methods for the plotting of a three- or five-day running average of the fission-rate which is more reliable than the histogram-plot used earlier. Instead of recording the number of animals for such a graph, one may use logarithms. Eckert and Feiler (1931) used a method of recording in which the number of divisions in a 24-hour period was recorded by a number. For instance, if eight paramecia were found in a dish where one had been isolated, the result would indicate three divisions and be so recorded; seven animals would equal two and three-fourths divisions, etc.

While Maupas, Calkins, and many others believed in a waning vitality in isolation lines resulting in natural death, others, namely Enriques (1907), Chatton and Chatton (1923), and especially Woodruff (1908–43), opposed this view and held that paramecium and other free-living ciliates were able to divide and reproduce indefinitely barring accidental death. It would appear that in the majority of cases where organisms died in culture, they were slowly killed

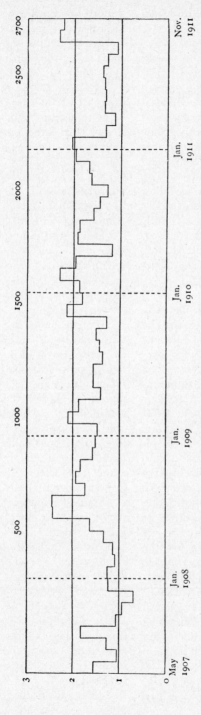

FIG. 130. Graph of the rate of reproduction of the lines of the pedigreed race of Woodruff (I) of *Paramecium aurelia* from start on May 1, 1907, to December 1, 1911, at the 2,705th generation. The ordinates represent the average daily rate of division of the four lines of the culture, again averaged for each month of its life to date. The vertical broken lines indicate the limits of the calendar years. The figures 1000, 1500, etc., represent generations and are placed above the months in which they were attained.

instead of terminating a naturally occurring life-cycle.

Woodruff has demonstrated convincingly that in *P. aurelia,* organisms could be maintained in uninterrupted multiplication (without the advent of conjugation) apparently indefinitely and at a high level of vitality. His experiment was terminated after 33 years.[2] The same conclusion of potential immortality was drawn by Metalnikow (1924) and Galadjieff (1932) and jointly (1916, 1933) for *P. caudatum.* On the other hand, many report that certain stocks of paramecium, even when carefully controlled cultural techniques are used, demonstrate a slow decline in vitality, as shown by a diminished fission-rate, then yield to death.

If one reads the technical methods employed by experimenters in the early studies, he will understand in part the causes for inconsistent results. Little or no attempt was made to employ bacteriologic methods. Dishes were wiped "clean" with a cloth; culture fluid was simply exposed to the air to become bacterized as the source of food. Paramecium was not washed or sterilized. Generally, a specimen was merely removed from a mixed or mass culture (or aquarium) and with it there were carried over many species of bacteria, some possibly deleterious. Information upon temperature was not recorded or even kept in many experiments.

However, if properly conducted, the method of isolation culture is valuable in determining fission-rate and demonstrating vitality. After the organism is sterilized (p. 106), it should be placed in a medium of known pH containing a known food source. Since bacteria are so frequently used as food, the medium should contain a given species of bacterium whose growth characteristics are known. Attempts should be made to standardize the culture fluid by the use of a uniform medium which may be buffered. Through the years, ground desiccated lettuce leaves have proved to be extremely useful, since it is inexpensive and a large amount of uniform quality may be prepared at one time and kept indefinitely. For buffering, Giese (1945) used a Sorensen buffer of approximately pH 7.0 which was made by mixing 3.8 parts of 0.5 M KH_2PO_4 (Sorensen) with 6.2 parts of 0.5 M Na_2HPO_4 and diluting tenfold. It may be necessary for very accurate work to test the pH and adjust the stock buffer by the addition of the appropriate salts. Directions for preparing the desiccated lettuce and the lettuce infusion are given on page 98. For division-rate determinations of paramecium, Giese suggests the use of a 0.5 Gm. sample of lettuce powder added to 50 cc. of distilled water and boiled for approximately one minute. To this combination is added 850 cc. of distilled water and 100 cc. of Sorensen buffer at pH 7.0, then stirred and poured into glass containers after filtering, then plugged and autoclaved after which the fluid may be kept indefinitely.

About 24 hours before using, the lettuce infusion is inoculated with a loopful of a strain of bacterium known to support growth of paramecium, i.e. *Pseudomonas ovalis.* This bacterium can be cultivated easily on 2 per cent agar containing 2 per cent yeast extract; the bacteria should be transferred weekly. For maintaining isolation cultures, a single sterilized paramecium is placed into a depression slide, or Bureau of Plant Industry Syracuse watch glass which holds 1 cc. or less of the bacterized lettuce infusion (Giese, 1945). Four of the small watch glasses are

---

[2] It is of interest to report that Woodruff's original race of *Paramecium aurelia*—now over 40 years old—is flourishing in several laboratories.

conveniently kept in a covered Petri dish, or more may be kept in a moist chamber if desired. The specimens should then be maintained at a given temperature.

Giese described an advantageous method using special culture tubes for the determination of the division rate in paramecium.

All dishes as well as the lettuce infusion used should be sterilized in an autoclave at 15 pounds pressure for 15–20 minutes.

In carefully controlled experiments upon the isolation method of studying the fission-rate for vitality, it will be seen that in a proper environment and with an adequately maintained supply of food (bacteria), growth and reproduction—at least in some races of paramecium and other Protozoa—are constant and potentially unlimited as has been shown by Woodruff and Baitsell (1911 a, b), Beers (1929), Darby (1930, 1930 a), and others. A straight line will be the result if the growth curve for the sum of the individuals is plotted on arithlog paper. If the food becomes diminished, the animals will starve and the fission-rate will decline. The fission-rate will also decline when the organisms are placed in a lower temperature or if the paramecia are placed in an environment with deleterious bacteria.

In their analysis of division rates for a period of three years, Richards and Dawson (1927) observed a secular trend and a seasonal rhythm for *P. aurelia* and two other ciliates. The seasonal rhythm was a yearly cycle with a maximum division rate in the summer and a minimum division rate during the winter. Their work and that of others demonstrates that the effect of the rhythm becomes less the longer the organism is maintained under laboratory conditions. It would appear that the seasonal cycle is associated with sunlight (Richards, 1929). A similar effect is noted by the writer in the culturing of the green paramecium, *P. bursaria* (unpublished).

Finally, it must be admitted that in the method of establishing isolation cultures in the laboratory, one reads reports of only successful cultivation after the series is once established. Inadvertently, selection plays a role in the study. Little or no information is available on the number of unsuccessful attempts made in the very beginning of the establishment of isolation cultures.

B. Causes and Effects of Conjugation and Its Relation to Vitality

In carefully controlled, long-term isolation cultures, it would appear that binary fission—at least in some races of paramecium—could continue endlessly without recourse to any other process. In these races, however, rhythms occur and in the preceding section, causes for them have been suggested.

In the investigation of the fundamental problem of protoplasmic old age in paramecium and other ciliates, the role of conjugation in the so-called life-cycle has figured prominently. For a long time, individuals have observed epidemics of conjugation in mass cultures of paramecium, and investigation has proceeded along the lines of finding methods to induce conjugation, to prevent it from occurring in mass cultures, and to determine the effects of the process.

Engelmann (1862, 1876) and Bütschli (1876) observed that ciliates in culture became reduced in size and showed signs of degeneration after a large number of divisions. Bütschli then suggested that conjugation intervened which had the effect of *Verjüngung* or rejuvenation, resulting in a restoration of the general vitality of the organisms, a view also held by Engelmann (1876) and Balbiani (1883).

The great French protozoölogist, Maupas (1888, 1889), concerned himself with the theory of rejuvenescence and studied experimentally vegetative reproduction, conjugation, and the relation between the two processes in the life-cycle. He began by investigating the life history of paramecium and other ciliates in detail from one generation to the next as a result of fission in order to determine whether or not successive generations of animals became senescent and degenerate. He was also interested in finding out if such animals could be restored after conjugation.

Maupas observed that the ciliates when followed through their successive generations slowly declined and showed the effects of senescence (as in higher organisms), then died. It is of interest to point out that as early as 1889, Fabre-Domergue, and many others since, attempted to explain that the degenerative changes ending in death of a race are due to poor environmental conditions. Concerning conjugation, Maupas held that after the process, ciliates do not have a more vigorous fission-rate than at other periods. He maintained that after conjugation, the organisms pass through a cycle of vegetative reproduction showing successively periods of youth, maturity, and old age. Differing from that of later workers, his concept of these periods was not by diverse fission-rates but in the readiness of organisms to enter into conjugation. He concluded from his experiments that after animals have become decidedly degenerate such as during the period of "old age," conjugation is not beneficial; that animals in this period die more quickly after conjugation than otherwise. However, when conjugation occurred at periods of nuclear maturity, the process was described as being beneficial; i.e. conjugation fails to cure degeneration but can prevent its occurrence.

Although he maintained that the rejuvenating effect of conjugation does not include an increased fission-rate, he was emphatic about the fact that the process resulted in rejuvenescence. He stated that the new nuclear apparatus, as a result of conjugation, acts as a kind of regenerating enzyme in which the organism is rejuvenated, due to the union of the two pronuclei of diverse origin acquired during the process.

Hertwig (1888–1902) confirmed the view of Maupas that conjugation does not increase the fission-rate by his utilization of the technique dealing with "split-pairs," a valuable method used by many since Hertwig. Soon after two individuals have joined to begin the sexual process, they are separated by the observer early enough to prevent nuclear exchange. By comparing such split-pairs with those which were allowed to complete conjugation, Hertwig found that the former multiplied more rapidly than the exconjugants. Although he found many different results in as many cases, Hertwig held that fertilization as a result of conjugation brings about rejuvenescence.

Résumés and the conclusions of early workers in regard to conjugation and rejuvenescence and their relation to the life-cycle are admirably discussed at length by Jennings (1929) in his monograph, "Genetics of the Protozoa," and this important work should be consulted for additional information.

The question of rejuvenescence as a result of conjugation has long been the subject of controversy among protozoologists. According to Bütschli, conjugation is for the purpose of preventing an occurrence of physiologic weakness and death and that after conjugation, a new period of reproductive activity is manifested by the increased fission-rate. Similarly, Calkins and Gregory (1913) claimed that their experiments indicated an increased vitality after conjugation in *P. caudatum*.

That this spectacular idea, based upon unsatisfactory or incomplete experimental data, has persisted even today in spite of other interpretations, is shown in many textbooks of general biology. After certain studies of conjugation in paramecium, Jennings (1912, 1913) concluded that one can scarcely speak of rejuvenescence when two-thirds of the ex-conjugants die. Galadjieff and Metalnikow (1933) reported that in *P. caudatum* multiplication *without conjugation* persisted in their cultures for 22 years and five months totalling 8704 generations (386 per year). At times they noticed periods of waning vitality, lasting in one case for four years; but they claim no relation between multiplication and the age of the cultures. Obviously, these workers conclude that conjugation is not obligatory in the life-cycle and assert that the concept of karyogamic rejuvenescence as a result of conjugation should be abandoned. However, they believe that conjugation is associated with creation and evolution in species; that it results in more perfect blends and combinations often at the price of certain others which do not thrive.

Maupas (1889) maintained that the conditions for conjugation are starvation, karyogamic maturity, and exogamy. Enriques (1902 a, 1906, 1907, 1907 a, 1908, 1909 a, 1912) and Zweibaum (1912, 1921, 1922) were among the first to maintain that it should be possible to induce epidemics of conjugation by external means, such as changes in the surrounding medium. Enriques claimed that at times he was able to induce conjugation in a non-conjugating culture of ciliates simply by the addition of fluid from a culture in which conjugation was occurring. Also, he was able to prevent conjugation in the latter culture by the addition of culture fluid from the former. According to Enriques, conjugation was also induced by the addition of fresh food to the culture.

According to Zweibaum (1912), cultures of *P. caudatum* demonstrate epidemics of conjugation when the specimens have passed through a period of five to six weeks of partial starvation or when specimens have been placed abruptly into a medium where there is a diminution of food at 20°–23° C. He claimed that certain salts, especially those of aluminum, gold, sodium, and iron, were most effective in inducing conjugation. Although he believed that his methods to induce conjugation could be applied to any race of *P. caudatum*, Jollos (1921, 1921 a) and Hopkins (1921) were unable to obtain similar results. Zweibaum reaffirmed Maupas in believing that paramecium must of necessity undergo conjugation while in lowered physiologic condition in order to avoid death, and the former concluded that the process of conjugation is in no way related to a life-cycle or dependent upon internal factors.[3]

In numerous papers, Calkins (1913, 1916, 1919, 1919 a, 1926, 1933) reported upon conjugation and its relation to vitality in the life-cycle of paramecium and other ciliates. He believed that youth, adolescence, and old age are characteristic stages of the protozoan life history with conjugation being a

[3] Willey and Lherisson (1930) observed the tendency of paramecia to conjugate in great numbers in wild cultures collected in the spring and autumn of the year. They reported a mean fission-rate higher in the summer than in the winter and concluded that climate may be a factor for this phenomenon. It was their opinion that there existed an annual cycle, beginning with vernal conjugation, passing through a high rate of summer fission to the autumn climax of conjugation prior to the coming winter.

part of adolescence. It was his opinion that conjugation or endomixis may postpone old age. He reported that old specimens of *P. caudatum* which had not conjugated for a long period of time were characterized by hypertrophy of the macronucleus, loss of trichocysts in the cortex, and abnormal division stages leading to monster formation. In later studies, he was of the opinion that conjugation does not always result in rejuvenescence but that increased vitality may be induced by means of parthenogenesis.

The causes, conditions, and effects of conjugation in paramecium were extensively studied by Jennings (1909–45) and many others including Raffel (1930), Chatton and Chatton (1931), and Sonneborn (1936). Jennings believed that external as well as internal conditions were responsible for conjugation. Concerning the former, Jennings stated that the conditions for inducing conjugation were different in different races but that essentially, conjugation occurred at the beginning of a decline in nutritive conditions following a period of exceptional richness which had induced rapid multiplication (hunger after satiety). In regard to internal characteristics, certain races showed a constitutional disposition to conjugate more readily than others.

Jennings observed that the intervals between successive conjugation periods varied from five days to a month in some races but in others, a still longer period of time elapsed. He as well as Joukowsky (1898), Enriques (1906), and Klitzke (1915, 1916) have shown that the descendants of an exconjugant, conjugated among themselves after eight or less divisions. Usually in mass cultures, there is an interval of at least several days—generally much more—between one epidemic of conjugation and the next (Jennings, 1910, 1910 a;

Calkins and Gregory, 1913; Hopkins, 1921; Ball, 1925 a). Hopkins (1921) reported that in his races the interval between epidemics of conjugation was four to 12 weeks for *P. caudatum* and two to six weeks for *P. aurelia*.

Jennings noted that conjugation differences in different races of paramecium showed no simple relation to relative size or other morphologic characteristics distinguishing the races. In a comparison between conjugants and non-conjugants in a given culture, the members of conjugating pairs are seen to be smaller and less variable than non-conjugants in the same culture. Several generations after conjugation, the progeny of the conjugants are somewhat larger than members of the same race that have not conjugated. This condition appears to be due to a slower fission-rate among the progeny of conjugants and after a few generations, the size differences of this type disappear. The differences in size between conjugants and non-conjugants would appear to be only temporary physiologic differentiations which are of no consequence to the later history of the stock. However, the progeny of conjugants are more variable in size and certain other respects than the progeny of the equivalent non-conjugants, thus conjugation increases variation. Hereditary differentiations arise as a result of conjugation within members of the same race. While it is true that the two members of a pair sometimes show hereditary differences, statistical examination of fission-rate shows a correlation between descendants of two conjugants nearly as close as the relationship of brothers and sisters and closer than that of cousins (Jennings).

According to Jennings, control experiments involving animals not allowed to conjugate while their culture mates conjugated showed that conjugation did not have a rejuvenating effect in general. Nevertheless, a few of the variations introduced by conjugation established stocks more vigorous than either the progeny of other conjugants or the progeny of non-conjugants.

Many earlier investigators observed that crowding of specimens in a single flourishing culture of paramecium frequently resulted in the production of many conjugating pairs. This phenomenon is called *selfing* and such conjugating pairs are also known as *selfing pairs*. Selfing pairs of conjugants in a given culture may be due to the hastening of the exhaustion of available food so that specimens are left in the physiologic state necessary for union, to increased chances for bodily contact in such a flourishing culture, and to the fact that there is present a sharply increased concentration of metabolites.

In a culture of paramecium as growing in the laboratory in mass culture, induction of hunger, as suggested by Maupas, is the method most successfully used in inducing conjugation (selfing). It appears that starvation in itself does not induce conjugation; but if paramecia are transferred from a culture richly supplied with food to one which is not, conjugation frequently occurs. By allowing the culture medium to evaporate, Hance (1917) was able to induce conjugation in paramecium. Jennings reported a race of *P. aurelia* that conjugated at the beginning of a decline in a rich food environment. In *P. caudatum,* Kasanzeff (1901) claimed that conjugation occurred during an increase instead of a decrease in the fission-rate. According to Calkins (1902 b), transfer of *P. caudatum* to distilled water tended to inhibit conjugation; Giese (1935) found the same condition to hold for *P. multimicronucleatum.* Hopkins (1921) and others have encountered several races of paramecium

which conjugated more frequently when distilled water was added to the culture and when cultures were well supplied with food.

In a study of the behavior of a conjugating race of *P. caudatum,* Ball (1925 a) observed that conjugation tended to occur in a culture before the maximum number of animals was reached and disappeared before the number of animals began to decline. The greatest number of conjugants was found in cultures from the fourth to the ninth day after transfer to fresh food. He concluded that there was no evidence for a cycle of conjugation and stated that the process appeared to reduce or destroy the conjugating power in some instances but not to affect it in others.

While it has been held by some that the progeny of a single individual could not conjugate among themselves without resulting in their destruction, a large body of information indicates that even with the occurrence of frequent reconjugation, the culture continues to flourish.

Hopkins (1921) investigated 24 races of paramecium in an effort to discover whether or not differences existed between the various races in respect to their ability to undergo conjugation. He observed diversities among the various races of *P. caudatum* and *P. aurelia* in their tendency to conjugate under natural and experimental conditions and was of the opinion that conditions for induction of conjugation in all races are not identical. His most effective method for induction of conjugation was to subject paramecia to a period of semi-starvation lasting four to eight weeks after which hay infusion and certain electrolytes were added to the culture to bring about a state of rapid multiplication. Sodium nitrate and calcium nitrate appeared to be most effective in this regard. Hopkins believed in the existence of an indirect connection between conjugation and the fission-rate since whenever conjugation was observed in a mass culture, it occurred after a period of accelerated division. He concluded that conjugation occurs in a culture when the division rate is at its maximum peak or shortly afterward.

In studying the role of starvation in the conjugation of P. multimicronucleatum, Giese (1935, 1938) was able to induce large numbers of paramecia to conjugate by environmental changes. Clones of paramecia were carefully cultivated in lettuce infusion on the bacterium Pseudomonas ovalis, and he noted that the onset of conjugation occurred with the exhaustion of the food supply. To determine whether or not the absence of food was a necessary condition for conjugation or merely coincidental with it, Giese added a loopful of bacteria to the culture when the food was exhausted and at the time conjugation usually occurred. With their food supply now replenished, paramecia proceeded to divide rapidly and failed to conjugate. In this manner and by repeated feedings, paramecia did not undergo conjugation until the food supply was diminished. He concluded that while feeding delays or postpones conjugation, lack of food after paramecia have been well fed leads to a chemophysical state of the protoplasm, resulting in the sticky state and conjugation.

The results of Giese are in accord with those of a large number of investigations wherein there is a period of food depletion or partial starvation after the animals have been well fed prior to conjugation.

Barbarin (1938) induced conjugation in P. caudatum as follows: approximately 3 cc. of a dense suspension of paramecia were placed in a small, specially constructed vessel in which bubbles of hydrogen were passed for an hour. The vessel was then hermetically sealed, and the paramecia were left overnight in the atmosphere of hydrogen at 25° C. Next morning, air was bubbled through the medium, and the paramecia were maintained for a few hours in this aerobic condition. As a result, numerous pairs of conjugants were claimed to be present.

The discovery of the existence of mating types in paramecium by Sonneborn (1937 c) has provided a key to the investigation of vitality, especially in regard to the effects of conjugation in the life of these organisms. This problem has been carefully and extensively studied with observations and experiments upon cultivation, mating, conjugation, and exconjugant behavior in P. bursaria by Jennings (1938–45). In a precise manner, he has presented, at least in part, explanations for what appeared to be conflicting claims of the earlier investigators in respect to the effects of conjugation.

Of P. bursaria and several other species which have been investigated, it has been observed that clones show a period of sexual immaturity during which the mating reaction fails to occur when specimens are mixed with clones of opposite mating type. There follows a transitional period during which the mating reaction is weak. Later, there occurs in clones, a period of sexual maturity in which the mating reaction is strongly marked resulting in the formation of conjugating pairs. These periods depend mainly upon cultural conditions and to a certain extent, upon inherent racial factors. Jennings reports that the period of immaturity may extend from a minimum of 12 days up to several years in P. bursaria. Also, that the period of maturity may last for several years only to be followed by a period of decline in which the fission-rate becomes reduced, ending finally with degeneration and death.[4]

Conjugation among clones commonly results in the production of four types of exconjugants as follows:[5] (Jennings, 1944 a)

1. Exconjugants may die without undergoing fission.

2. Exconjugants may undergo 1–4 stages of fission, then die.

3. Exconjugants may produce clones containing numerous specimens which are abnormal or pathologic.

4. Exconjugants may produce clones which consist of vigorous and actively dividing specimens.

It may be emphasized that in class 4, exconjugant clones may be produced which are even more vigorous than the parent clones. It is seen, therefore, that in this particular instance there is demonstrated in fact, rejuvenescence as a result of conjugation.

In studying the result of conjugation, the author (Wichterman, 1943) mated a clone of normally green *P. bursaria* with a clone of zoöchlorella-free, colorless specimens of the same species but of opposite mating type. Both of these clones had been maintained for several years in the laboratory. To determine the fate of exconjugants resulting from the cross, seven different mass matings were made over a long period. In observations made upon 57 isolated pairs of conjugants and their 114 exconjugants, it was found that 82.5 per cent of the

exconjugants failed to divide after conjugation. With only two exceptions, when a green exconjugant divided, its former colorless mate also divided. However, when seven different mass matings were made between two normally green clones of *P. bursaria*, 51 pairs of isolated conjugants and their 102 exconjugants disclosed the result that only 14.4 per cent of the exconjugants failed to divide after conjugation. Typical non-conjugating specimens were isolated and used as controls in each experiment.

Jennings (1944) reported that during the period of decline in a clone of *P. bursaria*, conjugation may occur up to the end of the period, but that conjugation of aged stocks results in the death of all or nearly all of the exconjugants. Regarding the fate of exconjugants, Jennings and Opitz (1944) cultivated 118 conjugating pairs as the result of a cross between a clone from the United States (variety II) and a Russian clone (variety IV). All died without multiplying. Of the lot, 87 pairs or 73.2 per cent died without separating while the remaining 31 pairs separated but the exconjugants died without dividing. Each clone by itself was capable of giving rise to normal, viable exconjugants; but when the two were brought together, death resulted. Thus it is seen not only that aged stocks when crossed may result in great mortality of the conjugant progeny but that racial factors are involved as well.

As a result of crossing diverse clones of *P. bursaria* from different geographical locations of Russia, Gause, Smaragdova, and Alpatov (1942) found that in all cases, some percentage of exconjugants was nonviable. Clones from the vicinity of Moscow were crossed; Moscow clones were crossed with those from southern Russia and with clones from northern Russia; also southern and

[4] Jennings (1944) reported that clones of a given mating type when kept in the laboratory up to eight years declined, ending with death of the organisms. The present writer received in 1940, clones of opposite mating types (C and D) of *P. bursaria* from Jennings and up to the present time (12 years from date of delivery), they are still being cultivated, mate and conjugate.

[5] Occasionally, conjugating pairs will be unable to complete the sexual process and die while joined.

northern clones were crossed. The least mortality of exconjugants was observed after mating Moscow clones with each other. The average percentage of non-viable individuals (exconjugants) rose somewhat in the mating of Moscow with southern clones and attained highest values in the Moscow matings with southern ones, and northern clones with southern ones.

In *P. bursaria,* animals belonging to any one variety do not, as a rule, mate with members of other varieties. However, Chen (1946 b) found an exception to this rule in the mating and subsequent conjugation of specimens belonging to certain Russian and American clones belonging to different varieties. In considerable detail, this investigator demonstrated that nuclear changes in such conjugations are decidedly abnormal, possibly due to physiologic incompatibility between the varieties. He reported that most of the conjugating pairs failed to separate even as late as 70 hours after the onset of conjugation.

It has been extensively reported by Jennings (1944 a, b, c) that conjugation of old clones of *P. bursaria* (from the standpoint of age as cultivated in the laboratory) results in a high mortality rate among the exconjugants or exconjugant clones (100 per cent or nearly so), while conjugation of young clones usually results in little or no mortality.[6] Even when specimens from an old clone conjugate with those from a young one, all or nearly all of the exconjugants die in the same manner as in the conjugation of specimens resulting from the mating of two old clones. Chen (1951 a) noted that when old clones of *P. bur-*

[6] In certain rare cases, however, conjugation between two recently collected clones resulted in a high mortality rate among the exconjugants. Clones of *P. bursaria* taken at random and mated, showed a mortality rate of 30–50 per cent among the exconjugants.

saria are mated with young clones, a cytological study reveals a high percentage of abnormal conjugants and exconjugants.

It is interesting to point out an observation reported by the writer (Wichterman, 1943) and confirmed by Jennings (1944 a) that if one member of a pair of conjugants of *P. bursaria* dies, the other generally dies also; if one lives, then the co-conjugant member is likely to live.

Jennings (1944 b, c) reported that successive inbreeding by conjugation of two clones results in a great mortality rate of the exconjugants. Sometimes it is seen to be five to 14 or more times as great (90–100 per cent) as the mortality of exconjugants from the conjugation of unrelated clones.

As a result of more exacting experimental work, the oft-quoted statement, *conjugation results in rejuvenescence,* is not wholly correct. Rejuvenescence through conjugation is a fact in some cases. In certain declining clones, conjugation, if it occurs before the decline in vitality is too great, may prevent further waning vitality and high vitality may be restored again. In other cases of conjugation, it is equally a fact that with the same type of carefully controlled experiments, exconjugants and their progeny become weak, abnormal, then die after possibly a few divisions.

In this regard, Sonneborn and Lynch (1932) demonstrated for *P. aurelia* that conjugation increases fission rate in some clones but decreases fission rate in others; it increases variation in some clones, but not in others; it increases mortality in some clones, but not in others. In a very definite manner, they showed the existence of racial differences in paramecium, as Jennings has done with *P. bursaria,* and suggested that the disagreement in results and theories of previous investigators may have been due to failure in examining the effects of conjugation in a sufficient number of different races within the species.

Finally, one important aspect of conjugation is that concerned with inheritance which is considered in detail in Chapter 10. The sexual union, with the interchange of nuclear material, produces new combinations of chromosomes resulting in individuals with new genetic constitutions. As in the Metazoa, including man, some of these combinations are good and certain are bad for the progeny. Jennings, Raffel, Lynch, and Sonneborn (1932) observed many "bad," i.e. short-lived stocks which arose after conjugation in *P. aurelia*. On the other hand, they also observed "good," or long-lived stocks after the sexual union.

In what appears to be the last paper before the death of this great student of paramecium, Jennings (1945), writing on age and mortality in paramecium (*P. bursaria*), comments finally upon conjugation as follows:

Inbreeding, as the conjugation of sibs (clones that are closely related through a common derivation from the previous conjugation of the same two clones) results in a high mortality, ranging usually from 50 to 100% of the inbred ex-conjugants. (Inbreeding by self-fertilization of a clone does not induce this high mortality.) Increased age of the conjugating clones increases the mortality among the ex-conjugants. Young and vigorous clones when mated show little or no mortality. Mortality (on the average) increases as the clones that conjugate are older. . . . Certain of the clones that result from conjugation show when again mated a high mortality that is independent of age and of inbreeding. (The cause of this presumably lies in their receiving a poor combination of chromosomes and cytoplasm in the conjugation by which they took origin.) . . . Death did not take origin in consequence of organisms becoming multicellu-

lar, as is sometimes set forth. Death occurs on a vast scale in the Protozoa, and it results from conditions that are intrinsic to the organisms. Most if not all clones ultimately decline and die out if they do not undergo some form of sexual reproduction (conjugation, clonal self fertilization, or autogamy). Rejuvenescence through sexual reproduction is a fact, and the perpetuation of the stock is mainly due to the formation of chains of clones, each clone the product of a sexual process in preceding clones. Yet conjugation produces, in addition to rejuvenated clones, vast numbers of weak, pathological, or abnormal clones, whose predestined fate is early death.

C. Relation of Endomixis, Autogamy, and Other Nuclear Phenomena to Vitality

In his long-continued experiments on the fission-rate, life and vitality of *P. aurelia* over many years and through many thousands of generations, Woodruff (1908–43) demonstrated conclusively that conjugation was not necessary for the continuation of his pedigreed strain. Indeed, his experiment clearly supported Weismann's concept that a protozoön was capable of multiplying indefinitely.

Begun in 1907, this famous culture was observed to pass through periodic rhythms. The paramecia were then examined cytologically by Woodruff and Erdmann (1914) during these rhythms, and there was discovered a periodic nuclear reorganization process occurring in the single animals. To this process in *P. aurelia,* they gave the name *endomixis* (Fig. 82 A). Later (1916), they reported the process as occurring in *P. caudatum,* and it has since been described by others as occurring in additional species of *Paramecium* as well as in other ciliates.

Endomixis was described as occurring in single specimens and characterized by a gradual disintegration of the macronucleus into chromatin bodies which are absorbed by the cytoplasm, the transformation of one or two of the products of the micronuclear divisions into new macronuclei to reconstitute the normal vegetative nuclear apparatus when distributed by cell division, the absence of a third micronuclear division, and the absence of synkaryon formation. Details of the process are given on page 260.

In a cytological reinvestigation of the phenomenon, Diller (1936) came to the conclusion that the so-called process of endomixis is in reality *autogamy,* or self-fertilization of the individual. In autogamy, three micronuclear divisions occur to produce the haploid gametic nuclei which fuse and form the diploid synkaryon. With the demonstration of the existence of autogamy in *P. aurelia* both cytologically by Diller and genetically by Sonneborn (1939), the process rests on a firm foundation. Wichterman (1939, 1940) reported a similar process for *P. caudatum* but in joined pairs to which he gave the name *cytogamy.* Until there is additional cytological evidence to support endomixis, it would appear that references to this process in the literature may be considered as autogamy.

It was stated that the interendomictic periods in both *P. aurelia* and *P. caudatum* varied in length. While it was originally announced that the periodicity of endomixis (autogamy) was approximately 25–30 days and 40–50 fissions for *P. aurelia,* and 50–60 days and 80–100 fissions for *P. caudatum,* later evidence discloses that environment and racial characters are involved which alter the periodicity of the rhythms.[7] Sonneborn (1937 b) reported

[7] Others who have studied the periodicity of the rhythms in paramecium are Young (1918), Jollos (1916, 1920), Erdmann (1919, 1920), Chejfec (1930), Galadjieff (1932), and Sonneborn (1937 b, 1938).

that considerable variation may exist in the interval, not only in different races but even in the same race under carefully standardized conditions. Indeed, it has been maintained by some that rhythms occur without so-called endomixis (autogamy). In a study of the relation of rhythms to nutrition and excretion in *P. aurelia, P. caudatum,* and *P. calkinsi,* Unger (1926) observed rhythms when nuclear reorganization was not occurring. He noted, however, that more food vacuoles were being formed at the top of the rhythm than at the bottom.

According to Erdmann (1920), one result of endomixis is the production of large variations in size of paramecia which were said to possess a survival factor. Caldwell (1933) was of the opinion that inherited diversities were produced as a result of endomixis in *P. aurelia.* Gelber (1938) and Pierson (1938) made statistical studies of the mortality rate in regard to endomixis and claimed that death occurred concomitant with the nuclear process. In a very young clone, i.e. when the time since the preceding endomixis was short, the mortality rate was said to be low. As the clone becomes older, i.e. the time since the preceding endomixis is long, the mortality rate at the next endomixis is high. When the time since the preceding endomixis is approximately 10 days, the mortality rate is 2.08 per cent; but if the time since the preceding endomixis is at least 125 days, the mortality rate is said to be 100 per cent.

Successful methods for the induction of endomixis (autogamy) in *P. aurelia* have been reported by Jollos (1916), Young (1917), and Sonneborn (1937 a). Paramecia are uniformly cultivated in isolation lines, and surplus animals, available after the daily transfer of the lines to fresh culture, are collected into approximately 10 drops of fresh culture medium in a depression dish and maintained at $31° \pm 1°$ C. without further change according to Sonneborn. Concerning the method, he reported that cultures set up soon after endomixis (autogamy) yielded conjugants, so that the induction of the process could not be studied quantitatively in these. Later cultures, however, showed increasing percentages of induction until 100 per cent was obtained after one day at $31°$ C. Earlier cultures which gave but low percentages of the process with this treatment could be induced to give larger percentages by subculturing or adding fresh culture medium to the original induction culture. Sonneborn and Lynch (1937) also reported that they were able to induce conjugation soon after endomixis but not after longer periods of time had elapsed.

According to Jennings and Sonneborn (1936) and others, if autogamy is omitted for long periods of time in *P. aurelia,* there is a decline in vitality as observed in the decreased fission-rate, and death is the result if the process fails to occur. However, if the fission-rate has declined appreciably and autogamy occurs, the normal fission-rate and vitality, in many instances, are restored. On the other hand, if autogamy is delayed too long, such as near the end of a period of decline when the vitality is at a very low ebb, the process will not restore vitality. Hence, it may be concluded that the greater the period since the last autogamic period, the greater the proportion of individuals succumb and that the recency of an autogamic period is a fair index of the vitality in *P. aurelia.*

The occurrence of nuclear reorganization in paramecium such as at autogamy and conjugation discloses that in some cases, but not all, there results an increased vitality of the organism as evidenced by increased fission-rate and prolongation of the life of the race. In these cases, the replacement of a new macronucleus appears to be of fundamental importance in continued vitality. That the macronucleus and not the micronucleus is important is attested by the existence of amicronucleate races. Such races have been described as living for a long time after passing through many generations. On the other hand, no instance of an amacronucleate race has ever been described. Until additional information is available, it would appear that the substitution of paramecium with a new macronucleus has the effect not only of preventing senescence and death but renewing vitality—at least in some races.

Concerning the role of autogamy and hemixis in vitality and the life-cycle, Diller (1936) concluded that independently from the hereditary effects which may issue from autogamy, the primary "purpose" of autogamy is a mechanism for readjusting the nucleo-cytoplasmic relationships of the cell. This would entail replacing a worn-out nucleus with a new one, or possibly freeing certain nuclear materials for cytoplasmic use. Possibly the same result may be achieved by hemixis in a direct way, independently of micronuclear activity. According to Diller, the role which hemixis plays in the life-cycle remains to be studied. See also Diller (1951).

In *P. aurelia,* Sonneborn (1940, 1947) observed that a small percentage of specimens, while undergoing autogamy and conjugation, develop a new macronucleus from each fragment of the old macronucleus, commonly giving rise to animals without micronuclei (Fig. 85).

To induce macronuclear regeneration, Sonneborn heated paramecia to $38°$ C. during the processes of fertilization and the subsequent reconstruction of the nuclear apparatus. As reported by him,

When exposed to this temperature during most of the period of meiosis and fertilization and up to the time of the first fission, the micronuclei are often aberrant, failing commonly to begin differentiation into macronuclei and often being lost entirely. However, when the exposure to 38° C. is limited to the period after fertilization, but starts before the second post-zygotic nuclear division (that is, before the division which normally results in the differentiation of the new macronuclei), the micronuclei remain normal and two of them as usual begin differentiating into macronuclei. These developing macronuclei, however, often have their growth arrested after a short time and do not resume growth and development until after two or more fissions have been completed. Arrested macronuclei do not divide until their growth has been resumed. Animals therefore arise which lack the new macronuclei, but, of course, contain many pieces of the old disintegrated macronucleus. The latter, when removed from the presence of the new macronuclei, proceed to regenerate into complete macronuclei in the way described above.

One important fact in macronuclear regeneration is that each piece of old macronucleus can regenerate a complete macronucleus which permits normal functioning of the animal in the absence of a micronucleus. This would indicate the macronucleus to be a compound structure consisting of many units each containing a full set of genes. It would appear that hereditary characters, including mating type, are determined by the macronuclei since micronuclei can be absent. New characters, such as size, form, fission-rate, etc., always appear in clones from macronuclear regenerates.

D. The Question of Cyst Formation in Paramecium

Although cysts have been described for various species of *Paramecium*, it is indeed surprising that they have not been encountered more frequently if cyst stages actually exist. Most investigators who have studied paramecium intensively under a variety of conditions and over a long period of years, are of the opinion that paramecia do not normally form cysts. Yet the following have described cysts or encystment for species of *Paramecium*: Prowazek (1899) for *P. bursaria;* Lindner (1899) for *"P. putrinum";* McClendon (1909) for *P. aurelia;* Simpson (1901), Ivanic (1926), and Michelson (1928) for *P. caudatum;* and Cleveland (1927, 1927 a) for an undetermined species of *Paramecium.* Certain of the methods set forth as inducing cyst formation have been attempted by the author with no success. In the study of a race of *P. trichium* over a period of many years, the author noticed active specimens to become spherical, which he called "pseudo-cysts" (Wichterman, 1939 a). These presented a striking resemblance to ciliate cysts in general appearance. They were somewhat smaller than active specimens, roughly spherical with a thick, irregular wall, conspicuous macronucleus and granular protoplasm with no differentiated cytoplasmic structures. Attempts at excystation proved unsuccessful and many could be found in various stages of degeneration.

An examination of Lindner's figure of a so-called cyst of *"P. putrinum"* proves the specimen to be not a paramecium but probably a *Colpoda*-like ciliate. Prowazek (1899) shows a figure of what he believed to be a cyst of *P. bursaria,* and the present author (unpublished) has on rare occasions seen specimens of the same species become rounded in masses of bacterial zoöglea seemingly forming cysts, but attempts at excystation failed.

Ivanic (1926) reported cysts of *P. caudatum* from an aquarium containing many other Protozoa. He described what he interpreted as encystment phenomena but failed to observe excystment. A precystic paramecium was described as short and thick which secreted a transparent layer of mucoid material that hardened. Cilia were said to become swollen and disappear, but the feeding apparatus was conspicuous.

McClendon (1909) repeatedly observed a thin membranous "cyst" wall surrounding a spherical stage of *P. aurelia* in which the organism was confined for a week but in which it was killed, apparently due to desiccation. Before drying, animals were seen to rotate in the "cysts" since the ciliary coat remained intact. He reported that when an animal left the "cyst," the

cially around the cytostome. Also visible in the so-called cyst was the macronucleus but not the micronucleus, while trichocysts and cytopharynx were resorbed. Such a fully formed "cyst" was said to resemble a sand grain (Fig. 131). Michelson claimed to have recovered motile specimens from the cyst. His work needs confirmation.

Cleveland (1927) injected large numbers of an undescribed species of *Paramecium* into the rectum of frogs and reported that a definite cyst membrane and cysts were produced. It seems possible that seen instead were encysted and excysting ciliates belonging possibly to the genus *Opalina* or *Nyctotherus*, both commonly found in frogs. Cleveland (1927 a) also reported upon natu-

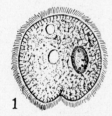 

FIG. 131. So-called stages in the encystment of *Paramecium caudatum*. (Michelson)

specimen appeared abnormal. So-called cyst stages were found either in masses of bacterial zoöglea or on the outside of such masses.

In what appears to be the most extensive report of cyst formation in paramecium, Michelson (1928) described encystment under conditions of slow desiccation entailing loss of peristome, vacuoles, and cilia. While he believed that drying of active specimens of *P. caudatum* in fine mud was the method of encystment in nature, he reported experimental encystment in agar. Paramecia were said to round up and become spherical or ellipsoidal with the dissolution of the cilia and with accentuation of the pellicular striations, espe-

ral and experimental ingestion of paramecia by cockroaches in which he claimed to have found actively motile specimens in the intestine. It seems possible that instead of observing paramecium, he encountered cyst and vegetative stages of *Nyctotherus ovalis*, a ciliate which is common in cockroaches. Hegner (1929) found that paramecia survive for about 24 hours and only in the forward part of the digestive tract of cockroaches.

From the foregoing accounts of so-called cyst stages of paramecium, it appears certain that some of the investigators have misinterpreted other ciliates which resemble paramecium. Before it can be accepted that cyst formation is an actuality in paramecium, accounts must include precise observation of encystment of the living animal, an adequate description of the cyst and especially the details of the excystment process. Also, it should be possible for other investigators to repeat the entire process. The fact that paramecium may be seen to become rounded and less active may not be cyst formation at all but simply phenomena associated with the slow death of the organism.

Considerably more experimental work must be done therefore before it can be established that paramecium is able to produce a viable cyst as is commonly the case with so many other free-living ciliates.

Chapter 12

Serologic Studies of Paramecium

A. Serology and Speciation

B. Serology in Respect to the Mating Reaction, Conjugation, and Inheritance

A. Serology and Speciation

Immunologic reactions and biochemical studies have proved to be valuable criteria for showing resemblances or differences in strains, races, or species of Protozoa. Serologic reactions may be used to compare the chemical composition with the morphologic structure. In general, it follows that if two species are closely related, the group reaction between them will be stronger than those not so closely related. Such investigation, therefore, may serve as a check for speciation. The essential purpose of immunologic classification—at least for the Protozoa—is primarily one of additional evidence for classifications now in use rather than the establishment of an entirely new classification based solely upon immunologic reactions.

Because serologic methods have been shown in a large number of cases to be more precise than certain other methods of demonstrating relationships between organisms, it is surprising that these highly specific tests have not been used more extensively with free-living Protozoa. Bacteria associated with ciliates in culture and used as food may have discouraged investigators from attempting such studies. However, it is now possible to cultivate an increasing number

of Protozoa free of bacteria normally found in the typical culture. Also the presence of the comparatively few bacteria on washed paramecia does not appear to interfere with the results.

In the field of parasitic protozoölogy, immunologic methods have been used extensively to demonstrate the validity of proposed species of *Leishmania*[1] and *Trypanosoma* where morphologic characters are identical in different species of the genus (see Taliaferro, 1929, 1941).

When paramecia are injected into the blood stream of a rabbit or other warm-blooded animals, they stimulate the rabbit's system to produce substances in the blood of the animal known as *antibodies*. Such antibodies are known as *immune antibodies* to differentiate them from natural antibodies which may exist in blood without immunization. The paramecia, or derivatives of them as well as other protein and possibly complex polysaccharide or lipid materials which can evoke antibodies, are known as *antigens*. Serum from the blood of such an animal containing antibodies is known as *antiserum*. Earlier, many immunologists believed that the introduction into the body of a single antigen resulted in the formation of a single antibody, but now it is generally held

[1] Noguchi (1924) by serological methods demonstrated the difference between morphologically identical species of *Leishmania* (*L. donovani*, *L. tropica* and *L. braziliense*).

379

that more than one kind of an antibody is formed in response to a single antigen.

Antibodies are generally classified according to the effect produced when mixed with the antigen. If the antibody produces a precipitate upon mixing with the antigen, the antibody is termed a *precipitin* and the reaction is called a precipitin reaction. If the antibody induces clumping or agglutination of the cells, the antibody is termed an *agglutinin* resulting in an agglutinative reaction. Another commonly employed test is the *complement fixation reaction,* which is based on the fact that a heat labile component called complement is bound when antigen and antibody interact.

Antiserum from a rabbit injected with a given species of *Paramecium* has a marked effect on the paramecia. The reaction has been described in detail by Bernheimer and Harrison (1940) for *Paramecium* and a somewhat similar one has been reported for the ciliate *Tetrahymena* by Kidder, Stuart, McGann, and Dewey (1945). When a small amount of the rabbit's blood serum containing these antibodies (antiserum) is introduced into a container of paramecia, the ciliates first respond by a gradual slowing of motion. The paramecia (*P. bursaria*) then become paralyzed or immobilized, and microscopic clots of a sticky substance form near the distal ends of the cilia; in many cases, several cilia become matted together in clumps. Typical ciliary activity may cease and paramecia left in this condition may die; on the other hand, there may be spontaneous recovery in some cases.

As early as 1902 Ledoux-Lebard studied the action of blood serum on paramecium. Later Rössle (1905) noted that the antiserum of a rabbit or guinea pig (when injected with a paramecium broth) made the pellicle of *Paramecium* sticky when tested with the antiserum a few days later. He compared the action of a five to 10 times dilution of normal and specific immune sera on *Paramecium* and found intensive paralyzing of paramecia to be the specific toxic action. Rössle found no definite phenomena which could be regarded as analogous to bacteriolysis. He used for the dilution of the normal and immune sera, the liquid from his culture of paramecia which was prepared without the addition of any particular electrolytes or salts. Later Rössle (1909), Sellards (1911), Takenouchi (1918), Schuckmann (1920), Masugi (1927), Rakieten (1928), Heathman (1932), Meyer and Shaffer (1936), and Robertson (1934, 1939 a, b) performed immunologic studies on paramecium or other Protozoa, but they were mainly concerned with an explanation of the antigen-antibody reactions rather than questions of speciation. It is of interest to report that Harrison, Fowler, et al. (1948) attempted to verify the claim of Roskin (1946, 1946 a) that the presence of cancers in human beings is accompanied by the presence of serum substances which are toxic for paramecia. In testing sera collected from 79 persons (28 with, and 51 without, cancer) upon *P. aurelia,* it was concluded that while human serum frequently contains fractions toxic for paramecia, this toxicity is not closely related to the presence of cancer, hence Roskin's claim is not substantiated.

We are indebted to Bernheimer and Harrison (1940) for their immunologic studies with *Paramecium* in which they investigated the problem of antigenic dissimilarities between morphologically undifferentiated and morphologically differentiated strains. Their study was concerned with the antigen-antibody reactions in the "aurelia" group of

Paramecium. In this group belong *P. aurelia*, *P. caudatum*, and *P. multimicronucleatum* (see p. 15 for specific characters). Fourteen clones of the three species, obtained from different sources, were investigated.

Paramecia from a rich culture were carefully washed several times by centrifugation then finally suspended in 3 or 4 ml. of distilled water. If the ciliates were found to be swimming actively, the suspension containing about 75,000 paramecia was injected in this volume into the peritoneal cavity of a rabbit.[2] Six or seven injections were given at weekly intervals until five rabbits were immunized. Complement of the sera were inactivated at icebox temperature for several months or at 56° C. for 30 minutes. These investigators demonstrated the presence of precipitin, complement-fixing, and immobilizing antibodies in the serum of rabbits immunized with paramecia. A standard technique was developed for the investigation of serologic specificity in *Paramecium* known as the *immobilization reaction*. They reported that when healthy paramecia were mixed with appropriate dilutions of homologous antiserum, disturbances in movement are first observed followed by a gradual decrease in rate of locomotion with dysfunction of organelles. Paramecia finally sink to the bottom of the container and die if allowed to remain in contact with the antiserum.

Their standard method for demonstrating the immobilization reaction is given as follows:

The serum, previously heated at 56°C. for 30 minutes, is diluted with measured volumes of a synthetic medium having the following composition:

$NaHCO_3$	20 mg
$MgH_2(CO_3)_2$	30 mg
$MgCl_2$	16 mg
$CaSO_4$	71 mg
$CaCl_2$	1 mg
Distilled water	1000 ml

The paramecia[3] to be used for testing are strained through bolting cloth and washed on filter paper with several portions of the same synthetic medium used for diluting serum. They are transferred to a 15 ml centrifuge tube and gently sedimented; the supernate is discarded, and the paramecia are resuspended in another portion of diluent. After a second or third centrifugation, the washed paramecia are suspended in such a volume of diluent as will yield a final concentration of approximately 500 paramecia per milliliter. In such a condition, the paramecia remain actively swimming for several days. Series of serum-dilutions are set up in rectangular plates of pyrex glass, each plate containing nine concavities, and suitable therefore, for nine combinations of antigen and serum. Two-tenth milliliter portions of serum-dilutions are placed in the concavities by means of 0.2 ml Kahn serological pipettes. The paramecia are added to the serum-dilutions in the concavities by means of a micro-pipette of such diameter as to deliver drops of approximately 0.02 ml in volume. Such drops contain, as a rule, between five and twenty paramecia. (It is unnecessary to use the same number of paramecia in each 0.2 ml portion of serum-dilution. The immobilizing titer[4] is found to be the same regardless of whether one or twenty paramecia are exposed to the action of 0.2 ml of serum-dilu-

---

[2] In the paramecia-suspensions for injection, an investigation was undertaken to determine whether bacteria were present in sufficient quantity to stimulate the production of antibacterial bodies. These investigators concluded that the sera did not contain antibacterial bodies detectable by agglutination in serum dilutions of 1:20 or greater, hence would not interfere with the reactions on paramecia.

[3] The paramecia to be used are obtained from a one- to two-week-old culture of lettuce infusion. The paramecia are simultaneously concentrated and washed of bacteria by sedimentation through a loose cotton plug in the tapered end of a 15 ml. centrifuge tube at a slow rate of centrifugation. In each test is included a control of spring water or normal serum.

[4] Titer is the greatest dilution of antibody which gives the immobilization effect.

Table 33

REACTIONS OF *P. multimicronucleatum*, *P. aurelia*, AND *P. caudatum* TO INACTIVATED HOMOLOGOUS ANTISERA AND THE REACTIONS OF VARIOUS STOCKS OF EACH SPECIES TO ANTISERA OF ONE STOCK OF EACH SPECIES (BERNHEIMER AND HARRISON)

	ANTI-P. MULTIMICRONUCLEATUM (STOCK A)												ANTI-P. AURELIA (STOCK N)												ANTI-P. CAUDATUM (STOCK A)											
	Undiluted	1:3	1:6	1:12.5	1:25	1:50	1:100	1:200	1:400	1:800	1:1600	Control	Undiluted	1:3	1:6	1:12.5	1:25	1:50	1:100	1:200	1:400	1:800	1:1600	Control	Undiluted	1:3	1:6	1:12.5	1:25	1:50	1:100	1:200	1:400	1:800	1:1600	Control
P. multimicronucleatum (stock A)	+	+	+	+	+	+	+	+	+	+	−	−	+	+	−	−	−	−	−	−	−	−	−	−	+	+	−	−	−	−	−	−	−	−	−	−
P. aurelia (stock N)	+	+	+	−	−	−	−	−	−	−	−	−	+	+	+	+	+	+	+	+	+	−	−	−	+	+	−	−	−	−	−	−	−	−	−	−
P. caudatum (stock A)	+	+	+	−	−	−	−	−	−	−	−	−	+	+	−	−	−	−	−	−	−	−	−	−	+	+	+	+	+	+	+	+	+	+	−	−
P. multimicronucleatum (stock B)	+	+	+	−	−	−	−	−	−	−	−	−																								
P. multimicronucleatum (stock D)	+	+	+	+	+	+	+	−	−	−	−	−																								
P. multimicronucleatum (stock E)	+	+	+	−	−	−	−	−	−	−	−	−																								
P. aurelia (stock V)													+	+	−	−	−	−	−	−	−	−	−	−												
P. aurelia (stock M)													+	+	+	+	+	+	+	+	+	−	−	−												
P. aurelia (stock B)													+	+	+	+	+	+	+	+	−	−	−	−												
P. aurelia (stock S)													+	+	−	−	−	−	−	−	−	−	−	−												
P. caudatum (stock B)																									+	+	−	−	−	−	−	−	−	−	−	−
P. caudatum (stock C)																									+	+	−	−	−	−	−	−	−	−	−	−
P. caudatum (stock D)																									+	−	−	−	−	−	−	−	−	−	−	−
P. caudatum (stock E)																									+	+	−	−	−	−	−	−	−	−	−	−

Plus indicates immobilization in two hours at 28°C; minus indicates motility. The effects of the lower serum-dilutions (up to 1:3, and in one serum, up to 1:6) are non-specific, but are included for purposes of completeness.

382

tion.) After two hours in a 28°C incubator (or at room temperature), the mixtures of paramecia and serum-dilutions are placed on the stage of a wide-field binocular microscope, and examined for the presence or absence of locomotion. A control drop containing only paramecia and diluent (no serum) is included with each set of tests.

Using this immobilization reaction, Bernheimer and Harrison reported that there is evidence suggesting the existence of two or more antigenic types in several different stocks of *P. multimicronucleatum, P. aurelia,* and *P. caudatum* (Table 33).

It is noteworthy to report that the reaction discloses well-defined species differences. Thus *P. multimicronucleatum* are immobilized by antimultimicronucleatum serum but *P. aurelia* and *P. caudatum* are not immobilized by this serum. Similarly, analagous specificity is exhibited by anticaudatum serum and by antiaurelia serum.

Later Bernheimer and Harrison (1941) reported investigations upon antigenic differentiation among 15 strains of *P. aurelia.* Fifteen rabbits were immunized against 15 strains of washed *P. aurelia.* The 15 sera were tested against the 15 strains in all possible combinations and against three other strains for which antisera had not been prepared. Table 34 shows the dif-

Table 34

SELECTED PROTOCOLS ILLUSTRATING IMMOBILIZING EFFECTS OF ANTISERA
IN *P. aurelia* (BERNHEIMER AND HARRISON)

| STRAIN | ANTISERUM | DILUTION OF ANTISERUM | | | | | | | | Diluent control | #1 CONCLUSION CONCERNING ACTION OF ANTI-SERUM UPON STRAIN TESTED |
		1:16	1:32	1:64	1:128	1:256	1:512	1:1024	1:2048		
G	N	—	—	—	—	—	—	—	—	—	No effect
K	B	r	r	—	—	—	—	—	—	—	Locomotion distinctly retarded by 1:16 dilution or greater
36	N	n/2	n/2	n/2	n/2	—	—	—	—	—	Approximately 50% immobilized up to and including 1:128 dilution
R	N	n—2	n—1	n—2	n—3	n—2	—	—	—	—	More than 50% immobilized up to and including 1:256 dilution
H	36	n	n	n	—	—	—	—	—	—	All immobilized up to and including 1:64 dilution
T	N	n	n	n	n	n	n	—	—	—	All immobilized up to and including 1:512 dilution

Minus indicates locomotion was unaffected; r indicates locomotion was distinctly retarded; n/2 indicates approximately half of the paramecia were immobilized; n—1, n—2 and n—3 indicate all but one, two and three paramecia were respectively immobilized; n indicates all the paramecia were immobilized.

ferent kinds of reactions observed in the 270 combinations of *P. aurelia* and antisera while the results of the combinations are summarized in Table 35. Their study showed that in some of the combinations, all the paramecia used were immobilized, yet in other combinations only part were immobilized while in still others none was immobilized. The work reveals that *P. aurelia* is antigenically heterogenous. Of the 18 strains of *P. aurelia*, 12 fall into three groups designated A, B, and C in the table. The remaining six strains placed in group D differ not only from those of group A, B, and C but also from each other, indicating that group D is heterogenous. The possibility remains that the complete absence of antigen reported by Bernheimer and Harrison might be a quantitative difference rather than complete absence of the antigen or antigens of the various groups.

The investigation also shows that some paramecia are more sensitive as test-antigens in the immobilization reaction than are others and that certain strains are very sensitive to immobilization by many heterologous antisera. Noted also was the frequent lack of reciprocal cross-reactivity. Bernheimer and Harrison reported the existence of some degree of parallelism between the geographic source of *P. aurelia* and the serologic group. They also pointed out the relationship existing between the serologic group and the mating variety.

Sonneborn (1943 a, b) reported the acquired immunity to a specific antibody and its development and inheritance in *P. aurelia*. He injected into rabbits a race known to be entirely pure (homozygous) for all its genes. The antibody thus formed in the rabbits' blood was shown to be specifically active on the race of paramecia even when the antiserum was diluted 1:4000; paramecia were quickly paralyzed and finally killed in concentrations greater than 1:1000. However, some paramecia that were observed to be temporarily resistant, were subjected to the antiserum and found to acquire complete immunity to it. Indeed, they were no longer affected by undiluted antiserum. As shown by Sonneborn, this induced immunity was inherited during reproduction by fission in the culture media entirely free from antiserum. In some lines, the immunity was quickly lost in approximately one week in passing through 25 to 30 generations. In other lines the immunity lasted for longer periods; in some cases it lasted for 75 days in passing through more than 279 generations.

This hereditary immunity in all immune cultures of *Paramecium* was quickly lost after fertilization regardless of the combinations brought together. When immune cultures underwent fertilization in dilute antiserum culture media, many lost their immunity at once, but others retained it for variable periods of time.

It is evident that this demonstration of acquired immunity by Sonneborn is not an ordinary persistent mutation but only temporarily inherited for a time during vegetative reproduction. It is a type of inheritance that was extensively studied by its discoverer Jollos (1921) under the name of *"Dauermodifikationen"* or long-lasting modifications. He produced resistant animals which in turn transmitted the resistance to their progeny by binary fission. This resistance lasted from several weeks to two months before it disappeared. After exposing paramecia of a sensitive clone to antiserum, he selected the few most resistant animals. These were cultured in the absence of antiserum and the progeny then exposed to a still greater concentration of antiserum. Using this method, lines of paramecia were ob-

Table 35

Results of Testing 15 Antisera Against 18 Strains of *Paramecium aurelia* in All Possible Combinations. Included in Table Are the Geographic Sources of the Strains, Their Mating Variety and the Suggested Serologic Groups (Bernheimer and Harrison)

STRAINS	SUGGESTED ANTIGENIC GROUPING	VARIETY (SEE TEXT)	ORIGINAL SOURCE	36	H	I	G	A	L	M	Y	C	V	X	I	B	T	N
N	A	1	Woodstock, Md.	—	—	—	—	—	—	—	—	—	—	—	16	64p	32	512p
T	A	1	Woodstock, Md.	—	—	—	—	—	—	—	—	—	—	—	16	64	32	512
B	A	1	Woodstock, Md.	—	—	—	—	—	—	—	—	—	—	—	16	64	64p	512h
1	A	1	Stanford, Cal.	—	—	—	—	—	—	—	—	—	—	r	128p	16p	r	32
R	A	1	Baltimore, Md.	—	—	r	—	—	—	—	—	—	—	16p	—	32	16p	256p
J	A	1	Baltimore, Md.	—	—	—	—	—	—	r	—	—	—	—	—	32p	32p	512p
X	B	2	Woodstock, Md.	—	r	r	r	r	r	—	r	r	64	256	r	—	—	—
V	B	2	Woodstock, Md.	—	r	—	r	r	16	—	r	r	128	256	r	r	—	—
C	C	2	Baltimore, Md.	128	128p	r	64p	r	1024	1024p	1024p	256	—	32	512p	64p	r	512
Y	C	3	Cockeysville, Md.	512p	64p	16	32p	r	1024	1024p	>2048p	256p	—	32	>2048p	64p	16p	1024p
M	C	3	Baltimore, Md.	512p	128p	r	64	r	1024	1024p	>2048p	512	—	32	>2048p	32p	16p	1024p
K	C	2	Baltimore, Md.	128	64p	256h	512h	128h	1024h	256	64h	128h	—	32h	512h	r	—	512h
L	D	2	Staunton, Va.	—	—	512	—	16	256	—	—	—	—	—	—	—	—	—
A	D	2	Strickersville, N. Y.	128p	32h	512	512h	512	—	—	—	r	r	—	—	—	—	—
G	D	2	Pineburst, N. C.	128p	16	512h	1024p	256h	—	—	—	—	r	16	—	—	—	—
I	D	2	Buffalo, N. Y.	64h	—	512	r	16	16	—	—	r	—	r	r	—	—	—
H	D	2	Halethorpe, Md.	64	64	128	64h	256	—	—	—	r	128	128	—	—	—	32p
36	D	2	Hamden, Conn.	16	32p	—	—	1024h	—	512p	>2048h	—	r	16	—	—	—	128h

tained which proved to be highly resistant to antiserum. According to Jollos the resistance, because of its impermanence, was believed to be due to modification of the cytoplasm and not nuclei. On the other hand, Harrison and Fowler (1945 a) report no such variation in extensive numbers of observations over a period of two years but this is questioned by Sonneborn (1947).

McClung (1943, 1944) experimented with five variations in the preparation of antigen for the production of immune sera against *Paramecium aurelia*. The most effective technique appeared to be the one in which six intravenous injections of the supernatant fluid were given two or three times a week. This supernatant fluid was obtained by centrifuging a culture which had been mechanically agitated. Before injection, the paramecia were broken into small fragments by drawing the suspension into a syringe fitted with a fine bore (#22 gauge) needle and then expelling the contents for a minimum of 25 times. This material was then centrifuged for one to two hours at 3500 r.p.m. and the supernatant fluid used for injection. Microscopic examination of this supernatant material revealed many separate cilia but relatively few if any intact cells. McClung was able to produce with regularity, sera giving a titer of approximately 1–1000; in some instances the value was as high as 1–4000. Evidence is given of a fractionation of antigens (in the supernatant material) comparable to the separation of somatic and flagellar antigens of the bacteria.

Harrison and Fowler (1945 a) and Harrison (1947) sought to develop and characterize derivative clones from individuals that were not paralyzed in the immobilization reaction when the paramecia were subjected to specifically related antiserum. By repeated selection, they found that serum-insensitive variants in serum-sensitive clones could give rise to clones of greater, lesser, or the same sensitivity as the original parent clone. They reported the establishment of one new clone of greater serum-sensitivity than its original parent clone as a result of continued selection of serum-sensitive individuals in successive clones. Throughout one year, this clone repeatedly showed some variation in sensitivity to antiserum for its original parent clone but was found to be usually of higher (never lower) sensitivity than the parent one.

Kimball (1947) found that when a line of *P. aurelia* was exposed for 24 hours to sublethal concentrations of antiserum or to the proteolytic enzyme trypsin, a high proportion of paramecia were resistant to the action of concentrated antiserum. He reported that the descendants of these resistant paramecia retained the resistance for periods ranging from one to more than 15 fissions; but when the lines were carried long enough, the resistance was finally lost without the occurrence of conjugation or autogamy. van Wagtendonk (1949) also investigated antigenic transformation in *P. aurelia* under the influence of proteolytic enzymes.

In a ciliate closely related to *Paramecium*, namely *Tetrahymena geleii* which was discovered and described by Furgason (1940), Tanzer (1941) studied the action of normal and immune serum. Tanzer, like Robertson (1939 a, b) found rather close relationship between the strains of ciliates from immobilization tests in homologous and heterologous antisera.

Kidder and Dewey (1945 b) examined six strains of the ciliate *Tetrahymena geleii* and two strains of *T. vorax* and reported their biochemical differences and similarities. The biochemical reactions refer to the ability or lack of ability of the ciliates to ferment certain carbohydrates, to synthesize certain amino acids, and to yield growth factors (Kidder, 1941 a). They observed that certain biochemical variations had occurred in their strains.

The same number of strains of bacteria-free *T. geleii* and *T. vorax* were examined by cross-adsorption techniques for antigenic relationships (Kidder, Stuart, McGann, and Dewey, 1945). They found that five of the strains of *T. geleii* possessed qualitatively identical antigens but that all differ quantitatively (amount, concentration, or distribution of antigenic materials). One strain possessed all the antigens of the other five strains plus additional antigens. Also noted were the marked antigenic differences found between *T. geleii* and *T. vorax*. It is worth noting that their serologic studies confirmed their earlier biochemical studies in showing that a variation had occurred between a parent strain and a daughter strain of *T. geleii*. They believe their evidence supports the hypothesis that one strain has varied by gaining antigens.

B. Serology in Respect to the Mating Reaction, Conjugation, and Inheritance

As mentioned earlier, Bernheimer and Harrison (1941) discovered a relationship existing between the serologic groups under investigation with the mating variety in *P. aurelia*. This relationship is shown in Table 35 in which all the strains belonging to group A also belong to mating variety 1 while both strains of group B belong to mating variety 2. Group C, however, is shown to contain two strains of mating variety 2. The six strains in group D belong to mating variety 2. It is of interest that Metz and Fusco (1948) were able to inhibit the mating reaction in *P. aurelia* with antiserum.

A serologic study of conjugation in

P. bursaria was undertaken by Harrison and Fowler (1945 b, 1946) in an attempt to compare the antigenic character of the daughters of exconjugants with the antigenic character of the clones prior to conjugation. Their experiments showed that the antigenic character in the cultures subsequent to conjugation was in many instances strikingly different from the individuals of the culture prior to conjugation. These investigators mated a typically green strain of *P. bursaria* with a colorless (zoöchlorellae-free) one. In addition to other than morphologic differences, the two strains were shown to be antigenically distinct. Each reacted quickly and extensively in homologous antiserum; neither reacted in antiserum for the other unless it had undergone conjugation with the other. The development of a gelatinous precipitate in and about the ciliary zone and the occurrence of ciliary tangling[5] with loss of ciliary activity was taken as positive evidence of antigen-antibody reaction when this phenomenon appeared within two hours after the paramecia were placed in antiserum. An examination of the exconju-

[5] This reaction of individual paramecia in the presence of effective antiserum at room temperature is given as follows: (1) The distal ends of a few cilia become enlarged as though they had picked up some amorphous substance on their surfaces. (2) Two or more altered cilia become attached at the enlargements and form tripod-like arrangements which are still capable of erratic movements. At first only one or two small areas of the surface of the animal are involved but later the reactions generally involve the whole surface of the animal. (3) A gelatinous precipitate is collected in the interciliary spaces; often in extensive reactions a part of the precipitate is extruded from the ciliary zone and is swept backward to form a "sea-anchor" trailing behind the slowly moving animal. Trichocyst extrusion is seen late in the observations of extensive reactions.

This reaction may be similar to the capsular swelling of encapsulated bacteria (*Neufeld quellung* reaction). An attempt should be made to isolate or characterize the substance found on the cilia.

gants revealed that within an hour after separation, 95 per cent of them contained some antigenic substance common to their mates which was not previously present within themselves. This new antigenic character appeared to persist for at least one month.

These investigators observed a sharp alteration in antigenic composition during the conjugation process of *P. bursaria* (Fig. 132). This change in antigenic character is not apparent early in the period of conjugation but, usually, only after the conjugants have been in fast union for periods approximating two thirds of the normal conjugation period.[6] They report that a detailed analysis of the results obtained with 64 conjugating pairs revealed the following: in 51 cases (80%) both individuals of the pairs had gained a new antigenic component from their mates while in nine cases (14%), one of the individuals had gained a new antigenic component while its mate had not. Neither of the individuals had gained new antigenic components in four cases (6%).

Many exconjugants showed that variable amounts of new antigenic components of cytoplasm were present in the majority of individuals immediately after separation from the conjugation process.

[6] Harrison and Fowler do not deny the possibility that the change in antigenic composition may commence with the beginning of conjugation and progress steadily through most of the union to completion toward the end of conjugation. The author (Wichterman, 1946 f, 1948 a) investigated the time relationships of nuclear phenomena during the conjugation process of green and colorless *P. bursaria*. It was found that paramecia remained joined for 24 hours at 25° C. and that transfer of pronuclei occurred 16–18 hours after the onset of the process. For a similar 24-hour period, Harrison and Fowler reported that changes in the antigenic configuration were completed within 16–18 hours which is coincidental with pronuclear transfer.

An important question that arises from the work of Harrison and Fowler is accounting for the method of *antigenic exchange* between the conjugants. These investigators list the following explanations for the transfer of antigenic components: (1) That an exudate from the individual which was sensitive to the serum prior to conjugation flowed over its mate and became attached nonspecifically to the tips of its cilia and interspersed within its ciliary zone. (2) That a chemical chain reaction began at the interface of attached paramecia and proceeded systematically through each molecule of protoplasm in turn until all reactive elements were affected. (3) That the protoplasmic elements involved in the immunity reactions passed through a rapid antigenic reorganization in response to, and under the influence of, the genes received in the newly gained pronuclear element. (4) That a transfer in volume of cytoplasm occurred some time during the conjugation process.

It is to be regretted that cytologists dealing with conjugation in *Paramecium* and other ciliates have been concerned nearly exclusively with nuclear phenomena. Although Hickson (1902) called attention to the fact that the question of cytoplasmic exchange between conjugants needs investigation, relatively little work has been done on the problem or even on the behavior of cytosomal structures during the process. Recently the author (Wichterman, 1937, 1937 b, c, 1939, 1940) investigated this facet of the mating process and was unable to observe an exchange of cytoplasm in the race of *P. caudatum* under investigation. The study led to the discovery of a new sexual process called *cytogamy* and revealed that the contiguous membranes of the joined individuals remained intact during the entire process and that there was no

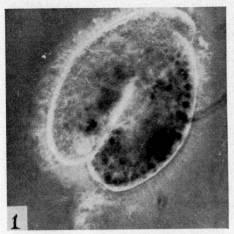 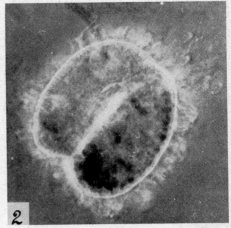

Fig. 132. Both figures are unaltered photomicrographs. The equipment used included a 4-mm objective with phase plate unit 0.25A + 0.25λ and a 5 × ocular in the microscope, and a Leica 35 mm camera. The factors of magnification and enlargements were uniform for both figures. (1) Twenty-four-hour (at 17° C.) conjugants of *P. bursaria* no. 225W (colorless), on the left, and no. 213 (green) on the right, in antiserum for strain no. 255W. Note ciliary tangling and light dispersion about the periphery of no. 255W and uninvolved cilia on no. 213. (2) Forty-eight-hour (at 17° C.) conjugants of *P. bursaria* no. 255W (colorless), on the left, and no. 213 (green), on the right, in antiserum for strain no. 255W. Note ciliary tangling and light dispersion about the periphery of both animals. (Harrison and Fowler)

transfer of pronuclei or cytoplasm. Instead, the pronuclei fused and formed a synkaryon each within the confines of a single individual. The study led him to question the classical interpretation of pronuclear transfer (as well as cytoplasmic transfer) in the conjugation process of *Paramecium* and other ciliates. Since then, a rather lively interest has developed in the problems of exchange of pronuclei and cytoplasm during the conjugation process. On the general problem, papers have appeared giving cytological, genetic, and serologic evidence of the transfer of pronuclei or cytoplasm or both during conjugation as well as proof of the existence of cytogamy. Chen (1940 a), using fixed and stained preparations of *P. bursaria*, presents cytological evidence of pronuclear transfer. In addition, he reports some cases in which the two pronuclei remain in the same conjugant and fuse to form a synkaryon (cytogamy). This suggests the persistence of the contiguous membranes between the members thus preventing cytoplasmic exchange. Sonneborn (1943 c, 1944) presents genetic evidence to show that there is nuclear and cytoplasmic transfer in *P. aurelia* during conjugation, and he also shows genetic evidence for cytogamy. He reports that the transfer of cytoplasm is crudely measured by the extent of the time interval between separation of conjugants at their anterior ends and separation at their paroral cones. If the interval is less than three and a half minutes, exchange of cytoplasm is not detected, regardless of what races were crossed, but when the interval is 20 minutes or more, "cytoplasmic factor" is invariably transferred. Further, he reports that exchange of cytoplasm at conjugation never occurs in crosses in certain races (although nuclei are exchanged in these crosses) but does occur in others. Concerning nuclear exchange, Sonneborn (1941) reported that there is a definite temperature factor involved in cross-fertilization and some indication that calcium increases the frequency of it and that sodium decreases it.

In presenting serologic evidence for the transfer of antigenic components between conjugants of *P. bursaria*, Harrison and Fowler are inclined to favor the view of cytoplasmic interchange as the mechanism by which this occurs.

The author is tempted to offer still another explanation of the transfer of antigenic components, for it occurs to him that a visible transfer of cytoplasm need not occur to account for antigenic transfer. It should be remembered that the joining in conjugation of two paramecia, as in other ciliates, results in a dedifferentiation of specialized cytosomal structures. The cilia and trichocysts dedifferentiate and disappear from the regions where the animals are joined, leaving only the thin membranes of each conjugant—one tightly applied to the other. These membranes may become dedifferentiated to the point where they may break down (permitting cytoplasmic interchange) or they may persist. Even though they persist, why is it not possible for the antigenic component, say the hapten,[7] to pass through the thin membranes just as one component of a solution passes through a semipermeable membrane in osmosis?

In the study of a stock of *P. aurelia* in reaction to antiserum, Sonneborn

(1943 a, b) obtained apparent *"Dauermodifikationen"* not unlike those reported earlier by Jollos. Approximately 20 per cent of the paramecia were normally resistant to homologous antiserum, but the resistance disappeared in a very few fissions. When such resistant animals, however, were selected from a serum test so that samples of their progeny could be tested from time to time to determine if they were resistant, the specimens demonstrated resistance which persisted up to 279 fissions before the resistance disappeared. It is seen from this work that a single exposure to antiserum appeared to transform a very transient resistance—like the work of Jollos—into a long lasting one, and Sonneborn concluded that these changes could not be due to an ordinary persistent mutation.

Sonneborn (1947 b, 1948, 1950 e), Sonneborn and Lesuer (1948) and others have studied the determination, inheritance and induced mutations of antigenic characters in certain races of *Paramecium aurelia*. As set forth by Sonneborn (Table 36), a single cell of serotype A in a given race (⅍51) gives rise to five distinctly different serotypes,[8] namely B, C, D, E and G. When each of these types of culture is injected into a different rabbit, a serum is obtained later from each. After the serum is heated and diluted, its action is tested by placing some paramecia in a drop of it. A given serum quickly paralyzes paramecia of the strain injected into the rabbit from which it is obtained, that is, a corresponding (homologous) serotype. The serum has no effect on paramecia of the remaining strains. Thus each of the six strains is identifiable by the serum which paralyzes it and the six sera can be used to identify any unknown culture of *P. aurelia*. According to Sonneborn, only one specific sub-

---

[7] Hapten is that portion of a molecule which has the specificities of the whole antigen but it itself is not capable of inciting antibody formation. It will bind antibody against the whole antigen.

[8] Serotypes = antigenic types.

Table 36

REACTIONS TO DIVERSE ANTISERA BY DIFFERENT HEREDITARY SEROTYPES OF PARAMECIA (A, B, C, D, E, and G) ALL DERIVED FROM AN ANCESTOR OF SEROTYPE A. EACH DILUTED ANTISERUM PARALYZES THE HOMOLOGOUS SEROTYPE, AS INDICATED BY +; BUT FAILS TO PARALYZE THE OTHER SEROTYPES, AS INDICATED BY — (SONNEBORN)

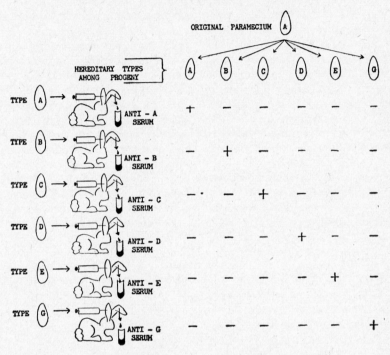

stance is involved in the paralysis reaction of most of the serotypes and indicates that the six specific substances in molecular form and composition are probably very similar. This specificity of substance to the paralysis reaction indicates that it is due to interaction between substances called antigens in the paramecia and specific complementary substances called antibodies which are formed in the rabbit in response to their presence. The six strains or serotypes are thus antigenically different. When animals of any of these types are exposed to its immobilizing antiserum for a short time, there follows a transformation of the antigenic type. These changes, from one type to another, yield in the trans-

formed animals cultures which remain permanently true to their new type through fission. Since any one of the six serotypes can transform to any of several of the others, it appears that transformation of serotype involves the replacement of one specific substance by another similar one which had not been present previously even in a detectible amount. Thus the six substances are mutually exclusive. When one is present, the others are absent.

It is interesting to note that in another race of *P. aurelia* (⧷29) which was collected 650 miles from race ⧷51, six serotypes were identified but two of these serotypes, namely F and H were not found in race ⧷51. It follows that only the four serotypes A, B, C, and D are common to both races.

Analysis of the genetic system involved in the control of these antigenic traits followed the plan used earlier by Sonneborn (1943 b); the breeding analysis reveals that the antigenic types are genically alike and differ only in some cytoplasmic character. In other words, the determination as to which antigen shall be primary is cytoplasmic. The opinion is expressed that the antigenic types differ only in cytoplasmic and not genic constitution and that when the antigenic traits are cytoplasmically determined, antibody induces permanent changes. It is concluded that the hereditary transformation of one antigenic type into another by means of exposure to specific antibodies against the primary antigen appears to be an example of a permanent, directed mutation of a particular trait.

Among other things, Kimball (1947, 1947 a, 1948) attempted to determine if the antiserum acts by selecting naturally occurring variants, as proposed by Harrison and Fowler, or if it induces changes such as reported by Jollos and Sonneborn. On the basis of his experiments, Kimball favored the interpretation as given by the latter workers and, like Sonneborn, believed that these changes which regularly disappeared during vegetative reproduction indicated that Mendelian recombinations of mutant genes could not be involved. Kimball is of the opinion that the resistance is produced by loss or inactivation of the antigens concerned in immobilization.

Masugi (1927) demonstrated that when in sufficient quantity, paramecia are able to remove the paralyzing antibodies from an antiserum thereby rendering it ineffective against any fresh paramecia that were added. With this in mind, Beale (1948) studied the process of transformation of antigenic type in *P. aurelia* by attempting to measure the antigens and antibodies involved in the reaction in which the ability to absorb antibody from a standard solution was used as the measure of the antigen in paramecium. The results of Beale's study lend support to the contention of Sonneborn that antigen type (*P. aurelia,* variety 4), depends upon competition between plasmagenes and that the plasmagenes may in reality be the antigens.

At the present time, serologic studies upon paramecium are currently active and we may look forward to important discoveries in the very near future. Newer work may place the whole subject in a different light as suggested by Sonneborn's (1950 a) most recent account. He has summarized his views and the main facts as follows:

(1) Each homozygous stock is characterized by a spectrum of possible serotypes. (2) Different stocks are characterized by different spectra of possible serotypes. (3) The specific immobilization substance of each serotype more or less completely excludes the presence or functioning (in the immobilization reaction) of any other specific immobilization substance. (4) Environmental conditions (including homologous antiserum) bring about directed transformations of serotype. (5) Many, possibly all, of the transformations are reversible, indicating that no potencies are lost in the process of transformation. (6) The differences among the serotypes in any one stock are due not to genic, but to cytoplasmic, differences which are inherited under the standard cultural conditions. (7) The specific differences between corresponding immobilization substances in different stocks are determined by allelic genes.

Additional information bearing upon genetic studies and serology is found in Chapter 10. The most recent review of serologic techniques and special methods in the genetics of antigenic types in

Paramecium is to be found in Sonneborn's (1950 b) "Methods in the general biology and genetics of *P. aurelia.*" This valuable paper should be consulted by those who are interested in serologic studies of *Paramecium*. In this active and fertile field of research, the reader is referred to the additional studies of Sonneborn (1949, 1949 a, 1950 a, c, d, e, f, 1951 a, b).

Chapter 13

Organisms Living In and Upon Paramecium

A. **Paramecium as a Source of Food for Metazoan Animals**
B. **Parasites and Enemies of Paramecium**
  1. Protozoan Parasites (Suctoria, Ciliata, Sarcodina)
  2. Parasitic Bacteria (Schizomycetes) and Related Microorganisms
  3. Fungi (Sphaerita)
C. **Symbiosis in Paramecium**

A. Paramecium as a Source of Food for Metazoan Animals

Many extremely young fishes are dependent upon *Paramecium* and other ciliates for food. However, it is a fact that comparatively little work has been done in a detailed study of this ecologic relationship. Because of their wide distribution and the speed of reproduction, species of *Paramecium* must play a vital role in this regard. It is a comparatively simple matter to obtain a rich, pure-line culture of a given species of *Paramecium* for experimental purposes. The green species—*P. bursaria*—should prove valuable in such feeding experiments because these animals contain the alga, *Chlorella*. Also, this species of *Paramecium* can be grown easily in a bacteria-free, synthetic medium (Loefer, 1936 b).

B. Parasites and Enemies of Paramecium

1. **Protozoan Parasites (Suctoria, Ciliata, Sarcodina).**

Species of *Paramecium* as well as other ciliates are frequently parasitized with microörganisms. To comprehend how widespread parasitism is among the Protozoa, one is referred to the monographic treatment of the subject by Kirby (1941). In discussing associations among organisms, it is occasionally difficult to draw the line between symbionts, commensals, and parasites. Symbionts (= symbiotes) are organisms living with a host in such a mutualistic relationship that both are benefited by the association. Commensals are organisms which live without benefit or injury to the host but to their own advantage. Parasites are organisms which cause injury in one form or another to the host while they benefit themselves.

SUCTORIA. The presence of certain living organisms in the cytoplasm of *Paramecium* and other ciliates was erroneously interpreted by Stein (1854) and formed the basis for his *Acineta* theory of embryo development. This theory, like Ehrenberg's famous Polygastric theory, fortunately enjoyed only a brief but nonetheless spectacular existence in the history of protozoölogy.

It is generally believed that Focke (1845) while studying *P. bursaria* was

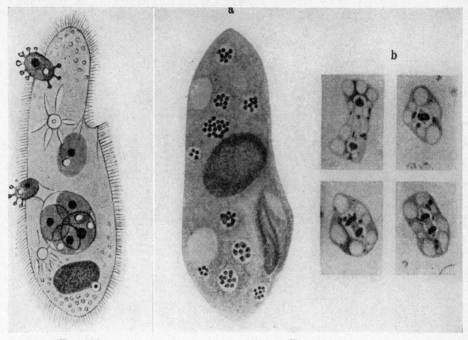

FIG. 133   FIG. 134

FIG. 133. *Paramecium caudatum* parasitized with young stages of the suctorian, *Sphaerophrya pusilla*. Note four specimens in large brood pouch with one specimen leaving it as a swarmer. (Bütschli)

FIG. 134. Paramecium parasitized with *Amoeba endophaga* (a). (b) Division stages of the ameba. (Fortner, after Kalmus)

the first to observe motile organisms in the cytoplasm of these ciliates. Subsequently, others made similar observations in *Paramecium* and in a number of different ciliates.[1] In reality, these small, motile, ciliated forms were developmental stages of *Suctoria (Acineta)*. In this class, cilia are present only in young individuals which are free-swimming. Later in development, the cilia are lost and as the suctorian becomes mature, it develops tentacles and becomes attached. In this group of Protozoa, internal or endogenous budding is characteristic of many species. Such buds are provided with cilia which, when liberated from the parent, may become parasitic in the cytoplasm of *Paramecium* or another ciliate. Stein, unaware of the life-cycle of the *Suctoria*, believed the ciliated young of these forms to be motile embryos. He claimed that these small, ciliated individuals were embryonic stages of vorticellids. Although he modified the theory later, Stein refused to admit that acinetids are independent organisms. It

[1] Claparède and Lachmann (1858–60) reported the presence of small droplets of macronuclear substance in the cytoplasm. These droplets were interpreted by them as early stages of "embryo formation" in a ciliate which they named "*P. putrinum.*" They claimed these stages of embryonic development differed from those described for *P. bursaria* and believed that this feature, although erroneous, was an important species character.

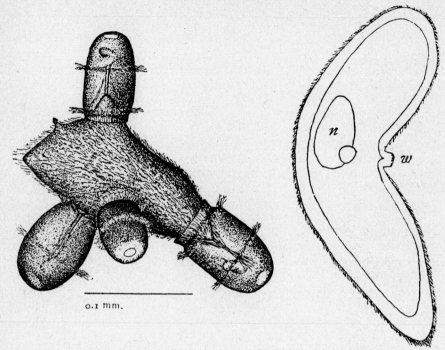

o.1 mm.

FIG. 135. (*Left*) *Paramecium* attacked by four didinia. Under such conditions the paramecium is usually torn in pieces and each *Didinium* obtains a portion. Sometimes one *Didinium* gets the entire Paramecium, forcing the others off during the process of swallowing. (*Right*) *Paramecium* which escaped after having been attacked by a *Didinium* showing the wound, *w*, the absence of cilia around the wound and a marked flexure in the body. (Mast)

was his belief that certain ciliates in their development must pass through phases resembling the *Acineta*—that podophryids represented developmental stages of *Paramecium*.

Stein's *Acineta* theory was successfully attacked by Cienkowsky (1855), Lachmann (1856), Balbiani (1860), Claparède and Lachmann (1860–61), Bütschli (1876), and Engelmann (1876). Balbiani correctly identified the so-called "embryos" in *Paramecium* as the young ciliated parasites belonging to the suctorian genus, *Sphaerophrya*. Experimentally, he was able to spread an infection in a *Paramecium* culture by the inoculation of several infected paramecia. We are indebted to Metch-

nikoff (1864) for his observations on the life-cycle of the ciliated parasitic stages in regard to entering and leaving host paramecia.

These parasitic stages of the suctorian live in the cytoplasm and their presence, especially when in large numbers, frequently results in the death of the host paramecia (Fig. 133).

CILIATA. The chief enemy of *Paramecium* is a voracious predator called *Didinium nasutum*. This very active, barrel-shaped ciliate possesses an expansible cytostome or mouth at its cone-shaped anterior end. *Didinium* attacks *Paramecium* and ingests the entire body of the latter through the enlarged cytostome—in some cases within a few seconds (Fig. 135). Strong trichites support the mouth of *Didinium* which apparently help the predator to bore through and pierce the soft body of *Paramecium*. Jennings (1931) observed that the searching of *Paramecium* by *Didinium* is one of trial and error. When *Didinium* comes in contact with a solid, the animal seemingly attempts to pierce and swallow it. The didinia even try to pierce the glass and small algae as well as other water plants. Paramecia appear to be excellent prey because they move more slowly than the didinia and also have very soft bodies. Reukauf (1930) reported that one *Didinium* can eat as many as 12 paramecia daily; a rich culture of paramecia can be completely depleted by the presence of these predaceous ciliates. One ordinarily thinks of the trichocysts as defense organelles but as Mast (1909) has demonstrated, the extruded trichocysts merely form a somewhat viscid mass or mat about the *Didinium* and hampers the predator only slightly. Reukauf, however, observed that occasionally the proboscis of *Didinium* is torn out in the encounter. In view of the behavior of *Paramecium* toward *Didinium*, the so-called defense function of the trichocysts seems highly questionable and that the extrusion of these structures represents perhaps a kind of injury reaction.

SARCODINA. Several members of the class *Sarcodina* have been described as being parasites of *Paramecium*. Wetzel (1925) reported temporary parasitism of the heliozoön *Raphidiocystis infestans* upon *Paramecium*. This species appears to be generally predatory upon flagellates and small ciliates. When it attacks *Paramecium* and other larger ciliates, the heliozoön behaves as a parasite (Fig. 136). The pseudopodia of the parasite become attached to various parts of the *Paramecium*, then the at-

tacker lies flat on the body of the host. According to Wetzel, many specimens of *Raphidiocystis* may engulf a single *Paramecium*, then extract cytoplasmic material from it. This association of the heliozoön with *Paramecium* is said to require 20 or 30 minutes or even up to two days. This relationship was discussed by Wetzel, who believed it to be a transitional stage between predatism on the one hand and true parasitism on the other.

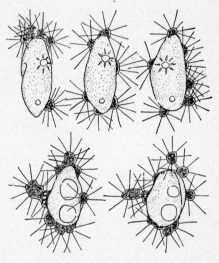

FIG. 136. The heliozoön, *Raphidiocystis infestans* attacking Paramecium. (Wetzel)

Kalmus (1931) reported that Fortner discovered a small ameba with division stages in the endoplasm of *P. caudatum* which the latter named *Ameba endophaga* (Fig. 134). In reality the parasite may be the ameboid thallus stages of the fungus, *Sphaerita* (page 405).

The common laboratory ameba, *Chaos diffluens* (*Amoeba proteus*) and the giant ameba of Roesel, *Chaos chaos* avidly eat paramecia when grown in the laboratory. *C. chaos* may, when starved, form a single food cup containing a hundred or more paramecia. While rich cultures of both these species can be grown in the laboratory exclusively on paramecia, it is believed that in nature many other organisms are used by these amebas as a source of food.

2. Parasitic Bacteria (Schizomycetes) and Related Microörganisms.

Bacterial parasites and closely related forms have been reported from the macronucleus, micronucleus, and cytoplasm of *Paramecium*. These small parasites were originally interpreted as fundamental structures of the protozoön and marked another spectacular but erroneous theory in the history of protozoölogy.

J. Müller (1856), before the Academy of Science of Berlin, discussed *Paramecium aurelia* (=*caudatum*) which was greatly enlarged and filled with dense interlacements of tortuous threads disposed in the form of curls. These threads were found mainly in hypertrophied macronuclei.

Claparède and Lachmann (1857, 1860–61, p. 259) reported the presence of thin, immobile rods in the nucleus (=macronucleus) of *Paramecium* and proposed that this nuclear body may, in some individuals, function as a testis while in other specimens, take on the role of an ovary. It is evident that he believed the parasites to be spermatozoa.

Later Stein (1859–78) observed fine, straight rods in hypertrophied macronuclei of *P. "aurelia"* (=*caudatum*). He agreed with the hypothesis of sexuality as expressed by Claparède and Lachmann but believed that the spermatozoa developed in the micronucleus and then fertilized the macronucleus.

Like Claparède and Lachmann, Balbiani (1861) observed the immobile rods in macronuclei of *P. caudatum* but correctly interpreted them as true parasites. However, he believed them to be parasites of the "ovary" (which was in reality the macronucleus) and that the spermatozoa originated in the testis of the ciliate which he called the nucleolus (=micronucleus). He proposed that *Paramecium* as well as other ciliates were hermaphroditic animals.

Engelmann (1862) found similar parasites in the macronuclei of *P. caudatum* but like Stein, regarded them as spermatozoa. Later (1876) he described microörganisms as being parasitic in the macronucleus of *Stylonychia mytilus*.

Bütschli (1876) observed rods in the macronucleus of *P. aurelia* and regarded them correctly as parasites. He then successfully attacked the hypothesis of sexual reproduction as set forth by the earlier workers and reviewed the subject in his important classification of 1889.

The work of Hafkine (1890 a) marks the beginning of careful and detailed study upon parasitism in *Paramecium*. This investigator named three different nuclear parasites all belonging to the genus *Holospora*. He was able to spread an infection by placing normal paramecia in a medium where he found the contagious agent and also by introducing infected paramecia into cultures of healthy specimens. Hafkine reported that in early stages of the infection, little effect is noted on the host, but a day or so after infection, all nuclear material is absorbed and the paramecia become almost filled with the parasites.

According to Hafkine, *Holospora undulata* and *H. elegans* invade the micronucleus and *H. obtusa* parasitizes the macronucleus (Figs. 137 A and B). Multiplication of *Holospora* is said to occur in two ways. During development, there is a period of rapid transverse division. At the end of this division period, the vegetative stages of the parasites take on the conspicuous and characteristic shape of spiral or straight filaments, depending upon the species. A bud is then formed at one of the ends of the slender parasite. From this bud a new organism develops, which in turn resembles its parent. Because of their method of reproduction, Hafkine believed the *Holospora* to be transitional forms between yeasts and bacteria.

Holospora undulata in the conspicuous, vegetative stage is said to be a spiralled organism which parasitizes the micronucleus. According to Hafkine, in the early stages of its life-cycle, the organism is a small fusiform body. During development, there is a period of rapid, transverse division which is followed by the spiral form so characteristic of the species. This is followed by an increase in the dimensions of the parasites which in turn become transformed into resistant spores. In this stage, *H. undulata* becomes less transparent but more refractive. At times, it becomes separated into three or four parts of different refractivity. Hafkine was able to see but few of the important stages in the life-cycle of this species. *H. elegans* is also a parasite of the micronucleus of *Paramecium* which, in the vegetative stage, is fusiform and more elongated and slender than in the other two species and not spiralled. *H. obtusa* is a parasite of the *macronucleus* which is slender and straight with both ends rounded. This species more recently has been studied by Bozler (1924), Fiveiskaja (1929), and Wichterman (1940).

Metchnikoff (1892), who also studied nuclear parasites of *Paramecium*, confirmed the views of Hafkine that the microörganisms penetrate directly to the macronucleus or micronucleus.

Bozler (1924 a) and Fiveiskaja (1929) studied the macronuclear parasites of *Paramecium* and noted in the host extensive vacuolation, the marked accumulation of fat droplets and excretion granules, as well as the abnormal displacement with loss of trichocysts. Bozler described clouds of quill-like, briskly moving microörganisms in the protoplasm of *Paramecium*. With the progression of the disease, he observed a migration of the micronucleus from its usual place near the macronucleus to the anterior end of the animal. The parasites were usually first found in vacuoles in the endoplasm of *Paramecium*. These appeared to be spindle-shaped forms, measuring from 3–6 microns in length, which rapidly developed in the macronucleus into the larger forms, measuring 20–30 microns in length.

Fiveiskaja, who studied *Holospora obtusa*, reported them as being long, thin, straight or slightly curved rods, measuring from 12–30 microns in length and 0.6–0.8 microns in diameter, showing frequently differential refractivity and stainability. An interesting feature reported by Fiveiskaja (which had been seen also by Bütschli, Balbiani, and Wichterman), was the presence of a dark region—evidently chromatin—in the parasite. She reported the pathologic changes that came about with the onset of death. This included diminution in number of formed food vacuoles, disappearance of mouth, cytopharynx and cytopyge, partial degeneration of cilia, production of large quantities of lipoids and crystals in the cytoplasm suggesting fatty degeneration, cessation of action of contractile vacuoles until eventually the host *Paramecium* is filled with large masses of the parasites resulting in death.

The author has encountered from time to time, different kinds of parasitic microörganisms in species of *Paramecium* which have been studied experimentally in both living and stained preparations (Wichterman, 1940, 1945,

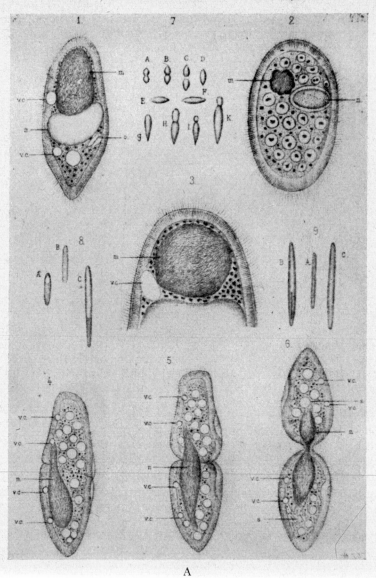

A

FIG. 137. (m) Micronucleus; (n) macronucleus; (v.c.) contractile vacuoles; (s) micro-organisms. (A) (1) Micronucleus (m) of Paramecium greatly enlarged with many micro-organisms; macronucleus (n) free of parasites. (2) Deformed, infected specimen. (3) Anterior end of Paramecium showing a large mass of parasites (*Holospora undulata*) in the greatly distended micronucleus. (4, 5, and 6) Three successive stages in the division of a parasitized Paramecium showing macronucleus (n) harboring the microörganisms.

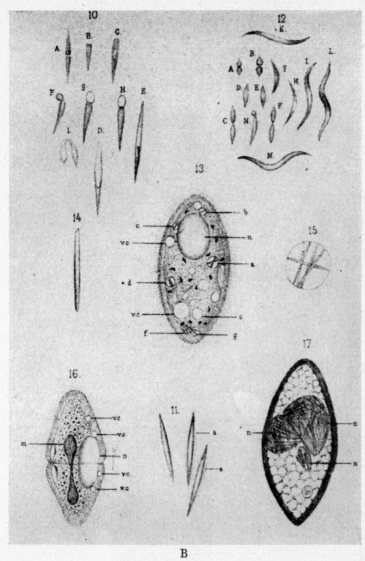

B

Note parasites free in endoplasm in 6. (7, 8 and 9) Different stages in the development and reproduction of *Holospora obtusa*. (B) (10 and 11) Stages in the development of *Holospora elegans*. (12) Stages in the development of *Holospora undulata*. (13) Paramecium showing examples of Holospora obtusa in the food vacuoles (a–g). (14) A specimen of *Holospora* from food vacuole in 13. (15) A food vacuole, greatly enlarged, containing parasites from specimen in 13. (16) Paramecium in division with *Holospora* in micronucleus (m). (17) Paramecium showing three infected macronuclear masses. (Hafkine)

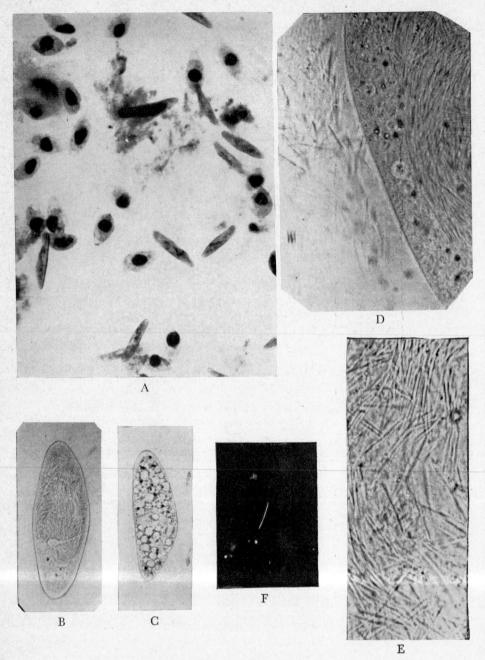

Fig. 138. (*For legend see facing page.*)

1946 d) (Fig. 138). In one such study of *Holospora* (Wichterman, 1940), enormous numbers of the parasites were found to reside and grow in the macronucleus of *P. caudatum* which distended that nuclear body until at times it filled the entire organism. Diseased paramecia were observed to have a very flexible body and fluid endoplasm. Heavily infected individuals moved about very slowly, settled to the bottom of the container and died (Fig. 138 A). The large masses of parasites in the host appeared opaque and hence could be detected easily with the low power of the microscope. In heavily parasitized paramecia, closer study of the macronucleus under high magnification showed the structure to be tightly packed with *Holospora*. It appears that the parasites thrive and grow on the macronuclear substance. Generally, the mass of parasites extended to the trichocyst layer and pressed against the sides of the host which resulted in a spindle-shaped organism. Normal cyclosis is either inhibited or stopped. Upon death of the host, the pellicle is ruptured and the parasites are liberated into the culture fluid. In experimental mass cultures of paramecia, fully 80 per cent of the organisms appeared to be infected.

Stained preparations of early stages of parasitized paramecia showed in many cases, a clear, small, wedge-shaped region in the center of the macronucleus. Its significance has not been ascertained. The micronucleus is not parasitized and is generally pushed to one end of the body. Fully grown, living parasites as seen under the micro-compression chamber appear cylindrical and measure approximately 20 microns in length and 1.3 microns in diameter. Under dark-field illumination they show thick walls and thin, blunt ends (Fig. 138 F). They possess no movement of their own. Stained specimens show a small oval mass of chromatin at one end of the parasite. Division of the parasite appears to be transverse, but smaller forms with pointed ends suggested other stages in the life-cycle and resembled those described and figured by Bütschli (1876) and Dehorne (1920). Attempts to grow the parasites in a paramecium-broth and on agar plates proved unsuccessful although the parasitized paramecia were easily maintained in hay-infusion cultures.

To study fission in infected paramecia, daily isolation cultures were maintained, in some instances for nearly two months. It was found that heavily parasitized specimens failed to divide and died. A moderately infected *Paramecium* may take up to several days in the division process. Frequently large L-shaped monsters were produced which lasted for several days before death. On the other hand, moderately infected paramecia were observed to undergo fission in which one daughter appeared

---

FIG. 138. *Paramecium caudatum* parasitized by the microorganism, *Holospora obtusa*. (A) Diseased, dead, and dying paramecia as found on bottom of culture flask in typical epidemic (low power). The spherical black mass in each organism consists of great numbers of tightly packed parasites. (Photographed immediately upon removal from culture). (B) Heavily parasitized Paramecium showing most of body occupied by *Holospora obtusa* (living specimen). (C) Diseased, vacuolated specimen prior to death. (D) Photograph (high power) of part of living infected Paramecium showing parasites in animal and in culture medium outside body. (E) Photograph (higher power) of living, rod-shaped parasites from animal in D. (F) Photograph (oil imm.) of living parasite (dark-field). Parasite measured 20 microns. (Wichterman)

to receive all or nearly all of the parasites while the remaining daughter appeared free of them. In one such experiment, the heavily infected daughter died while the remaining daughter passed through 42 generations in 56 days.

Paramecia were observed to become infected by ingesting parasites enclosed in food vacuoles.

Petschenko (1911) studied a parasite of *Paramecium* which he named *Drepanospira mülleri*[2] and placed with the *Spirillaceae*. According to this investigator, the vegetative stage of this parasite is long and narrow with two gentle bends or turns in its body. One end of the parasite is rounded while the other end gracefully tapers into a point. He was able to study the entire life-cycle of diseased paramecia in epidemics which lasted for a month. Petschenko reported that endospores of the parasite are ingested by *Paramecium* in the form of small vacuoles which are easily confused with food vacuoles once in the cytoplasm. The endospores then develop in the cytoplasm into curved rods, which form a large mass nearly filling the body of *Paramecium*. The slightly spirally twisted parasite possesses a nucleated region near one (anterior) end. There is a stage called the resting period, during which the parasite is smaller and curved only once and has the nucleated portion occupy from one-half to nearly all of the stage. Parasitized paramecia have alveolar cytoplasm showing the macronucleus undergoing degenerative changes. The macronuclei may be abnormally small, vacuolated, fragmented, or completely lost from view. In such parasitized animals, the micronucleus is usually never seen.

[2] The generic name, *Drepanospira* (Gr.) which means sickle or reaping hook, is characteristic of its shape. The species is named in honor of Jean Müller who was the first to call attention to these parasites in 1856.

It is reported that intoxication of the cell, which acts directly on the cytoplasm—indirectly on the nuclear material—is a result of chemical action due to waste products and secretions of the parasite. Petschenko noted that externally, *Drepanospira mülleri* is the same as *Holospora undulata* and *H. elegans;* and since Hafkine did not observe nuclear elements in *Holospora,* it is possible that they all belong to the same genus.

Differing from *Holospora* is a parasite reported by Calkins (1904) which he called *Caryoryctes cytoryctoides.* He observed an infection in which 80 per cent of *Paramecium caudatum* in culture contained parasites in the macronucleus.

In cultures of a race of zoöchlorellae-free *Paramecium bursaria,* fairly large numbers of individuals were discovered parasitized with *Schizomycetes* (Wichterman, 1945). These spherical microorganisms, which stain with the Feulgen reaction, measure less than a micron and exist in globular aggregates from approximately 3–45 microns in diameter in the endoplasm of paramecia. Individual specimens contain from one to 12 of these masses. Frequently four or fewer large masses or parasites are seen in which the body of the protozoön is misshapen. In severe cases of parasitism, the host assumes an abnormal pyriform shape with the aggregate of parasites at the wide end of the body.

In heavily parasitized paramecia the micronucleus is always present and appears normal but is not found in its typical position next to the macronucleus. Instead, the micronucleus is located at either end of the ciliate. On the other hand, the macronucleus appears to be attacked. In many cases the bacterial aggregate appears pressed against the macronucleus forming a cup-shaped depression and resulting in its hypertrophy or structural alteration. In such specimens the macronucleus is considerably smaller than in typical animals and suggests that macronuclear material is used in the metabolism of these microorganisms. It is easy to understand that the larger masses of parasites interfere not only with cyclosis but other trophic functions of the cell as well.

Generally, parasitized *P. bursaria* do not conjugate with members of the opposite mating type, but one case was found in which a specimen with a bacterial aggregate measuring 32 microns was joined in conjugation at the stage of pronuclear exchange. Also a fission stage was found in which the macronucleus and micronucleus had completed division, but the body of the protozoön had failed to constrict or divide. There is shown in these cases, the persistence of nuclear division in spite of parasitism.

In another study of a race of *P. caudatum* characterized by unstable micronuclear behavior (Wichterman, 1946 d), cytological investigation disclosed that most specimens from pure-line cultures showed a single micronucleus, but many were present with two micronuclei and fewer with three or four. Some appeared to be amicronucleate.

The method of producing supernumerary micronuclei was discovered by studying dividing individuals. Instead of a micronuclear product passing to each daughter at the end of fission as in normal animals, the late telophase stage of the dividing micronucleus is found in an anterior or posterior daughter at the time of constriction of the cell. The result is that one daughter is bimicronucleate and the other daughter amicronucleate. The bimicronucleate forms are capable of dividing normally. Division stages show two micronuclei in late telophase stages that result in four micronuclei, two going to each daughter.

However, aberrations may again occur with normal fission to produce variations in number of micronuclei. Generally two contractile vacuoles are present, but specimens with three vacuoles, even four, are not uncommon. In some, the macronucleus possesses a large refractive glass-like body free of chromatin. In others, macronuclei appear in various stages of fragmentation or normal. A parasitic microörganism, perhaps a virus, appears to be present in the macronucleus and cytoplasm that causes this derangement. This induction of micronuclear variation in *Paramecium* by parasites may be of considerable evolutionary significance.

Viruses are not uncommon in multicellular plants and animals, and in view of the fact that many different microorganisms are present in paramecium, it is not unreasonable to expect their occurrence in these ciliates. See page 344 for the possible relation of *kappa* to a virus or related microörganism.

3. Fungi (Sphaerita).

Parasitic fungi belonging to the order *Chytridiales*, family *Olipiaceae*, have been reported from a large number of both free-living and parasitic Protozoa. Such parasitic chytrids belong to two genera, namely *Sphaerita* and *Nucleophaga*, but only the former has been described as a parasite of *Paramecium* (Cejp, 1935). In his study of autogamy in *P. calkinsi*, Diller (1948 a) reported the presence in many specimens of a fungoid symbiont of the macronucleus.

Members of the genus *Sphaerita* measure from approximately 3 microns to less than a micron. In the cytoplasm of the host, each is a one-celled, uninucleate, ameboid thallus enclosed by a delicate membrane. This stage grows into a multinucleate thallus of considerable size called a sporangium after which growth ceases and sporulation begins. The size and number of spores or zoöspores produced by the sporangium are variable. Generally, they are spherical in shape, and they have been reported as possessing one, two, or no flagella. Cejp described as many as 11 sporangia in a single paramecium and zoöspores with two flagella. According to him, each zoöspore possesses a central or eccentrically placed nucleus.

Rupture of the large sporangium occurs in the cytoplasm of the host and generally results in its death and disintegration. Zoöspores are thus liberated in the water to infect other specimens.

C. Symbiosis in Paramecium

The green paramecium, *P. bursaria*, is the only one of the eight well-defined species which exhibits a truly symbiotic relationship because of the presence of green unicellular algae in the cytoplasm. These algae belong to the order *Chlorococcales*, family *Oöcystaceae*, and are members of the genus *Chlorella*. They are spherical to broadly ellipsoidal in shape and measure approximately 4–9 microns in diameter. *Chlorella* possesses a large, single, parietal chloroplast, which is usually cup-shaped. On the basis of morphology, many species of *Chlorella* are indistinguishable from one another, but physiologic differences exist. The algae are free-living or grow within the cells of invertebrates. *C. vulgaris* is the commonest, free-living species.

Reproduction in *Chlorella* is asexual by means of two, four, or more autospores which are formed within a mother-cell. The autospores are liberated by the rupture of the wall of the mother-cell.

Although the algae live symbiotically in the cytoplasm of *P. bursaria*, the chlorellae are autotrophic (phototrophic, holophytic) and not necessarily dependent upon the host; they have been isolated and grown in pure culture. They are able to decompose carbon dioxide by means of the contained chlorophyll in the presence of sunlight, thus liberating the oxygen and combining the carbon with other elements derived from water and inorganic salts.

Von Siebold (1849) first suggested the relationship of the green organisms living in animals to chlorophyll-bearing forms. Shortly after, De Bary (1879) introduced the term *symbiosis*[3] into biologic literature, Brandt (1882) established the generic name of *Zoöchlorella* to specify these algae. However, after Entz (1881–82) reported that the algal cells from crushed *Stentor* could be grown independently of the host, Brandt (1883) attempted to withdraw the name *Zoöchlorella*. Bütschli (1882) and Pringsheim (1915, 1928) were of the opinion that the algal cultures of Entz represented free-living algae which had been ingested by the host. Beijerinck (1890) demonstrated the similarity of the zoöchlorellae of *Hydra, Stentor,* and *Paramecium* to his newly described free-living *Chlorella vulgaris*.[4] Because he was unsuccessful in growing them independently of their hosts, he believed the algae had become changed as a result of their symbiotic relationship. Le Dantec (1892) was of the opinion that the zoöchlorellae of *P. bursaria* were identical with the strain of free-living *Chlorella vulgaris* as observed by Beijerinck, although he too was unable to culture the algae independently of the paramecia. In spite of

[3] As first used by its originator, De Bary, the word symbiosis was employed as a collective term to include parasitism and mutualism. He recognized antagonistic and mutualistic symbiosis as the two main groups in the association.

[4] In his "Fresh-water Algae of the United States," Smith (1933) reports that *Chlorella (Zoöchlorella) conductrix,* Brandt, is found in *Hydra, Stentor,* and *Paramecium.*

the attempt by Brandt to withdraw his own term of *Zoöchlorella,* the name is in wide usage today. It should be interpreted as a non-taxonomic word having reference to green, unicellular algae which live symbiotically within the protoplasm of host organisms.

Pringsheim (1915) cultivated *P. bursaria* in a medium free of organic food material. He claimed to have washed thoroughly the paramecia free from bacteria. Into a sterile inorganic nutrient solution, he placed two specimens and soon had a rich culture of hundreds of individuals. Pringsheim concluded that since no organic food was available to the paramecia, there was shown a mutual benefit of one organism upon the other.

Other investigators, namely Dangeard (1900), Reichensberger (1913), Buchner (1921), and Oehler (1922 b), have contributed knowledge concerning the symbiosis between the chlorellae and *P. bursaria.* More recently we are indebted to Parker (1926), Gelei (1927), Goetsch and Scheuring (1927), Pringsheim (1928), and Loefer (1934, 1936 a, b, c, 1938) for their comprehensive experimental studies.

Oehler (1922 b) investigated the symbiosis of algae in *P. bursaria* by experimentally infecting a colorless (zoöchlorellae-free) strain with a species of free-living *Chlorella,* then eliminating the alga from the host and finally growing the alga independently again. Pringsheim (1928) was unsuccessful in repeating this work and came to the conclusion that Oehler had created an artificial infection. Indeed, Pringsheim was unable to culture the algae independently from any of his strains of *P. bursaria.* Oehler was successful also in establishing certain *Ulotrichales*— namely *Hormidium nitens, Stichococcus bacillaria,* and *Stichococcus mirabilis* in the host paramecia. That the

algae from *P. bursaria* can be cultured independently was demonstrated by Loefer (1936 a). This investigator first cultivated *P. bursaria* in a bacteria-free medium then cultivated their algae independently on agar slants. It is of interest to report that a culture of the alga was sent to Pringsheim's laboratory at Prague for identification. Concerning this strain, Pringsheim (1946) noted that the strain was morphologically indistinguishable from other species of *Chlorella,* although physiologically different. Loefer (1936 a) described it as not corresponding to the free-living *Chlorella vulgaris* but more closely resembling *C. ellipsoidea.*

While the symbiotic algae have been grown independently of their host paramecia, races of colorless, zoöchlorellae-free *P. bursaria* have in turn been cultivated by a number of investigators.[5] Jennings (1938) found that the symbiotic green algae in *P. bursaria* can be reduced in number and removed entirely from animals by a series of rapid fissions of host paramecia. The author has studied a large number of pedigreed races of *P. bursaria* from many different sources. Most of these races have been maintained continuously for over 10 years with observations made regularly throughout the period. Certain races are shown to contain specimens which may be characterized by the approximate number of algae and the intensity of their chlorophyll. Sev-

[5] Famintzin (1891) reported that wild, green *P. bursaria* can divide only once or twice if deprived of chlorellae, then die. On the other hand, the author has had growing in his laboratory for many years, two such colorless races of *P. bursaria* which were originally green with zoöchlorellae (Wichterman, 1941, 1943). These colorless races require more frequent subculturing than typically green ones yet appear perfectly healthy and readily demonstrate the mating reaction and undergo subsequent conjugation with green races of the opposite mating type (Wichterman, 1946 a, b, c).

eral races of colorless (zoöchlorellae-free) *P. bursaria* were also cultivated. It soon became evident that cultures of typically green specimens could be maintained in flasks with tightly fitting glass covers, for periods of many months throughout the years without the need of frequent subculturing.[6] Such cultures are characterized by the lack of bacterial growth in them. On the other hand, colorless *P. bursaria* (or even those races of paramecia with very few zoöchlorellae) require subculturing every two or three weeks for survival. In such cultures, there must be present an adequate bacterial flora which is utilized as food. The author (Wichterman, 1947 b, 1948 b) has produced permanently colorless races of *P. bursaria* devoid of all zoöchlorellae by irradiation with high dosages of x-rays (up to 600,000 *r*). Since the zoöchlorellae are more sensitive to x-rays than the paramecia, certain dosages will kill the algae but not the protozoön.

Colorless *P. bursaria* behave quite differently from green specimens. Instead of concentrating to the side of the glass container exposed to light as with the latter, colorless ones appear to be rather uniformly distributed throughout a culture and they are more active swimmers. Le Dantec (1892) reported that if green and colorless *P. bursaria* live together for a time, the number of colorless ones become less. He believed that the green specimens left chlorellae which infected the colorless ones. One should not overlook the interpretation that the colorless ones may disappear

from such a mixed culture and leave only green specimens.

An observation of the author (Wichterman, 1941) seems worthy of note in respect to possible toxicity of zoöchlorellae. Zoöchlorellae suspensions from green individuals of the mating type opposite from the colorless *P. bursaria* were made. When a colorless specimen was placed in such a suspension of zoöchlorellae, the ciliate ingested the algae rapidly. Food vacuoles contained from one to five zoöchlorellae. However, in less than a day, the colorless *P. bursaria* became darkly granular, sluggish, then died apparently as a result of an antibiotic substance produced by the suspension. Yet when zoöchlorellae from another race which also mated with the colorless paramecia were placed with them, no such lethal effect took place, showing perhaps different chemical or physiologic characteristics of the algae.

The work of Parker (1926) on symbiosis in *P. bursaria* is of interest. He demonstrated that the paramecia contributed to the food of the algae by measuring the amount of CO_2 produced by the ciliates under the two diverse conditions of light and darkness. In seeking to determine whether the CO_2 given off by the paramecia in respiration was utilized by the algae, he found that the amount of the gas given off in darkness was greatly in excess to that given off in the light. As a result of photosynthetic activity, the greater part of the CO_2 liberated by the paramecia in the light was utilized at once by the algae. He noted that in darkness, apparently no appreciable absorption of CO_2 by the algae occurs.

That the algae contributes to the food of *P. bursaria* was also demonstrated by this investigator who eliminated the CO_2 from the air which was supplied to the culture and by removing the CO_2

[6] In the author's laboratory, two plugged test tubes, each containing a pedigreed race of *P. bursaria,* were suspended by wire and close to a north window. Up to the present time—which is 11 years after their establishment—the cultures are still flourishing in the original test tubes. The only attention given them is the addition of sterile lettuce infusion a few times a year to compensate for that lost by evaporation.

from the culture itself. Green algae require CO_2. Parker reported that in the culture of *P. bursaria* exposed to CO_2-free air, all life was extinct at the end of five days, but specimens of *P. aurelia* subjected to the CO_2-free air were in good condition after five days.

While some authors maintain that under certain conditions, *P. bursaria* digests a certain number of chlorellae, Pringsheim and Parker are of the opinion that this occurrence is rare. From the observations and reports of a large number of protozoölogists, there is shown to be a truly mutualistic symbiosis between the chlorellae and *P. bursaria*.

It is of interest to report that when green *P. bursaria* are irradiated with 300,000 to 600,000 Roentgen units, the algae gradually disappear, paramecia become colorless, and optically active crystals appear in the cytoplasm where they persist indefinitely. The crystals, which vary in size from a few to 12 microns, are found largely in the posterior end of the body. This appearance of crystals is correlated with a gradual change in the type of nutrition, i.e. from a holophytic to a holozoic one in which the colorless, x-rayed, crystal-bearing forms are now dependent upon the bacteria as a source of food. Unlike the green forms, food vacuoles packed with living bacteria are quickly formed—in some instances one every 15 to 20 seconds.

When recently collected "wild" *P. bursaria* are examined immediately, crystals and zoöchlorellae are present. This is probably due to the fact that considerable organic material is ingested from the immediate environment in nature hence the animals may not be completely dependent upon a holophytic type of nutrition in their natural habitat.

In freshly made cultures of green, crystal-free *P. bursaria* growing in the laboratory and rich in bacteria, crystals occur for a short time; appearing the day of inoculation into a new culture, the crystals persist for about one week. However, as the bacteria disappear in cultures the crystals also disappear and the zoöchlorellae and paramecia come into a truly symbiotic association. It is then that the green paramecia utilize the photosynthetic products of the algae while the algae on the other hand utilize certain metabolic products of the ciliates. A similar effect is produced when green crystal-free *P. bursaria* are placed in total darkness for periods of three or more weeks. As the zoöchlorellae gradually decrease in number, crystals make their appearance. When such forms are again placed in the light, the zoöchlorellae increase in number while the crystals disappear (Wichterman, 1948).

Chapter 14

Research Technique in the Study of Paramecium

A. Study of Living Specimens
   1. General and Special Methods
   2. Use of Anesthetics and Dyes
B. Methods for Killing and Staining
   Paramecium
   1. Temporary Preparations
      NUCLEI
      PELLICLE
      CONTRACTILE VACUOLES
      CYTOPYGE
      TRICHOCYSTS AND CILIA
      CYTOPLASMIC INCLUSIONS
   2. Permanent Preparations
      a. KILLING AND FIXING FLUIDS
      b. TECHNIQUES OF FIXATION
      c. SECTIONING OF PARAMECIA
      d. STAINING METHODS FOR
         PERMANENT PREPARATIONS
      e. RELIEF STAINING
      f. SILVER AND OSMIC IMPREG-
         NATION METHODS

A. Study of Living Specimens

For the collection, cultivation, sterilization, and the general technique in
the handling of paramecia and glassware, the reader is referred to Chapter
4. Good textbooks of physiology are
available which give the basic techniques for the investigation of certain
physiologic problems dealing with paramecium and other Protozoa. However,
some techniques of this nature are so
detailed that the original papers of certain experiments must be consulted for
the method or technique concerned.

1. General and Special Methods.

Insofar as possible, paramecium
should be studied in the living condition. It seems that in recent years, there
has been a tendency to overlook the
possibilities of obtaining valuable information from a close study of the living
animal. Instead, dyes or permanently
stained preparations are resorted to
without exhausting the possibilities of
observing paramecium alive. The new
phase microscope is a valuable instrument for observing in the living condition fine details of nuclei, cilia, etc., not
readily seen in the ordinary microscope
(Bennett, Jupnik, Osterberg, and Richards, 1946). The use of the electron
microscope has proved invaluable for
the detailed study of cilia and trichocysts (Jakus, Hall, and Schmitt, 1942;
Jakus, 1945; Jakus and Hall, 1946; and
Anderson, 1951).

To observe the normal swimming and
spiralling habits, the organisms should
be placed in a shallow dish and examined with a wide-field, low-power microscope such as a binocular dissecting microscope. The common practice of observing paramecium under
higher power is to place a small amount
of culture fluid containing a large number of specimens on a microscope slide.
To prevent crushing the animals, a few
pieces of glass, i.e. finely drawn glass
rods, should first be placed around the

drop before the cover-glass is put on the preparation. At first the paramecia will be seen to move about very actively but frequently; after about five minutes, they will become less active especially if there are small clusters of bacteria present (Fig. 52). Cotton fibers or teased lens paper may also be placed in the drop of fluid containing para- mecia to impede active locomotion. Viscous non- or slightly toxic substances also can be utilized, such as a solution of 1–3 per cent agar gelatin, quince- seed jelly, and a solution of completely hydrolyzed, medium viscosity polyvinyl alcohol (Moment, 1944). A 5–10 per cent solution of methyl cellulose, be- cause of its high viscosity, low tonicity, and non-toxic properties has proved to be excellent for slowing locomotion (Marsland, 1943; Brown, 1944).[1] A ring of the solution is made on a slide, the ring being small enough to be covered by a cover-glass. A drop of cul- ture medium containing paramecia is placed inside the ring and the cover- glass added. As the methyl cellulose dif- fuses inward, slowing down of the organisms occurs progressively. Buck (1943) and Bullough (1946) give addi- tional methods for arresting movement in paramecium.

By means of a new device (Fig. 139), it is possible to obtain quickly and easily, large numbers of dividing stages of paramecium and certain other nega- tively geotropic ciliates (Wichterman, 1952). The method is based on the fact that dividing specimens not only move more slowly than typical vegetative specimens but once division has pro- gressed to the point where a definite

transverse constriction furrow is formed, the fission stages, especially with their characteristically uncoördinated man- ner of swimming, tend to settle toward the bottom of the culture container.

The device consists of a 4000 ml. pyrex beaker from which the bottom has been cut off and ground smooth. To this opening is fused a pyrex funnel having the same diameter as the open- ing at the bottom of the cut beaker. To

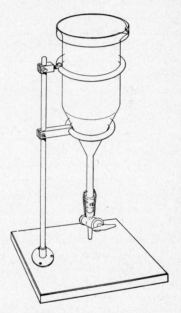

Fig. 139. Device for obtaining large num- bers of fission stages of Paramecium. (1/12 actual size.) (Wichterman)

the stem of the funnel there is joined the inner member of a 19/38 standard taper, ground-glass joint; the outer member is fastened to a glass stopcock having a 5 mm. opening. The ground- glass joint is held firmly in place with two coiled springs. The glass device is set on a stand consisting of a wooden base and upright steel arm which holds two adjustable, rubber-covered brass rings.

The device is used in the following

[1] In 50 cc. of boiling distilled water, add and mix thoroughly methyl cellulose (Dow Methocel, pharmaceutical grade, viscosity 15 cps.). Allow methyl cellulose to soak for 20– 30 minutes, then add 50 cc. of cold water. When the mixture reaches room temperature, stir until smooth.

manner: a rich culture (approximately 4 liters or less) of species pure *Paramecium caudatum* is placed in the unit. Next, a rich suspension of suitable bacteria is added to the culture. The actively feeding paramecia then divide at a fairly rapid rate. By simply opening the stopcock from time to time and drawing off a few cubic centimeters of culture fluid, large numbers of fission-stages may be obtained.

Although not holding as large a volume of culture fluid, the standard 2-liter separatory funnel may be used successfully in place of the device.

A "hanging drop" preparation is desirable for long-continued study of one or a number of paramecia in a drop of culture fluid prepared as follows: a drop of culture fluid is placed on a cover-glass, and with the use of a low-power microscope, the specimen is introduced into the drop by means of a capillary pipette. The cover-glass is then inverted over the concavity of a depression slide which is then sealed with petroleum jelly or some other sealant to make an airtight compartment.

For studying fission-rates, etc., specimens may be placed directly in the concavities of depression slides in which a cover-glass is not used. To prevent evaporation and contamination, such preparations are placed in moist chambers. For studying growth and obtaining fission-rates, the moist chambers should be placed in a constant temperature chamber at the optimum temperature (24°–28° C). The establishment of isolation cultures to determine the fission-rate is found on page 361.

The author (Wichterman, 1940) has studied nuclear behavior and fine structures of paramecium in the living condition with a precision microcompression chamber (Fig. 106S). The instrument consists of a base plate measuring 45 x 75 mm. with a threaded circular shell.

Into the threaded shell is screwed the compressor cylinder on the upper edge of which is engraved a calibrated scale. The scale reads 0–100 on the vertical lines. The compressor screw threads are 0.5 mm. wide, giving 500 microns compression with one rotation of the compression cylinder (each division on the scale equals 5 microns). The base plate has a hole in its center to which a flat circular glass slide is cemented with a thin film of beeswax. Inside the compressor cylinder is the rotator cylinder controlled by a pin, shown projected at an angle of 45° from the compressor pins in the illustration. A No. 1 cover-glass 25 mm. in diameter is held in the bottom of the rotator cylinder by means of a screw ring. Rotation of the preparation without further compression is possible. At any time the distance between the two pieces of glass can be determined.

For the study of a single specimen of paramecium or a joined pair, either may be placed with the aid of a capillary pipette and low-power binocular microscope into a small drop of culture fluid on the circular glass slide of the base plate of the chamber. The joined metal cylinders, the lower of which contains the cover-glass, are screwed down while observations continue under the microscope. When the cover-glass approaches the meniscus of the drop of culture fluid containing the specimen, condensation of water vapor is observed on the cover-glass. This serves notice that it requires very slow and critical turning of the rotocompressor to only a few more microns compression to make contact with the drop. Extremely fine adjustments to within a few microns make it possible to prevent gradually joined paramecia from spiralling between the cover-glass and the circular glass slide, after which time, detailed observations may be made on nuclear behavior. The drop of culture fluid, of course, does not come in contact with the metal of the microcompression chamber. Evaporation is practically eliminated when a ring of moist filter paper is placed in a circular groove in the base plate.

For detailed study, any objective can be used including oil immersion. In using the precision microcompression chamber, it is possible to observe in the living condition, the entire behavior of nuclear phenomena and cytoplasmic behavior over a long period of time. In addition, accurate time relationships can be determined during the union. Indeed, in this manner, the divisions of the micronuclei and their behavior in sexual phenomena have been photographed on motion-picture film (Wichterman, 1940 a).

To prevent unnecessary confinement of joined paramecia which are to be observed over long intervals of time, a single pair may be studied at definite intervals and placed back again in a small amount of culture fluid. At other times during critical nuclear stages, the paramecia may be studied in the chamber continuously.

For observation of cilia and certain other structures, the *dark-field condenser* can be used with great advantage. Polarizing prisms placed in the regular microscope are valuable in studying optically active crystals of paramecium.

2. Use of Anesthetics and Dyes.

If anesthetization is desired, alcohol, ether, or chloroform may be sprayed lightly on the preparation or small amounts of aqueous solutions of one of the following added: 0.1 per cent nicotine, 10 per cent methyl alcohol, 0.1–0.2 per cent chloral hydrate, 1.0 per cent cocaine hydrochlorate, 1.0 per cent cocaine hydrochloride, 1–33 per cent magnesium sulfate, 1.0 per cent magnesium chloride, or 0.1 per cent Chloretone. According to Cole and Richmond (1925), one drop of 0.12 per cent Chloretone when added to a drop of equal size of culture containing paramecia will anesthetize the organisms in about 10 minutes and keep them so up to 10 days.

Also useful for retarding movements are the following: saturated Sodium Amytal; Rousselet solution consisting of 3 cc. of 2 per cent cocaine, 1 cc. of 90 per cent alcohol and 6 cc. of distilled water; Corri solution consisting of 10 cc. of 96 per cent methyl alcohol, 90 cc. water and 3 drops of chloroform.

Concentration of paramecia in the field has been described on page 95 but if concentrated paramecia are needed in the laboratory, electrophoresis or centrifugation may be employed.

Intravitam Dyes.

For certain observations, vital staining is recommended because when used in extremely weak concentrations, parts of the living organism can be stained with little or no injury to paramecium. The dye may be dissolved in distilled water, original culture medium, or in absolute alcohol. Occasionally the dye may be used without first placing it in solution.

Since they more effectively stain the organism, basic dyes are more commonly used than acid ones. Janus green B (1:10,000–100,000) stains mitochondria; methylene blue (1:10,000) stains nuclei, cytoplasmic granules and other structures; neutral red (1:3000–50,000) which stains nuclei also serves as an indicator showing a yellowish red color in an alkaline region, cherry-red in a weak acid region and blue in a strongly acid region. Other useful intravitam dyes are Nile blue sulfate (1:30,000); indulin (1:100,000–200,000); brilliant cresyl blue (1:50,000); Victoria blue (1:100,000–200,000), and rhodamine (1:20,000).

Ball (1927) tested a number of basic dyes in which he determined the minimum concentration needed to stain paramecium in regard to their toxicity and his results are given in Table 37.

Food Vacuoles.

To study ingestion, cyclosis, the path taken by food vacuoles in the protoplasm and egestion, powdered carmine[2]

selection in paramecium refer to Chapter 6. Food granules in living paramecia may be effectively demonstrated by adding 0.1 per cent brilliant cresyl blue mixed with equal parts of the culture fluid. This method results in the food vacuoles becoming brightly colored.

The use of *indicators* is extremely valuable in studying the acidity and alkaline changes (pH) of the food vacuoles during digestion and their use and results are treated in detail on page 166. These changes can be demonstrated

Table 37

STAINABILITY AND TOXICITY OF SOME BASIC DYES UPON PARAMECIUM (AFTER BALL, 1927)

Dyes	Minimum Concentration That Will Stain Paramecium	Toxicity: Per Cent Dead in One Hour
Bismarck brown	1:150,000	0
Methylene blue	1:100,000	5
Methylene green	1:37,500	5
Neutral red	1:150,000	3
Toluidine blue	1:105,000	5
Basic fuchsin	1:25,000	30
Safranin	1:9000	30
Aniline yellow	1:5500	0
Methyl violet	1:500,000	20
Janus green B	1:180,000	40

or "Chinese ink" are commonly used. A small amount of finely powdered carmine or the powder from a block of "Chinese ink" is distributed over a drop of culture fluid containing paramecia. These substances are readily ingested by paramecium and enter into the food vacuole similar to ingestion of nutritious food. Many substances can be used but the particles should be insoluble in the medium and not too heavy or too large. For a list of these and other substances which may be used to test food

[2] Bragg (1937) observed that certain batches of carmine proved to be extremely toxic to paramecium. He observed that the toxic substances could be removed easily by first washing 1 Gm. of carmine in 1 liter of hot or cold tap water.

readily by the use of neutral red (1–50,000) in which this indicator becomes cherry-red at the acid end and yellow in the alkaline region of the color range. The following dyes dissolved in 100 cc. of distilled water give good results as indicators: neutral red, 50 mg.; phenol red, 100 mg.; litmus, 150 mg.; Congo red, 50 mg. A drop of the indicator-solution is mixed with a drop of culture containing paramecia and observed under the microscope.

Feeding and hydrogen-ion concentration changes may be studied advantageously by the use of heat-killed yeast cells stained with Congo red (Buck, 1943). Also the ciliates are quieted by this method. A small amount of compressed yeast is mixed with a very dilute solution of Congo red, brom thymol blue, brom phenol blue, or brom cresol purple. The mixture is boiled for 10 minutes after which a small amount is added to the slide containing paramecia. Food vacuoles stained orange will have a pH of 5 or more while those stained blue, a pH of 3 or less.

B. Methods for Killing and Staining Paramecium

1. Temporary Preparations.

NUCLEI. Acetocarmine and methyl green with acetic acid are two excellent and widely used nuclear stains. Aceto-carmine may be prepared by boiling carmine in 45 per cent acetic acid until no more will dissolve (usually 0.5 Gm. of carmine in 100 cc. of the acid), cooling and filtering. To use effectively, a drop of culture fluid containing paramecia is placed on a slide to which is added a small drop of aceto carmine. After about one minute, a drop of 45 per cent glacial acetic acid (or acid of the same concentration as used in the stain) is applied, then the preparation is covered with a glass slip. For observation with the oil immersion lens, the cover slip should be ringed with petroleum jelly. The acetic acid tends to make the cytoplasm transparent and leaves the nuclei stained red. Methyl green acetic is a solution of methyl green in 0.5–1.0 per cent acetic acid.

Although these two dyes are commonly employed in which 1.0 per cent of the solution of the dye is used in 1.0 per cent acetic acid, stronger or weaker concentrations may be used effectively.

PELLICLE. A 0.5–10 per cent solution of nigrosin, when mixed in a smear preparation or drop of paramecia-culture and dried in air, reveals clearly the pellicular pattern. A permanent preparation may be made by simply covering the thoroughly dry smear with dammar or balsam and adding a coverglass.

Upon application of a 35 per cent solution of alcohol to living specimens, the pellicle is raised in blister formation and may then be studied.

CONTRACTILE VACUOLES. Paramecia placed in a heavy suspension of "Chinese ink" powder or fine carbon particles will readily show discharge of the vacuolar contents to the outside of the body.

CYTOPYGE. Egestion and the cytopyge may be observed by placing paramecia in a medium rich with bacterial food. A suspension of carmine or carbon particles also may be used. Under the higher powers of the microscope, one should focus on the subterminal region of the body below the ingestatory structures.

TRICHOCYSTS AND CILIA. Trichocyst extrusion may be demonstrated upon application of 0.5 per cent methylene blue, methyl green with acetic acid, acetocarmine, or a number of different acids with or without dyes in solution. The use of colored inks is highly recommended in which a small amount of red ink (Sanford's) is introduced into a drop of culture fluid containing paramecia, then covered with a cover-glass. After about five minutes, a drop of blue ink (Waterman's) is allowed to run under the cover-glass. While the red ink is slowly toxic to the paramecia, the animals are killed instantly when they come in contact with the blue ink. Dead, stained specimens show the cilia bright red, cytoplasm pink and the trichocysts bright blue. Occasionally the macronucleus will be stained also (Halter, 1925).

Cilia may also be observed clearly in the living condition after the animal has had its movement slowed down with one of the reagents described earlier.

Currents created by the cilia may be observed clearly if the organisms are placed in dense suspensions of finely powdered carbon or "Chinese ink."

CYTOPLASMIC INCLUSIONS. Cytoplasm and cytoplasmic inclusions stain with the use of strongly concentrated indicator-solutions and many basic dyes. Lugol's solution (composed of 3 Gm. potassium iodide, 50 cc. distilled water, and 2 Gm. iodine), if diluted 1:5–1:10 stains glycogen bodies and cilia.

2. Permanent Preparations.

A permanent preparation involves the killing and fixing of paramecia, followed by staining and dehydration in such a manner that certain structural details are seen more distinctly than in the living animal. A large variety of fixing agents, stains, counterstains, and special techniques have been employed by protozoölogists in an effort to show cytological detail. Since paramecia are relatively small, practically all fixing agents penetrate rapidly. As a rule, fixatives containing picric acid do not permit as clear a staining differentiation as those containing mercuric chloride. This is especially the case with carmine stains. When in concentration of 1–5 per cent, acetic acid generally results in distinctly clear cytoplasmic differentiation. The following are recommended for the study of paramecium and Protozoa in general.

a. KILLING AND FIXING FLUIDS.

Sublimate Mixtures.

Schaudinn's Fluid.

This is perhaps the most widely used fixative and many variations exist; for them, see Wenrich (1937). The writer has used the following for many years with excellent results and recommends

it most highly. The fixative should not be prepared until ready for fixation.

Saturated aqueous solution,
  mercuric chloride[3] . . . . 2 parts
95 per cent ethyl alcohol . . . . . 1 part

Immediately before using, add 5 cc. of glacial acetic acid to each 100 cc. of the above mixture. The fluid is then heated to 45° C. or less and applied to the organisms where they are left for 20–30 minutes. The fluid should be allowed to cool to room temperature immediately after the organisms are introduced into the fixing reagent.

After fixation, specimens should be transferred to 50 per cent then 70 per cent alcohol for approximately 10 minutes each. To remove all traces of corrosive sublimate crystals, a few drops of strong iodine or Lugol's solution should be added to the 70 per cent alcohol containing the specimens at a ratio of about three drops to each 50 cc. of alcohol. Specimens, which may be kept indefinitely in the 70 per cent alcohol, are now ready for a staining procedure.

Worcester's Fluid.

This is prepared by using 90 cc. of 10 per cent Formalin which is saturated with corrosive sublimate and 10 cc. of glacial acetic acid. Follow directions as given for Schaudinn's fluid.

Sublimate-acetic Fluid.

Saturated aqueous
  sublimate solution . . . . 100 cc.
Glacial acetic acid . . . .   2 cc.

Follow directions as given for Schaudinn's fluid.

[3] Made by dissolving 7 Gm. mercuric chloride in 100 cc. distilled water and should be absolutely clear before using. Follow with iodine wash, any fixative containing this reagent. Because of its corrosive property, avoid bringing any metal instruments in contact with it.

Sublimate-alcohol-acetic Fluid.

This is prepared by using 95 per cent alcohol saturated with corrosive sublimate. To each 100 cc. of the mixture, add 5 cc. of glacial acetic acid immediately before using and heat as with Schaudinn's fluid. After fixation, transfer paramecia to 85 per cent alcohol and treat with iodine solution.

Formol-mercuric Chloride.

Add formaldehyde to a saturated aqueous solution of mercuric chloride in the concentration of 5 per cent. Fix for 20–30 minutes then transfer through 35, 50 and 70 per cent alcohol allowing 10 minutes for each stage.

Bouin's Fluid.

Glacial acetic acid	5 cc.
100 per cent Formalin (40 per cent formaldehyde)	15 cc.
Saturated aqueous solution, picric acid	75 cc.

Fix at room temperature for 10–30 minutes, then transfer and wash specimens in several changes of 50 per cent alcohol allowing about 10 minutes for each change. Place in 70 per cent alcohol and retain there until wash alcohol is no longer deeply tinged with yellow color.

Carnoy's Fluid.

Absolute ethyl alcohol	75 cc.
Glacial acetic acid	25 cc.

Fix specimens for 10–30 minutes, wash in 95 per cent alcohol then hydrate to 85 per cent and 70 per cent alcohol allowing about 10 minutes for each change.

Flemming's Fluid.

1 per cent chromic acid	30 cc.
2 per cent osmium tetroxide	8 cc.
Glacial acetic acid	2 cc.

Fix specimens for 30–60 minutes, wash for at least one hour with running water or many changes of water after which material may be stained or dehydrated to 70 per cent alcohol for preservation.[4]

Perenyi's Fluid.

95 per cent alcohol	3 parts
10 per cent nitric acid	4 parts
0.5 per cent chromic acid	3 parts

b. TECHNIQUES OF FIXATION. In the study of cytological detail, it is advisable to use a number of different fixing methods and techniques in order to obtain the fullest possible information from the preparations. Immediately following are a number of techniques all of which have been used successfully by the author and which are highly recommended.

Cover-glass Method.

This method consists of causing the paramecia to adhere to the cover-glass upon fixation in such a manner that the specimens may be transferred through all subsequent reagents by simply handling the cover-glass. Chemically clean cover-glasses, preferably circular ones of #1 thickness and measuring 7/8 of an inch in diameter, are smeared with Mayer's egg-albumen fixative. This consists of equal parts of the albumen of a fresh egg and glycerin to which is added a crystal of thymol to prevent or inhibit mold growth. A very small drop is placed on the center of the cover-glass and smeared over the entire surface to make as thin a film as possible. This may be done with a finger. Using a

---

[4] After fixation, washing and prior to staining, some prefer to dehydrate the specimens to 95 per cent alcohol for "hardening." This is not always necessary since fixatives like Schaudinn's fluid will harden the paramecia. However, certain techniques require dehydration to 95 per cent alcohol.

micropipette, concentrated paramecia, with a minimal amount of culture fluid, are then placed over the thin film of albumen fixative and distributed over the entire surface. If the meniscus of the fluid on the cover-glass is too high, allow the fluid to evaporate until only a thin film is present. This must be done carefully since too thorough evaporation will result in killing of the organisms before fixation. Next, the cover-glass preparation containing the living paramecia is inverted over the dish (Petri) containing fixing fluid and dropped with the film surface down. Generally the cover-glass will float on the surface of the fixing fluid for a few seconds thereby permitting one to hold the cover-glass with forceps before it drops to the bottom of the container. The cover-glass should then be inverted again to prevent the smeared surface from coming in contact with the bottom of the glass dish so that the cover-glass comes to rest with the smeared surface up. After fixation, the cover-glasses with paramecia affixed to them may be transferred to subsequent fluids, stains, dehydrating and clearing agents. For mounting, a drop of dammar or other mounting medium is placed in the center of a microscope slide, and the cover-glass is then inverted and placed over the dammar to make a permanent preparation.

Another convenient method requiring the use of albumen-smeared cover-glasses, is one in which the organisms are concentrated—preferably by centrifugation—then squirted into a fixing fluid by means of a capillary pipette. Paramecia should then be washed and dehydrated through graded series of alcohols (50, 70, 85 per cent) allowing about 10 minutes for each change. It is best to avoid centrifuging killed and preserved specimens since in their hardened condition, they are easily

damaged. Instead, the author uses small vials and replaces the fluid rather than changing the specimens. Preserved paramecia always settle to the bottom of the vial in a short time. To cause the preserved paramecia to adhere to the cover-glass, simply smear with Mayer's albumen fixative as previously described, then add a drop of concentrated paramecia now in 85 per cent alcohol. It will be found that the paramecia become fairly well distributed over the surface of the cover-glass. At the same

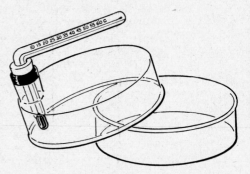

Fig. 140. Modified Petri dish with clinical-type thermometer in lid for continuous temperature observation. (Wichterman)

time, the alcohol will coagulate the albumen and cause the paramecia to become affixed to the glass. After a short time and being careful not to allow drying of the preparation, place a few drops of 95 per cent alcohol over the smear, then transfer the cover-glasses, with smeared surfaces up, into 95 per cent alcohol for further hardening of the aubumen. Preparations are now ready for hydration and a staining procedure.

In the above process, microscope slides may be used instead of cover-glasses (Smith, 1944).

Special Devices for Cover-glass Method.

For the fixation and staining of paramecia and other Protozoa, a number of special devices have been described as aids in the technique. Some are of distinct use; others so unnecessary, cumbersome, and expensive that they are of dubious value for either the beginner or expert.

Cover-glass preparations are customarily fixed, stained, dehydrated, etc. in Petri dishes. The author (Wichterman, 1945 a) has described a simple modified Petri dish for accurately maintaining and observing the temperature of the fluid in the dish (Fig. 140). It has proved to be useful for fixation and staining procedures requiring heat, such

Fig. 141. Columbia staining well for staining smeared cover glass preparations. (Courtesy, A. H. Thomas Co.)

as Schaudinn's fluid and the Feulgen reaction. Extremely useful containers for the smeared, cover-glass technique are the Columbia staining wells (Fig. 141). Chen (1942) and the writer (Wichterman, 1946 d) have described staining racks for the convenient handling of large numbers of cover-glass preparations through the various reagents.

Fixing and Staining en Masse.

The writer uses a beaker of hot Schaudinn's fluid in which the concentrated paramecia, obtained by centrifugation, are dropped by means of a pipette. The killed paramecia become snow-white, and when a piece of black paper is placed beneath the beaker, the organisms can be seen easily with the naked eye as they settle to the bottom of the container.

Washing and dehydration are done in the same beaker by simply siphoning off the reagents and adding fresh fluids. Staining may be continued in the same manner, except that it is more convenient to use a tall, narrow vial. After staining and dehydrating in vials, the paramecia are cleared and placed in xylol where extreme care must be exercised since the specimens may be easily damaged in this reagent. Xylol is then replaced with dammar, and specimens in a drop of the dammar are placed on a slide and covered with a cover-glass.

Occasionally, after the usual fixation methods, one finds that trichocysts have been extruded. The following procedure as described in the bulletin of the Carolina Biological Supply Company (1949, V. 12, p. 32) is useful in killing paramecia in bulk since it results in good fixation without extrusion of trichocysts.

1. A rich culture of paramecium is strained through four thicknesses of handkerchief linen.

2. Six to 12 drops of pure ethyl alcohol are added to the culture (around edges of finger bowl). This amount is added to about 150–200 cc. or two thirds of the small finger bowl capacity of culture medium.

3. Place culture in freezing unit of refrigerator for about 10 minutes, then add six to 12 drops of alcohol. Replace in freezing unit.

4. After intervals of about 10 minutes, add alcohol as before and replace in refrigerator until examination with a binocular microscope shows that the paramecia are found to be nearly immobilized. At this stage, most of the specimens will be found on the bottom of the finger bowl.

5. While watching the specimens at this stage through the binocular microscope, hand centrifuge the dish using a gentle, steady, circular motion until the paramecia are congregated in the center of the bowl.

6. Draw off paramecia with a pipette and

squirt them into about 50 per cent alcohol which has been brought to the same temperature as the culture.

7. Continue centrifuging and pipetting off specimens until most of the paramecia are recovered.

8. After about five to 10 minutes, add as much Bouin's (or other fixative) as there is 50 per cent alcohol. Allow paramecia to settle and draw off upper surface of fixing mixture.

9. Concentrated specimens may now be stained, cleared, and mounted.

c. SECTIONING OF PARAMECIA. Paramecia may be embedded, sectioned, and stained individually or en masse. The fundamental procedure is the same as that for the preparation of metazoan tissues, hence no attempt will be made here to describe the elementary technique. Instead, any book on histological technique may be consulted.

Sectioning of a Single Specimen.

Killed, unstained paramecia may be easily followed with the naked eye with the aid of black paper placed under the container. Prior to infiltration and embedding in paraffin, the specimen should be intensely stained with eosin in 95 per cent alcohol to enable the investigator to make identification in the paraffin; the excess eosin will be removed in later stages. Next, the specimen is removed by means of a micropipette then placed through several changes of absolute alcohol and xylol. Infiltration and embedding are done under a low-power binocular microscope. A small glass dish with a deep concavity is heated, then melted paraffin added after the embedding dish is placed in a Petri dish. The specimen is then added to the paraffin, which is kept melted for approximately 15 minutes with the heat generated by an electric lamp placed close to the dish. Prior to embedding, the specimen is oriented with a blunt but finely drawn glass rod which is first

heated. When paramecium is in the desired position, ice water is poured into the Petri dish to solidify the paraffin in the embedding dish.

Collodion and combined collodion and paraffin embedding methods are described by Wenrich (1937).

Sectioning en Masse.

Fixed and preserved paramecia may be completely dehydrated in alcohol by sedimentation, in preference to centrifugation, then through absolute alcohol followed by xylol. Specimens are next removed with a minimum amount of xylol and placed in melted paraffin contained in a Lefevre watch glass where infiltration and embedding are done. This dish has at the bottom a small rectangular well or depression in which the paramecia should settle. The remainder of the procedure is essentially the same as that in embedding single specimens.

As with metazoan tissues, concentrated paramecia may be embedded by simply transferring them to a small mold or paper boat of melted paraffin. Beers (1937) has successfully used a transparent piece of fresh mesentery in which a bag is formed and tied to hold the preserved Protozoa. The entire bag with contained Protozoa is dehydrated, cleared, infiltrated, embedded, and sectioned.

Another method requires the use of a thin-walled glass tube 1 cm. or less in diameter and 5 cm. or more in length. A piece of fine gauze (#20) is neatly tied around one end and immersed in alcohol after which concentrated fixed and preserved paramecia are dropped into the other end. One simply lifts the tube from one reagent to the other including melted paraffin for infiltration and embedding. The gauze, of course, prevents the paramecia from leaving the tube but permits the reagents to enter and penetrate the organisms. It is a simple matter to remove the gauze and the hardened paraffin from the glass tube for sectioning.

d. STAINING METHODS FOR PERMANENT PREPARATIONS. The selection of a stain will depend upon the structures one wishes to study and to a certain extent upon the fixative employed. Of a large number used by the author, the following are highly recommended.

Heidenhain's Iron Hematoxylin Method.

This is the most widely used method for staining chromosomes and nuclear structures. It also stains fibrils and basal granules of cilia. Many modifications of the basic method exist and for them, see Wenrich (1937). This method requires that the organisms be placed in a mordant, iron alum (ammonioferric sulfate) and a dye, hematoxylin. Since crystals of iron alum become yellow to brown when left standing, select only clear violet crystals in preparing the solution. The hematoxylin solution should be "ripened" or oxidized for best results. This may be done by pouring the solution in large beakers and letting it stand overnight. The beakers should first be covered with layers of cheesecloth to prevent the entrance of dust but not air. It is convenient to dissolve the required amount of dye in absolute alcohol after which distilled water is added to make the desired percentage.

The basic method and a modification follow:

1. Fixation: Use any of the fixing fluids described earlier.
2. Washing: Follow directions given for the fixing fluid selected.
3. Dehydration: Use a graded series of alcohols, i.e. 95, 85, 70, 50, 35, and 10 per cent alcohol then water, allowing the specimens to remain in each stage for five to 10 minutes. If material is washed

in 95 per cent alcohol begin dehydration with 85 per cent alcohol; if washed in 50 per cent alcohol, next use 35 per cent alcohol, etc. until water is reached.

4. Mordanting: Place specimens in 4 per cent aqueous solution of iron alum for 12–24 hours.[5]
5. Washing: Wash carefully in several changes of distilled water for five to eight minutes.
6. Staining: Place specimens in 0.5 per cent aqueous solution of Heidenhain's hematoxylin for the same length of time they remained in mordant.
7. Washing: Rinse specimens well in water. Material may be left here during the destaining or differentiation.
8. Destaining: *Iron alum method.* Differentiate carefully with a 1 or 2 per cent solution of iron alum under the microscope. As soon as the desired results are obtained, place specimens in tap water and wash carefully for one hour then dehydrate.

Picric acid method (Tuan, 1930). Although this method requires more time, the exceptionally good results warrant its use, and the writer recommends it in preference to iron alum destaining. The destaining is done with a saturated aqueous solution of picric acid. To check destaining, remove specimens from the picric acid and *wash them well in a large volume of tap water* to remove excess picric acid. Then rinse in dilute ammonia water (approximately 50 cc. distilled water to which is added several drops of concentrated ammo-

[5] Longer methods, i.e. 24 hours are generally more reliable.

nium hydroxide). Next rinse in a container of tap water and examine under the microscope. If not completely differentiated, place specimens back in picric acid, and repeat the process until desired results are obtained; then wash in tap water for one hour and dehydrate.

9. Dehydration: Specimens after washing should be gradually dehydrated by placing them in the graded series of alcohols through 95 per cent allowing approximately five minutes for each stage.

10. Clearing: Place specimens in clearing oil composed of four parts of thyme oil and one part clove oil. Use two changes of oil allowing specimens to remain in each for about six minutes. This is followed by two changes of xylol with six minutes for each change.

11. Mounting: Finally specimens should be mounted in dammar, balsam, euparal, clarite, or other mounting medium.

Other Hematoxylin Stains.

Mayer's hemalum and Delafield's hematoxylin stains are useful in the cytological study of paramecium since nuclei are stained clearly and permanently with little time needed for the procedure.

For each stain, specimens, after fixation, should be washed and finally hydrated to water after which the stain is applied for four to eight minutes at room temperature. Specimens are then removed from the stain, washed well in tap water for 30 minutes, dehydrated, cleared, and mounted. Alcoholic eosin (0.5 per cent solution in 95 per cent alcohol) makes an effective counterstain.

Mayer's Hemalum.

Hematoxylin .... 1 Gm.
Distilled water .... 1 liter

After the hematoxylin is dissolved, add 0.2 gr. of sodium iodate and 50 Gm. alum, dissolve and filter.

Delafield's Hematoxylin.

To 400 cc. of a saturated aqueous solution of ammonium alum add slowly, 25 cc. of a solution consisting of 95 per cent alcohol in which 4 Gm. of hematoxylin crystals are dissolved. Expose to light and air for several days then filter. Next add 100 cc. of glycerin and 100 cc. of methyl alcohol, expose as before to darken stain, then filter.

Since the staining power of Delafield's hematoxylin tends to increase with age, first stain some of the specimens experimentally. If too strong, dilute approximately 1:3 with distilled water.

Purified Methyl Green.

Pollister and Leuchtenberger (1949) have described a method using purified methyl green for the critical staining of chromatin which is believed to be as selective as the Feulgen reaction. Their work should be consulted for details.

Feulgen's Nucleal Reaction.

The Feulgen reaction, which is specific for the presence of thymonucleic acid in the cell, is an extremely valuable technique for demonstrating sharply and clearly, fine detail of chromosomes and nuclei in paramecium. The reaction is carried out in two principal steps, the first of which is a mild acid hydrolysis splitting the purine bases and the carbohydrate of the thymonucleic acid, thereby liberating aldehyde groups of the aldopentose sugars. The second step is a reaction between these liberated aldehyde groups and the fuchsin-sulfurous acid reagent (Schiff's reaction). Basic fuchsin is reduced and decolorized by the action of sulfur dioxide in the formation of this reagent. Nuclear material is colored a deep violet and the cytoplasm is left colorless unless a counterstain is used. The writer has used a number of variations and modifications of the original method and highly recommends the following two techniques for paramecium.

Feulgen Method with Sodium Bisulfite (NaHSO₃).

1. Fix paramecia in a sublimate fixative (Schaudinn's fluid preferred).[6]
2. Wash and dehydrate through alcohol series and allow specimens to remain in 95 per cent alcohol for approximately 24 hours.
3. Hydrate through alcohol series to water.
4. Place in *normal HCl* at room temperature for two minutes.
5. Transfer (hydrolyze) to *normal HCl maintained at exactly 60° C.* for five minutes.
6. Rinse in *normal HCl* at room temperature for one minute.
7. Place in *fuchsin-sulfurous acid reagent* one half to one hour.
8. Wash by passing specimens through three changes or dishes (covered) each containing *dilute sulfurous acid reagent* (10 per cent NaHSO₃), one and a half minutes in each dish, agitating the specimens during the period of washing.
9. Wash in running tap water for five to 10 minutes.
10. Dehydrate in graded series of alcohol in usual manner to 95 per cent alcohol.
11. Counterstain with fast green dissolved in 95 per cent alcohol; then

destain with 95 per cent alcohol, if counterstain is desired.
12. Clear in clearing oil and xylol (as described earlier) and mount.

Reagents for the Feulgen Nucleal Reaction

Normal HCl Solution

This is prepared by adding 82.5 cc. of concentrated hydrochloric acid with a specific gravity of 1.17–1.185 to 1 liter of distilled water.

Fuchsin-sulfurous Acid Reagent

Dissolve 2 Gm. of basic fuchsin in 400 cc. of hot distilled water by shaking. Then cool to about 50° C. and filter. Continue cooling to about 25° C., add 40 cc. of normal HCl solution; then add and dissolve 2 Gm. of anhydrous sodium bisulfite (NaHSO₃).[7] Place in glass-stoppered bottles (brown glass if available), and fill to top so as to exclude air and keep in dark. This reagent will turn to pale straw color after several hours but should not be used for at least 24 hours after being made.

Dilute Sulfurous Acid Reagent (Feulgen Wash)

Water with an excess of sulfur dioxide may be prepared as follows:

Distilled water	200 cc.
10 per cent anhydrous sodium bisulphite[7]	10 cc.
Normal HCl solution	10 cc.

Feulgen Method with Potassium Metabisulfite (K₂S₂O₅).

Investigators have experienced considerable difficulty with the Feulgen reaction due principally to unsatisfactory basic fuchsin in which the cytoplasm became stained because of im-

---

[6] Weak Flemming's fluid or virtually any fixative may be used providing the specimens are washed *very thoroughly* after fixation.

[7] Substitute potassium metabisulfite in place of sodium bisulfite in the preparation of these reagents if the second Feulgen method is used.

purities in the dye. It has been reported that by treating the bleached solution with activated charcoal, impurities of the dye may be removed. Also the use of potassium metabisulfite instead of sodium sulfite as a source of sulfur dioxide in the fuchsin-sulfurous reagent results in a more dependable and stable solution (Stowell, 1945). Additional important references dealing with the Feulgen technique follow: Feulgen (1926), Wyckoff, Ebeling, and Ter Louw (1932), Dodson (1946), and Rafalko (1946).

The writer has used the potassium metabisulfite substitution in treating paramecium with excellent results and the method follows:

1. Fix paramecia in a sublimate fixative (Schaudinn's fluid preferred).
2. Wash and dehydrate through alcohol series and allow specimens to remain in 95 per cent alcohol for 24 hours.
3. Hydrate through alcohol series to distilled water.
4. Place in normal HCl at room temperature for 1 minute.
5. Hydrolyze in normal HCl at 50° C. for 20 minutes (or 60° C. for five minutes).
6. Rinse in normal HCl at room temperature for one minute.
7. Place in fuchsin-sulfurous acid reagent (now containing potassium metabisulfite) for two hours.
8. Pass specimens through three changes or dishes of the sulfurous acid bleaching solution (now containing 10 per cent potassium metabisulfite) allowing 10 minutes for each change.
9. Wash in running tap water for five minutes.
10. Rinse in distilled water for one minute.
11. Dehydrate in graded series of alcohol in usual manner to 95 per cent.

12. Counterstain with fast green, if desired, clear and mount as in other methods.

Borax-carmine, Indulin Method.

This method, accidentally discovered by Lynch (1929), is one not commonly employed by protozoölogists, but because of its excellence it merits greater usage for the cytological study of paramecium and other Protozoa. Nuclei are stained a brilliant red or vermilion and the cytoplasm and cilia, a transparent blue. When properly used it not only results in a strikingly beautiful preparation but also shows fine detail and is permanent.

After using it upon paramecium and other Protozoa, the following method is recommended:

1. After fixation,[8] dehydrate through the alcohol series to 95 per cent, then hydrate to 50 per cent alcohol, allowing specimens to remain in each stage for five to 10 minutes.
2. Place specimens in Grenacher's borax-carmine for six to 24 hours. (Best results are obtained with Grübler's "rubrum opticum.")
3. In the same dish containing the stain and specimens carefully and slowly add enough drops of concentrated hydrochloric acid while agitating the dish to produce a *cloudy* or flocculent pink precipate. Distribute the precipitate in the container with a pipette or by shaking. Avoid adding too much acid since this will dissolve the precipitate and offset the desired results.
4. Allow specimens to remain in this precipitated borax-carmine 12–24 hours.

---

[8] Sublimate fixatives are especially recommended but most fixatives can be used. However, Flemming's or similar chromic acid fixatives do not give good results.

5. Transfer specimens to slightly acidified 50 and 70 per cent alcohol allowing five to 10 minutes for each stage.

6. Specimens now greatly overstained with a deep red color are destained under the microscope by adding and thoroughly mixing drops of concentrated hydrochloric acid to the 70 per cent alcohol until the cytoplasm of paramecia is colorless and nuclei brilliant red color.

7. Place in 80 per cent alcohol for 10 minutes when suitably destained.

8. Counterstain with a 5 per cent solution of Grübler's indulin in 80 per cent alcohol to stain cytoplasm and cilia a light blue color. Counterstaining must be done carefully since indulin is not easily removed from the cytoplasm.

9. Transfer specimens through several dishes of 95 per cent alcohol for a total of 10 minutes.

10. Clear with clearing oil, xylol, and mount as described earlier for other methods.

Osmium-toluidin Method.

A valuable method for staining cilia, the pellicular structure and contractile vacuolar apparatus is the osmium-toluidin technique of Gelei which follows:

1. Fix for one to 12 hours in 10 parts of a 2 per cent solution of osmic acid and one part formol.

2. Without washing, place specimens directly for one to 12 hours into a solution composed of 1 Gm. of alum, 2 Gm. potassium bichromate and 100 cc. distilled water.

3. Wash in distilled water.

4. Place in 1 per cent solution of ammonium molybdate for one to 12 hours.

5. Wash in distilled water.

6. Stain in 0.1–0.3 per cent aqueous solution of toluidin blue at 50°–60° C. for two to four minutes.

7. Dehydrate through series of alcohols, clear and mount in dammar.

Borrel's Staining Method.

Fix ciliates in Schaudinn's fluid or sublimate acetic mixture; then remove fixative by washing. Stain five to 20 minutes or longer in a saturated aqueous solution of magenta. In place of magenta, a mixture of equal parts of sufranin saturated in water and saturated in absolute alcohol, may be used. After staining, rinse preparation, then stain in a mixture of two parts of saturated aqueous indigo-carmine and one part saturated aqueous picric acid for two to five minutes. Rinse again until green stain fails to leave specimens. Dehydrate rapidly to 95 per cent alcohol, differentiate nuclear stain, complete dehydration, clear and mount in dammar.

Nuclei are stained red and cytoplasmic structures are stained green.

e. RELIEF STAINING. This is a fairly rapid and excellent method for the demonstration of the pellicular pattern and general surface structure, cilia, mouth, cytopharynx and contractile vacuolar apparatus. The dye in the staining solution is precipitated in and upon the organisms, and the method has been used advantageously by Bresslau (1921), Coles (1927), King (1928, 1935), and others wherein a variety of stains was employed.

Paramecia or other Protozoa are concentrated in a drop of clear culture fluid, placed on a slide or cover-glass to which is added a drop of the staining solution. The two drops are then mixed and the paramecia distributed evenly over the surface after which the preparations are allowed to dry thoroughly in air and mounted directly in dammar. The staining solutions should be of fairly high concentration and low toxicity.

King (1928) has demonstrated fine detail of paramecium in using separately 10 per cent aqueous solutions of nigrosin, Chinese blue and opal blue.

f. SILVER AND OSMIC IMPREGNATION METHODS.

Silver Impregnation Methods.

Klein (1926 a, b, 1927 a, b, 1927–28, 1928, 1931) is responsible for a valuable technique in which colloidal silver from silver nitrate is reduced and deposited in certain structures of paramecium and other ciliates by the action of sunlight.

Dry Method.

Concentrated paramecia in a drop of clear culture fluid are distributed on a cover-glass or slide and allowed to dry thoroughly in air. Preparations are next placed in 2 per cent aqueous solution of silver nitrate for six to eight minutes then washed thoroughly in distilled water. Smears are then placed in a dish (preferably smoothly glazed white porcelain) and exposed to daylight for two to eight hours depending upon the light intensity. This step, the reduction of silver on paramecium, should be controlled under the microscope. When the "silverline" pattern or network is clearly differentiated, remove from daylight, rinse preparation well in distilled water, dry thoroughly in air and mount directly in dammar.

Wet Method.

Gelei and Horváth (1931) developed a "wet" method in which the organisms are fixed in a sublimate-formaldehyde solution consisting of 95 cc. of saturated corrosive sublimate and 5 cc. of formaldehyde. After fixation, paramecia are washed with several changes of non-chlorinated water, once with distilled water, then placed in 2 per cent solution of silver nitrate for approximately 15 minutes. Specimens are then transferred to distilled water and exposed to daylight for an hour or less depending upon the intensity of the sunlight. Next, paramecia are washed in five changes of distilled water one minute for each change, dehydrated in the alcohol series, cleared and mounted on a slide.

The advantage of this method is proper fixation and absence of distortion since drying of the specimens is not allowed to occur during the procedure.

Osmic Impregnation Method.

For the osmic impregnation of paramecium, King (1935) recommends for fixation the osmo-sublimate mixture of Mann or the modified Champy fixative after Nassonov. With this method, King was able to show the detailed structure of the contractile vacuoles of paramecium. The organisms, after fixation are washed in running water, then placed in 2 per cent osmic acid at 38° C. for two to four days. After impregnation, the paramecia are washed, dehydrated in the series of alcohols, cleared and mounted.

Impregnated specimens also can be infiltrated with paraffin and sectioned 4–10 microns in thickness.

Additional information concerning special techniques may be found in the section on protozoölogical methods by Wenrich in McClung's (1937) "Microscopical Technique," and in the books by Bêlâr (1928), and Calkins and Summers (1941). The book, "Methods in the Study of Protozoa" by Kirby (1947) is extremely useful for many and varied specialized technical procedures. General techniques including fundamental processes, fixatives, stains, etc., may be found in the textbooks dealing with histological technique.

Methods and procedures in the isolation, cultivation, collection, concentration, sterilization, establishment of pure-line, pedigreed clones, induction of mating phenomena and conjugation, etc., will be found in their respective sections of the book. The comprehensive account by Sonneborn (1950 b) dealing with methods in the general biology and genetics of *P. aurelia* should be consulted by students planning research with these and related organisms.

Chapter 15

Usefulness of Paramecium in the Study of Fundamental Problems in Biology

A. Vitality Studies and Immortality
B. Action of Drugs, Dyes, Organic Extracts, etc., upon Paramecium and Its Usefulness in General Physiology
C. Cytological Studies
D. Studies on Inheritance
E. Serologic Studies
F. Growth and Population Studies

To the experimenter interested in studying basic and fundamental problems in biology, paramecium proves to be an exceptionally useful organism. No other protozoön has been so extensively investigated from so many different aspects.

Specimens can be collected in nature and cultivated easily in the laboratory. Beginning with a single specimen, the experimenter can obtain readily a rich, genetically pure pedigreed culture of progeny. For detailed microscopic observations, the animal or animals may be removed easily and quickly from the culture medium and placed into the fluid to be tested while observations continue. The animals then may be quickly removed from the test fluid and placed back again in their normal culture medium if necessary. Reproducing asexually by binary fission, the organisms are hereditarily alike, thereby insuring uniformity in test animals and controls. Also, long-continued experi-

ments may be performed upon the same race of such pedigreed animals.

Compared with other experimental animals, the maintenance of cultures of paramecium requires little space, little cost and once cultural conditions have been standardized, comparatively little time.

It should be remembered that besides consisting of a single cell, paramecium is an entire organism, performing the same vital activities of nutrition, excretion, irritability, etc. which are fundamental attributes of living protoplasm. Unlike tissue cells which are predominantly specialized for fairly well-defined physiologic functions such as secretion, contraction, etc., we find in paramecium structural complexities due to intracellular specializations. Hence, it is difficult to conceive of the organism as simply a single cell since division of labor in a physiologic concept reaches its highest expression in ciliates as exemplified by paramecium.

A. Vitality Studies and Immortality

Weismann's belief of potential immortality as exemplified in the Protozoa was first substantiated by Woodruff (1932, 1943) in his famous, long-continued pedigreed race of P. aurelia which had been maintained through many thousands of generations over a period of many years. First isolated and grown in the laboratory in 1907, it is of interest to report that this race is at present being maintained in several laboratories.

Eliminating the possibility of accidental death, it has been shown that one specimen will produce two daughters, each of the two daughters four granddaughters, etc. provided there is ample food and space. Under optimal conditions, there may be as many as five to seven generations per day, but in nature, it is very likely less.

Paramecium, of course, must ingest a certain amount of food—principally bacteria—to maintain its vegetative and reproductive activities. Generally speaking, the more suitable bacterial food ingested by paramecium, the greater will be its fission or reproductive rate. It also follows that when paramecium is supplied with little bacterial food, its reproductive rate decreases sharply. Finally when the organism is deprived of food, inanition phenomena occur ending in death.

Chejfec (1929) attacked the problem of immortality in paramecium not from the standpoint of the race but of a single individual. He succeeded in keeping single specimens alive for 120 days without the occurrence of division by the daily feeding with reduced amounts of bacterial food. Chejfec estimated that *P. caudatum* requires about 100,000 *B. coli* per hour while Ludwig reported that about 30,000 *B. subtilis* were sufficient (see Chapter 6 for food requirements). Chejfec observed the specimens daily and gave them only one-half the quantity of bacterial food according to signs of inanition. It was his aim to provide paramecium with barely enough food to maintain the vegetative activities but not quite enough to permit reproduction. With more refined methods, the intriguing experiments of Chejfec should be extended.

The subject of vitality is more extensively studied in Chapter 11.

Suggested problems in respect to vitality, the life-cycle, and immortality follow:

1. Comprehensive and carefully executed experiments using various species of *Paramecium* in an effort to induce encystment and if that is possible, excystment as well.

2. The roles of the described nuclear processes in respect to the life history of paramecium.

3. Mortality experiments concerned with reduced bacterial feeding to determine the length of time individuals of the various species of *Paramecium* may be kept alive without resorting to fission.

B. Action of Drugs, Dyes, Organic Extracts, etc., upon Paramecium and Its Usefulness in General Physiology

The fact that paramecium is highly sensitive to the action of drugs, dyes, and many other toxic substances reaffirms the conviction held by E. Duclaux who held that life manifestations of a living organism are capable of an infinitely more sensitive reaction than in any of our most perfected chemical reagents.

The action of hundreds of drugs and many other substances is treated in Chapter 5. Experiments dealing with cellular physiology are treated in Chapters 6, 7, and 8. It is not the object here to give repetitious information on the subject but to emphasize the fact that paramecium is an important and extremely useful tool for the solving of fundamental problems in the field of general biology, physiology, and pharmacology.

In spite of the large amount of investigation and numerous publications in this field, much of the work is of little value. Many investigators have failed to familiarize themselves with the normal physiologic activities of paramecium. Indeed, more than a few investigators have even failed to identify properly the species used. It is now possible to obtain rich cultures of a reliably identified species of *Paramecium* from biologic supply houses or from the many investigators presently engaged in work upon paramecium. Such cultures of correctly identified paramecia may be easily cultivated throughout the length of the experiment.

Since paramecium is a bacterial feeder, one studying the effects of certain substances—say hormonal extracts—must exercise great care and caution in respect to bacterial concentrations in the controls and experimental animals. The results of a number of investigators are worthless because this fact was disregarded.

Criteria for Determining the Action and Death Point of Chemicals, Drugs, and Other Agents.

A valuable measure of the effect of the environment and factors in it is the determination of the fission-rate as an index to vitality. The subject is treated in detail in Chapter 11, and should be consulted. Vitality studies by means of isolation cultures are useful for long-continued experiments. However, immediate and specific effects on the protoplasm of paramecium can be determined by the placing of specimens into the substance to be tested. Obviously, agents of low toxicity or concentration may produce influences or injuries upon paramecium which are not as easily detected under the microscope as more potent agents of greater toxicity. In establishing the moment or instant of death, one should look for the complete cessation of movement of the organism or of the cilia, complete stoppage of the contractile vacuolar activity or, in other instances, the disintegration of the cell-body.

Injuries or effects due to agents causing changes less than instant cessation of locomotion or bursting of the cell-body require greater observational care. One should look for the following:

   a. Changes in ciliary activity or locomotion
   b. Changes in rate of cyclosis
   c. Changes in volume of the cell-body
   d. Changes in rate of anterior and posterior contractile vacuole activity
   e. Trichocyst extrusion
   f. Slow peripheral disintegration of the cell-body

Some problems in general physiology which may be investigated with paramecium are listed below:

1. Specific action of drugs, dyes, organic extracts, snake antiserum, antibiotics, carcinogenic compounds, etc. (see Chapter 5).

2. Photodynamic influences after treatment with eosin, rose bengal, and other substances.

3. Effects of x-rays, ultraviolet light, and radioactive isotopes upon vegetative and reproductive activities of paramecium.

4. Respiration studies upon single specimens during fission and non-fission stages; during conjugation and mating.

5. Repetition and amplification of the spectacular experiments of Jollos. By slowly raising the temperature and increasing the poison concentrations with arsenic acid, he was able to produce resistant long-lasting but impermanent strains ("Dauermodifikation").

6. Effects of various salts and other agents upon ciliary action.

7. Microdissection experiments involving behavior of fragments during

fission, non-fission, and conjugation stages in respect to problems of morphogenesis.

8. Nutritional experiments involving the selectivity and utilization of certain species of bacteria as food. A comparison of monobacterial feeding with mixed species of bacteria. Attempts at growth of paramecia in culture without the presence of any living microörganisms to serve as food (pure cultures).

9. Behavior of paramecia as a result of electric stimulation not only during vegetative activity but during fission, conjugation, and mating phenomena.

10. Continued investigations upon the mating phenomena from a biochemical approach in an effort to discover the substance responsible for the "clumping reaction" leading to conjugation.

C. Cytological Studies

Problems suggested below should yield important information not only in the life of paramecium but in the field of biology as well.

1. Morphogenic studies during fission and conjugation.

2. Selection of certain mating types of *P. bursaria* and *P. trichium* for a detailed study of conjugation in respect to chromosome structure and behavior. In these species, chromosomes are large, hence can be studied easily.

3. Investigations of the various nuclear processes not only with fixed and stained specimens but studied in the *living condition* in the precision, microcompression chamber.

4. Behavior, origin, and fate of cilia and trichocysts during fission, non-fission, and conjugation stages.

5. Greater detailed studies of the macronucleus, especially during fission and during development and degeneration at time of conjugation.

6. Additional information obtained with the electron microscope in regard to the finer structure of paramecium.

D. Studies on Inheritance

Paramecium has played an important role in the problems dealing with inheritance in the Protozoa due primarily to the extensive research work of Jennings and Sonneborn and the latter's associates. Details of inheritance in uniparental and biparental reproduction may be found in Chapter 10. In spite of the recent notable studies made upon inheritance in paramecium, it appears that only a beginning has been made thus far. Continued investigations with such excellent subjects as the species of *Paramecium* must eventually yield abundant information on the nature of the gene and of gene action.

Some problems concerned with inheritance in paramecium follow:

1. The relationships of genic characters of parents and progeny as effected by environmental conditions.

2. Studies concerned with acclimatization and immunity to unfavorable environmental conditions and their possible persistence to future generations.

3. Investigations dealing with genic factors that may exist outside of the chromosomes (plasmagenes) and the role of cytoplasm in inheritance.

4. The role of the macronucleus and macronuclear regeneration in inheritance.

5. The inheritance of mating types at conjugation in the various species of *Paramecium*.

6. The extent to which cytoplasm and/or pronuclei are exchanged in joined pairs of paramecia and its relation to inheritance.

7. Studies on the origin of diverse macronuclei which arise from identical micronuclei.

8. Investigations dealing with "cytoplasmic lag" phenomena.

9. Additional information on the cytoplasmic killer factor, kappa.

E. Serologic Studies

It has been shown that immunologic reactions have been of considerable value in demonstrating resemblances or differences in strains, races, and species of *Paramecium* and other Protozoa. Serologic reactions may be used to compare the chemical composition with the morphologic structure. As mentioned earlier (Chapter 12), serologic methods have been shown to be more precise than certain other methods of demonstrating relationships between organisms. It is therefore surprising that these highly specific tests have not been used more extensively with paramecium and other free-living Protozoa.

Some problems concerned with serology in paramecium follow:

1. The relationship between serologic groups and mating varieties in the species of *Paramecium*.

2. The determination, inheritance, and induced mutations of antigenic characters in the species of *Paramecium*.

3. A serologic study of conjugation in the various species of *Paramecium* in an attempt to compare the antigenic character of the daughter exconjugants with the antigenic character of the clones prior to conjugation.

4. Elucidation of the fact that serum-resistant clones arise following exposure of temporarily resistant animals to antiserum in which these clones produce two kinds of lines, viz., some resistant and others susceptible.

F. Growth and Population Studies

Paramecium has been used extensively in growth studies and to a lesser extent, in population studies as described in Chapter 6. It is easily possible to conduct growth experiments in relation to food concentration, pH of the culture medium, temperature, and other external factors. Some problems concerned with studies in this field follow:

1. The relation to growth of paramecium after prolonged cultivation upon a single species of bacterium used as the single food source.

2. Whether or not changes in the species of bacteria used as food are necessary for continued existence.

3. A study of the effects upon growth and nuclear phenomena of paramecium with so-called "deleterious bacteria" which may not be suitable as food.

4. Experiments designed to cultivate species of paramecium free of their principal food source, bacteria, as well as other microörganisms.

5. The relation of food and growth to nuclear phenomena in paramecium.

6. Continuation of the theoretical and experimental investigations of Gause (1934 f) and his associates concerning the struggle for existence in which paramecium is used as the test animal.

Bibliography

Abderhalden, E. and O. Schiffmann (1922), Weitere Untersuchungen über die von einzelmen Organen hervorgebrachten Substanzen mit spezifischer Wirkung. VII, Chemotaktische Versuche an Paramaecien und Untersuchungen über die Geschwindigkeit ihrer Teilung unter dem Einfluss von Optonen aus verschiedenen Organen, *Arch. ges. Physiol.,* **194,** 206–217.

Ackerman, E. (1950), Optimum frequencies for sonic disintegration of paramecia, *Federation Proc.,* **9,** 3.

—————— (1951), Vibrating plate transducers for frequency studies of the breakdown rate of biological cells, *Rev. Scient. Instruments,* **22,** 649–651.

—————— (1951 a), Resonances of biological cells at audible frequencies, *Bull. Math. Biophysics,* **13,** 93–106.

—————— (1952), Cellular fragilities and resonances observed by means of sonic vibrations, *J. Cellular Comp. Physiol.,* **39,** 167–190.

Ackert, J. E. (1916), On the effect of selection in Paramecium, *Genetics,* **1,** 387–405.

Acton, H. W. (1921), On the behavior of *Paramecium caudatum* towards the cinchona alkaloids, *Indian J. M. Research,* **9,** 339–358.

Adolph, E. F. (1922), The physiological action of excretory products, *Anat. Record,* **24,** 396.

—————— (1925), Some physiological distinctions between fresh water and marine organisms, *Biol. Bull.,* **48,** 327–334.

—————— (1926), The metabolism of water in Amoeba as measured in the contractile vacuole, *J. Exp. Zoöl.,* **44,** 355–381.

—————— (1931), The Regulation of Size as Illustrated in Unicellular Organisms, Springfield, Ill., Charles C Thomas, 1–233.

—————— (1943), Physiological Regulations, Lancaster, Pa., Jaques Cattell Press.

Alexandrowa, W. and C. Istomina (1903), Einige Beobachtungen an Infusorien, *Trav. soc. naturalistes Leningrad,* **34,** 159–160.

Allee, W. C. (1931), Animal Aggregations, Chicago, Univ. of Chicago Press, 1–431.

—————— (1934), Recent studies in mass physiology, *Biol. Rev.,* **9,** 1–48.

—————— (1938), The Social Life of Animals, New York, W. W. Norton & Company, 1–293.

—————— (1941), Integration of problems concerning protozoan populations with those of general biology, *Bio. Symposia,* **4,** 75–89.

Allescher, M. (1912), Uber den Einfluss der Gestalt des Kernes auf die Grössenabnahme hungernder Infusorien, *Arch. Protistenk.,* **27,** 129–171.

Allman, G. J. (1855), On the occurrence among the Infusoria of peculiar organs resembling thread-cells, *Quart. J. Microscop. Sci.,* **3,** 177–179.

Allsopp, C. B., et al. (1951), Symposium on radiation chemistry. 1. Radiation chemistry in relation to radiobiology, by C. B. Allsopp. 2. Elementary processes in the radiation chemistry of water and implications for radiobiology, by M. Burton. 3. Hydrogen peroxide formation in water exposed to ionizing radiations, by P. Bonet-Maury. 4. A review of the evidence for the production of free radicals in water consequent on the absorption of ionizing radiations, by F. S. Dainton. 5. Some aspects of the biochemical effects of ionizing radiations, by W. M. Dale. 6. Radiation chemistry of organic halogen compounds, by W. Minder, *Brit. J. Radiol.*, **24**, 413–440.

Alpatov, V. V. (1937), Heat resistance and its changes under the influence of salts, narcotics, and electrical stimulation, *Bull. soc. naturalistes Moscow*, Sect. biol., **46**, 133.

——— and O. K. Nastjukova (1932), Effect of various ultraviolet ray exposures on the division rate in Paramecium, *Compt. rend. acad. sci. U.R.S.S.*, Sér. A., **1932**, 287–291.

——— and ——— (1934), Differences in susceptibility in ultraviolet radiation of Paramecium caudatum and Paramecium bursaria, *Proc. Soc. Exptl. Biol. & Med.*, **32**, 99–101.

Altenburg, E. (1946), The symbiont theory in explanation of the apparent cytoplasmic inheritance in Paramecium, *Am. Naturalist*, **80**, 661–662.

——— (1946), The "Viroid" theory in relation to plasmagenes, viruses, cancer and plastids, *Am. Naturalist*, **80**, 559–567.

——— (1948), The role of symbionts and autocatalysts in the genetics of the ciliates, *Am. Naturalist*, **82**, 252–264.

Alverdes, F. (1922 a), Studien an Infusorien über Flimmerbewegung, Lokomotion und Reizbeantwortung, *Arb. a. d. Geb. d. exp. Biol.*, **3**, 123.

——— (1922 b), Lebendbeobachtungen an beflimmerten und begeisselten Organismen, *Verhandl. d. Zool. Ges.*, **27**, 37–39.

——— (1922 c), Zur Lokalisation des chemischen und thermischen Sinnes bei Paramecien und Stentor, *Zool., Anz.*, **55**, 19–21.

——— (1922 d), Zur Lehre von den Reaktionen der Organismen auf äussere Reize, *Biol. Zentr.*, **42**, 218–222.

——— (1922 e), Untersuchungen über Flimmerbewegung, *Arch. ges. Physiol. Pflügers*, **195**, 227–244.

——— (1923), Der Sondercharakter der von den Ciliaten gezeigten Galvanotaxis, *Arch. ges. Physiol. (Pflügers)*, **198**, 513–542.

——— (1923 a), Beobachtungen an *Paramecium putrinum* und *Spirostomum ambiguum*, *Zool. Anz.*, **55**, 277–287.

——— (1925), Spezielle Physiologie der Flimmer- und Geisselbewegung, *Handb. d. norm. u. path. Physiol.*, **8**.

——— (1937), Das Lernvermögen der einzelligen Tiere, *Ztschr. Tierpsychol.*, **1**: 35–38.

——— (1939), Weiteres über die Marburger Dressurversuche an niederen Tieren, *Zool. Anz. Suppl.* **12**, 103–110.

Amberson, W. R. (1928), The influence of oxygen tension upon the respiration of unicellular organisms, *Biol. Bull.*, **55**, 79–91.

Amster, Göttingen (1922), Ein neus Züchtungs-verfahren für Protozoen, *Zentr. Bakt. Parasitenk.*, **89**, 166–168.

Anderson, T. F. (1951), Techniques for the preservation of three-dimensional structure in preparing specimens for the electron microscope, *Trans. N. Y. Acad. Sci.*, **13**, 130–134.

Andrejewa, E. (1928), Über den Einfluss des Bodens als einen Regulierungsfaktor des Mediums auf die Kultur von *Paramecium aurelia*, *Arch. Protistenk.*, **63**, 94–104.

——— (1930), Die elektrische ladung und die Bewegungsgeschwindigkeit der Infusorien Paramaecium caudatum, *Kolloid-Z.*, **51**, 348–356.

——— (1931), Zur Frage über die physikalisch-chemische Bestimmung der Korrelationen einiger physiologischer Prozess bei *Paramecium caudatum*, *Arch. Protistenk.*, **73**, 346–360.

Anonymous (1703), An extract of some letters sent to Sir C. H. relating to some microscopical observations. *Phil. Tr. Roy. Soc.*, **23**, 1368.

Aunap, E. (1927), Eine Methode. Infusioren durch den Objektträger zu fixieren und zu färben, *Arch. Protistenk.*, **60**, 193–196.

Austin, M. L. (1946), Contributions towards an analysis of the killing action of variety 4 killers in *Paramecium aurelia, Anat. Record*, **96**, 18.

—— (1948), The killing substance, Paramecin: activity of single particles, *Am. Naturalist*, **82**, 51–59.

—— (1951), Sensitivity to paramecin in *Paramecium aurelia* in relation to stock, serotype, and mating type, *Physiol. Zoöl.*, **24**, 196–204.

Back, A. (1939), Sur un type de lésions produites chez *Paramecium caudatum* par les rayons x, *Compt. rend. soc. biol.*, **131**, 1103–1106.

—— and L. Halberstaeder (1945), Influence of biological factors on the form of Roentgen-ray survival curves. Experiments on *Paramecium caudatum, Am. J. Roentgenol.*, **54**, 290–295.

Balamuth, W. (1940), Contributions to the problem of regeneration in Protozoa, *Am. Naturalist*, **74**, 528–541.

—— (1940 a), Regeneration in Protozoa: A problem of morphogenesis, *Quart. Rev. Biol.*, **15**, 290–337.

Balbiani, E. G. (1858), Note relative a l' existence d' une génération sexuelle chez les Infusoires, *Compt. rend.*, **46**, 628–632.

—— (1858 a), Note relative a l'existence d' une génération sexuelle chez les infusoires, *J. physiol.*, **1**, 347–352.

—— (1860), Note sur un cas de parasitisme improprement pris pour un mode de reproduction des Infusoires cilies, *C. R. Acad. Sci. Paris*, **51**, 319–322.

—— (1860 a), Etudes sur la reproduction des protozoaires. Du role des organes générateurs. Dans la division des Infusoires ciliés, *J. physiol. Homme et Anim.*, **3**, 71–87.

—— (1860 b), Observations et expériences sur les phénomènes de réproduction fissipare chez les Infusoires ciliés, *Comp. rend. acad. sci.*, **1**, 1191–1195.

—— (1861), Recherches sur les phénomènes sexuels des infusoires, *J. physiol. Homme et Anim.*, **4**, 465–520.

—— (1881–82), Les Protozoaires, *J. Microg.*, **5** and **6**.

—— (1883), Bütschli et la conjugaison des infusoires, *Zol. Anz.*, **6**, 192–196.

—— (1888), Recherches experimentales sur la merotomie des Infusoires cilies, *Rec. zool. suisse*, **5**, 1–72.

—— (1891), Sur la formation des monstres doubles chez les infusoires, *J. de l'anat. et physiol.*, **27**, 169–196.

—— (1892), Nouvelles recherches expérimentales sur la mérotomie des Infusoires ciliés, *Ann. de Micrographie*, **4**, 369–449.

—— (1893), Mérotomie des infusoires ciliés, *Ann. de Micrographie*, **5**, 49–84.

—— (1898), Etudes sur l'action des sels sur les infusoires, *Arch. d'anat. micr.*, **2**, 318–600.

Baldwin, W. M. (1920), A study of the combined action of x-rays and of vital stains upon Paramecia, *Biol. Bull.*, **39**, 59–66.

Ball, G. H. (1925), Studies on Paramecium I. Experiments on the action of various endocrine substances of liver, and of glycogen on the division rate of *Paramecium, Univ. Calif. Pubs. Zoöl.*, **26**, 353–383.

—— (1925 a), II. The behavior of a conjugating race of *Paramecium caudatum, Univ. Calif. Pubs. Zoöl.*, **26**, 385–433.

—— (1927), III. The effects of vital dyes on *Paramecium caudatum, Biol. Bull.*, **52**, 68–78.

Bancroft, F. W. (1905), On the validity of Pflügers Law for the galvanotropic reactions of Paramecium, *Univ. Calif. Pubs. Physiol.*, **2**, 193.

—— (1905 a), Über die Gültigkeit des Pflüger'schen Gesetzes fur die galvanofropischen Reaksionen von Paramecium, *Pflügers Arch. ges. Physiol.*, **107**, 535–556.

Bancroft, F. W. (1906 a), On the influence of the relative concentration of calcium ions on the reversal of the polar effects of the galvanic current in Paramecium, *J. Physiol.* (*London*), **34**, 444–463.

—— (1906 b), The control of galvanotropism in Paramecium by chemical substances, *Univ. Calif. Pubs. Physiol.*, **3**, 21–31.

Banta, A. M. (1914), A new form of collecting pipette, *Science*, **40**, 98–99.

Barbarin, V. V. (1938), Agents déterminant la balance de graisse et de glycogène chez le *Paramaecium caudatum*, *Biol. Zhur.*, **7**, 391–398.

—— (1939), Factors determining the fat-glycogen balance in *Paramecium caudatum* III. The influence of suffocation on the accumulation of fat and glycogen, *Chem. Zentralbl.*, **1**, 1193.

—— (1940), Alternation of sensibility in *Paramecium caudatum* during starvation, *Leningradskii Gosudarstvennyi Pedagogichgskii. Uchenye Zapiski*, **30**, 51–64.

Bardeen, C. R. (1906–8), The action of the X-rays on Paramecia, *Anat. Rec.*, **1**, 59–60.

Barnes, T. C. (1937), Textbook of General Physiology, Philadelphia, Blakiston and Co., 1–554.

—— and H. Z. Gaw, The chemical basis for some biological effects of heavy water, *J. Am. Chem. Soc.*, **57**, 590–591.

Baron, M. A. (1926), Über mitogenetische Strahlung bei Protisten, *Arch. Entwicklungs-mech. mechan.*, **108**.

Barratt, J. O. W. (1904 a), Die Wirkung von Sauren und Basen auf lebende Paramecien, *Ztschr. f. allg. Physiol.*, **4**, 438–484.

—— (1904 b), The lethal concentration of acids and bases in respect to *Paramecium aurelia*, *Proc. Roy. Soc., London*, **74**, 100–104.

—— (1905 a), Die Addition von Säuren und Alkalien durch lebendes Protoplasma, *Ztschr. f. allg. Physiol.*, **5**, 10–34.

—— (1905 b), Die Kohlensäureproduktion von *Paramecium aurelia*, *Ztschr. f. allg. Physiol.*, **5**, 66–73.

—— (1905 c), Der Einfluss der Konzentration auf die Chemotaxis, *Ztschr. f. allg. Physiol.*, **5**, 73–94.

Baskina, V. P. (1924), Changes in the physiological state of *Paramecium caudatum* in cultures, measured by their resistance against poisons, *Bull. inst. recherches biol. Perm*, **10**, 423–434.

Bauer, K. (1926), Über die Wirkung von Histamin und Adrenalin auf Protozoen und Leukozyten, *Verhandl. d. Zool. Ges.*, **31**, 172–177.

Bayer, G. and T. Wense (1936 a), Über den Nachweis von Hormonen in einzelligen Tieren, I. Cholin und Acetylcholin im Paramecium, *Pflügers Arch. ges. Physiol.*, **237**, 417–422.

—— (1936 b), Über den Nachweis von Hormonen in einzelligin Tieren, II. Adrenalin (Sympathin) im Paramaecium, *Pflügers Arch. ges. Physiol.*, **237**, 651–654.

Beale, G. H. (1948), The process of transformation of antigenic type in *Paramecium aurelia*, Variety 4, *Proc. Nat. Acad. Sci. U.S.*, **34**, 418–423.

Beck, L. V. and A. C. Nichols (1937), Action of fluorescent dyes on Paramecium as affected by pH, *J. Cellular Comp. Physiol.*, **10**, 123.

Becker, G. R. (1926), Vital staining and reduction of vital stains by protozoa, *Biol. Bull.*, **50**, 235–238.

Becquerel, P. (1936), La vie latente de quelques Algues et Animaux inferieurs aux basses températures et la conservation de la vie dans l'univers, *Compt. rend.*, **202**, 978–981.

Beers, C. D. (1926), The life-cycle in the ciliate *Didinium nasutum* with reference to en-cystment, *J. Morphol. and Physiol.*, **42**, 1.

—— (1928), The relation of dietary insufficiency to vitality in the ciliate *Didinium nasutum*, *J. Exp. Zoöl.*, **51**, 121–133.

———— (1929), On the possibility of indefinite reproduction in the ciliate *Didinium nasutum* without conjugation or endomixis, *Am. Naturalist,* **63,** 125–129.

———— (1931), Some effects of conjugation in the ciliate *Didinium nasutum, J. Exp. Zoöl.* **58,** 455–470.

———— (1935), Structural changes during encystment and excystment in the ciliate *Didinium nasutum, Arch. Protistenk.,* **84,** 133–155.

———— (1937), A method for the sectioning of protozoa en masse, *Science,* **86,** 381–382.

———— (1942), An endorsement of the use of generic names as common nouns, *Science,* **96,** 403–404.

———— (1946), *Tillina magna:* Micronuclear number, encystment and vitality in diverse clones; capabilities of amicronucleate races, *Biol. Bull.,* **91,** 256–271.

Behrend, K. (1916), Über die Wirkung des Glyzerins auf Protisten und Pflanzenzellen, *Arch. für Protistenk.,* **36,** 174–187.

Bêlâr, K. (1926), Der Formwechsel der Protistenkerne, *Ergeb. u. Fortschr. der zool.,* **6,** 235–654.

———— (1928), Methoden zur Untersuchung der Protozoen, Methodik d. wiss, *Biologie,* Berlin, Springer, **1,** 735.

Bennett, A. H., H. Jupnik, H. Osterberg, and O. W. Richards (1946), Phase Microscopy, *Trans. Am. Microscop. Soc.,* **65,** 99–131.

Bernheimer, A. W. (1938), A comparative study of the crystalline inclusions of protozoa, *Trans. Am. Microscop. Soc.,* **58,** 336–343.

———— and J. A. Harrison (1940), Antigen-antibody reactions in Paramecium: The aurelia group, *J. Immunol.,* **39,** 73–83.

———— (1941), Antigenic differentiation among strains of *Paramecium aurelia, J. Immunol.* **41,** 201–208.

Beutler, R. (1929), Liefert das Glykogen die Energie für den Flimmerschlag?, *Ztschr. f. vergl. Physiol.,* **10,** 540–545.

Bhatia, B. L. (1923), On the significance of extra contractile vacuoles in *Paramecium caudatum, J. Roy. Microscop. Soc.,* **262,** 69–72.

Biancani, E. H. and A. Dognon (1935), Les ultra-sons et leurs action biologiques, *J. physiol. et path. gen.,* **32,** 1083–1106.

Bichniewicz, S. (1913), Die Beeinflussung der Giftigkeit des Chinins durch Fremdstoffe gegenüber *Colpidium colpoda, Z. allgm. Physiol.,* **15,** 133–183.

Bidder, G. P. (1923), The relation of the form of a sponge to its currents, *Quart. J. Microscop. Sci.,* **67,** 293.

Bills, C. E. (1922), Inhibition of locomotion in Paramecium and observations on certain structures and internal activities, *Biol. Bull.,* **42,** 7–13.

———— (1923), A pharmacological comparison of six alcohols, singly and in admixture, on Paramecium, *J. Pharmacol. Exp. Therap.,* **22,** 49.

———— (1924), Some effects of the lower alcohols on *Paramaecium, Biol. Bull.,* **47,** 253–264.

Binet, A. (1889), Psychic Life of Microorganisms, Chicago, Open Court Publishing Co.

Binz, C. (1867–94), Über die Wirkung antiseptischer Stoffe auf Infusoria von Pflanzenjauche, *Zentralbl. f. d. med. Wissensch.,* Real-Enzyklopadie, **4.**

Birge, E. A. and C. Juday (1911), The inland lakes of Wisconsin. The dissolved gases of the water and their biological significance, *Wisconsin Geol. Natural Hist. Surv. Bull.,* **22,** 259.

Birukoff, B. (1899), Untersuchungen über Galvanotaxis, *Arch. ges. Physiol. (Pflügers),* **77,** 555–585.

———— (1904), Zur Theorie der Galvanotaxis, *Arch. f. Anat. u. Physiol.,* 271–296.

———— (1906), Zur Theorie der Galvanotaxis, *Arch. ges. Physiol. (Pflügers),* **111,** 95–143.

Blum, H. F. (1941), Photodynamic Action and Diseases Caused by Light, New York, Reinhold Publishing Co., 1–309.

Bodine, J. H. (1921) Hydrogen-ion concentration of protozoan cultures, *Biol. Bull.*, **41**, 73–77.

Boell, E. J. The effect of respiratory inhibitors on the oxygen consumption of *Paramecium calkinsi, Anat. Record*, **84**, 493–494.

––––– (1946), The effect of sodium azide on *Paramecium calkinsi, Biol. Bull.*, **91**, 238–239.

––––– and Woodruff, L. L. (1941), Respiratory metabolism of mating types in *Paramecium calkinsi, J. Exp. Zoöl.*, **87**, 385–402.

Bokorny, T. (1895), Einige vergleichende Versuche über das Verhalten von Pflanzen und niederen Tieren gegen basische Stoffe, *Pflügers Arch. ges Physiol.*, **59**, 557–562.

––––– (1896), 1. Vergleichende Studien über die Giftwirkung verschiedener chemischer Substanzen bei Algen und Infusorien. 2. Toxikologische Notizen über Orthound Paraverbindungen, *Pflügers Arch. ges Physiol.*, **64**, 262–306, 306–312.

––––– (1905), Nochmals über die Wirkung stark verdünnter Lösungen auf lebende Zellen, *Pflügers Arch. ges Physiol.*, **110**, 174–226.

––––– (1905 d), Das Kupfer und die Giftwirkung des destillierten Wassers. Übereinstimmendes Verhalten der Metalle der Kupfergruppe gegen Zellen der nierderen Pflanzen, *Chemikerz.*, **29**, 687 and 1201.

––––– (1911), Verhalten von Infusorien und anderen niederen Organismen sowie Pflanzen gegen stark verdünnte wässerige Auflösungen von Basen, *Arch. Zellforsch*, **7**, 1–26.

Boland, J. C. (1928), Immobilization of paramecia, *Science*, **67**, 654.

Bordas, L. (1906), Moyens de defense et d'attaque de quelques infusoires, *Naturaliste*, Paris, **28**, 161–62.

Borowski, W. M. (1922), Zur Biologie der Infusorien, *Arch. russ. Protist.*, **1**, 136–140.

Bovie, W. T. and D. M. Hughes (1918), The effects of quartz ultraviolet light on the rate of division of *Paramecium caudatum, J. Med. Research*, **39**, 223–231.

Bozler, E. (1924 a), Über die Morphologie der Ernährungsorganelle und die Physiologie der Nahrungsaufnahme von *Paramaecium caudatum Ehrenberg, Arch. Protistenk.*, **49**, 163–215.

––––– (1924 b), Über die physikalische Erklärung der Schlundfadenströmungen, ein Beitrag zur Theorie der Protoplasmaströmungen, *Z. vergleich. Physiol.*, **2**, 82–90.

––––– (1926), Reizphysiologische Untersuchungen an Paramaecien, *Zool. Anz.*, **2**, 124–129.

Bragg, A. N. (1935 a), A new method of distinguishing species of Paramecium, *Proc. Oklahoma Acad. Sci.*, **15**, 56–57.

––––– (1935 b), The initial movements of the food vacuole of *Paramecium trichium Stokes, Arch. Protistenk.*, **85**, 421–425.

––––– (1936), Some morphological variations in *Paramecium bursaria* (Ehrenberg), *Arch. Protistenk.*, **88**, 69–75.

––––– (1936 a), Observations on the initial movements of the food vacuoles of *Paramecium multimicronucleata* P. and M. with comments on conditions in other species of the genus, *Arch. Protistenk.*, **88**, 76–84.

––––– (1936 b), Selection of food in *Paramecium trichium, Physiol. Zoöl.*, **9**, 433–442.

––––– (1937), Toxicity of certain carmine preparations to Protozoa, *Proc. Oklahoma Acad. Sci.*, **17**, 57.

––––– (1939 a), Selection of food by Protozoa, *Turbox News*, **17**, 41–44.

––––– (1939 b), Some methods useful in studying selection of food by Protozoa, *Turbox News*, **17**, 41–44.

––––– and H. R. Hulpieu (1925), A method of demonstrating acidity of food-vacuoles in Paramecium, *Science*, **61**, 392.

Brahmachari, P., U. Brahmachari and R. Banerjea (1932), Chemotherapy of quinoline compounds III. The action of certain quinoline compounds on Paramecium, *J. Pharmacol. Exp. Therap.*, **44**, 445–448.

—— *et al.* (1933), Chemotherapy of quinoline compounds IV. The action of certain quinoline compounds on Paramecia, *J. Pharmacol. Exp. Therap.*, **48**, 149–150.

Bramstedt, F. (1935), Dressurversuche mit *Paramecium caudatum* und *Stylonchia mytilus*, *Z. vergleich. Physiol.*, **22**, 490–516.

—— (1937), Wirkung des hypophysenhinterlappenhormons auf *Paramecium caudatum*, *Zool. Anz.*, **117**, 97–103.

—— Über die Dressurfähigkeit der Ciliaten, *Zool. Anz.* Suppl. (*Verhandl. deutsch. zool. Gesellsch.*, 41) **12**, 111–132.

Brandt, K. (1881), Färbung lebender einzelliger Organismen, *Biol. Zentralbl.*, **1**, 202–205.

Brandwein, P. F (1935), The culturing of fresh-water protozoa and other small invertebrates, *Am. Naturalist*, **69**, 628–632.

Bresslau, E. (1913), Über das spezifische Gewicht des Protoplasmas und die Wimperkraft der Turbellarien und Infusorien, *Verhandl. deut. zool. Ges.*, Bremen.

—— (1921), Die Gelatinierbarkeit des Protoplasmas als Grundlage eines Verfahrens zur Schnellanfertigung gefärbter Dauerpräparate von Infusorien, *Arch. f. Protistenk.*, **43**, 467–480.

—— (1924), Ein einfacher, insbesondere für kleine Flüssigkeitsmengen geeigneter Apparat zur Bestimmung der Wasserstoffionenkonzentration (Hydrionometer) mit den Michaelisschen Indikatoren, *Deutsche med. Wchnschr.*, **6**.

—— (1928), Die Stäbchenstruktur der Tektinhüllen, III., Arb. aus dem Staatsinst. f. experim., *Therap. u. d. Georg Speyer-Hause*, **21**, 26.

Bretschneider, L. H. and G. Hirsch (1927), Nahrungsaufnahme, intraplasmatische Verdauung und Ausscheidung bei *Balantidium giganteum*, *Z. vergleich. Physiol.*, **6**, 3–4.

Brodsky, A. (1908), Sur la structure intime du Frontonia leucas, *Rev. suisse zool.*, **16**, 75–128.

—— (1924), Die Trichocysten der Infusorien, *Arch. russ. Protistol.*, **3**, 23–37.

Browder, A. (1915), The effect of lecithin and cholesterol upon the division rate in *Paramecium*, *Univ. Calif. Pubs. Physiol.*, **5**, 1–3.

Brown, J. (1934), Pure cultures of Paramecium, *Science*, **80**, 409–410.

Brown, M. G., J. M. Luck, G. Sheets, and C. V. Taylor (1933), The action of x-rays on *Euplotes taylori* and associated bacteria, *J. Gen. Physiol.*, **16**, 397–406.

Brown, R. B. (1944), On quieting Paramecium with methyl cellulose, *Science*, **100**, 62.

Brown, R. H. J. (1940), The protoplasmic viscosity of Paramecium, *J. Exp. Biol.*, **17**, 317–324.

Brown, V. E. (1930), The neuromotor apparatus of Paramecium, *Arch. zool. exp. et gén.*, **70**, 469–481.

Brücke, E. (1861), Die Elementarorganismen, *Abhandl. Wiener Akad. nat.-math.*, **44**, 381–406.

Buck, J. B. (1943), Quieting Paramecium for class study, *Science*, **97**, 494.

Budgett, S. P. (1898), On the similarity of structural changes produced by lack of oxygen and certain poisons, *Am. J. Physiol.*, **1**, 210–214.

Budington, R. A., and H. F. Harvey (1915), Division rate in ciliate protozoa as influenced by thyroid constituents, *Biol. Bull.*, **28**, 304–314.

Bullington, W. E. (1925), A study of spiral movement in the ciliate infusoria, *Arch. Protistenk.*, **50**, 219–75.

—— (1930), A further study of spiraling in the ciliate Paramecium, with a note on morphology and taxonomy, *J. Exp. Zoöl.*, **56**, 423–51.

Bullough, W. S. (1946), Agar technique for arresting movement in Protozoa, *Science*, **104**, 227.

Buonanni, F. (1691), Observationes circa Viventia (etc.), Rome.

Burbanck, W. D. (1950), Growth of pedigreed strains of *Paramecium caudatum* and *Paramecium aurelia* on a non-living medium, *Biol. Bull.*, **99**, 353–354.

Burbanck, W. D. and M. P. Burbanck (1951), Autogamy of *Paramecium aurelia* (stock 51.7) grown on non-living media, *Biol. Bull.*, **101**, 206–206.

—— and M. P. Burbanck (1951 a), Autogamy in *Paramecium aurelia* (stock 51.7) grown in isolation culture on living and non-living media, *Proc. Am. Soc. Protozoöl.*, **2**, 13.

Burge, W. E. (1924), The effect of different anaesthetics on the catalase content and oxygen consumption of unicellular organisms, *Am. J. Physiol.*, **69**, 304–306.

—— and A. M. Estes (1926), The effect of insulin, thyroxin and temperature on the sugar metabolism of Paramecium, *J. Metab. Research*, **78**, 183–86.

—— and —— (1928), A study of the inorganic constituents of the body with respect to their stimulating effect on sugar metabolism, *Am. J. Physiol.*, **85**, 103–105.

——, ——, G. C. Wickwire, and M. Williams (1927), The effect of the internal secretions and temperature on the metabolism of amino acids and simple sugars by animal cells, *Am. J. Physiol.*, **81**, 468–469.

Bürger, O. (1908), Nuevos estudios sobre Protozoos Chilenos del agua dulce, *Anales univ. Chile*, **122**, 137–203.

Burian, R. (1924), Die Exkretion, *Wintersteins Handb. d. vergl. Physiol.*, **2**, 2.

Burt, R. L. (1945), Narcosis and cell division in *Colpoda steinii*, *Biol. Bull.*, **88**, 12–29.

——, G. W. Kidder, and C. L. Claff (1941), Nuclear reorganization in the family Colpodidae, *J. Morphol.*, **69**, 537–561.

Bütschli, O. (1873), Vorläufige Mittheilung einiger Resultate von Studien über die Conjugation der Infusorien und die Zelltheilung, *Z. wiss. Zoöl.*, **25**, 426–441.

—— (1873 a), Einiges über Infusorien, *Arch. mikroskop. Anat.*, **9**, 657–678.

—— (1876), Studien über die ersten Entwicklungsvorgänge der Eizelle, die Zelltheilung und der Konjugation der Infusorien, *Abhandl. Senckenb. naturforsch. Ges.*, **10**, 1–250.

—— (1885), Glykogen in Protozoen, *Z. Biol.*, **21**, 603–612.

—— (1887–89), Protozoa; Bronn's Klassen und Ordnungen des Tierreichs, Leipzig, **1**, 1–2035.

Buytendijk, F. J. (1919), Acquisition d'habitudes par des êtres unicellulaires, *Arch. néerland. physiol.*, **3**, 455–468.

Caldwell, L. (1933), The production of inherited diversities at endomixis in *Paramecium aurelia*, *J. Exp. Zoöl.*, **66**, 371–407.

Calkins, G. N. (1902 a), Marine Protozoa from Woods Hole, *Bull. U. S. Fish Comm.*, **21**, 413.

—— (1902 b), Studies on the life-history of Protozoa: I. The life-cycle of *Paramecium caudatum*, *Arch. Entwickelungsmech. Organ.*, **15**, 139–186.

—— (1902 c), Studies on the life-history of Protozoa: The six hundred and twentieth generation of *Paramecium caudatum*, *Biol. Bull.*, **3**, 192–205.

—— (1904), Studies on the life history of Protozoa: IV. Death of the A series, *J. Exp. Zoöl.*, **1**, 423–461.

—— (1906), *Paramecium aurelia* and *Paramecium caudatum*, Biol. Stud. Pupils of W. T. Sedgwick, Chicago, 1–10.

—— (1906 a), The protozoan life cycle, *Biol. Bull.*, **11**, 229–244.

—— (1911), Effects produced by cutting Paramecium cells, *Biol. Bull.*, **21**, 36–72.

—— (1913), Further light on the conjugation of paramecium, *Proc. Soc. Exp. Biol. & Med.*, **10**, 36.

—— (1915), Cycles and rhythms and the problem of "immortality" in Paramecium, *Am. Naturalist*, **49**, 65–75.

—— (1916), General biology of the protozoan life cycle, *Am. Naturalist*, **50**, 257–270.

—— (1919), The restoration of vitality through conjugation, *Proc. Natl. Acad. Sci.*, **5**, 95–102.

—— (1919 a), *Uroleptus mobilis Engelm.* II. Renewal of vitality through conjugation, *J. Exp. Zoöl.*, **29**, 121–156.

—————— (1926), Organization and variation in protozoa, *Sci. Monthly*, **22**, 341–351.

—————— (1933), The Biology of the Protozoa, Philadelphia, Lea and Febiger, 1–607.

—————— (1934), Factors controlling longevity in protozoan protoplasm, *Biol. Bull.*, **67**, 410–431.

—————— and S. W. Cull (1907), The conjugation of *Paramecium aurelia* (*caudatum*), *Arch. Protistenk.*, **10**, 375–415.

—————— and W. H. Eddy (1916–17), The action of pancreatic vitamin upon the metabolic activity of Paramecium, *Proc. Soc. Exp. Biol. Med.*, **14**, 162–164.

—————— and L. H. Gregory (1913), Variations in the progeny of a single exconjugant of *Paramecium caudatum*, *J. Exp. Zoöl.*, **15**, 467–525.

—————— and C. C. Lieb (1902), Studies on the life-history of Protozoa. 2. The effect of stimuli on the life-cycle of *Paramecium caudatum*, *Arch. Protistenk.*, **1**, 355–71.

—————— and F. M. Summers (1941), Protozoa in Biological Research, New York, Columbia University Press, 1–1148.

Cantacuzène, A. (1925), Sensibilité comparée de diverses infusoires a certains alcaloïdes du Quinquina, *Compt. rend. soc. biol.*, **93**, 1600–1601.

Carben, H. (1914), Über die Wirkung von Kombinationen einiger Gifte mit Methylenblau und Eosin, *Aus d. Pharm. Inst. Univ. München.*, 25.

Carlgren, O. (1899–1900), Uber die Einwirkung des konstanten galvanischen Stromes auf niedere Organismen, *Arch. Anat. u. Physiol.* (Physiol. section) (Supplement of 1900), 49–76, 465–480.

—————— (1905), Der Galvanotropismus und die innere Kataphorese, *Z. allgem. Physiol.*, **5**, 123–130.

Carter, H. J. (1861), Notes and corrections on the organization of Infusoria, etc., *Ann. and Mag. Natural Hist.*, **8**, 281–290.

Catcheside, D. G. (1951), The Genetics of Microorganisms, New York, Pitman Pub. Corp., 1–223.

Catterina, G. (1897), Contribuzione allo studio sull'importanze dei Protozoi nella purificazione delle acque, *Atti Soc. Veneto-Trentine Padova* (2), **3**, 153–166.

Causey, D. (1926), Mitochondria in ciliates with especial reference to *Paramecium caudatum*, *Univ. Calif. Pubs. Zoöl.*, **28**, 231–250.

Cejp, K. (1935), *Sphaerita,* parasit Paramecií, *Publ. Fac. Sci. Univ. Charles*, **141**, 3–7.

Certes, A. (1880), Sur la glycogénèse chez les infusoires, *Compt. rend. soc. biol.*, **90**, 77–80.

—————— (1884), Note relative à l'action des hautes pressions sur la vitalité des micro-organismes d'eau douce et d'eau de mer, *Compt. rend. mém. soc. biol.* Paris, 3.

—————— (1885), Dell uso delle materie coloranti nello studio fisiologico e istologico degli infusorii viventi, *Boll. sci. Pavia*, **7**, 46; *J. Microbiol.*, Paris, **9**, 212.

—————— (1885 a), De l'emploi des matières colorantes dans l'etude physiol. et hist. des infusoires vivants, *Compt. rend. soc. biol.*, **37**, 1.

Chalkley, H. W. (1930 a), Resistance of Paramecium to heat as affected by changes in hydrogen-ion concentration and in inorganic salt balance in surrounding medium, *U.S. [A.] Pub. Health Service. Pub. Health Depts.*, **45**, 481–489.

—————— (1930 b), On the relation between the resistance to heat and the mechanism of death in Paramecium, *Physiol. Zoöl.*, **3**, 425–440.

Chambers, M. H. (1918), The effects of some food hormones and glandular products on the rate of growth of *Paramecium caudatum*, *Biol. Bull.*, **35**.

—————— (1919), The effect of some food hormones and glandular products on the rate of growth of *Paramecium caudatum*, *Biol. Bull.*, **36**, 82–91.

Chambers, R. (1924), The physical structure of protoplasm as determined by micro-dissection and injection, General Cytology, Sec. V, (Edited by E. V. Cowdry), Chicago, 234–309.

—————— and J. A. Dawson (1925), The structure of the undulating membrane in the ciliate *Blepharisma*, *Biol. Bull.*, **48**, 240–242.

Chao, P. K. (1952), Kappa size and concentration per cell in relation to stages of life cycle, genotype and mating type in *Paramecium aurelia, Proc. Nat. Acad. Sci.,* (in press).

Chase, A. M. and O. Glaser (1930), Forward movement of Paramecium as a function of th hydrogen ion concentration, *J. Gen. Physiol.,* **13,** 627–636.

Chatton, E. and S. Brachon (1933), Sur une Parámécie a deux races: *Paramoecium Duboscqui,* N. sp., *Compt. rend. soc. biol.,* **114,** 988.

―――― and ―――― (1936), Les deux formes du *Paramaecium Dubosqui Chatton* et *Brachon* apres trois annees de culture. Passage du type *sphaerocaryum* au type *bactrocaryum, Compt. rend. soc. biol.,* **121,** 711.

―――― and M. Chatton (1923), L' influence des facteurs bactériens sur la nutrition, la multiplication et la sexualité des Infusoires, *Compt. rend. acad. sci. Paris,* **176,** 1262–1265.

―――― and ―――― (1931), La conjugaison du *Paramaecium caudatum* déterminée expérimentalement par modification de la flore bactérienne associée. Races dites conjugantes et non conjugantes, *Compt. rend. acad. sci. (Paris),* **193,** 206–208.

―――― and Lwoff, A. (1930), Imprégnation, par diffusion argentique, de l'infraciliature des ciliés marins et d'eau douce, après fixation cytologique et sans dessication, *Compt. rend. soc. biol.,* **104,** 834–936.

――――, ―――― and M. Lwoff (1931), L' origine in fraciliaire et la genèse des trichocystes et des trichites chez les *Ciliés Foettingeriidae, Compt. rend. acad. sci.,* **193,** 670.

―――― and L. Tellier (1927), Sur les limites de résistance de quelques infusoires d'eau douce aux solutions de chlorures. Limits of resistance of some fresh water infusoria to solutions of chlorides., *Compt. rend. soc. biol.,* **97,** 285–288.

Chejfec, M. (1928), On the nuclear reorganization of *Paramecium caudatum, Acta Biol. Exptl.,* **2,** 89–121.

―――― (1929), Die Lebensdauer von *Paramecium caudatum* in Abhängigkeit von der Nährungsmenge, *Acta Biol. Exptl.,* **4,** 73–118.

―――― (1930), Zur Kenntnis der Kernreorganisationsprozesse bei *Paramecium caudatum, Arch. Protistenk.,* **70,** 87–118.

―――― (1932), Regulacja i regeneracja u *Paramecium caudatum, Acta Biol. Exptl.,* **7,** 115–134.

―――― (1937), Das Verhalten von *Paramecium caudatum* in Chininlösungen, *Acta Biol. Exptl.,* **11,** 220–228.

Chen, T. T. (1940), Conjugation of three animals in *Paramecium bursaria, Proc. Nat. Acad. Sci.,* **26,** 231–238.

―――― (1940 a), Evidences of exchange of pronuclei during conjugation in *Paramecium bursaria, Proc. Nat. Acad. Sci.,* **26,** 241–243.

―――― (1940 b), Conjugation in *Paramecium bursaria* between animals with very different chromosome numbers and between animals with and without micronuclei, *Proc. Nat. Acad. Sci.,* **26,** 243–246.

―――― (1940 c), Conjugation in *Paramecium bursaria* between animals with diverse nuclear constitutions. Significance of conjugation between animals with very different chromosome numbers and between animals with and without micronuclei, *J. Heredity,* **3,** 185–196.

―――― (1940 d), Polyploidy in *Paramecium bursaria, Proc. Nat. Acad. Sci.,* **26,** 239–240.

―――― (1940 e), Polyploidy and its origin in Paramecium, *J. Heredity,* **31,** 175–184.

―――― (1940 f), A further study on polyploidy in Paramecium, *J. Heredity,* **31,** 249–251.

―――― (1942), A staining rack for handling cover-glass preparations, *Stain Technol.,* **17,** 129–130.

―――― (1944), Staining nuclei and chromosomes in Protozoa, *Stain Technol.,* **19,** 83–90.

―――― (1945), Induction of conjugation in *Paramecium bursaria* among animals of one mating type by fluid from another mating type, *Proc. Nat. Acad. Sci.,* **31,** 404–410.

—— (1946), Conjugation in *Paramecium bursaria*. I Conjugation of three animals, *J. Morphol.*, **78**, 353–395.

—— (1946 a), Temporary pair formation in *Paramecium bursaria*, *Biol. Bull.*, **91**, 112–117.

—— (1946 b), Conjugation in *Paramecium bursaria*. II Nuclear Phenomena in lethal conjugation between varieties, *J. Morphol.*, **79**, 125–262.

—— (1946 c), A technique for counting numerous chromosomes, *J. Morphol.*, **78**, 221–230.

—— (1946 d), Varieties and mating types in *Paramecium bursaria*. I New variety and types from England, Ireland, and Czechoslovakia, *Proc. Nat. Acad. Sci.*, **32**, 173–181.

—— (1949), Conjugation between double monsters and single animals in *Paramecium bursaria*, *Proc. Nat. Acad. Sci.*, **35**, 108–111.

—— (1949), Killer substances produced by *Paramecium bursaria*, *Western Soc. Naturalists*, 19th meeting, 18.

—— (1951), Conjugation in *Paramecium bursaria*. III. Nuclear changes in conjugation between double monsters and single animals, *J. Morphol.*, **88**, 245–292.

—— (1951 a), Conjugation in *Paramecium bursaria*. IV. Nuclear behavior in conjugation between old and young clones, *J. Morphol.*, **88**, 293–360.

Chen, Y. T. (1944), Mating types in *Paramecium caudatum*, *Am. Naturalist*, **78**, 334–340.

Child, C. M. (1913), Demonstration of the axial gradients by means of potassium-permanganate, *Biol. Bull.*, **36**, 133–146.

—— (1914), The axial gradient in Ciliate infusoria, *Biol. Bull.*, **26**, 36.

—— (1934 a), Differential reduction of methylene blue by living organisms, *Proc. Soc. Exp. Biol. Med.*, **32**, 34–36.

—— (1934 b), The differential reduction of methylene blue by Paramecium and some other ciliates, *Protoplasma*, **22**, 377–394.

—— (1941), Patterns and Problems of Development, Chicago, Univ. of Chicago Press, 1–811.

—— and E. Deviney (1926), Contributions to the physiology of *Paramecium caudatum*, *J. Exp. Zoöl.*, **43**, 257–312.

Chopra, R. N. and J. S. Chowhan (1931), The action of the venom of the Indian cobra (*N. naia vel tripudians*) on certain Protozoa, *Indian J. Med. Research*, **18**, 1103–1111.

Cienkowsky, L. (1855), Bemerkungen über Stein's Acinetenlehre, *Bull. acad. imp. sci. St. Petersb., physmath.*, **13**, 297–304, *Quart. J. Microscop. Sci.*, (1857), **5**, 96–103.

Claff, L. C. (1940), A migration-dilution apparatus for the sterilization of Protozoa, *Physiol. Zoöl.*, **13**, 334–343.

—— and T. N. Tahmisian (1948), Cartesian diver technique: a simplified mixing method in a new type of cartesian diver vessel, *Biol. Bull.*, **95**, 253.

—— and —— (1949), Cartesian diver technique, *J. Biol. Chem.*, **179**, 577–583.

Claparede, E. and J. Lachmann (1857), Note sur la reproduction des Infusoires, *Ann. sci. nat. Zool.*, **4**, 221–244.

—— and —— (1858–61), Études sur les Infusoires et les Rhizopodes, **1, 2**, Genève.

Clark, J. (1888), Über den Einfluss niederer Sauertoffpressungen auf die Bewegungen des Protoplasmas, *Ber. deut. botan. Ges.*, Berlin, **6**, 273.

Cleveland, L. R. (1927), The encystment of Paramecium in the recta of frogs, *Science*, **66**, 221–222.

—— (1927 a), Natural and experimental ingestion of Paramecium by cockroaches, *Science*, **66**, 222.

Coehn, A. and W. Barratt (1905), Über Galvanotaxis vom Standpunkte der physikalischen Chemie, *Z. allgem. Physiol.*, **5**, 1–9.

Cohen, B. M. (1934), The effect of conjugation within a clone of *Euplotes Patella*, *Genetics*, **19**, 25–39.

Cohen, B. M. (1934 a), On the inheritance of body form and of certain other characteristics, in the conjugation of *Euplotes Patella, Genetics,* **19,** 40–61.

Colas-Belcour and A. Lwoff (1925), L'utilisation des glucides par quelques protozoaires, *Compt. rend. soc. biol.,* **93,** 1421.

Cold Spring Harbor Symposia on Quantative Biology. (1946), Heredity & Variation in Microorganisms, **11.**

Cole, E. C. (1934), Acid fuchsin for demonstration of ingestion in *Paramecium caudatum, Proc. Soc. Exp. Biol. Med.,* **32,** 138–139.

Cole, F. J. (1926), The History of Protozoology, London, Univ. of London Press, 1–64.

Cole, W. H. (1925), Pulsation of the contractile vacuole of Paramecium as affected by temperature, *J. Gen. Physiol.,* **7,** 581–587.

——— and E. Richmond (1925), The use of chloretone as an anesthetic for paramecium, *Proc. Soc. Exp. Biol. Med.,* **22,** 231–233.

Coles, A. C. (1927), Relief staining of bacteria, protozoa, infusoria, *Watson Microscope Record,* **10,** 23–25.

Collett, M. E. (1919–21), I. The toxicity of acids to ciliate infusoria. II. The rôle of molecule and of ions. III. Antagonism of the toxic action of acids by inorganic chlorides, *J. Exp. Zoöl.,* **29,** 443–472, **34,** 67–74, 75–100.

Colombo, C. (1904), L'azione biologica e terapeutica dei campi magnetici variabili, *Gazz. med. ital.,* **55,** 471–472.

Conn, H. W. (1905), A preliminary report on the protozoa of the fresh waters of Connecticut, *Connecticut Geol. Natural His. Survey.*

Cori, G. T. (1923), The influence of thyroid extracts and thyroxin on the rate of multiplication of Paramaecia, *Am. J. Physiol.,* **65,** 295–299.

Corti, B. (1774), Osservazioni microscopiche sulla Tremella e sulla circolazione del fluido in una Pianta acquajuola (Chara), Lucca.

Cosmovici, N. L. (1892), Le qu'il faut entendre par système aquifère, organes segmentaires, organes excrèteurs, néphridies, *Congrès intern. zoöl.,* **2,** 16–40.

——— (1931), Les phenomènes mécaniques de la digestion chez les infusoires, *Compt. rend. soc. biol.,* **106,** 745.

——— (1931 a), Les phenomenes de la digestion chez les infusoires. Qu'est-ce qu'une vacuole digestive et que faut-il entendre par cette expression?, *Compt. rend. soc. biol.,* **106,** 749.

——— (1933), La nutrition et la rôle physiologique du vacuome chez les infusoires. La théorie canaliculaire du protoplasma, *Ann. sci. univ. Jassy,* **17,** 294–336.

Costamagna, S. (1899), Ricerche interno alla digestione nei Cigliati mediante il rossoneutro, *Atti accad. Torino,* **34,** 1035–1044.

Crampton, G. C. (1912), Experiments performed upon Protozoa confined in capillary tubes, *Arch. Protistenk.,* **27,** 9–15.

——— (1912 a), Inhibition of cell division in Paramaecium, *Science,* **35,** 634–635.

Crane, M. M. (1921), The effect of hydrogen ion concentration on the toxicity of alkaloids for Paramecium, *J. Pharmacol. Exp. Therap.,* **18,** 319.

Crowther, J. A. (1926), The action of X-Rays on *Colpidium colpoda, Proc. Roy. Soc. (London) B,* **100,** 390–404.

Crozier, W. J. (1923), A note on the reaction of protoplasm, *Proc. Soc. Exp. Biol. Med.,* **21,** 58.

Cull, S. W. (1907), Rejuvenescence as the result of conjugation, *J. Exp. Zoöl.,* **4,** 85–89.

Cunningham, B. and P. L. Kirk. (1941), The chemical metabolism of *Paramecium caudatum, J. Cellular Comp. Physiol.,* **18,** 299–316.

——— and ——— (1942), The oxygen consumption of single cells of *Paramecium caudatum* as measured by a capillary respirometer, *J. Cellular Comp. Physiol.,* **20,** 119–134.

Cunningham, E. (1910), On the velocity of steady fall of spherical particles through fluid medium, *Proc. Roy. Soc. (London) A.*, **83**, 357.

Cutler, D. W. and L. M. Crump (1923), The rate of reproduction in artificial cultures of *Colpidium colpoda*, *Biochem. J.*, **17**, 878–886.

―――― and ―――― (1923 a), III. The rate of reproduction in artificial cultures of *Colpidium colpoda*, *Biochem. J.*, **18**, 905–912.

―――― and ―――― (1925), The influence of washing upon the reproduction rate of *Colpidium colpoda*, *Biochem. J.*, **19**, 450–453.

Dale, D. (1913), On the action of electrolytes on Paramecium, *J. Physiol.*, **46**, 129.

Dale, H. H. (1901), Galvanotaxis and chemotaxis of ciliate infusoria Part 1, *J. Physiol.*, **26**, 291–361.

Damerow, A. P. (1931), An attempt to grow Paramoecia in pure cultures of tubercle bacilli, *Am. Rev. Tuberc.*, **24**, 363–366.

Dangeard, P. A. (1901), Les Zoochlorelles du *Paramecium bursaria*, *Le Botaniste*, **7**, 161.

Daniel, G. E., R. R. Spencer, and D. Calnan (1945), Methylcholanthrene and the environment of Paramecium, *J. Natl. Cancer Inst.*, **6**, 157–160.

Daniel, J. F. (1908), The adjustment of Paramecium to distilled water and its bearing on the problem of the necessary inorganic salt content, *Am. J. Physiol.*, **23**, 48–63.

―――― (1909), Adaptation and immunity of lower organisms to ethyl alcohol, *J. Exp. Zoöl.*, **6**, 571–611.

Danielsohn, P. (1899), Über die Einwirkung verschiedener Akridinderivate auf Infusorien, Diss. München, 21 pp.

Darby, H. H. (1929), The effect of the hydrogen concentration on the sequence of protozoan forms, *Arch. Protistenk.*, **65**, 1–37.

―――― (1930), Studies on growth acceleration in Protozoa and yeast, *J. Exp. Biol.*, **7**, 308–316.

―――― (1930 a), The experimental production of life cycles in ciliates, *J. Exp. Biol.*, **7**, 132–142.

Davenport, C. B. (1897), Experimental Morphology, **1**, 1–280.

―――― and W. E. Castle (1895), Studies on morphogenesis III. On the acclimatization of organisms to high temperatures, *Arch. Entwicklungsmech. mech.*, **2**, 227–249.

―――― and H. V. Neal (1896), Studies on morphogenesis. V. On the acclimatization of organisms to poisonous chemical substances, *Arch. Entwicklungsmech. mech.*, **2**, 564–583.

Dawson, J. A. (1924), Inheritance of an abnormality of form in *Paramecium aurelia*, *Proc. Soc. Exp. Biol. Med.*, **22**, 104–106.

―――― (1926), A mutation in *Paramecium aurelia*, *J. Exp. Zoöl.*, **44**, 133–157.

―――― (1928), The mutant *Paramecium aurelia*, *Science*, **68**, 258.

―――― (1928 a), A comparison of the life 'cycles' of certain ciliates, *J. Exp. Zoöl.*, **51**, 199–208.

―――― and O. W. Richards (1927), The analysis of the division rates of ciliates, *J. Gen. Physiol.*, **10**, 853–858.

Day, H. C. (1930), Studies on the contractile vacuole of Spirostomum and Paramecium, *Physiol. Zoöl.*, **3**, 56–71.

Day, L. M. and M. Bentley (1911), A note on learning in Paramaecium, *J. Anim. Behav.*, **1**, 67–73.

DeBarros, R. (1940), Colchicine and Paramecia, *Unim. São Paulo, Bol.*, **17**, 97–116.

deFromentel, E. (1874), Études sur les Microzoaires ou Infusoires Proprement Dits, Paris, 1–364.

DeGaris, C. F. (1927), The accession of contractile vacuoles during fission in *Paramecium caudatum*, *Anat. Record*, **37**, 136.

―――― (1927 a), Experimental studies on retarded development of the single cell—*Paramecium caudatum*, *Anat. Record*, **35**, 33–34.

DeGaris, C. F. (1927 b), A genetic study of *Paramecium caudatum* in pure lines through an interval of experimentally induced monster formation, *J. Exp. Zoöl.*, **49**, 133–148.

―――― (1928), The effects of anterior and posterior selections on fission rate in pure lines of *Paramecium caudatum*, *J. Exp. Zoöl.*, **50**, 1–14.

―――― (1935), The use of double monsters as means of identification in crossing pure lines of *Paramecium caudatum*, *Am. Naturalist*, **69**, 84–86.

―――― (1935 a), Lethal effects of conjugation between *Paramecium aurelia* and double monsters of *Paramecium caudatum*, *Am. Naturalist*, **69**, 87–91.

―――― (1935 b), Heritable effects of conjugation between free individuals and double monsters in diverse races of *Paramecium caudatum*, *J. Exp. Zoöl.*, **71**, 209–256.

Degen, A. (1905), Untersuchungen über die kontraktilen Vakuolen, *Botan. Ztg.* **63**, 163–226.

Dehorne, A. (1911), La non-copulation du noyau échangé et du noyau stationaire et la disparition de ce dernier dans la conjugaison de *Paramecium caudatum*, *Compt. rend. acad. sci. Paris*, **152**, 922–925.

―――― (1920), Contribution à l'étude comparée de l'appareil nucléaire des infusoires ciliés, *Arch. zool. exp.* **60**, 47.

―――― and F. Morvillez (1926), Contribution à l'étude de l'action d'alcaloïdes sur les infusoires. Facteurs des variations de résistance, *Compt. rend. soc. biol.*, **94**, 704–706.

DeJong, D. A. (1922), Micro-organismes et basses températures, *Arch. néerland. sci.*, **7**, 588–591.

Delage-Herouard (1896), Traité de Zoologie Concrète, La Cellule et les Protozoaires, Paris, Schleicher Frères, **1**.

DeLamater, A. J. (1939), Effect of certain bacteria on the occurrence of endomixis in *Paramecium aurelia*, *Biol. Bull.*, **76**, 217–225.

Delboeuf, J. (1891), Une loi mathématique applicable à la dégénerescence qui affecte les Infusoires ciliés a la suite des fissiparations constamment répétées, *Rev. sci. Paris*, **47**, 386–371.

Dembowski, J. (1921), Über die Nahrungswahl und die sogenannten Gedächt-niser-scheinungen bei *Paramecium caudatum*, *Trav. Lab. biol. Nencki Varsovie*, **1**, 1–37.

―――― (1922), Weitere Studien über die Nahrungswahl bei *Paramecium caudatum*, *Trav. Lab. biol. Nencki Varsovie*, **1**, 1–16.

―――― (1922 a), Über den Einfluss der Suspensionskonzentration auf die Anzahl der gebildeten Nahrungsvakuolen bei *Paramecium caudatum*, *Trav. Lab. biol. Nencki Varsovie*, **1**, 1–16.

―――― (1923), Über die Bewegungen von *Paramecium caudatum*, *Arch. Protistenk.*, **47**, 25–54.

―――― (1923 a), Untersuchungen über die Bewegung von *Paramecium caudatum* in Tropfen verschiedener geom. Gestalt, *Trav. Lab. biol. Nencki Varsorie*, **1**, 1–32.

―――― (1928–29), Ruchy pionowe *Paramaecium caudatum*. 1. Wzgledne polozenie śpodka ciezkości w ciele wymoczka. 2. Wplyw niektórych warunków zewnetrznych. (Vertical movement of *Paramecium caudatum*. 1. The relative location of equilibrium center in the body of the infusorian. 2. Influence of certain outside conditions), *Acta Biol. Exptl.* **3**, 19–45, 195–240.

―――― (1929) Die Vertikalbewegungen von *Paramecium caudatum*. 1. Die Lage des Gleichgewichtszentrums im Körper des Infusors, *Arch. Protistenk.*, **66**, 104–132.

―――― (1929 a), Vertikalbewegung von *Paramecium caudatum*. I. Die relative Lage des Gleichgewichtszentrums im Körper des Infusors, *Trav. Lab. biol. Nencki Varsovie*, **5**, 1947.

―――― (1929 b), Die Vertikalbewegungen von *Paramecium caudatum*. II. Einfluss einiger Aussenfaktoren, *Arch. Protistenk.*, **68**, 215–261.

―――― (1931), Die Vertikalbewegungen von *Paramecium caudatum*. III. Polemisches und Experimentelles, *Arch. Protistenk.*, **73**, 153–187.

———— (1931 a), Dalse studja nad geotropizmem Paramecium. (Further studies on geotropism in Paramecium). *Acta Biol. Exptl.* (*Warsaw*), **6**, 59–87.

———— (1938), Über die Rhytmik der *Paramecium teilungen, Acta Biol. Exptl.* (*Warsaw*), **12**, 22–33.

Diebschlag, E. (1940), Über die Lernfähigkeit von *Paramecium caudatum, Zool. Anz.,* **130**, 257–271.

Diller, W. F. (1934), The conjugation of *Paramecium trichium, Anat. Record,* **60**, 92–93.

———— (1934 a), Autogamy in *Paramecium aurelia, Science,* **79**, 57.

———— (1936), Nuclear reorganization processes in *Paramecium aurelia,* with descriptions of autogamy and "hemixis," *J. Morphol.,* **59**, 11–67.

———— (1940), Nuclear variation in *Paramecium caudatum, J. Morphol.,* **66**, 605–633.

———— (1942), Re-conjugation in *Paramecium caudatum, J. Morphol.,* **70**, 229–259.

———— (1947), Conjugation in *Paramecium trichium, Anat. Record,* **99**, 72.

———— (1948), Nuclear behavior of *Paramecium trichium* during conjugation, *J. Morphol.,* **82**, 1–52.

———— (1948 a), Induction of autogamy in single animals of *Paramecium calkinsi* following mixture of two mating types, *Biol. Bull.,* **95**, 265.

———— (1948 b), An extra post-zygotic division in *Paramecium caudatum, Biol. Bull.,* **95**, 265.

———— (1949), An extra postzygotic nuclear division in *Paramecium caudatum, Anat. Record,* **105**, 62–63.

———— (1949 a), An abbreivated conjugation process in *Paramecium trichium, Biol. Bull.,* **97**, 331–343.

———— (1950), An extra postzygotic nuclear division in *Paramecium caudatum, Trans. Am. Microscop. Soc.,* **69**, 309–316.

———— (1950 a), Cytological evidence for pronuclear interchange in *Paramecium caudatum, Trans. Am. Microscop. Soc.,* **69**, 317–323.

———— (1951), The relationship of autogamy in single animals to conjugation in *Paramecium aurelia, Anat. Record,* **111**, 58.

Dimitrowa, A. (1928), Untersuchungen über die überzähligen pulsierenden Vakuolen bei *Paramecium caudatum, Arch. Protistenk.,* **64**, 462–478.

———— (1930), Zur Frage der Teilungs-geschwindigkeit bei *Paramaecium caudatum Ehrenberg, Arch. Protistenk.,* **72**, 554–558.

———— (1932), Die fördernde Wirkung der Exkrete von *Paramecium caudatum Ehrbh.* auf dessen Teilungsgeschwindigkeit, *Zool. Anz.,* **100**, 127–132.

Dippell, R. V. (1948), Mutations of the killer Plasmagene, kappa, in variety 4 of *Paramecium aurelia, Am. Naturalist,* **82**, 43–50.

———— (1950), Mutation of the killer cytoplasmic factor in *Paramecium aurelia, Heredity,* **4**, 165–187.

Di Tomo, M. (1932), Ricerche sul comportamento di *Paramecium caudatum* in un dato volume di cultura liquida, *Boll. d. zool.,* **3**, 137–140.

Dobell, C. (1911), The principles of protistology, *Arch. Protistenk.,* **23**, 269–310.

———— (1914), A commentary on the genetics of the ciliate Protozoa, *J. Genetics,* **4**, 131–190.

———— (1932), Antony van Leeuwenhoek and his "Little Animals", New York, Harcourt, Brace and Co., 1–435.

Dodson, E. O. (1946), Some evidence for the specificity of the Feulgen Reaction, *Stain Technol.,* **21**, 103–105.

Doflein, F. (1902), Das System der Protozoen, *Arch. Protistenk.,* **1**, 169.

———— (1907), Beobachtungen und Ideen über die Konjugation der Infusorien, *Sitzber. Ges. Morph. phys. München,* **23**, 107–114.

———— (1916), Lehrbuch der Protozoenkunde, Jena, G. Fischer, ed. 4, 1–1189.

———— (1918), Theilung und Tod der Einzelligen, *Zool. Anz.* **49**, 306–308.

Doflein, F. and E. Reichenow (1928), Lehrbuch der Protozoenkunde, Jena, G. Fischer, ed. 5.

Dogiel, V. (1925), Die Geschlechtsprozesse bei Infusorien, *Arch. Protistenk.*, **50**, 283.

—— (1929), Experimente über die Ernährungsphysiologie der Infusorien, *Congr. intern. Zool., Budapest, Part II*, 879–886.

—— and M. Issakowa-Keo (1927), Physiologische Studien an Infusorien. II. Der Einfluss der Salzlösungen auf die Ernährung von Paramecium, *Biol. Zentr.*, **47**, 577–586.

Dognon, A. (1927), Étude sur la photosensibilisation biologique: La fluorescence et la penetration des photosenbilisateurs. (Biological photosensitization: Fluorescence and penetration of sensitizing agents), *Compt. rend. soc. biol.*, **97**, 1590–1592.

—— (1928), La Photosensibilisation biologique. Influence de la concentration du sensibilisateur et de l'intensité lumineuse. (Biological photosensitization. Influence of concentration of the sensitizing agents and light intensity), *Compt. rend. soc. biol.*, **98**, 283–285.

——, E. Biancani, and H. Biancani (1932), Action des ultra-sons sur les Paramécies. Influence de divers facteurs extérieurs, *Compt. rend. soc. biol.*, **3**, 754–755.

Donatelli, L. and P. Pratesi (1937), The pharmacology of sodium glycerophosphate and tetramethylammonium compounds, *Boll. soc. ital. biol. sper.*, **12**, 349.

Doniach, I. (1939), A comparison of photodynamic activity of some carcinogenic with non-carcinogenic compounds, *Brit. J. Exp. Path.*, **20**, 227–235.

Dornfeld, E. J. (1939), The ultracentrifuge in cellular biology, *Marquette M. Rev.*, **3**, 51–55.

Doudoroff, M. (1936), Studies in thermal death in Paramecium, *J. Exp. Zoöl.*, **72**, 369–385.

Downing, W. L. (1951), Structure and morphogenesis of the cilia and the feeding apparatus in *Paramecium aurelia, Proc. Am. Soc. Protozoöl.* **2**, 14–15.

Duggar, B. M. (1936), Biological Effects of Radiation, New York, McGraw Hill Book Co., 2 vols.

Dujardin, F. (1841), Histoire Naturelle des Zoöphytes: Infusoires, Paris.

Dunihue, F. W. (1931), The vacuome and the neutral red reaction in *Paramaecium caudatum, Arch. Protistenk.*, **75**, 476–497.

Ebbecke, U. (1935), Das Verhalten von Paramaecien unter der Einwirkung hoher Druckes, *Pflügers Arch. ges. Physiol.*, **236**, 658–661.

—— (1936), Über plasmatische Kontraktionen von roten Blütkorperchen, Paramäcien und Algenzellen unter der Einwirkung hoher Drucke, *Pflügers Arch. ges. Physiol.*, **238**, 452–466.

Eckert, F. and M. Feiler (1931), Studien über die Bedeutung des umsatzsteigernden Einflusses hochverdünnter Chininlösungen auf Paramecium, *Z. vergleich. Physiol.*, **14**, 93–121.

Eddy, S. (1928), Succession of Protozoa in cultures under controlled conditions, *Trans. Am. Microscop. Soc.*, **47**, 283–319.

Edmondson, C. H. (1920), Protozoa of the Devil's Lake Complex, North Dakota, *Trans. Am. Microscop. Soc.*, **39**, 167–198.

Efimoff, A. and W. Efimoff (1922), Überkältung und Ausfrieren von Protozoen, *Russ. Arch. Protistol*, **1**, 151–167.

—— and —— (1924), Über Ausfrieren und Überkältung der Protozoen, *Arch. Protistenk.*, **49**, 433–446.

—— and —— (1925), Vitale Färbung und photodynamische Erscheinungen, *Biochem. Z.*, **155**, 376.

Efimoff, W. (1922), Die photodynamische Sensibilisierung der Protozoen und der Satz von Talbot, *Russ. Arch. Protistol*, **1**, 148–150.

——, N. J. Nekrassow, and A. W. Efimoff (1928), Die Einwirkung des Oxydationspotentials und der H-Ionenkonzentration auf die Vermehrung der Protozoen und Abwechslung ihrer Arten, *Biochem. Z.,* **197,** 105–118.

——, ——, and —— (1928 a), Les influences de la nourriture et des conditions physico-chimiques sur l'augmentation des Protozoaires, *Intern. Congr. Zool.,* **10,** 912.

Ehrenberg, C. G. (1830), Organization systematik und geographisches Verhältniss der Infusionsthierchen: Akad. der Wissen. Berlin, Druckerei der Königlichen Akad. Wissen. Berlin.

—— (1833), Abhandl. d. Akad. d. Wissensch. zu Berlin, Druckerei der Königl. Akad. der Wissensch. Berlin.

—— (1838), Die Infusionsthierchen als vollkommene Organismen, Leipzig.

Ehret, C. F. (1951), The effects of visible, ultraviolet, and x-irradiation on the mating reaction in *Paramecium bursaria, Anat. Record,* **111,** 112.

Ehrmann, P. (1895), Über die kontraktile Vakuole der Infusorien, *Sitzber. Naturw. Ges. Leipzig,* 19/21 Jg., 89–102.

Eisenberg-Hamburg, E. (1925), Sur le fonctionnement de la vésicule pulsatile chez les infus. Contribution aux recherches sur la perméabilité de la membrane cellulaire, *Trav. lab. biol. Nencki Varsovie,* **2,** 1–30.

—— (1926), Recherches sur le fonctionnement de la vesicule des infusoires dans les conditions normales et sous l'actions de certains agents experimentaux: pression osmotique et electrolites, *Arch. biol.,* **35,** 441–464.

—— (1929), Recherches comparatives sur le fonctionnement de la vacuole pulsatile chez les infusoires paracites de la grenouille et chez les infusoires d'eau douce: Influence de la pression osmotique, des électrolytes et du pH, *Arch. Protistenk.,* **68,** 451–470.

—— (1930), l'influence des sels de strontium sur les mouvements du *Paramecium caudatum.* Le rôle des sels de calcium et de la concentration en ion hydrogène, *Acta Biol. Exp.,* **4,** 261–278.

—— (1932), Einfluss der Sr-Salze auf die Bewegung von *Paramecium caudatum.* Die Rolle des Ca und der Konzentration der Wasserstoffionen, *Arch. Protistenk.,* **77,** 108–124.

Eismond, J. (1890), Eine einfache Untersuchungsmethode für lebende Infusorien, *Zool. Anz.,* **13,** 723–727.

Elliott, A. M. (1935), The influence of pantothenic acid on growth of protozoa, *Biol. Bull.,* **68,** 82–92.

—— (1937), Vitamin B_1 and the growth of Protozoa, *Anat. Record Suppl.,* **70,** 127.

—— (1939), The vitamin B complex and the growth of *Colpidium striatum, Physiol. Zoöl.,* **12,** 363–373.

—— (1950), The growth-factor requirements of *Tetrahymena geleii E., Physiol. Zoöl.,* **23,** 85–91.

Ellis, J. (1769) Observations on a particular manner of increase in the animalcula of vegetable infusion, with the discovery of an indissoluble salt arising from Hemp-seed put into water until it becomes putrid, *Phil. Trans. Roy. Soc.,* **69,** 138–152.

Emery, F. E. (1928), The metabolism of amino-acids by *Paramecium caudatum, J. Morphol.,* **45,** 555–577.

Engelmann, T. W. (1862), Zur Naturgeschichte der Infusionsthiere, *Z. wiss. Zoöl.,* **11,** 347–393.

—— (1868), Über die Flimmerbewegung, Leipzig, Engelmann.

—— (1876), Über Entwicklung und Fortpflanzung von Infusorien, *Morphol. Jahrb.,* **1,** 573–634.

—— (1878), Zur Physiologie der kontraktilen Vakuolen der Infusionsthiere, *Zool. Anz.,* **1,** 121.

—— (1879), Physiologie der Protoplasma-und Flimmerbewegung, *Hermann's Handb. der Physiol.,* **1,** 341–408.

Engelmann, T. W. (1882), Über Licht- und Farbenperception niederster Organismen Onderzoek. Physiol. Lab., *Utrecht*, 7, 234, *Pflügers Arch. ges. Physiol.*, 29, 387.

Enriques, P. (1902), Ricerche osmotiche sugli Infusori, *Atti acc. Lincei Rend.*, 11, 340–344.

——— (1902 a), Ricerche sui Protozoidei infusioni, *Atti acc. Lincei Rend.*, 11, 392–397.

——— (1906), Sulla condizione che determinano la coniugazione negli Infusori, e del differenziamento sessuale nei Vorticellidi, Bologna, 1–60.

——— (1907), Il dualismo nucleare negli Infusori e il suo significato morphologico e funzionale, *Biol. Torina*, 1, 326–351.

——— (1907 a), La coniugazione e il differenziamento sessuale negli Infusori, *Arch. Protistenk.*, 9, 195–296.

——— (1908), Die Conjugation und sexuelle Differenzierung der Infusorien. 2. Abth., *Arch. Protistenk.*, 12, 213–276.

——— (1909), La teoria di Spencer sulla divisione cellulare studiata con ricerche biometriche negli infusori, *Arch. fisiol.*, 7, 113–136.

——— (1909 a), La sexualité chez les Protozoaires, *Scientia* (Milan), 6.

——— (1912), Il dualismo nucleare negli Infusori e il suo significato morfologico e funzionale. 2. Die Nahrung and Struktur des Makronukleus, *Arch. Protistenk.*, 26, 420–434.

Epshtein, G. V. and O. N. Babikova (1936), The action of arichin and plasmocid on protozoa, *Z. mikrobiol. Epidemiol. immunitätsforsch. (U.S.S.R.)*, 16, 840.

Erdmann, R. (1919), Endomixis and size variations in pure lines of *Paramecium aurelia*, *Proc. Soc. Exp. Biol. Med.*, 16, 60.

——— (1920), Endomixis and size-variations in pure bred lines of *Paramecium aurelia*. *Arch. Entwicklungsmech. mechan.*, 46, 85–148.

——— (1922), Art und artbildung bei protisten, *Biol. Zentr.* 42, 49.

——— (1925), Endomixis bei *Paramecium bursaria, Sitzber. Ges. naturforsch. Freunde Berlin*, 24–27.

——— and L. L. Woodruff, (1914), Vollständige Erneuerung des Kernapparates ohne Zellverschmelzung bei reinlinigen Paramecien, *Biol. Zentr.*, 34, 484–496.

——— and ——— (1916), The periodic reorganization process in *Paramecium caudatum, J. Exp. Zoöl.*, 20, 59–97.

Erhard, H. (1910), Studien über Flimmerzellen, *Arch. Zellforsch.*, 4, 309.

Essex, H. E. and J. Markowitz (1930), The physiologic action of rattlesnake venom (crotalin). I. Effect on blood pressure: symptoms and post-mortem observations. II. The effect of crotalin on surviving organs. III. The influence of crotalin on blood, in vitro and in vivo. IV. The effect on lower forms of life. V. Some experiments on immutiny to croatin, *Am. J. Physiol.*, 92, 317–328, 329–334, 335–341, 342–344, 345–348.

Estabrook, A. H. (1910 a), Effect of chemicals on growth in Paramecium, *J. Exp. Zoöl.*, 8, 489–534.

——— (1910 b), Effect of external agents on growth in Paramecium, *Science*, 31, 467.

Ewer, R. F. (1948), The genetics of the ciliate Protozoa, *Science Progress*, 36, 450–459.

Fabre, P. (1942), Galvanotropisme d'une paramécie en voie de division et transport électrique de substance d'un individu à l'autre, *Compt. rend. soc. biol.*, 136, 332–334.

Fabre-Domergue, P. (1888), Recherches anatomiques et physiologiques sur les infusoires cilies, *Ann. sci. nat. Zool.*, 5, 1–140.

——— (1889), *Ann. Micrograph.*, 2, 237.

——— (1897), A propos des "Trichiten" et des "Stutzfasern" des Infusoires ciliés, *Zool. Anz.*, 20, 3–4.

Famintzin, A. (1891), Beitrag zur Symbiose von Algen und Thieren, *Mém. Acad. Petersburg.*, 38, 15.

Fantham, H. B. and A. Porter (1945), The microfauna, especially the Protozoa found in some Canadian mosses, *Proc. Zool. Soc. London*, 115, 97–174.

Fauré-Fremiet, E. (1909), Constitution du macronucleus des Infusoria ciliés, *Compt. rend. acad. sci. Paris,* **148,** 659–661.

—— (1909 a), La structure physicochimique du macronucleus des Infusoria ciliés, *Bull. Soc. zool.,* **34,** 55–56.

—— (1911), Production experimentale de "trichites" chez Didinium, *Compt. rend. soc. biol. Paris,* **71,** 146–147.

——, C. Léon, A. Mayer, and L. Plantefol (1929), Recherches sur le besoin d'oxygene libre. I. L'oxygène et les mouvements des Paraméciens, *Ann. physiol.,* **5,** 633–641.

——, ——, ——, and —— (1929 a), L'oxygène libre et les mouvements des paramécies. *Compt. rend. soc. biol. Paris,* **101,** 627–628.

—— (1945), Symétrie et polarité chex les Ciliés bi- ou multicomposites, *Bull. Biol.,* **79,** 106–150.

—— (1948), Les mecanismes de la morphogénèse chez les cilies, *Folia Biotheoretica,* **3,** 25–58.

Feiler, M. (1927), Über die Chininwirtung auf die Tirzelle (Effect of quinine on animal cells), *Arch. Protistenk.,* **59,** 562–581.

—— (1928 a), Über die Einwirkung des Plasmochins auf *Paramecium caudatum, Arch. Protistenk.,* **61,** 133–143.

—— (1928 b), Über die Chininwirkung auf die Tierzelle I und II, *Arch. Protistenk.,* **61,** 119–132.

—— (1929 a), Weitere Untersuchungen über die oligodynamen Wirkungen der Alkaloide auf *Paramecium caudatum, Arch. Protistenk.,* **67,** 157–204.

—— (1929 b), Über neue Versuche, betreffend die oligodyname Einwirkung von Alkaloiden auf *Paramecium caudatum, Zool. Anz.,* **80,** 323–330.

—— (1931), Über den Einfluss der höheren Verdünnungsstufen von Chinin auf das Wachstum der Bakterienflora des Heuaufgusses, *Zentr. Bakt. Parasitenk.,* **83,** 63–68.

Ferguson, F. F., J. R. Holmes, and E. Lavor (1942), The effect of some sulfonamide drugs upon several free-living micro-organisms, *J. Elisha Mitchell Sci. Soc.,* **58,** 53–59.

Fermor-Adrianowa, X. (1913), Die Bedeutung der Enzystierung bei *Stylonychia pustulata, Zool. Anz.,* **42,** 380–384.

—— (1925), Die Variabilität von Paramecium, *Arch. Protistenk.,* **52,** 418–426.

Fetter, D. (1926), Determination of the protoplasmic viscosity of Paramecium by the centrifuge method, *J. Exp. Zoöl.,* **44,** 279–283.

Feulgen, R. (1926), Die Nuclearfärbung, Abderhalden: *Handb. d. biol.* Arbeitsmeth. Liefrg. 213

Figge, F. H. J. and R. Wichterman (1952), The effects of hematoporphyrin and other substances upon x-ray sensitivity of *Paramecium, Biol. Bull.,* **103,** 302.

Finley, H. E. (1930), Toleration of fresh water protozoa to increased salinity, *Ecology,* **11,** 337–347.

—— (1946), Patterns of sexual reproductive cycles in ciliates, *Biodynamica,* **6,** 31–79.

Fischer, S. (1885), Beobachtungen über den kontraktilen Behälter der Infusorien, *Die Welt,* Warschau, 691–727.

—— and Kemnitz (1916), Über die Einwirkung einiger Porphyrine auf Paramaecien, *Z. physiol. Chem.,* **96,** 309–313.

Fitzgerald, P. J., M. L. Eidinoff, J. E. Knoll and E. B. Simmel (1951), Tritium in radioautography, *Science,* **114,** 494–498.

Fiveiskaja, A. (1929), Einfluss der Kernparasiten der Infusorien auf den Stoffwechsel, *Arch. Protistenk.,* **65,** 275–298.

Flather, M. D. (1919 a), The effects of a diet of polished and of unpolished rice upon the metabolic activity of Paramecium, *Biol. Bull.,* **36,** 54–62.

—— (1919 b), The influence of the glandular extracts upon the contractile vacuoles of *Paramecium caudatum, Biol. Bull.,* **37,** 22–39.

Florentin, R. (1899), Études sur la faune des mars salées de Lorraine, *Ann. sci. nat. Zool.*, **10**.

Focke, G. W. (1836), Uber einige Organisationsverhältnisse bei polygastrischen Infusorien und Rädertieren, *Isis*.

────── (1845), Ergebnisse ferneren Untersuchungen dere polygastrischen Infusorien, *Verhandl. Ges. deut. naturforsch. Artzte*, **22**, 109–110.

Fortner, H. (1924), Über die physiologisch differente bedeutung der kontraktilen Vakuolen bei *Paramecium caudatum*, *Zool. Anz.*, **60**, 217.

────── (1925), Über die Gesetzmässigkeit der Wirkungen des osmotischen Druckes physiologisch indifferenter Lösungen auf einzellige, tierische Organismen, *Biol. Zentr.*, **45**, 417–446.

────── (1926 a), Der Intoxikationsexponent, *Biol. Zentr.* **46**, 185–190.

────── (1926 b), Zur Frage der diskontinuierlichen Exkretion bei Protisten, *Arch. Protistenk.*, **56**, 295–320.

────── (1926 c), Uber die Gesetzmaessigkeit der Wirkungen des Osmotischen Druckes physiologisch indifferenter Loesungen auf einzellige, tierische Organismen, *Biol. Zentr.*, **45**, 417–446.

────── (1927), Beiträge zur Praxis der Protistenuntersuchung, *Z. wiss. Mikroskop.*, **44**, 463.

────── (1928 a), Über die Vakuolentätigkeit und ihre Beziehungen zu Plasmakolloiden, *Protoplasma*, **3**, 602.

────── (1928 b), Das Intoxikationstheorem, *Protoplasma*, **3**, 536.

────── (1928 c), Zur Kenntnis der Verdauungsvorgänge bei Protisten. Studie an *Paramecium caudatum*, *Arch. Protistenk.*, **61**, 282–292.

────── (1933), Uber den Einfluss der Stoffwechselendprodukte der Futterbakterien auf die Verdauungsvorgänge bei Protozoen, *Arch. Protistenk.*, **81**, 19–56.

────── (1933 a), Über den Einfluss der Stoffwechselendprodukte der Futterbakterien auf die Verdauungsvorgänge bei Protozoen (Untersuchungen) an *Paramecium caudatum*, *Arch. Protistenk.*, **81**, 19–56.

Fortunato, F. (1940), The behavior of *Paramecium caudatum* in solid agar medium, *Proc. Penn. Acad. Sci.*, **14**, 124–125.

Fox, H. M. (1925), The effect of light on the vertical movement of aquatic organisms, *Proc. Cambridge Philo. Soc.*, **1**, 219–224.

Frantz, V. (1912), Was ist höherer Organismus? *Biol. Zentr.*, **31**, 1–21, 33–41.

French, J. W. (1940), Trial and error learning in Paramecium, *J. Exp. Psychol.*, **26**, 609–613.

────── (1940 a), Individual differences in Paramecium, *J. Comp. Psychol.*, **30**, 451–456.

Frey, H. (1858), Das einfachste thierische Leben, *Monatsschr. d. wiss. Vereins in Zürich*.

Frings, H. (1948), Dried skim milk powder for rearing protozoa, *Turbox News*, **26**, 33–37.

Frisch, J. A. (1935), Experimental adaptation of fresh-water ciliates to sea water, *Science*, **81**, 537.

────── (1937), The rate of pulsation and the function of the contractile vacuole in *Paramecium multimicronucleatum*, *Arch. Protistenk.*, **90**, 123–161.

────── (1939), The experimental adaptation of Paramecium to sea water, *Arch. Protistenk.*, **93**, 38–71.

Fühner, H. (1917), Beiträge zur Toxikologie des Arsenwasserstoffs. I. Die Wirkung auf Protozoen, *Arch. exptl. Path. Pharmakol.*, **82**, 44–50.

Furgason, W. H. (1940), The significant cytostomal pattern of the "*Glaucoma-Colpidium*" group, and a proposed new genus and species, *Tetrahymena geleii*, *Arch. Protistenk.*, **94**, 224–266.

Fürth, R. (1920), Über die Anwendung der Theorie der Brownschen Bewegung auf die ungeordnete Bewegung niederer Lebewesen, *Pflügers Arch. ges. Physiol.*, **184**, 294–299.

Fyg, W. (1929), Vitalfärbung von Paramaecien zur Verauschaulichung des Verdauungsvorganges. (Vital staining of paramaecia to demonstrate its digestive processes,) *Mikroskop. Naturfreunde*, **7**, 125–126.

Galadjieff, M. A. (1932), K probleme bessmertiia proteïshikh. (Dvadtsat' let kol'-tury infuzorii *Paramaecium caudatum* bez kon'iugatsti.) [Immortality in Protozoa. (twenty years of Paramecium caudatum cultured without conjugation),] *Bull. acad. sci. U.R.S.S. Classe sci. math. et Nat.*, **9**, 1269–1300, **10**, 1531–1557.

—— and E. Malm (1929), *Compt. rend. acad. sci. U.R.S.S. Ser. A.*, 433.

—— and S. Metalnikow (1933), L'immortalité de la cellule. Vingt-deux ans de culture d'Infusoires sans conjugaison, *Arch. zool. exp. et gen.*, **75**, 331–352.

Gale, C. K. (1935), Penetrative and selective heat effects of short and ultrashort waves, *Arch. Phys. Therapy, X-Ray, Radium*, **16**, 271–277.

Galiano, E. F. (1914), Beitrag zur Untersuchung der Chemotaxis der Paramecien, *Z. allgem. Physiol.*, **16**, 359–372.

—— (1929), Sobre el Concepto de Quimotaxis de las células, *Mém. soc. Española hist. natural.*, **15**, 867–871.

Galina, R. (1914), Über den Einfluss äusserer und innerer Faktoren auf die Pulsationsfrequenz der kontraktilen Vakuole von Vorticella nebulifera, mit besonderer Berücksichtigung der narkotischen Agentien, *Z. allgem. Physiol.*, **16**, 419–473.

Galtsoff, P. S., F. E. Lutz, P. S. Welch, and J. G. Needham (1936), Culture methods for invertebrate animals, Ithaca, Comstock Publishing Co., **32**, 590 pp.

Garbrowski, L. (1907), Gestaltsänderung und Plasmoptyse, *Arch. Protistenk.*, **9**, 53–83.

Garner, M. R. (1934), The relation of numbers of *Paramecium caudatum* to their ability to withstand high temperatures, *Physiol. Zoöl.*, **7**, 408–434.

Garnjobst, L., E. L. Tatum, and C. V. Taylor (1943), Further studies on the nutritional requirements of *Colpoda duodenaria*, *J. Cellular Comp. Physiol.*, **21**, 199–212.

Gause, G. F. (1934 a), Eksperimental'noe issledovanie bor' by za suchchestvovanie mezhdu *Paramecium caudatum, Paramecium aurelia* i *Stylonychia mytilus.* (An experimental investigation of the struggle for existence between *Paramecium caudatum, Paramecium aurelia,* and *Stylonychia mytilus,*) *Zool. Zhur.*, **13**, 1–17.

—— (1934 b), O protsessakh unichtozheniia adnogo vida drugim v populiatsiiakh infuzoriï (Destruction of one species by another in populations of Protozoa,) *Zool. Zhur.*, **13**, 18–26.

—— (1934 c), Über die Konkurrenz zwischen zwei Arten, *Zool. Anz.*, **105**, 219–222.

—— (1934 d), Untersuchungen über den kampf ums Dasein bei Protisten, *Biol. Zentr.*, **54**, 536–547.

—— (1934 e), The Struggle for Existence, Baltimore, Williams and Williams Co., 1–163.

—— (1935 a), Experimentelle Untersuchungen über die Konkurrenz zwischen *Paramecium caudatum* und *Paramecium aurelia*, *Arch. Protistenk.*, **84**, 207–224.

—— (1935 b), Exposés de biométrie et de statistique biologique. 9. Vérifications expérimentales de la theorie mathématique de la lutte pour la vie, *Actualités sci. et ind.*, **277**, 1–63.

—— (1937), The toxic action of cinchonine isomers upon ciliates, *Arch. Protistenk.*, **88**, 180.

—— (1939), Studies on natural selection in Protozoa. I. The adaptation of *Paramecium aurelia* to the increased salinity of the medium, *Zool. Zhur.*, **18**, 631.

—— (1942), The relation of adaptability to adaptation, *Quart. Rev. Biol.*, **17**. 99–114.

——, O. K. Nastukova, and W. W. Alpatov (1934 a), The influence of biologically conditioned media on the growth of a mixed population of *Paramecium caudatum* and *Paramecium aurelia*, *J. Animal Ecol.*, **3**, 222–230.

——, ——, and —— (1934 b), Vliianie biologicheski izmenennoï sredy na rost smeshannoï populiatsii *Paramecium caudatum* i *P. aurelia* (Influence of biologically conditioned media on the growth of the mixed population of *Paramecium caudatum* and *Paramecium aurelia*), *Zool. Zhur.,* **13**, 629–638.

Gause, G. F. and N. P. Smaragdova (1939), Studies on natural selection in Protozoa. II. A comparative investigation of adaptation of *Paramecium caudatum* to the increased salinity of the medium and to quinine solutions, *Zool. Zhur.,* **18**, 642.

———, ———, and W. W. Alpatov (1942), Geographic variation in Paramecium and the rôle of stabilizing selection in the origin of geographic differences, *Am. Naturalist,* **76**, 63–74.

Gaw, H. Z. (1936), Physiology of the contractile vacuole in ciliates. I. Effects of osmotic pressure. II. Effects of hydrogen ion concentration. III. Effect of temperature. IV. Effect of heavy water, *Arch. Protistenk.,* **87**, 185–224.

Geckler, R. P. (1949), Nitrogen mustard inactivation of the cytoplasmic factor kappa, in *Paramecium aurelia,* variety 4, *Science,* **110**, 89–90.

——— (1950), Inheritance of changes induced by nitrogen mustard in *Paramecium aurelia,* variety 4, *Genetics,* **35**, 108.

——— (1950 a), Genetic changes induced by exposure to nitrogen mustard and their inheritance in *Paramecium aurelia,* Variety 4, *Genetics,* **35**, 253–277.

——— and R. F. Kimball (1951), Effects of nitrogen mustard on cell division in *Paramecium, Anat. Record,* **111**, 107–108.

——— and ——— (1952), Effects of nitrogen mustard on cell division in Paramecium, *Nuclear Sci. Abs.,* **6**, 141.

Gelber, B. (1952), Investigations of the behavior of *Paramecium aurelia:* I. Modification of behavior after training with reinforcement, *J. Comp. Physiol. Psychol.,* **45**, 58–65.

Gelber, J. (1938), The effect of shorter than normal interendomictic intervals on mortality after endomixis in *Paramecium aurelia, Biol. Bull.,* **74**, 244–246.

Gelei, G. (1937), Ein neues Fibrillensystem im Ectoplasma von Paramecium, *Arch. Protistenk.,* **89**, 133–162.

——— (1938), Über die Isolation der erregungsleitenden Bahnen bei Ciliaten, *Biol. Zentr.,* **58**, 219–228.

——— (1939), Neuere Beiträge zum Bau und zu der Funktion des Exkretions-systems vom Paramecium, *Arch. Protistenk.,* **92**, 384–400.

Gelei, J. v. (1925), Új Paramecium szeged környékéröl. *Paramecium nephridiatum.* nov. sp., *Állattani Közlemények,* **22**, 121–162.

——— (1925 a), Nephridialapparat bei den Protozoen, *Biol. Zentr.,* **45**, 676–683.

——— (1926–27), Eine neue Osmium-Toluidinmethode für Protistenforschung, *Mikrokosmos,* **20**, 97–103.

——— (1926 a), Cilienstruktur und Cilienbewegung, *Verhandl. deut. zool. Ges.,* 202–13.

——— (1926 b), Zur Kenntnis des Wimperapparates, *Z. Anat. Entwicklungsgeschichte.,* **81**, 530–553.

——— (1927), Angaben zu der Symbiosefrage von Chlorella, *Biol. Zentr.,* **47**, 449–461.

——— (1928), Nochmals über Nephrodialapparat bei den Protozoen, *Arch. Protistenk.,* **64**, 479–94.

——— (1929), A Véglények Indegrendszere, *Állattani Közlemények,* **26**, 164–190.

——— (1932), Die reizleitenden Elemente der Ciliaten in nafs hergestellten Silber-bzw. Goldpräparaten, *Arch. Protistenk.,* **77**, 152–174.

——— (1933), Neue Silberbilder vom Nephridialapparat des Parameciums, *Arb. biol. Forsch. Inst.,* **6.**

——— (1933 a), Neue Beiträge zum Feinbau der Zilien, *Biol. Zentr.,* **53**, 512–515.

——— (1934), Eine interessante Anomalie im Cytopharynx von Paramecium, *Zool. Anz.,* **105**, 123–124.

——— (1934 a), Der feinere Bau des Cytopharynx von Paramecium und seine systematische Bedeutung, *Arch. Protistenk.,* **82**, 331–362.

—— (1934 b), Das Verhalten der ectoplasmatischen Elemente des Parameciums während der Teilung, *Zool. Anz.*, **107**, 161–177.

—— (1935), Infusorien im Dienste der Forschung und des Unterrichtes, *Biol. Zentr.*, **55**, 57–74.

—— (1936), Die Bildung der Porus excretorius und sein Verhältnis zum Neuronemensystem, *Állattani Közlemények*, **56**.

—— (1937–38), Schraubenbewegung und Körperbau bei Paramecium, *Arch. Protistenk.*, **90**, 165–177.

—— (1937 a), Die zusammengesetzte Pulsationsblase bei Paramecium, *Mikrokosmos*, **30**.

—— (1938), Beiträge zur Ciliatenfauna der Umgebung von Szeged. VII. *Paramecium nephridiatum*, *Arch. Protistenk.*, **91**, 343–356.

—— (1939), Das äussere Stützgerüstsystem des Parameciumkörpers, *Arch. Protistenk.*, **92**, 245–272.

—— and P. Horváth (1931), Eine nass Silber-bzw. Goldmethode für die Herstellung der reizleitenden Elemente bei den Ciliaten, *Z. wiss. Mikroskop.*, **48**, 9–29.

Gelfan, S. (1927), The electrical conductivity of protoplasm and a new method of its determination, *Univ. Calif. Pubs. Zoöl.*, **29**, 453–456.

Geppert, L. J. and E. H. Shaw Jr. (1937), The kinetics of the toxic action of antiseptic agents toward *Paramecium caudatum*, *Proc. S. Dakota Acad. Sci.*, **17**, 34–38.

Gerard, R. W. and L. H. Hyman (1931), The cyanide sensitivity of Paramecium, *Am. J. Physiol.*, **97**, 524–525.

Gersch, M. (1937), Vitalfärbung als Mittel zur Analyse physiologischer Prozesse, *Protoplasma*, **27**, 412.

Giemsa, G. B. C. (1923), Methoden zur Färbung der Protozoen. Kraus-Uhlenhuth, *Hdb. d. mikrobiol. Die Technik*, Berlin, Urban and Schwarzenberg, **1**, 359–380.

—— and S. Prowazek (1908), Wirkung des Chinins auf die Protistenzelle, *Beihefte z. Arch. Schiffs-u. Tropen-Hyg.*, **12**, 88.

Giese, A. C. (1935), The role of starvation in conjugation of Paramecium, *Physiol. Zoöl.*, **8**, 116–125.

—— (1938), Race and conjugation of Paramecium, *Physiol. Zoöl.*, **11**, 326–332.

—— (1938 a), Size and conjugation in Blepharisma, *Arch. Protistenk.*, **91**, 125–135.

—— (1939), Ultraviolet radiation and cell division. I Effects of λ2654 and 2804A upon *Paramecium caudatum*, *J. Cellular Comp. Physiol.*, **13**, 139–150.

—— (1939 a), Studies on conjugation in *Paramecium multimicronucleatum*, *Am. Naturalist*, **73**, 432–444.

—— (1941), Mating types in *Paramecium multimicronucleatum*, *Anat. Record*, **81**, 131–132.

—— (1942), Action of ultraviolet on cells, *The Collecting Net*, **17**.

—— (1945), A simple method for division-rate determination in Paramecium, *Physiol. Zoöl.*, **18**, 158–161.

—— (1945 a), The ultraviolet action spectrum for retardation of division of Paramecium, *J. Cellular Comp. Physiol.*, **26**, 47–55.

—— (1945 b), Ultraviolet radiations and life, *Physiol. Zoöl.*, **18**, 223–250.

—— (1946), An intracellular photodynamic sensitizer in Blepharisma, *J. Cellular Comp. Physiol.*, **28**, 119–128.

—— (1952), Thermal studies on protozoans, *Anat. Record*, **113**, 109.

—— (unpublished manuscript) Mating types in *Paramecium multimicronucleatum*.

—— and M. A. Arkoosh (1939), Tests for sexual differentiation in *Paramecium multimicronucleatum* and *Paramecium caudatum*, *Physiol. Zoöl.*, **12**, 70–75.

—— and E. B. Crossman (1945), The action spectrum of sensitization to heat with ultraviolet light, *J. Gen. Physiol.*, **29**, 79–87.

Giese, A. C. and E. B. Crossman (1946), Sensitization of cells to heat by visible light in presence of photodynamic dyes, *J. Gen. Physiol.*, **29**, 193–202.

—— and H. D. Heath (1948), Sensitization to heat by X-rays, *J. Gen. Physiol.*, **31**, 249–258.

—— and P. A. Leighton (1933), Fluorescence of cells in the ultraviolet, *Science*, **77**, 509–510.

—— and —— (1935 a), Quantitative studies on the photolethal effects of quartz ultraviolet radiation upon Paramecium, *J. Gen. Physiol.*, **18**, 557–571.

—— and —— (1935 b), The long wave-length limit of photolethal action in the ultraviolet, *Science*, **81**, 53–54.

—— and E. A. Reed (1940), Ultraviolet radiation and cell division variation in resistance to radiation with stock, species and nutritional differences in Paramecium, *J. Cellular Comp. Physiol.*, **15**, 395–408.

—— and C. V. Taylor (1935), Paramecia for experimental purposes in controlled mass cultures on a single strain of bacteria, *Arch. Protistenk.*, **84**, 225–231.

Giglio-Tos, E. (1908), Sull interpretazione morfologica e fisiologica degli Infusori, *Biol. Torino*, **2**, 1–79.

Gilman, L. C. (1939), Mating types in *Paramecium caudatum, Am. Naturalist,* **73**, 445–450, *Biol. Symposia,* **1**, 177–182.

—— (1941), Mating types in diverse races of *Paramecium caudatum, Biol. Bull.*, **80**, 384–402.

—— (1946–47), Mating reactions in *Paramecium caudatum, Proc. S. Dakota Acad. Sci.*, **26**, 43.

—— (1946–47 a), Mating types in *Paramecium caudatum* from South Dakota, *Proc. S. Dakota Acad. Sci.*, **26**, 112.

—— (1949), Intervarietal mating reactions in *Paramecium caudatum, Biol. Bull.*, **97**, 239.

—— (1950), The position of Japanese varieties of *Paramecium caudatum* with respect to American varieties, *Biol. Bull.*, **99**, 348–349.

Giroud, A. (1929), Recherches sur la nature chimique du chondriome, *Protoplasma,* **7**, 72–98.

Glaser, O. (1924), Temperature and forward movement of Paramecium, *J. Gen. Physiol.*, **7**, 177–189.

Glaser, R. W. (1932), Culture of certain Protozoa, bacteria free, *J. Parasitol.*, **19**, 23.

—— and N. A. Coria (1930), Methods for the pure culture of certain Protozoa, *J. Exp. Med.*, **51**, 787–806.

—— and —— (1933), The culture of *Paramecium caudatum* free from living organisms, *J. Parasitol.*, **20**, 33–37.

—— and —— (1935), The culture and reactions of purified Protozoa, *Am. J. Hyg.*, **21**, 110–120.

Gleichen, W. F. v. (1778), Abhandlung über die Saamen- und Infusionsthierchen etc., Nürnberg.

Goetsch, W. (1924), Lebensraum und Körpergrösse, *Biol. Zentr.*, **44**, 529–560.

—— and L. Scheuring (1926), Parasitismus und Symbiose der Algengattung Chlorella, *Z. Morphol. Ökol. Tiere,* **7**, 220–53.

Goldfuss, G. A. (1782–1848), Handbuch der Zoologie, Nürnberg.

Goldschmied-Hermann, A. (1935), Vergleichende Untersuchungen uber die Wirkung verschiedener Narkotika auf Paramecien, *Biol. Generalis,* **11**, 254–260.

Gourret, P. and P. Roeser (1886), Le Protozoaries du Vieux-Port de Marseille, *Arch. zool. exp. et gen.*, **4**, 443–534.

Goyal, R. K. (1935), Action of conissine on certain protozoa in vitro, *Compt. rend. soc. biol.*, **120**, 296.

Grabowski, U. (1939), Experimentelle Untersuchungen über das angebliche Lernvermögen von Paramaecium, *Z. Tierpsychol.*, **2**, 265–282.

Gray, J. (1928), Ciliary Movement, New York, The Macmillan Co., 1–162.

—— (1929), The mechanism of ciliary movement, *Am. Naturalist*, **63**, 68–81.

—— (1930), The mechanism of ciliary movement. VI Photographic and stroboscopic analysis of ciliary movement, *Proc. Roy. Soc. (London) B.*, **107**.

Greeley, A. W. (1903), The reactions of Paramoecia and other Protozoa to chemical and electrical stimuli, *Science*, **17**, 980–982.

—— (1904), Experiments on the physical structure of the protoplasm of paramœcium. and its relation to the reactions of the organism to thermal, chemical and electrical stimuli, *Biol. Bull.*, **7**, 3–32.

Greenleaf, W. E. (1924), The influence of volume of culture medium and cell proximity on the rate of reproduction of Protozoa, *Proc. Soc. Exp. Biol. Med.*, **21**, 405–406.

—— (1926), The influence of volume of culture medium and cell proximity on the rate of reproduction of Infusoria, *J. Exp. Zoöl.*, **46**, 143–167.

Greenwood, M. (1894), On the constitution and mode of formation of "food vacuoles" in infusoria etc., *Trans. Roy. Soc. (London)*, **185**.

—— and E. R. Saunders (1894), On the rôle of acid in protozoan digestion, *J. Physiol.*, **16**, 441–467.

Grenet, H. (1903), Action du champ magnétique sur les infusoires, *Compt. rend. soc. biol. Paris*, **55**, 957–958.

Grethe, G. (1895), Über die Wirkung verschiedener Chininderivate auf Infusorien, *Deut. Arch. klin. Med.*, **56**.

Griffiths, A. B. (1888), A method of demonstrating the presence of uric acid in the contractile vacuoles of some of the lower organisms, *Proc. Roy. Soc. Edinburgh*, **16**, 131–135.

Grinwald, E. (1929), Untersuchungen über die Entwicklungsfaktoren der Mikrobenkulturen. Wirkung der allelokatalytischen Substanz in der Kultur von *Colpidium colpoda*, *Acta Biol. Exp.*, **3**, 81–100.

Grobicka, J. and J. Wasilewska (1925), Essai d'analyse chimique quantitative de l'infusoire *Paramecium caudatum*, *Trav. inst. Nencki Warszawa.*, **3**, 1–23.

Gromova, E. N. (1941), Vliianie vneshnikh faktofov na strukturu makronuklevsa *Paramecium caudatum*. (The effect of external factors on the structure of the macronucleus in *Paramecium caudatum*), *Zool. Zhur.*, **20**, 187–197.

Grover, W. W. and E. H. Shaw Jr. (1947), A preliminary report on the mechanism of the toxic action of thiourea to *Paramecium caudatum*, *Proc. S. Dakota Acad. Sci.*, **26**, 85–86.

Gruber, A. (1884–85), Über künstliche Teilung der Infusorien, *Biol. Zentr.*, **4**, 771, **5**, 137.

—— (1886), Beiträge zur Kenntnis der Physiologie und Biologie der Protozoen, *Ber. naturforsch. Ges. Freiburg Breisgau*, **1**, 1–33.

—— (1886 a), Der Konjugationsprozess von *Paramecium aurelia*, *Ber. naturforsch. Ges. Freiburg Breisgau*, **2**, 31–32, 43–60.

—— (1889), Biologische Studien an Protozoen, *Biol. Zentr.*, **9**, 14–23.

—— (1890), Die Konjugation der Infusorien, *Biol. Zentr.*, **10**, 136–50.

Gruithuisen, F. von P. (1812), Beyträge sur Physiognosie und Eautognosie für Freunde der Naturforschung, München.

Gunn, J. A., and R. St. A. Heathcote (1931), The pharmacological actions of some alkyl derivatives of Harmol: 1. ethyl harmol, *Quart. J. Pharm. Pharmacol.*, **4**, 549–565.

Gurwitsch, A. (1904), Morphologie und Biologie der Zelle, Jena.

Guyer, M. F., and F. Daniels (1928), Cancer irradiation with cathode rays, *J. Cancer Research*, **12**, 166–187.

Haeckel, E. (1873), Zur Morphologie der Infusorien, *Jena. Z.*, **7**.

Hafkine, W. M. (1890), Recherches sur l'adaptation au milieu chez les infusoires et les bactéries. Contribution à l'étude de l'immunité, *Ann. inst. Pasteur*, **4**, 363–379.

———— (1890 a), Maladies infectieuses des Paramécies, *Ann. inst. Pasteur*, **4**, 148–162.

Halberstaedter, L. and A. Back (1943), Influence of colchicine alone and combined with x-rays on Paramecium, *Nature*, **152**, 275–276.

Hall, R. P. (1939), The trophic nature of the plant-like flagellates, *Quart. Rev. Biol.*, **14**, 1–12.

———— (1941), Vitamin deficiency as one explanation for inhibition of protozoan growth by conditioned medium, *Proc. Soc. Exp. Biol. Med.*, **47**, 306–308.

———— (1942), Incomplete proteins as nitrogen sources and their relation to vitamin requirements in *Colpidium campylum*, *Physiol. Zoöl.*, **15**, 96–107.

———— (1944), Comparative effects of certain vitamins on populations of *Glaucoma piriformis*, *Physiol. Zoöl.*, **17**, 200–209.

———— and W. B. Cosgrove (1944), The question of the synthesis of thiamin by the ciliate *Glaucoma piriformis*, *Biol Bull.*, **86**, 31–40.

————, D. F. Johnson, and J. B. Loefer (1935), A method for counting Protozoa in the measurement of growth under experimental conditions, *Trans. Am. Microscop. Soc.*, **54**, 298–300.

———— and R. F. Nigrelli (1937), A note on the vacuome of *Paramecium bursaria* and the contractile vacuole of certain ciliates, *Trans. Am. Microscop. Soc.*, **56**, 185–190.

Halter, C. R. (1925), Staining Paramecium in the class-room, *Science*, **61**, 90.

Hamburger, C. (1904), Die Konjugation von *Paramecium bursaria* (*Focke*), *Arch. Protistenk.*, **4**, 199–239.

Hamilton, A. (1904), The toxic action of scarlatinal and pneumonic sera on paramoecia, *J. Infectious Diseases*, **1**, 211–228.

Hammett, F. S. (1929), The chemical stimulus essential for growth by increase in cell number, *Protoplasma*, **7**, 297–332, *Proc. Am. Phil. Soc.*, **68**, 151–161.

———— (1930), The natural chemical regulation of growth by increase in cell number, *Proc. Am. Phil. Soc.*, **69**, 217–223.

Hammond, J. C. (1935), Stroboscopic observation of ciliary movement in the Protozoa, *Science*, **82**, 68–70.

Hance, R. T. (1915), The inheritance of extra contractile vacuoles in an unusual race of *Paramecium caudatum*, *Science*, **42**, 461–462.

———— (1917), Studies on a race of Paramecium possessing extra contractile vacuoles. I. An account of the morphology, physiology, genetics and cytology of this new race, *J. Exp. Zoöl.*, **23**, 287–333.

———— (1918), Paramoecia with extra contractile vacuoles, *Science*, **48**, 295–296.

———— (1925), A combined culture medium and indicator for Paramecium, *Science*, **62**, 351.

———— (1927–28), X-rays—new tools for biologists, *Proc. Penna. Acad. Sci.*, **2**, 1–4.

———— (1931), Certain x-ray effects upon *Paramecium*, *Proc. West Va. Acad. Sci.*, **5**, 56–57.

———— and H. Clark (1926), Studies on X-ray effects. XIV. The effect of X-ray on the division rate of Paramecium, *J. Exp. Med.*, **43**, 61–70.

Hansen, K. (1927), Some studies of Paramecium concerning their isolation, sterilization and nutritive requirements, *Acta Path. Microbiol. Scand.*, **4**, 1–38.

Hardin, G. (1944), Symbiosis of Paramecium and Oikomonas, *Ecology*, **25**, 304–311.

Hargitt, G. T. and Fray, W. W. (1917), The growth of Paramecium in pure cultures of bacteria, *J. Exp. Zoöl.*, **22**, 421.

———— and R. L. Phillips (1921), The rate of metabolism of Paramecium with controlled bacterial food supply, *Anat. Record*, **23**, 123.

Harnisch, O. (1926), Kritische Studien über die Gewöhnung freilebender Protozoen an Gifte, *Zool. Anz. Supp.*, **2**, 99–108.

Harper, E. H. (1911), The geotropism of Paramoecium, *J. Morphol.*, **22**, 993–1000.

——— (1912), Magnetic control of geotropism in Paramoecium, *J. Animal. Behavior*, **2**, 181–189.

——— (1912 a), Magnetic control of the movements of Paramoecium which have ingested iron, *Science*, **35**, 939–940.

Harris, F. I. and H. S. Hoyt (1919), The action of ultraviolet light on certain bacteria in relation to specific absorption by amino acids, *Univ. Calif. Pubs. Path.*, **2**, 245–250.

Harrison, J. A. (1947), Antigenic variation in protozoa and bacteria, *Ann. Rev. Microbiol.*, **1**, 19–42.

——— and E. H. Fowler (1944), Serologic differentiation of the three species of paramecia in the aurelia group, *Anat. Record*, **89**, 20.

——— and ——— (1945), Antigenic variation in pure cultures of Paramecia, *J. Bact.*, **49**, 519.

——— and ——— (1945 a), Antigenic variation in clones of *Paramecium aurelia*, *J. Immunol.*, **50**, 115–125.

——— and ——— (1945 b), Serologic evidence of cytoplasmic interchange during conjugation in *Paramecium bursaria*, *Science*, **102**, 377–378.

——— and ——— (1945 c), An antigen-antibody reaction with Tetrahymena which results in dystomy, *Science*, **102**, 65–66.

——— and ——— (1946), A serologic study of conjugation in *Paramecium bursaria*, *J. Exp. Zoöl.*, **101**, 425–444.

——— and ——— (1946 a), The reaction of Paramecia in specific antiserum, *J. Bact.*, **51**, 626.

———, M. E. Sano, E. H. Fowler, R. H. Shellhamer, and C. A. Bocher (1948), Toxicity for paramecia of sera from cancerous and non-cancerous persons, *Federation Proc.*, **7**.

Hartmann, M. (1907), Das System der Protozoen, *Arch. Protistenk.*, **10**, 139.

——— (1919), Allegemeine Morphologie und Physiologie der Protozoen, *Lehrb. d. Mikrobiol.*, Jena, G. Fischer, **1**.

——— (1921), Ergebnisse und Probleme der Protistenkunde, *Festschr. d. Kaiser-Wilhelm-Ges.*, Springer.

——— (1926), Über experimentelle Unsterblichkeit von Protozoenindividuen, *Naturwissenschaften*, **14**, 433–435.

——— (1928), Praktikum der Protozoologie, Jena, G. Fischer.

Hartog, M. (1888), Preliminary note on the functions and homologies of the contractile vacuole in plants and animals, *Rep. Brit. A. Adv. Sc.*, 714.

——— and A. E. Dixon (1893), On the digestive ferments of a large protozoon, *Rep. Brit. A. Adv. Sc.*, **63**, 801–802.

Harvey, E. N. (1911), Studies on the permeability of cells, *J. Exp. Zoöl.*, **10**, 507–556.

——— (1931), Observations on living cells, made with the microscope-centrifuge, *J. Exp. Biol.*, **8**, 267–274.

——— (1934), Biological effects of heavy water, *Biol. Bull.*, **66**, 91–96.

——— (1934 a), The air turbine for high speed centrifuging of biological material, together with some observations on centrifuged eggs, *Biol. Bull.*, **66**, 48–54.

Hausmann, G. (1927), Über die Bewegung einiger ziliater Protozoen im Wechselstrom, *Biol. Generalis*, **3**, 463–474.

Hausmann, W. (1911), Die sensibilisierende Wirkung des Hämatoporphyrins, *Biochem. Z.*, **30**, 276–316.

——— (1912), Über optische Sensibilisation im Tier- und Pflanzenreiche, *Fortschr. d. Naturw. Forsch.*, **6**, 243.

——— and W. Kolmer (1907), Colchicine and Paramecium, *Biochem. Z.*, **3**, 503.

——— and ——— (1908), Über die sensibilisierende Wirkung pflanzlicher und tierischer Farbstoffe auf Paramaecien, *Biochem. Z.*, **15**, 13–18.

Haye, A. (1930), Über den Exkretionsapparat bei den Protisten, nebst Bemerkungen über einige andere feinere Strulturverhältnisse der untersuchten Arten, *Arch. Protistenk.*, **70**, 1–86.

Hayes, F. R. (1930), The physiological response of Paramaecium to sea-water, *Z. vergleich. Physiol.*, **13**, 214–222.

Heathman, L. (1932), Studies of the antigenic properties of some free living and pathogenic amebas, *Am. J. Hyg.*, **16**, 97–123.

Hegner, R. W. (1926), The interrelations of Protozoa and the utricles of utricularia, *Biol. Bull.*, **50**, 239–270.

—— (1926 a), The Protozoa of the pitcher plant *Sarracenia purpurea*, *Biol. Bull.*, **50**, 271–276.

—— (1929), The viability of Paramecia and Euglenae in the digestive tract of cockroaches, *J. Parasitol.*, **15**, 272–275.

Heilbrunn, L. V. (1921), Protoplasmic viscosity changes during mitosis, *J. Exp. Zoöl.*, **34**, 417–447.

—— (1926), The absolute viscosity of protoplasm, *J. Exp. Zoöl.*, **43**, 255–278.

—— (1928), The Colloid Chemistry of Protoplasm, Berlin, Borntraeger.

—— (1947), An Outline of General Physiology, Philadelphia, W. B. Saunders Co., 1–748.

Heinrich K. (1903), Über die Wirkung einiger noch nicht untersuchter fluores zierender Stoffe auf Paramaecien, Diss. München, 17 pp.

Henderson, V. E. (1930), The present status of the theories of narcosis, *Physiol. Rev.*, **10**, 171–220.

Henri, V. and Mme. Henri (1912), Excitation des organismes par les rayons ultra-violets, *Compt. rend. soc. biol.*, **72**, 1083–1085.

—— and —— (1912 a), Action photodynamique du sélénium colloïdal, *Compt. rend. soc. biol. Paris*, **72**, 326–328.

Herfs, A. (1922), Die pulsierende Vakuole der Protozoen ein Schutzorgan gegen Aussüssung. Studien uber Anpassung der Organismen an das Leben im Süsswasser, *Arch. Protistenk.*, **44**, 227–260.

Hermann, J. (1784), *Der Naturforscher*, **20**, 157.

Hertel, E. (1904), Über Beeinflussung des Organismus durch Licht, speziell durch die chemisch wirksamen Strahlen, *Z. allgem. Physiol.*, **4**, 1–43.

—— (1905), Über physiologische Wirkung von Strahlen verschiedener Wellenlänge. Vergleichende physiologische Untersuchungen, *Z. allgem. Physiol.*, **5**, 95–122.

Herter, K. (1927), Taxien und Tropismen der Tiere, *Tabulae biol.*, **4**, 348–376.

Hertwig, R. (1888), Ueber Kernteilung bei Infusorien, *Sitzber. Ges. Morphol. Physiol. München*, **3**, 127–128.

—— (1888 a), Ueber Kernteilung und ihre Bedeutung für Zellteilung und Befruchtung, *Sitzber. Ges. Morphol. Physiol. München*, **4**, 83–87.

—— (1889), Ueber die Conjugation der Infusorien, *Abhandl. bayer. Akad. Wiss.*, **17**, 150–233.

—— (1889 a), Ueber die Conjugationen der Infusorien, *Sitzber. Ges. Morphol. Physiol. München*, **5**, 35–38.

—— (1892), Über Befruchtung und Konjugation, *Verhandl. deut. zoöl.*, **2**, 95–112.

—— (1900), Was veranlasst die Befruchtung der Protozoen? *Sitzber. Ges. Morphol. Physiol. München*, **15**, 62–69.

—— (1902), Die Protozoen und die Zelltheorie, *Arch. Protistenk.*, **1**, 1.

—— (1903), Über das Wechselverhältnis von Kern und Plasma, *Sitzber. Ges. Morphol. Physiol. München*, **18**, 77–100.

—— (1914), Über Parthenogenesis der Infusorien und die Depressionszustände der Protozoen, *Biol. Zentr.*, **34**, 557–581.

Herwerden, M. A. van (1926), Omkeerbare gelvorming en weefselfixatie, *Verslag Gewone Vergader. Afdeel. Natuurk., Nederland Akad. Wetenschap.*, **35**, 574–577.

Herzfeld, E. (1926), Über das Vorkommen von Missbildungen und Monstrositäten bei Paramaecium sp. nebst einigen experimentellen Untersuchungen über deren Bedeutung, *Jena. Z. Naturw.*, **62**, 81–124.

Hetherington, A. (1934), The pure culture of Paramecium, *Science*, **79**, 413–414.

———— (1934 a), The role of bacteria in the growth of *Colpidium colpoda*, *Physiol. Zoöl.*, **7**, 618–641.

Heubner, W. (1925), Zur Pharmakologie der Reizstoffe, *Arch. exp. Path.*, **107**, 129–154.

Hibbard, H. (1945), Current status of our knowledge of the golgi apparatus in the animal cell, *Quart. Rev. Biol.*, **20**, 1–19.

Hickson, S. J. (1902), *Dendrocometes paradoxus* Part I—Conjugation, *Quart. J. Microscop. Sci.*, **45**, 325–359.

Hill, J. (1752), History of Animals, in General Natural History, London, **3**.

Hinrichs, A. (1927), Ultraviolet radiation and living processes, *Trans. Illinois State Acad. Sci.*, **20**, 316–319.

———— (1928), Ultraviolet radiation and division in *Paramecium caudatum*, *Physiol. Zoöl.*, **1**, 394–415.

Hirsch, E. (1914), Untersuchungen über die biologische Wirkung einiger Salze, *Zool. Jahrb., Abstr. Zool. Physiol.*, **34**, 559–677.

Hirschler, J. (1924), Sur les composants lipoidiferes du plasma des Protozoaires, *Compt. rend. soc. biol.*, **10**, 891–893.

———— (1927), Studien über die sich mit Osmium schwärzenden Plasmakomponenten einiger Protozoenarten, *Z. Zellforsch.*, **5**, 704–786.

Hiwatashi, K. (1949), Studies on the conjugation of *Paramecium caudatum:* I. Mating types and groups in the races obtained in Japan, *Science Repts. Tôhoku Imp. Univ.*, **18**, 137–140.

———— (1949 a), Studies on the conjugation of *Paramecium caudatum:* II. Induction of pseudoselfing pairs by formalin killed animals, *Science Repts Tôhoku Imp. Univ.*, **18**, 141–143.

———— (1950), Studies on the conjugation of *Paramecium caudatum:* III. Some properties of the mating type substances, *Science Repts Tôhoku Imp. Univ., Fourth Ser.*, **18**, 270–275.

Hoare, C. A. (1927), Schewiakoff's keys for the determination of the Holotrichous ciliates, Translated from the Russian. *Proc. Zool. Soc. London*, **1**, 1173–1174.

Hofmeister, W. (1865), Über die Mechanik der Bewegungen des Protoplasmas, *Flora*, 7–12.

Hollis, H. and W. L. Doyle (1937), Enzyme studies on protozoa, *J. Cellular Comp. Physiol.*, **12**, 306.

Holter, H. and W. L. Doyle (1938), Studies on enzymatic histochemistry. XXVIII. Enzymatic studies on protozoa, *J. Cellular Comp. Physiol.*, **12**, 295–308.

Hopkins, H. S. (1921), The conditions for conjugation in diverse races of Paramecium, *J. Exp. Zoöl.*, **34**, 338–384.

Horning, E. S. (1926), Studies on the mitochondria of Paramecium, *Australian J. Exp. Biol. Med. Sci.*, **3**, 89–95.

———— (1927), On the relation of mitochondria to the neuclus, *Australian J. Exp. Biol. Med. Sci.*, **4**, 75–78.

———— (1927 a), On the orientation of mitochondria in the surface cytoplasm of infusorians, *Australian J. Exp. Biol. Med. Sci.*, **4**, 187–190.

Horton, F. M. (1935), On the reactions of isolated parts of *Paramecium caudatum*, *J. Exp. Biol.*, **12**, 13–16.

Hosoi, T. (1937), Protoplasmic streaming in isolated pieces of paramecium, *J. Faculty Sci., Imp. Univ. Tokyo Sect. IV*, **4**, 299–305.

Howland, R. B. (1924), On excretion of nitrogenous waste as a function of the contractile vacuole, *J. Exp. Zoöl.*, **40**, 231–250.

Howland, R. B. (1924 a), Experiments on the contractile vacuole of *Amoeba verrucosa* and *Paramecium caudatum, J. Exp. Zoöl.,* **40,** 251–262.

―――― and A. Bernstein (1931), A method for determining the oxygen consumption of a single cell, *J. Gen. Physiol.,* **14,** 339–348.

Hughes, D. M. and W. T. Bovie (1918), The effects of fluorite ultraviolet light on the rate of division of *Paramecium caudatum, J. Med. Research,* **39,** 233–238.

Humphrey, B. A. and G. F. Humphrey (1947), Succinic dehydrogenase in Protozoa, *Nature,* **159,** 374.

―――― and ―――― (1948), Studies in the respiration of *Paramecium caudatum, J. Exp. Biol.,* **25,** 123–134.

Hunt, R. (1907), Studies on experimental alcoholism, *Bull. 33, Hyg. Lab. Mar. Hosp. Washington.*

Hutchinson, A. H. and M. R. Ashton (1929), The specific effects of monochromatic light on the growth of Paramecium, *Can. J. Research,* **1,** 292–304.

―――― and ―――― (1931), Specific effect of monochromatic light upon plasmolysis in Paramecium, *Can. J. Research,* **4,** 614–623.

―――― and ―――― (1933), The effect of radiant energy on diastase activity, *Can. J. Research,* **9,** 49–64.

Hutchison, R. H. (1915), The effects of certain salts, and of adaption to high temperatures on the heat resistance of *Paramecium caudatum, J. Exp. Zoöl.,* **19,** 211–224.

Huygens, C. (1888), Oeuvres Complètes, La Haye.

Hyman, L. H. (1925), Methods of securing and cultivating Protozoa: I General statements and methods, *Trans. Am. Microscop. Soc.,* **44,** 216–221.

―――― (1931), Methods of securing and cultivating Protozoa: II Paramecium and other ciliates, *Trans. Am. Microscop. Soc.,* **50,** 50–57.

―――― (1940), The Invertebrates Protozoa through Ctenophora, New York, McGraw-Hill Co.

―――― (1941), Lettuce as a medium for the continuous culture of a variety of small laboratory animals, *Trans. Am. Microscop. Soc.,* **60,** 365–370.

―――― (1942), The transition from the unicellular to the multicellular individual, *Biol. Symposia,* **8,** 27–42.

Ibara, Y. (1926), Culture medium for the ciliate Lacrymaria, *Science,* **63,** 212.

Iida, T. T. (1940), Electrical conductivity of cytolyzed Paramecium, *Japan. J. Zoöl.,* **8,** 407–414.

Ilowaisky, S. A. (1926), Über die Kernprozesse der getrennten Konjuganten der *Stylonychia mytilus* und *Paramecium caudatum, Arch. Protistenk.,* **53,** 249–252.

Imel, H. G. (1915), Some preliminary observations on the oxygenless region of Center Lake, Kosciusko County, Indiana, *Proc. Indiana Acad. Sci.,* 345–356.

Ishikawa, H. (1911), Wundheilungs- und Regenerationsvorgänge bei den Infusorien, *Arch. Entwicklungsmech. mechan.,* **35,** 1–29.

Israel, O. and T. Klingmann (1897), Oligodynamische Erscheinungen an pflanzlichen und tierischen Zellen, *Arch. path. Anat.,* **147,** 293–340.

Ivanic, M. (1926), Über die mit den Reorganisationsprozessen der Bewegungs und Nahrungsaufnahmeorganellen verbundene Ruhestadien von *Paramecium caudatum, Zool. Anz.,* **68,** 1–9.

Jacobs, M. H. (1912), Studies on the physiological characters of species. 1. The effects of carbon dioxide on various Protozoa, *J. Exp. Zoöl.,* **12,** 519–542.

―――― (1919), Acclimatization as a factor affecting the upper thermal death points of organisms, *J. Exp. Zoöl.,* **27,** 427–442.

―――― (1922), The effect of carbon dioxide on the consistency of protoplasm, *Biol. Bull.,* **41,** 14–30.

Jacobson, I. (1931), Fibrilläre Differenzierungen bei Ciliaten, *Arch. Protistenk.,* **75,** 31–100.

Jacobson, W. E. (1948), Non-mating-type-specific "activation" in *Paramecium aurelia:* induction of macronuclear breakdown in sensitive stock P of variety 1 by killer stock G of variety 2, *Anat. Record,* **101,** 708.

Jahn, T. L. (1934), Problems of population growth in the Protozoa, *Cold Spring Harbor Symposia Quant. Biol.,* **II,** 167–180.

—— (1935 a), Studies on the physiology of the euglenoid flagellates. V. The effect of certain carbohydrates on the growth of *Euglena gracilis, Arch. Protistenk.,* **86,** 238–250.

—— (1935 b), Studies on the physiology of the euglenoid flagellates. VII. The effect of salts of certain organic acids on growth of *Euglean gracilis, Arch. Protistenk.,* **86,** 258–262.

—— (1941), Respiratory metabolism, Protozoa in Biological Research, New York, Columbia Univ. Press, 1–1148.

Jakus, M. A. (1945), The structure and properties of the trichocysts of Paramecium, *J. Exp. Zoöl.,* **100,** 457–485.

—— and C. E. Hall (1946), Electron microscope observations of the trichocysts and cilia of Paramecium, *Biol. Bull.,* **91,** 141–144.

——, ——, and F. O. Schmitt (1942), Electron microscope studies of the structure of Paramecium trichocysts, *Anat. Record,* **84,** 474–475.

Jellinek, S. (1936), Biology of ultrashort waves, *Arch. Phys. Therapy, X-Ray, Radium,* **17,** 512.

Jennings, H. S. (1897), Studies on reactions to stimuli in unicellular organisms. I. Reactions to chemical, osmotic and mechanical stimuli in the ciliate infusoria, *J. Physiol.,* **21,** 258–322.

—— (1899), The behavior of unicellular organisms, in Reprint, Biological Lectures from Marine Biological Laboratories, Woods Hole, Mass., Boston, Ginn and Co., 93–112.

—— (1899 a), Studies in reactions to stimuli in unicellular organisms. II. The mechanism of the motor reactions of Paramecium, *Am. J. Physiol.,* **2,** 311–341.

—— (1899 b), Studies on reactions to stimuli in unicellular organisms. IV. Laws of chemotaxis in Paramecium, *Am. J. Physiol.,* **2,** 355–379.

—— (1899 c), The psychology of a protozoan, *Am. J. Psychol.,* **10,** 1–13.

—— (1900), Studies on reactions to stimuli in unicellular organisms. V. On the movements and motor reflexes of the Flagellata and Ciliata, *Am. J. Physiol.,* **3,** 229–260.

—— (1901), On the significance of the spiral swimming of organisms, *Am. Naturalist,* **35,** 369–378.

—— (1904), A method of demonstrating the external discharge of the contractile vacuole, *Zool. Anz.,* **27,** 656–658.

—— (1904 a), The behavior of Paramecium. Additional features and general relations, *J. Comp. Neurol.,* **14,** 441–510.

—— (1904 b), Contributions to the study of the behavior of lower organisms: 1. Reactions to heat and cold in the ciliate infusoria, *Carnegie Inst. Wash. Pub.,* **16,** 1–28.

—— (1904 c), Contributions to the study of the behavior of lower organisms: 2. Reactions to light in ciliates and flagellates, *Carnegie Inst. Wash. Pub.,* **16,** 29–71.

—— (1904 d), The theory of tropisms, *Carnegie Inst. Wash. Pub.,* **16,** 89–107.

—— (1904 e), Physiological states as determining factors in the behavior of lower organisms, *Carnegie Inst. Wash. Pub.,* **16,** 109–127.

—— (1904 f), The method of trial and error in the behavior of lower organisms, *Carnegie Inst. Wash. Pub.,* **16,** 235–252.

—— (1905), The basis for taxis and certain other terms in the behavior of Infusoria, *J. Comp. Neurol.,* **15,** 138–143.

Jennings, H. S. (1905 a), Papers on reactions to electricity in unicellular organisms, *J. Comp. Neurol. Psychol.*, **15**, 528–534.

―――― (1908), The interpretation of the behavior of the lower organisms, *Science*, **27**, 698–710.

―――― (1908 a), Heredity, variation and evolution in Protozoa. 1. The fate of new structural characters in Paramecium in connection with the problem of inheritance of acquired characters in unicellular organisms, *J. Exp. Zoöl.*, **5**, 577–632.

―――― (1908 b), Heredity, variation and evolution in Protozoa. II. Heredity and variation of size and form in Paramecium with studies of growth, environmental action and selection, *Proc. Am. Phil. Soc.*, **47**, 393–546.

―――― (1909), Diverse races of Paramecium and their relation to selection and to conjugation, *Science*, **29**, 424–425.

―――― (1909 a), Heredity and variation in the simplest organisms, *Am. Naturalist*, **43**, 321–337.

―――― (1910), What conditions induce conjugation in Paramecium?, *J. Exp. Zoöl.*, **9**, 279–298.

―――― (1910 a), Experiments on the effect of conjugation on the life history in Paramecium, *Science*, **31**, 466–467.

―――― (1911), Assortative mating, variability and inheritance of size, in the conjugation of Paramecium, *J. Exp. Zoöl.*, **11**, 1–134.

―――― (1912), Age, death and conjugation in the light of work on lower organisms, *Pop. Sc. Monthly*, **80**, 563–577.

―――― (1913), The effect of conjugation in Paramecium, *J. Exp. Zoöl.*, **14**, 279–391.

―――― (1920), Life and Death, Heredity and Evolution in Unicellular Organisms, Boston, 233 pp.

―――― (1922), Variation in uniparental reproduction, *Am. Naturalist*, **56**, 5–15.

―――― (1923), Inheritance in unicellular organisms, Eugenics, Genetics and the Family, **1**, 59–64.

―――― (1929), Genetics of the Protozoa, *Bibliographia Genetica*, **5**, 105–330.

―――― (1931), Behavior of the Lower Organisms, New York, Columbia University Press, 1–366.

―――― (1938), Sex reaction types and their interrelations in *Paramecium bursaria*. I and II. Clones collected from natural habitats, *Proc. Nat. Acad. of Sci.*, **24**, 112–120.

―――― (1939), Genetics of *Paramecium bursaria*. I. Mating types and groups, their interrelations and distribution; mating behavior and self sterility, *Genetics*, **24**, 202–233.

―――― (1939 a), *Paramecium bursaria:* Mating types and groups, mating behavior, self-sterility; their development and inheritance, *Am. Naturalist*, **73**, 414–431.

―――― (1940), Chromosomes and Cytoplasm in Protozoa, *Am. Assoc. Advancement Sci.*, **14**, 44–55.

―――― (1940 a), The beginning of social behavior in unicellular organisms, *Science*, **92**, 539–546.

―――― (1941), The transition from the individual to the social level, *Science*, **94**, 447–453.

―――― (1941 a), Genetics of *Paramecium bursaria*. II. Self-differentiation and self-fertilization of clones, *Proc. Am. Phil. Soc.*, **85**, 25–48.

―――― (1941 b), Inheritance in Protozoa, chap. 15 in Protozoa in Biological Research, New York, Columbia University Press.

―――― (1942), Senescence and Death in Protozoa and Invertebrates, Problems of Ageing, ed. 2, 29–48.

―――― (1942 a), Genetics of *Paramecium bursaria*. III. Inheritance of mating type in crosses and in clonal self fertilizations, *Genetics*, **27**, 193–211.

―――― (1944), *Paramecium bursaria:* Life history. I. Immaturity, maturity and age, *Biol. Bull.*, **86**, 131–145.

—— (1944 a), *Paramecium bursaria:* Life history. II. Age and death of clones in relation to the results of conjugation, *J. Exp. Zoöl.,* **96,** 17–52.

—— (1944 b), *Paramecium bursaria:* Life history. III. Repeated conjugations in the same stock at different ages, with and without inbreeding, in relation to mortality at conjugation, *J. Exp. Zoöl.,* **96,** 243–273.

—— (1944 c), *Paramecium bursaria:* Life history. IV. Relation of inbreeding to mortality of ex-conjugant clones, *J. Exp. Zoöl.,* **97,** 165–197.

—— (1945), *Paramecium bursaria:* Life history. V. Some relations of external conditions, past or present, to ageing and to mortality of exconjugants, with summary of conclusions on age and death, *J. Exp. Zoöl.,* **99,** 15–31.

—— (1945 a), Social life and interrelationships in certain Protozoa, *Sociometry,* **8,** 9–20.

—— [Chairman], et al. (1939 b), Symposium on mating types and their interactions in the ciliate infusoria. I. Introduction, by H. S. Jennings. II. *Paramecia aurelia:* Mating types and groups; lethal interactions; determination and inheritance, by T. M. Sonneborn. III. *Paramecium bursaria:* Mating types and groups, mating behavior, self-sterility; their development and inheritance, by H. S. Jennings. IV. Studies on conjugation in *Paramecium multimicronucleatum,* by A. C. Giese. V. Mating types in *Paramecium caudatum,* by L. C. Gilman. VI. Mating types in *Euplotes,* by R. F. Kimball, *Am. Naturalist,* **73,** 385–456.

—— and G. T. Hargitt (1910), Characteristics of diverses races of Paramecium, *J. Morphol.,* **21,** 495–561.

—— and C. Jamieson (1902), Studies on reactions to stimuli in unicellular organisms. X. The movements and reactions of pieces of ciliate infusoria, *Biol. Bull.,* **3,** 225–234.

—— and K. S. Lashley (1913), Biparental inheritance and the question of sexuality in Paramecium, *J. Exp. Zoöl.,* **14,** 393–466.

—— and —— (1913a), Biparental inheritance of size in Paramecium, *J. Exp. Zoöl.,* **15,** 193–199.

—— and E. M. Moore (1902), Studies on reactions to stimuli in unicellular organisms. VIII. On the reactions of infusoria to carbonic and other acids, with especial reference to the causes of the gatherings spontaneously formed, *Am. J. Physiol.,* **6,** 233–250.

—— and P. Opitz (1944), Genetics of *Paramecium bursaria.* IV. A fourth variety from Russia. Lethal crosses with an American variety, *Genetics,* **29,** 576–583.

——, D. Raffel, R. S. Lynch, and T. M. Sonneborn (1932), The diverse biotypes produced by conjugation within a clone of *Paramecium aurelia, J. Exp. Zoöl.,* **62,** 363–408.

—— and T. M. Sonneborn (1936), Relation of endomixis to vitality in *Paramecium aurelia, Extrait compt. rend. 12th Intern. Congr. Zool. Lisbon, 1935.*

Jensen, P. (1893), Die absolute Kraft einer Flimmerzelle, *Arch. ges. Physiol.,* **54,** 537–551.

—— (1893 a), Über den Geotropismus niederer Organismen, *Pflügers Arch. ges. Physiol.,* **53,** 428.

Jickeli, C. F. (1884), Über die Kernverhältnisse der Infusorien, *Zool. Anz.,* **7,** 468–473, 491–497.

Jirovec, O. (1926), Protozoenstudien I, *Arch. Protistenk.,* **56,** 280–290.

Joblot, L. (1718), Descriptions et usages de plusieurs nouveaux microscopes . . . . avec de nouvelles observations (etc.), part 2, (Sec. Ed. with additions by pub. 1754).

Jodlbauer, A. (1904), Über die Wirkung photodynamischer -fluoreszierender- Substanzen auf Paramaecien aun Enzyme bei Röntgen- und Radiumbestrahlung, *Deut. Arch. kiln. Med.,* **80,** 488–491.

—— (1908), Die sensibilisierende Wirkung fluoreszierender Stoffe, *Jahrb. Leistung. Fortschr. Physik. Med.,* **1,** 280–294.

—— (1913), Die Sensibilisierung durch fluoreszierende Stoffe, *Strahlentherapie,* **2,** 71–83.

Joff, N. A. (1923), Über Regeneration bei Infusorien, *Arch. soc. Russ. Protistenk. Moskau,* **2**, 220–229.

Johnson, W. H. (1929), The reactions of Paramecium to solutions of known hydrogen ion concentration, *Biol. Bull.,* **57**, 199–224.

—— (1936), Studies on the nutrition and reproduction of Paramecium, *Physiol. Zoöl.,* **9**, 1–14.

—— (1937), Experimental populations of microscopic organisms, *Am. Naturalist,* **71**, 5–20.

—— (1946), Growth of Paramecium in sterile culture, *Anat. Record,* **96**, 17–18.

—— (1950), Further studies on the sterile culture of Paramecium, *Proc. Am. Soc. Protozool.,* **1**, 9.

—— and E. G. S. Baker (1941), The sterile culture of Paramecium, *Anat. Record,* **81**, suppl. 68.

—— and —— (1942), The sterile culture of *Paramecium multimicronucleata, Science,* **95**, 333–334.

—— and —— (1943), Effects of certain B vitamins on populations of *Tetrahymena geleii, Physiol. Zoöl.,* **16**, 172–185.

—— and G. Hardin (1938), Reproduction of Paramecium in old culture medium, *Physiol. Zoöl.,* **11**, 333–346.

—— and E. L. Tatum (1945), The heat labile growth factor for Paramecium in pressed yeast juice, *Arch. Biochem.,* **8**, 163–168.

Jollos, V. (1913 a), Experimentelle Untersuchungen an Infusorien, *Biol. Zentr.,* **33**, 222–236.

—— (1913 b), Über die Bedeutung der Konjugation bei Infusorien, *Arch. Protistenk.,* **30**, 328–334.

—— (1914), Variabilität und Vererbung bei Mikro-organismen, *Z. indukt.,* **12**, 14–35.

—— (1916), Die Fortpflanzung der Infusorien und die potentielle Unterblichkeit der Einzelligen, *Biol. Zentr.,* **36**, 495–515.

—— (1920), Experimentelle Verebungsstudien an Infusorien, *Z. indukt. Abstammungs-Verblehre,* **25**, 77–97.

—— (1921), Experimentelle Protistenstudien: I. Untersuchungen über Variabilität und Vererbung bei Infusorien, *Arch. Protistenk.,* **43**, 1–222.

—— (1921 a), Experimentelle Protistenstudien, Jena, G. Fischer.

—— (1934), Dauermodifikationen und Mutationen bei Protozoen, *Arch. Protistenk.,* **83**, 197–219.

Jones, A. W. and W. J. Cloyd (1950), The adhesive function of the trichocysts of Paramecium, *Turtox News,* **28**, 231–232.

Jones, E. P. (1928), Population curves of Paramecium, *Proc. Penna. Acad. Sci.,* **2**, 1927–1928.

—— (1930), Paramecium infusion histories: I. Hydrogen ion changes in hay and hay-flour infusions, *Biol. Bull.,* **59**, 275–284.

—— (1932), Concentrating Paramecium and rotifers without centrifuging, *Science,* **75**, 52.

—— (1933), Paramecium infusion histories: III. Population and size changes within a pure line, *Univ. Pitt. Bull.,* **29**, 1–8.

—— (1936), Comparative measurements of living and fixed Protozoa, *Trans. Am. Microscop. Soc.,* **55**, 516–518.

—— (1937), *Paramecium multimicronucleatum;* mass-culturing, maintaining and rehabilitating mass-cultures, and securing concentrations, Culture Methods for Invertebrate Animals, Ithaca, Comstock.

Jones, R. M. (1950), McClung's Handbook of Microscopical Technique (by 36 authors), ed. 3, New York, P. B. Hoeber, 803.

Joseph, H. (1914), Zur Klarstellung gegenüber Herrn Dr. P. Kammerers Gegenkritik, *Naturwissenschaften*, 5.

────── and S. Prowazek (1902), Versuche über die Einwirkung von Röntgenstrahlen auf einige Organismen, besonders auf deren Plasmatätigkeit, *Z. allgem. Physiol.*, 1, 142.

Joukowsky, D. (1898), Beiträge zur Frage nach den Bedingungen der Vermehrung und des Eintritts der Konjugation bei den Ciliaten, *Verhandl. Nat.-med. Ver. Heidelberg*, 26, 17–42.

Juday, C. (1908), Some aquatic invertebrates that live under anaerobic conditions, *Trans. Wisconsin Acad. Sci.*, 16, 10–16.

Junker, H. (1925), Über die Wirkung extrem verdünnter Substanzen auf Paramaecien, *Biol. Zentr.*, 45, 26–34.

────── (1928), Die Wirkung extremer Potenzverdünngen auf Organismen, *Pflügers Arch. ges. Physiol.*, 219, 647–672.

Kaas, J. (1909), Structure et fonctions biologiques endoplasmatique du *Paramecium aurelia*, *Mem. Pontific. Accad. Nuovi Lincei*, 26, 109–156.

Kahl, A. (1928), Die Infusorien (Ciliata) der Oldesloer Salzwasserstellen (*Paramecium ficarium.*), *Arch. Hydrobiol.*, 19, 50–123.

────── (1930), Wimpertiere oder Ciliata (Infusoria), Jena, Dahl-Bischoff, 1–886.

Kahler, H., W. H. Chalkley, and C. Voegtlin (1929), The nature of the effect of a high frequency field on *Paramecium*, *Pub. Health Repts.*, 44, 339–347.

Kalmus, H. (1927), Das Kapillar-Respirometer, eine neue Versuchsanordnung zur Messung des Gaswechsels von Mikroorganismen, *Biol. Zentr.*, 47, 595–600.

────── (1928 a), Die Messungen der Atmung Gärung und CO_2-Assimilation Kleiner Organismen in der Kapillare, *Z. vergleich. Physiol.*, 7, 304.

────── (1928 b), Untersuchungen über die Atmung von *Paramecium caudatum*, *Z. vergleich. Physiol.*, 7, 314–322.

────── (1929), Beobachtungen und Versuche über die Tätigkeit der kontraktilen Vakuole eines marinen Infusors: Amphileptus gutta Cohn, nebst morphologischen und systematischen Vorbemerkungen, *Arch. Protistenk.*, 66, 409.

────── (1929 a), Versuche über die Teilung von Paramaecien in der Kapillare nebst Bemerkungen über den sogenannten Raumfaktor, *Arch. Protistenk.*, 66, 402–408.

────── (1931), Paramecium, Das Pantoffeltierchen, Jena, G. Fischer, 1–188.

Kamada, T. (1928), The time-intensity factors in the electro-destruction of the membrane of Paramecium, *J. Faculty Sci. Imp. Univ. Tokyo*, IV, 2, 41–49.

────── (1928a), Current strength and anodal galvanotropism in Paramecium, *J. Faculty Sci. Imp. Univ. Tokyo*, 2, 29–40.

────── (1929), Control of galvanotropism in Paramecium, *J. Faculty Sci. Imp. Univ. Tokyo*, 2, 123–139.

────── (1934), Some observations on potential differences across the ectoplasm membrane of Paramecium, *J. Exp. Biol.*, 11, 94–102.

────── (1935), Contractile vacuole of Paramecium, *J. Faculty Sci. Imp. Univ. Tokyo*, 4, 49–61.

────── (1940), Ciliary reversal of Paramecium, *Proc. Imp. Acad.* (*Tokyo*), 16, 241–247.

Kanda, S. (1914 a), On the geotropism of Paramecium and Spirostomum, *Biol. Bull.*, 26, 1–25.

────── (1914b), Further studies of Paramecium and Spirostomum, *Biol. Bull.*, 34, 101.

────── (1918), Further studies on the geotropism of *Paramecium caudatum*, *Biol. Bull.*, 34, 108–119.

Kanitz (1907), Der Einfluss der Temperatur auf die pulsierenden Vakuolen der Infusorien und die Abhängigkeit biologischer Vorgänge von der Temperatur überhaupt, *Biol. Zentr.*, 27, 11–25.

Kasanzeff, W. (1901), Experimentelle Untersuchungen über *Paramecium caudatum*, Inaug.-Diss., Zürich, 60 pp.

Kasanzeff, W. (1928), Beitrag zur Kenntnis der Grosskerne der Zilliaten, *Trav. lab. zool. Stat. Sebastopol (Acad. Sci. Leningrad)*, Ser. 2, Nr. 11, 1–30.

Kavanagh, A. J. and O. W. Richards (1937a), The unusual mortality which characterizes a Paramecium culture, *Anat. Record,* **70** (Suppl.), 39.

—— and —— (1937b), The potential longevity of a Paramecium culture, *Anat. Record,* **70** (Suppl.), 39.

Kay, P. and E. H. Shaw Jr. (1941), The influence of bacteria on the toxicity of sodium laurylsulfonate to several strains of *Paramecium caudatum, Proc. S. Dakota Acad. Sci.,* **21**, 80.

Kent, W. S. (1880–82), A Manual of the Infusoria, London, 3 vols.

Khainsky, A. (1910), Physiologische Untersuchungen über *Paramecium caudatum, Biol. Zentr.,* **30**, 267–278.

—— (1910 a), Zur Morphologie und Physiologie einiger Infusorien (*Paramecium caudatum*) auf Grund einer neuen histologischen Methode, *Arch. Protistenk.,* **21**, 1–60.

Khawkine, M. W. (1888), Le Principe de l'hérédité et les lois de la mécanique en application à la morphologie de cellules solitaires, *Arch. zool. exp.,* **6.**

Kidder, G. W. (1933), Studies on *Conchophthirius mytili De Morgan:* II. Conjugation and nuclear reorganization, *Arch. Protistenk.,* **74**, 27–49.

—— (1933a), On the Genus *Ancistruma Strand* (=*Ancistrum Maupas*): II. The conjugation and nuclear reorganization of *Ancistruma isseli Kahl, Arch. Protistenk.,* **81**, 1–18.

—— (1933b), On the Genus *Ancistruma Strand:* I. The structure and division of *Ancistruma mytili Quenn.* and *Ancistruma isseli Kahl, Biol. Bull.,* **64**, 1–20.

—— (1934), Studies on the ciliates from fresh water mussels: I The structure and neuromotor system of *Conchophthirus anodontae Stein., Conchophthirus curtus Engl.* and *Conchophthirus magna sp. nov., Biol. Bull.,* **66**, 69–90.

—— (1938), Nuclear reorganization without cell division in *Paraclevelandia simplex* (Family Clevelandellidae), an endocommensal ciliate of the wood-feeding roach, *Panesthia, Arch. Protistenk.,* **91**, 69–77.

—— (1941), The technique and significance of control in Protozoan culture, Protozoa in Biological Research, New York, Columbia University Press, Chap. 8.

—— (1941a), Growth studies on ciliates: VII Comparative growth characteristics of four species of sterile ciliates, *Biol. Bull.,* **80**, 50–68.

—— (1946), Studies on the biochemistry of Tetrahymena: VI Folic acid as a growth factor for *Tetrahymena Geleii W., Arch. Biochem.,* **9**, 51–53.

—— and V. C. Dewey (1945), Studies on the biochemistry of Tetrahymena: I. Amino acid requirements, *Arch. Biochem.,* **6**, 425–432.

—— and —— (1945a), Studies on the biochemistry of Tetrahymena: VII. Riboflavin, pantothen, biotin, niacin and pyridoxine in the growth of *Tetrahymena geleii W., Biol. Bull.,* **89**, 229–241.

—— and —— (1945b), Studies on the biochemistry of *Tetrahymena:* III Strain differences, *Physiol. Zoöl.,* **18**, 137–156.

—— and —— (1947), Studies on the biochemistry of Tetrahymena: X. Quantitative response to essential amino acids, *Proc. Nat. Acad. Sci.,* **12**, 347–356.

—— and —— (1949), Studies on the biochemistry of Tetrahymena. XIII. B vitamin requirements, *Arch Biochem.,* **21**, 66–73.

—— and Diller, W. F. (1934), Observations on the binary fission of four species of common free-living ciliates, with special reference to the macronuclear chromatin, *Biol. Bull.,* **67**, 201–219.

——, D. M. Lilly, and C. L. Claff (1940), Growth studies on ciliates: IV. The influence of food on the structure and growth of *Glaucoma vorax, sp. nov., Biol. Bull.,* **78**, 9–23.

—— and C. A. Stuart (1939), Growth studies on ciliates: I. The role of bacteria in the growth and reproduction of Colpoda, *Physiol. Zoöl.,* **12**, 329–340.

——, ——, V. G. McGann, and V. C. Dewey (1945), Antigenic relationships in the genus Tetrahymena, *Physiol. Zoöl.*, **18**, 415–425.

Kijenskij, G. (1925), Beiträge zur amitotischen Teilung bei den Infusorien, *Vestnik Král. České Spolec Nauk*, 2. Kl.

Kimball, R. F. (1937), The inheritance of sex at endomixis in *Paramecium aurelia*, *Proc. Nat. Acad. Sci.*, **23**, 469–474.

—— (1939), Change of mating type during vegetative reproduction in *Paramecium aurelia*, *J. Exp. Zoöl.*, **81**, 165–179.

—— (1939 a), A delayed change of phenotype following a change of genotype in *Paramecium aurelia*, *Genetics*, **24**, 49–58.

—— (1939 b), Mating types in Euplotes, *Am. Naturalist*, **73**, 451–456.

—— (1941), Double animals and amicronucleate animals in *Euplotes patella* with particular reference to their conjugation, *J. Exp. Zoöl.*, **86**, 1–33.

—— (1942), The nature and inheritance of mating types in *Euplotes patella*, *Genetics*, **27**, 269–285.

—— (1943), Mating types in the ciliate Protozoa, *Quart. Rev. Biol.*, **18**, 30–45.

—— (1947), The induction of inheritable modification in reaction to antiserum in *Paramecium aurelia*, *Genetics*, **32**, 486–499.

—— (1947a), Heritable resistance to antiserum produced by antiserum and by trypsin in *Paramecium aurelia*, *Genetics*, **32**, 93.

—— (1947b), Induction of mutations in *Paramecium aurelia* by beta radiation, *Anat. Record*, **99**, 47.

—— (1948), The role of cytoplasm in inheritance in variety 1 of *Paramecium aurelia*, *Am. Naturalist*, **82**, 79–84.

—— (1949), The effect of ultraviolet light upon the structure of the macronucleus of *Paramecium aurelia*, *Anat. Record*, **105**, 63.

—— (1949 a), Inheritance of mutational changes induced by radiation in *Paramecium aurelia*, *Genetics*, **34**, 412–424.

—— (1949b), The induction of mutations in *Paramecium aurelia* by beta radiation, *Genetics*, **34**, 210–222.

—— and N. Gaither (1950), Photorecovery of the effects of ultraviolet radiation on *Paramecium aurelia*, *Genetics*, **35**, 118.

—— and —— (1951), The influence of light upon the action of ultraviolet on *Paramecium aurelia*, *J. Cellular Comp. Physiol.*, **37**, 211–233.

—— and —— (1952), Role of externally produced hydrogen peroxide in damage to *Paramecium aurelia* by x-rays, *Proc. Soc. Exp. Biol. Med.*, **80**, 525–529.

Kimura, Y. (1935), The biological action of the rays from radioactive substances: I. The effects of small dose radiation upon the reproductive activity of unicellular organisms, *Sci. Papers Inst. Phys. Chem. Research* [*Tokyo*], **28**, 27–47.

King, E. (1693), *Phil. Trans. Roy. Soc.*, **17**, 861–865.

King, R. L. (1928), The contractile vacuole in *Paramecium trichium*, *Biol. Bull.* **55**, 59–68.

—— (1933), Contractile vacuole of Euplotes, *Trans. Am. Microscop. Soc.*, **52**, 103–106.

—— (1935), The contractile vacuole of *Paramecium multimicronucleata*, *J. Morphol.*, **58**, 555–571.

—— (1950), Multiple pores and contractile vacuoles in Paramecium, *Anat. Record*, **108**, 53.

—— (1951), Origin and position of the excretory pores in *Paramecium aurelia*, *Proc. Am. Soc. Protozoöl.*, **2**, 12.

—— (1952), The pores of the contractile vacuoles of *Paramecium multimicronucleatum* and *P. caudatum*, *Proc. Soc. Protozoöl.*, **3**, 7–8.

—— and H. W. Beams (1937), The effect of ultracentrifuging on Paramecium, with special reference to recovery and macronuclear reorganization, *J. Morphol.*, **61**, 27–49.

King, R. L. and H. W. Beams (1940), A comparison of the effects of colchicine on division in Protozoa and certain other cells, *J. Cellular Comp. Physiol.*, **15**, 1–3.

—— and —— (1941), Some effects of mechanical agitation on *Paramecium caudatum*, *J. Morphol.*, **68**, 149–159.

King, S. D. (1927), The Golgi apparatus of protozoa, *J. Roy. Microscop. Soc.*, **47**, 342–355.

Kinosita, H. (1936), a. Time-intensity relation of the cathodal galvanic effect in Paramecium; b. Electric excitation and electric polarization in Paramecium, *J. Faculty. Sci. Imp. Univ. Tokyo*, **4**, 155–163.

—— (1936a), Electric polarization of hydrogen ions in Paramecium, *J. Faculty Sci. Imp. Univ. Tokyo Sect. IV*, **4**, 137–148.

—— (1936b), Effect of external pH on the electric polarization of OH ions in Paramecium, *J. Faculty Sci. Imp. Univ. Tokyo Sect. IV*, **4**, 149–154.

—— (1936c), Electric excitation and electric polarization in Paramecium, *J. Faculty Sci. Imp. Univ. Tokyo Sect. IV*, **4**, 155–161.

—— (1936d), Time-intensity relation of the cathodal galvanic effect in Paramecium, *J. Faculty Sci. Imp. Univ. Tokyo Sect. IV*, **4**, 163–170.

—— (1936e), Supernormal phase of electric response in Paramecium, *J. Faculty Sci. Imp. Univ. Tokyo Sect. IV*, **4**, 171–184.

—— (1936f), Addition of subliminal stimuli in Paramecium, *J. Faculty Sci. Imp. Univ. Tokyo Sect. IV*, **4**, 195.

—— (1936g), Effect of change in orientation on the electrical excitability of Paramecium, *J. Faculty Sci. Imp. Univ. Tokyo Sect. IV*, **4**, 189–194.

—— (1938), Electrical stimulation of Paramecium with two successive subliminal current pulses, *J. Cellular Comp. Physiol.*, **12**, 103–117.

—— (1939), Electrical stimulation of Paramecium with linearly increasing current, *J. Cellular Comp. Physiol.*, **13**, 253–261.

Kirby, H. (1941), Organisms living on and in Protozoa, in Calkins, G. N. and F. M. Summers, Protozoa in Biological Research, New York, Columbia University Press, Chap. 20.

—— (1942), A parasite of the macronucleus of Vorticella, *J. Parasitol.*, **28**, 311–314.

—— (1947), Methods in the Study of Protozoa, Berkeley, Univ. of Calif. Press, 1–116.

Kisch, B. (1913), Untersuchungen über Narkose, *Z. Biol.*, **60**, 401.

Kissa, H. (1914), Die Wirkung kombinierter Narkotika der Fettreihe auf Colpidien, *Z. allgen. Physiol.*, **16**, 320–340.

Kitching, J. A. (1934), The physiology of contractile vacuoles: I. Osmotic relations, *J. Exp. Biol.*, **11**, 364–381.

—— (1936), The physiology of contractile vacuoles: II. Body volume control in marine Peritricha, *J. Exp. Biol.*, **13**, 11–27.

—— (1938), The physiology of contractile vacuoles. III. The water balance of fresh water Peritricha, *J. Exp. Biol.*, **15**, 143–151.

—— (1939), The effects of a lack of oxygen and of low oxygen tensions on Paramecium, *Biol. Bull.*, **77**, 339–353.

—— (1939a), The effects of lack of oxygen, and of low oxygen tensions, on the activities of some Protozoa, *Biol. Bull.*, **77**, 304–305.

—— (1939b), On the activity of Protozoa at low oxygen tensions, *J. Cellular Comp. Physiol.*, **14**, 227–236.

Kite, G. L. (1913), Studies on the physical properties of protoplasm: I. The physical properties of the protoplasm of certain animal and plant cells, *Am. J. Physiol.*, **32**, 146–164.

—— (1915), Studies on the permeability of the internal cytoplasm of animal and plant cells, *Am. J. Physiol.*, **37**, 282.

Klein, A. P. (1943), The nitrogen compounds necessary for growth in *Colpidium striatum* Stokes with special reference to the amino acids, *Physiol. Zoöl.*, **16**, 405–417.

Klein, B. M. (1926 a), Über eine neue Eigentumlichkeit der Pellicula von *Chilodon unicinatus Ehrenberg, Zool. Anz.*, **67**, 160–162.

—— (1926b), Ergebnisse mit einer Silbermethode bei Ciliaten, *Arch. Protistenk.*, **56**, 243–279.

—— (1927–28), Morphologisches und Physiologisches vom Pantoffeltierchen (Paramecium), *Naturforscher*, **1**.

—— (1927a), Über die Darstellung der Silberliniensysteme des Ciliatenkörpers, *Mikrokosmos*, **20**, 233–235.

—— (1927b), Die Silberliniensysteme der Ciliaten. Ihr Verhalten während der Teilung und Konjugation, neue Silberbilder, Nachträge, *Arch Protistenk.*, **58**, 55–142.

—— (1928), Die Silberliniensysteme der Ciliaten Weitere Resultate. (The silver-line system of ciliates), *Arch. Protistenk.*, **62**, 177–260.

—— (1931), Über die Zugehörigkeit gewisser Fibrillen bzw. Fibrillenkomplexe zum Silber-liniensystem, *Arch. Protistenk.*, **74**, 401–416.

Klitzke, M. (1914), Über Wiederkonjuganten bei *Paramecium caudatum, Arch. Protistenk.*, **33**, 1–20.

—— (1915), Über Wiederkonjuganten bei *Paramecium caudatum, Arch. Protistenk.*, **36**, 67.

—— (1916), Ein Beitrag zur Kenntnis der Kernentwicklung bei den Konjuganten, *Arch. Protistenk.*, **36**, 215.

Klokaciova, Z. (1924), Sui Parameci con più di due vacuoli contrattili, *Boll. ist. zool. Roma*, **2**, 47–54.

—— (1924 a), A proposito dei vacuoli contrattili supplementari di *Paramecium caudatum, Boll. ist. zool. Roma.*, **2**.

—— (1927), Azione dei cloruri di Li, K, Na, Ca sulla divisione e sui vacuoli contrattili di *Paramecium caudatum, Boll. ist. zool. Roma*, **4**, 151–170.

Knoch, M., and H. König (1951), Zur Struktur der Paramaecien-Trichocysten, *Naturwiss.*, **38**, 531.

Koch, A. (1916), Moderne Probleme der Tierphysiologie, *Naturwissenschaften*, **4**, 101–04, 109–14.

Koehler, O. (1922), Über die Geotaxis von Paramecium, *Arch. Protistenk.*, **45**, 1–94.

—— (1922a), Über die Geotaxis von Paramecium, *Verhandl. zool. Ges.*, **26**, 69–71.

—— (1928), Galvanotaxis, *Handb. d. norm. u. path. Physiol.*, **11**.

—— (1930), Über die Geotaxis von Paramecium, *Arch. Protistenk.*, **70**, 279–307.

—— (1939), Ein Filmprotokoll zum Reizverhalten querzertrennter Paramecien, *Zool. Anz. Suppl.* (*Verhandl. deut. Zool. Ges. 41*) **12**, 132–142.

Koehring, V. (1930), The neutral-red reaction, *J. Morphol.* **49**, 45–137.

Kofoid, C. A. and L. E. Rosenberg (1940), The neuromotor system of *Opisthonecta henneguyi* (*Fauré-Fremiet*), *Proc. Am. Phil. Soc.*, **82**, 421–437.

Kölsch, K. (1902), Untersuchungen über die Zerfliessungerserscheinungen der ciliaten Infusorien, nebst Bemerkungen über Protoplasmastruktur, Bewegungen und Vitalfärbungen, *Z. Jahrb., Abt. Morphol.*, **16**, 273–422.

Kōno, T. (1930), Untersuchungen zur Frage der Vitalfärbung und deren Beeinflussung durch Gifte, *Protoplasma*, **11**, 118–177.

Kopaczewski, W. (1928), Pénétration électrocapillaire des matieres colorantes dans la cellule. (Electrocapillary penetration of colored materials into the cell), *Compt. rend. acad. sci.* (*Paris*), **186**, 1758–1761.

Korentschewsky, W. (1902), Vergleichend-pharmakologische Untersuchungen über die Wirkung von Giften auf einzellige Organismen, *Arch. exptl. Path. Pharmakol.*, **49**, 7–31.

Köster, W. (1933), Untersuchungen über Teilung und Konjugation bei *Paramecium multimicronucleatum*, *Arch. Protistenk.*, **80**, 410–433.

Kostka, G. (1926), Die Kultur der Mikroorganismen, *Mikrokosmos* suppl.

Kriz, R. A. (1924), Selective action of strychnine and nicotine on a single cell, *Am. Naturalist*, **54**, 464–469.

Krizenecky, J. (1926), Untersuchungen über die biologische Wirkung des Biocleins, *Věstnik. Ceskoslov. Akad. Zemědělské*, **2**, 1.

Krüger, F. (1929), Dunkelfelduntersuchungen an den Trichocysten von *Paramecium caudatum*, *Verhandl. deut. zool. Ges.*, **33**, 267–272.

—— (1930), Untersuchungen über den Bau und die Funktion der Trichocysten von *Paramecium caudatum*, *Arch. Protistenk.*, **72**, 91–134.

—— (1931), Dunkelfelduntersuchungen über den Bau der Trichocysten von *Frontonia leucas*, *Arch. Protistenk.*, **74**, 207–235.

—— (1936), Die Trichocysten der Ciliaten im Dunkelfeldbild, *Zoologica*, **34**, 1–82.

Kruse, W. (1910), Die Lehre von Stoff- und Kraftwechsel der Kleinwesen für Ärzte und Naturforscher dargestellt, Leipzig, F. C. W. Vogel, 1200 pp.

Kruszyński, J. (1939), Mikrochemische Untersuchungen des veraschten *Paramecium caudatum*. Verteilung von Kalk und Eisen in der Zelle, *Arch. Protistenk.*, **92**, 1–9.

Kudelski, A. (1899), Note sur la métamorphose partielle des noyaux chez les Paramecies, *Bibliographie Anat. Paris*, **6**, 270–272.

Kudo, R. R. (1947), Protozoology, Springfield, Ill., C. C Thomas, ed. 3.

Kulagin, N. (1899), Zur Biologie der Infusorien, *Physiologiste Russe*, **1**, 269–275.

Lacaillade, C. W. Jr. (1933), The determination of the potency of bee venom in vitro, *Am. J. Physiol.*, **105**, 251–256.

Lachovskij, P. (1926), Ultramikroskopische Untersuchungen der Paramecien, *Russ. Arch. Protistologii*, **5**, 111–132.

Lackey, J. B. (1932), Oxygen deficiency and sewage Protozoa: with descriptions of some new species, *Biol. Bull.*, **63**, 287–295.

Lampiris, N. A. (1915), Untersuchungen über die Wirkung der Gold- und Platinsalze auf Paramecien, mit besonderer Berücksichtigung ihrer sensibilisierenden Fähigkeit, *Diss. Aus dem pharm. Inst. Univ. München*, 35 pp.

Landis, E. M. (1920), An amicronucleate race of *Paramecium caudatum*, *Am. Naturalist*, **54**, 453–457.

—— (1925), Conjugation of *Paramecium multimicronucleata Powers and Mitchell*, *J. Morphol.*, **40**, 111–167.

Lang, A. (1901), Lehrbuch der vergleichenden Anatomie der wirbellosen Tiere, Jena, G. Fischer.

Lankester, E. R. (1903), A Treatise on Zoology. Part I. Introduction and Protozoa, London, A. and C. Black.

Lansing, A. I. (1938), Localization of calcium in *Paramecium caudatum*, *Science*, **87**, 303–304.

Lauterborn, R. (1916), Die sapropelische Lebewelt, *Verhandl. Naturalist. med. Ver.*, Heidelberg, N. F., **13**, 395–481.

Lea, D. E. (1947), Actions of Radiations on Living Cells, New York, The Macmillan Co., 1–402.

Le Dantec, F. (1890), Recherches sur la digestion intracellulaire chez les protozoaires, *Ann. inst. Pasteur*, **4**, 776–791, **5**, 163–170.

—— (1892), Recherches sur la symbiose des algues et des Protozoaires, *Ann. inst. Pasteur*, **6**, 190–198.

—— (1897), La régénération du micronucleus chez quelques Infusoires ciliés, *Compt. rend. acad. sci. Paris*, **125**, 51–52.

Ledermüller, M. F. (1760–65), Mikroskopische Gemüths- und Augenergötzungen, Nürnberg.

Ledoux-Lebard (1902), Action du sérum sanguin sur les Parámécies, *Ann. inst. Pasteur,* **16**, 510–521.

—— (1902 a), Action de la lumiere sur la toxicité de l'eosine et de quelques autres substances pour les parámécies, *Ann. inst. Pasteur,* **16**, 587–594.

Lee, H. (1949), Change of mating type in *Paramecium bursaria* following exposure to X-rays, *J. Exp. Zoöl.,* **112**, 125–130.

Lee, J. S. (1941), A technique for continuous or intermittent observation of the contractile vacuoles of Paramecium, *Science,* **94**, 332.

—— (1941 a), A combined fixative and stain for the cilia and trichocysts of Paramecium, *Science,* **94**, 352.

Lee, J. W. (1942 a), The effect of temperature on food vacuole formation in Paramecium, *Physiol. Zoöl.,* **15**, 453–458.

—— (1942 b), The effect of pH on food-vacuole formation in Paramecium, *Physiol. Zoöl.,* **15**, 459–465.

Leeuwenhoek, A. van (1674), *Phil. Trans. Royal Soc.,* **9**, 178–182.

—— (1677), Concerning little animals by him observed in Rain-Well-Sea- and Snow water; as also in water wherein pepper had lain infused, *Phil. Trans. Royal Soc.,* **12**, 821–831.

—— (1875), Onderzoekingen over Ontwikkeling en Voortplanting van Infusoria, Utrecht, P. W. van der Weijer.

Lehmann, F. M. (1927), Über das Verhalten von Paramaecien in reinen Linien gegenüber Zellgiften, *Sitzber. Ges. naturforsch. Freunde Berlin,* 23–24.

Leichsenring, J. M. (1925), Factors influencing the rate of oxygen consumption in unicellular organisms, *Am. J. Physiol.,* **75**, 84–92.

Leontjew, H. (1923–24), Zur Biophysik der niederen Organismen, *Arch. russ. Prot.,* **3**, 201–206, **4**, 121–125.

—— (1926), I. Über das spezifische Gewicht des Protoplasmas, *Biochem. Z.,* **170**, 326–329.

Lepeschkin, W. W. (1925), Untersuchungen über das Protoplasma der Infusorien, Foraminiferen und Radiolarien, *Biol. Generalis,* **1**, 368–395.

Lepsi, J. (1926), Infusorien des Süsswassers und Meeres, *Mikroskopie F. Naturfreunde,* Berlin.

Leslie, L. D. (1940 a), Nutritional studies of *Paramecium multimicronucleata:* I. Quantitative and qualitative standardization of the food organism, *Physiol. Zoöl.,* **13**, 243–250.

—— (1940 b), Nutritional studies of *Paramecium multimicronucleata:* II. Bacterial foods, *Physiol. Zoöl.,* **13**, 430–438.

Levander, K. M. (1894), Materialien zur Kenntnis der Wasserfauna in der Umgebung von Helsingfors, mit besonderer Berücksichtigung der Meeresfauna: I. Protozoa, *Acta pro Fauna et Flora Fennica,* **12**.

Lewin, K. R. (1910), Nuclear relations of *Paramecium caudatum* during the asexual period, *Proc. Cambridge Phil. Soc.,* **16**, 39–41.

—— (1911), The behavior of the Infusoria, *Proc. Roy. Soc. Biol.,* **84**, 332–344, 572.

Lherisson, C. (1930), Experimental ingestion of Paramecium, *Trans. Roy. Soc. Can.,* **24**, 155–156.

Lieberkühn, N. (1856), Contributions to the anatomy of the Infusoria, *Müller's Arch.*

Liebermann, P. R. (1929), Ciliary arrangement in different species of Paramecium, *Trans. Am. Microscop. Soc.,* **48**, 1–11.

Liebmann, H. (1936), Auftreten, Verhalten und Bedeutung der Protozoen bei der Selbstreinigung stehenden Abwassers, *Z. Hyg. Infektionskrank.,* **118**, 29–63.

Lindeman, R. L. (1942), Experimental simulation of winter anaerobiosis in a senescent lake, *Ecology,* **23**, 1–13.

Lindner, G. (1899), Die Protozoënkeime im Regenwasser (*Paramecium putrinum*), *Biol. Zentr.*, **19**, 421–432, 456–463.

Linné, Karl von (1735–78), Systema Naturae, (Editions 1–12).

Lipschütz, A. (1915), Allgemeine Physiologie des Todes, Graunschweig, F. Wieweg.

Lipska, I. (1910), Recherches sur l'influence de l'inanition chez *Paramecium caudatum*, *Rev. suisse zool.*, **18**, 591–646.

—— (1910 a), Les effets de l'inanité chez les infusoires, *Compt. rend. soc. phys. hist. nat. Genève*, **17**, 15–17.

Little, P. A., J. J. Oleson, and J. M. Schaefer (1951), Growth-promoting effect of aureomycin on *Stylonychia pustulata*, *Paramecium caudatum* and *Paramecium bursaria*, *Proc. Am. Soc. Protozoöl.*, **2**, 10.

Lloyd, F. E. (1928), The contractile vacuole, *Biol. Rev. Biol. Proc. Cambridge Phil. Soc.*, **3**, 329–358.

—— and J. Beattie (1928), The pulsatory rhythm of the contractile vesicle in Paramecium, *Biol. Bull.*, **55**, 404–419.

Lloyd, L. (1946 a), Movements in culture of some sewage-filter organisms, *Nature*, **157**, 844.

—— (1946 b), Attraction of narrow clefts for some organisms inhabiting sewage filters, *Proc. Leeds Phil. Lit. Soc., Sci. Sect.*, **4**, 323–342.

—— (1947), Unco-ordinated growth in Paramecium induced by "Gammexane," *Nature*, **159**, 135.

—— (1948), Induced and spontaneous abnormality in ciliates, *Nature*, **162**, 188.

—— (1949), Induced and spontaneous abnormalities in Paramecium, *Proc. Leeds Phil. Lit. Soc., Sci. Sect.*, **5**, 140–154.

Loeb, J. (1912), Über die Hemmung der Giftwirkung von NaJ, $NaNO_3$, NaCNS und anderen Salzen, *Biochem. Z.*, **43**, 181–202.

—— and S. P. Budgett (1897), Zur Theorie des Galvanotropismus, *Pflügers Arch. ges. Physiol.*, **65**, 518–534.

—— and I. Hardesty (1895), Über de Lokalisation der Atmung in der Zelle, *Pflügers Arch. ges. Physiol.*, **61**, 583–594.

—— and H. Wasteneys (1911), Die Entgiftung von Säuren durch Salze, *Biochem. Z.*, **33**, 489–502.

—— and —— (1913), Narkose und Sauerstoffverbranch, *Biochem. Z.*, **56**, 295.

Loefer, J. B. (1934), Bacteria-free culture of Paramecium, *Science*, **80**, 206–207.

—— (1936), Effect of certain 'peptone' media and carbohydrates on the growth of *Paramecium bursaria*, *Arch. Protistenk.*, **87**, 142–150.

—— (1936 a), A simple method for maintaining pure-line mass cultures of *Paramecium caudatum* on a single species of yeast, *Trans. Am. Microscop. Soc.*, **55**, 255–256.

—— (1936 b), Bacteria-free culture of *Paramecium bursaria* and concentration of the medium as a factor in growth, *J. Exp. Zoöl.*, **72**, 387–407.

—— (1936 c), Isolation and growth characteristics of the "Zoochlorella" of *Paramecium bursaria*, *Am. Naturalist*, **70**, 184–188.

—— (1938), Effect of hydrogen-ion concentration on the growth and morphology of *Paramecium bursaria*, *Arch. Protistenk.*, **90**, 185–193.

—— and R. P. Hall (1936), Effect of ethyl alcohol on the growth of eight protozoan species in bacteria-free cultures, *Arch. Protistenk.*, **87**, 123–130.

Löffler, F. (1889), Eine neue Methode zum Färben der Mikroorganismen, im besonderen ihrer Wimperhaare und Geisseln, *Zentr. Bakt. Parasitenk.*, **6**, 209–224.

Löhner, L. (1913), Verleichende Untersuchungen über Erstickung, Wärmelähmung und Narkose mit Protozoen, *Z. allgem. Physiol.*, **15**, 199.

—— and B. E. Markovits (1922), Zur Kenntnis der oligodynamischen Metallgiftwirkungen auf die lebendige Substanz: 1. Paramaecienversuche, *Pflügers Arch. ges. Physiol.*, **195**, 417–431.

Loisele, G. (1903), Sur la sénescence et la conjugaison des Protozoaires, *Zool. Anz.*, **26**, 484.

Lotka, A. J. (1925), Elements of Physical Biology, Baltimore.

―――― (1932), The growth of mixed populations: Two species competing for a common food supply, *J. Wash. Acad. Sci.*, **22**, 461.

Lozina-Lozinsky, L. (1929 a), Le choix de la nourriture dans les différents milieux par le *Paramecium caudatum*, *Compt. rend. soc. biol.*, **100**, 321–323.

―――― (1929 b), Zur Physiologie der Ernährung der Infusorien: I. Nahrungsauswahl und Vermehrung bei *Paramecium caudatum*, *Izvest. Nauch-Issledovatel. Inst. Lesgafta*, **15**, 91–136.

―――― (1929 c), Chemotaxis in relation to the choice of food stuffs by Infusoria, *Compt. rend. acad. sci. U.R.S.S. Sér. A*, **17**, 403–408.

―――― (1929 d), Physiology of nutrition of Infusoria: 1. Food selection and multiplication of *Paramecium caudatum*, (*Bull. inst. sci. Lesshaft*,) **15**, 91–136.

―――― (1929 e), Le choix de la nourriture chez *Paramoecium caudatum*, *Compt. rend. soc. biol.*, **100**, 722–724.

―――― (1931), Zur Ernährungsphysiologie der Infusorien: Untersuchungen über die Nahrungsauswahl und Vermehrung bei *Paramecium caudatum*, *Arch. Protistenk.*, **74**, 18–120.

Luboska, A. and J. Dembowski (1950), Cyclosis in Paramecium, *Acta Biol. Exptl.*, **15**, 19.

Lucas, M. S. (1930), Results obtained from applying the Feulgen reaction to Protozoa, *Proc. Soc. Exp. Biol. Med.*, **27**, 258–260.

Luck, J. M., G. Sheets, and J. O. Thomas (1931), The role of bacteria in the nutrition of Protozoa, *Quart. Rev. Biol.*, **6**, 46–58.

Ludloff, K. (1895), Untersuchungen über den Galvanotropismus, *Pflügers Arch. ges. Physiol.*, **59**, 525.

Ludwig, W. (1927), Die Ursachen der extremen Giftwirkung der Schwermetallionen, sowie der Verunreiniung von Wasser und Glas auf *Paramecium aurelia*, *Z. vergleich. Physiol.*, **6**, 623–687.

―――― (1928 a), Der Betriebsstoffwechsel von *Paramoecium caudatum*, *Arch. Protistenk.*, **62**, 12–40.

―――― (1928 b), Über den funktionellen Zusammenhang zwischen Populationsdichte, Nahrungsdichte und Teilungsrate bei Protisten und über die Zunahme der Bevölkerungsdichte überhaupt, *Biol. Generalis*, **4**, 351–376.

―――― (1929 c), Über die Empfindlichkeitsgrenzen des Plasmas der Protozoen X[6], *Congr. intern. zool.*, *Budapest 1927, Part I*, 583–590.

―――― (1929 d), Untersuchungen über die Schraubenbahnen niederer Organismen, *Z. vergleich. Physiol.*, **9**, 734–801.

―――― (1929 e), Vergleichende Untersuchungen über das Wachstumsgesetz, *Biol. Zentr.*, **49**, 735–758.

―――― (1930 a), Zur Nomenklatur und Systematik der Gattung Paramecium, *Zool. Anz.*, **92**, 33–41.

―――― (1930 b), Zur Theorie der Flimmerbewegung (Dynamik, Nutzeffekt, Energuebukabz), *Z. vergleich. Physiol.*, **13**, 397–504.

―――― (1931), Über vergleichende Wachstumsuntersuchungen, *Biol. Zentr.*, **51**, 116–119.

―――― and C. Boost (1939), Ueber das Wachstum von Protistenpopulationen und des allelokatalytischen Effekt, *Arch. Protistenk.*, **92**, 453–484.

Lund, B. L. (1918), The toxic action of KCN and its relation to the state of nutrition and age of the cell as shown by Paramecium and Didinium, *Biol. Bull.*, **35**, 211–231.

Lund, E. E. (1933), A correlation of the silverline and neuromotor systems of Paramecium, *Univ. Calif. Pubs.*, **39**, 35–76.

―――― (1941), The feeding mechanisms of various ciliated Protozoa, *J. Morphol.*, **69**, 563–573.

Lund, E. J. (1918 a), Quantitative studies on intracellular oxidation: I. Relation of oxygen concentration and the rate of intracellular oxidation in *Paramecium caudatum,* *Am. J. Physiol.,* **45**, 351–364.

——— (1918 b), Quantitative studies on intracellular oxidation: II. The rate of oxidation in *Paramecium caudatum* and its independence of the toxic action of KCN, *Am. J. Physiol.,* **45**, 365–373.

——— (1918 c), Quantitative studies on intracellular oxidation: III. Relation of the state of nutrition of Paramecium to the rate of intracellular oxidation, *Am. J. Physiol.,* **47**, 166–177.

——— (1921), Quantitative studies on intracellular respiration: V. The nature of the action of KCN on Paramecium and Planaria, with an experimental test of criticism and certain explanations offered by Child and others, *Am. J. Physiol.,* **57**, 336–349.

Luyet, B. J. (1933), Current intensity factor in electrocution of paramaecia, *Proc. Soc. Exp. Biol. Med.,* **30**, 924–926.

——— (1935), La structure nucléaire étudiée in vivo par la méthode de la photographie ultra-violette, *Compt. rend. acad. sci.* [Paris], **200**, 2035–2036.

——— (1936), Mesure directe de la dose léthale de courant électrique chez les Paramécies, *Compt. rend. acad. sci.* [Paris], **202**, 511–512.

——— and P. M. Gehenio (1935), Comparative ultra-violet absorption by the constituent parts of Protozoan cells (Paramaecium), *Biodynamica,* **7**, 1–14.

Lwoff, A. (1923), Sur la nutrition des Infusoires, *Compt. rend. acad. Sci. Paris,* **176**, 928–931.

——— (1924), Le pouvoir de synthèse d'un protiste hétérotrophe: *Glaucoma piriformis,* *Compt. rend. soc. biol.,* **91**, 344–345.

——— (1925), 1. La nutrition des infusoires aux dépens de substances dissoutes. 2. Influence d'extraits de glandes et d'organes sur la vitesse de multiplication des infusoires, *Compt. rend. soc. biol.,* **93**, 1272, 1352.

——— (1929), La Nutrition de Polytomella uvella et la pouvoir de synthese des Protistes heterotrophes. Les Protistes mesotrophes, *Compt. rend. acad. sci.,* **188**, 114–116.

——— (1929 a), Milieux de culture et d'entretien pour *Glaucoma piriformis, Compt. rend. soc. biol.,* **100**, 635–636.

——— (1932), Recherches Biochimiques sur la Nutrition des Protozoaires, *Monographies inst. Pasteur.*

——— (1938), Les facteurs de croissance pour les microorganismes, *Ann. Inst. Pasteur,* **61**, 580–617.

——— (1950), Problems of Morphogenesis in Ciliates, New York, J. Wiley and Sons, 103 pp.

——— (1951), Biochemistry and Physiology of Protozoa, New York, Academic Press, **1, 2, 3.**

——— and M. Lwoff (1937), L'Aneurine, facteur de croissance pour le cilié *Glaucoma piriformis, Compt. rend. soc. biol.,* **126**, 644–646.

———, and ——— (1938), La Specificité de l'aneurine, facteur de croissance pour le cilié *Glaucoma piriformis, Compt. rend. soc. biol.,* **127**, 1170–1172.

——— and N. Roukelman (1926), Variations de quelques formes d'azote dans une culture pure d'infusoires, *Compt. rend. acad. Sci.,* **183**, 156–159.

Lwoff, M. (1938), L'Hematine et l'acide ascorbique, facteurs de croissance pour le flagellé *S. chizotrypanum cruzi, Compt. rend. acad. sci. Paris,* **206**, 540–542.

Lynch, J. E. (1929), Eine Neue Karminmethod Für Total Praparate, *Z. wiss. Mikroskop.,* **46**, 465–469.

——— (1930), Studies on the ciliates from the intestine of *Strongylocentrotus, Univ. Calif. Pubs. Zoöl.,* **33**, 307–350.

Lyon, E. P. (1905), On the theory of geotropism in Paramecium, *Am. J. Physiol.,* **14**, 421–432.

—— (1918), Note on geotropism of Paramecium, *Biol. Bull.*, **34**,120.

McCleland, N. and R. A. Peters (1920), Toxicity of some inorganic substances to the paramecium, *J. Physiol.*, **53**, 12.

McClendon, J. F. (1909), Protozoan studies, *J. Exp. Zoöl.*, **6**, 265–283.

McClung, C. E. et al (1937), Microscopical Technique, New York, P. B. Hoeber, Inc., 1–698.

McClung, L. S. (1943), A technique for the production of antisera for *Paramecium aurelia*, *J. Bact.*, **46**, 576.

—— (1944), A technique for the production of immune sera for *Paramecium aurelia*, *Proc. Indiana Acad. Sci.*, **53**, 47–49.

MacDougall, M. S. (1931), Another mutation of *Chilodon Uncinatus* produced by ultraviolet radiation, with a description of its maturation processes, *J. Exp. Zoöl.*, **58**, 229–236.

MacLennan, R. F. (1933), The pulsatory cycle of the contractile vacuoles in the Ophryoscolecidae, ciliates from the stomach of cattle, *Univ. Calif. Pubs. Zoöl.*, **39**, 205–250.

—— (1936), Dedifferentiation and redifferentiation in *Ichthyophthirius*. II. The origin and function of cytoplasmic granules, *Arch. Protistenk.*, **86**, 404–426.

—— and H. K. Murer (1934), Localization of mineral ash in the organelles and cytoplasmic components of Paramecium, *J. Morphol.*, **55**, 421–433.

Maier, H. N. (1903), Über den feineren Bau der Wimperapparate der Infusorien, *Arch. Protistenk.*, **2**, 73–179.

Makarov, P. V. (1940), On relation between the vital staining of protozoan macronucleus and its microscopic structure, *Arch. Russes Anat. Hist. Emb.*, **25**, 105–112.

Malm, E. N. (1930), The action of alkaloids on protozoa in relation to the H ion concentration and the semipermeable character of the cell membranes, *Trav. sta. biol. Sébastopol*, **2**, 17–30.

Marcovits, E. (1922), Cytologische Veränderungen von Paramaecium nach Bestrahlung mit Mesothorium, *Arch. exp. Zellforsch.*, **16**, 238–248.

Markova, T. G. (1945), Fiziologicheskie osobennosti *Paramecium caudatum* razlichnogo vozrasta (Individual noe stranenie prosreishikh.) [Physiological features of *Paramecium caudatum* of different ages. (Individual aging of protozoans)], *Zoologicheskii Zhurnal*, **24**, 32–41.

Markovits, J. (1928), Strahlenwirkung auf die Zellteilung. (Effect of radiation on cell division), *Arch. exp. Zellforsch*, **6**, 315–321.

Marsland, D. A. (1943), Quieting Paramecium for the elementary student, *Science*, **98**, 414.

Massart, J. (1889), Sensibilité et adaptation des organismes à la concentration de solutions salines, *Arch. biol.*, **9**, 515–570.

—— (1901), Recherches sur les organismes inférieurs: IV. Le lancement des trichocystes (chez *Paramecium aurelia*), *Bull. Acad. roy. Belg. Classe Sci.*, **2**, 91–106.

—— (1902), Versuche einer Einteilung der nicht-vervösen Reflexe, *Biol. Zentra.*, **22**, 9–23, 41–52, 65–79.

Mast, S. O. (1906), Light reactions in lower organisms: I. *Stentor coruleus*, *J. Exp. Zoöl.*, **3**, 359–399.

—— (1909), The reactions of *Didinium nasutum* with special reference to the feeding-habits and the function of trichocysts, *Biol. Bull.*, **16**, 91–118.

—— (1946), The food-vacuole in Paramecium, *Anat. Record.*, **94**, 29.

—— (1947), The food vacuole in Paramecium, *Biol. Bull.*, **92**, 31–72.

—— and W. J. Bowen (1944), The food-vacuole in the Peritricha, with special reference to the hydrogen-ion concentration of its content and of the cytoplasm, *Biol. Bull.*, **87**, 188–222.

Mast, S. O. and K. S. Lashley (1916), Observations on ciliary current in free-swimming Paramecia, *J. Exp. Zoöl.,* **21,** 281–293.

───── and J. E. Nadler (1926), Reversal of ciliary action in *Paramecium caudatum, J. Morphol. and Physiol.,* **43,** 105–117.

─────, M. Pace, and L. R. Mast (1936), The effect of sulfur on the rate of respiration and on the respiratory quotient in *Chilomonas paramecium, J. Cellular Comp. Physiol.,* **8,** 125–139.

Masugi, M. (1927), Über die Wirkung des Normal-sowie des spezifischen Immunserums auf die Paramäzien. Über die Immunität derselben gegen die beiden Serumwirkungen, *Krankheitsforsch.,* **5,** 375–402.

Matheny, W. A. (1910), Effects of alcohol on the life cycle of *Paramecium, J. Exptl. Zoöl.,* **8,** 193–205.

Maupas, E. (1883), Contribution a l'étude morphologique et anatomique des Infusoires ciliés, *Arch. zool. exp. et gén.,* **1,** 427–664.

───── (1885), Sur le glycogène chez les Infusoires ciliés, *Compt. rend. soc. biol.,* **101,** 1504.

───── (1886 a), Sur la conjugaison des Paramécies, *Compt. rend. soc. biol.,* **103,** 482–484.

───── (1886 b), Sur la puissance de multiplication des Infusoires ciliés, *Compt. rend. soc. biol.,* **103,** 1006–1008.

───── (1886 c), Sur la conjugaison du *Paramecium bursaria, Compt. rend. soc. biol.,* **103,** 955–957.

───── (1888), Recherches expérimentales sur la multiplication des Infusoires ciliés, *Arch. zool. exp. et gén.,* **6,** 165–277.

───── (1889), Le rajeunissement Karyogamique chez les ciliés, *Arch. zool. exp. et gén.,* **7,** 149–517.

Mayeda, T. (1928), De l'effet de la réaction (pH) du milieu sur le galvanotropisme de la paramécie, *Compt. rend. soc. biol.,* **99,** 108–110.

───── and S. Date (1929), De l'influence de la concentration en ions H sur le galvano-tropisme de la paramécie, *Compt. rend. soc. biol. Paris,* **101,** 633–635.

Mayer, A. G. (1909), The relation between ciliary and muscular movements, *Proc. Soc. Exp. Biol. Med.,* **7,** 19.

───── (1910), The converse relation between ciliary and neuromuscular movements, *Carnegie Inst. Wash. Pub.,* **132,** 1–25.

Medes, G. and A. K. Stimson (1942), Attempts at culturing *Paramecium multimicronu-cleatum* in cell-free media, *Growth,* **6,** 145–150.

Medrkiewiczowna, H. (1921), Einfluss des durch die Oberfläche der Kultur eindringenden O_2 auf die Teilungsrate, *Soc. Sci. Varsovie,* **1.**

Meissner, M. (1888), Beiträge zur Ernährungsphysiologie der Protozoen, *Z. wiss. Zoöl.,* **46,** 498–516.

Melnikov, N. H., A. M. Avetesyan, and M. S. Rokitskaya (1941), On the protozoic ac-tivity of some phenols, *Compt. rend. acad. sci. U.R.S.S.,* **31,** 123–124.

Mendelssohn, M. (1895), Über den Thermotropismus einzelliger Organismen, *Pflügers Arch. ges. Physiol.,* **60,** 1–27.

───── (1902), Recherches sur la thermotaxie des organismes unicellulaires, *J. physiol. et path. gén.,* **4,** 393–409.

───── (1902 a), Recherches sur l'interférence de la thermotaxie avec d'autres tactismes et sur le mécanisme du mouvement thermotactique, *J. physiol. et path. gén.,* **4,** 475–488.

───── (1902 b), Quelques considérations sur la nature et la rôle biologique de la thermotaxie, *J. physiol. et path. gén.,* **4,** 489–496.

Merton, H. (1925), Experimentelle Untersuchungen des Kinoplasmas der Flimmererzeu-gung, *Z. Zellforsch. u. mikroskop. Anat.,* **2,** 382–407.

——— (1928), Untersuchungen über die Entstehung amöbenähnlicher Zellen aus absterbenden Infusorien: 1. Entstehung, Morphologie und Züchtung der Autoplasmen, *Sitzber. heidelberg. Akad. Wiss. Math.- naturw. Klasse*, 1–29.

——— (1932), Die Verwendung von Kupfersalzen zur Herstellung von Paramaecium -Präparaten, *Arch. Protistenk.*, **76**, 171–197.

——— (1935), Versuche zur Geotaxis von Paramaecium, *Arch. Protistenk.*, **85**, 33–60.

Mesnil, F. and H. Mouton (1903), Sur une diastase protéolytique extraite des Infusoires ciliés, *Compt. rend. soc. biol.*, **55**, 1016–1019.

Metalnikow, S. (1904), Über die intrazelluläre Verdauung, *Bull. Acad. Sci. Pétersburg*, **19**, 187–193.

——— (1907), Über die Ernähung der Infusorien und deren Fähigkeit, ihre Nahrung zu wählen, *Trav. soc. naturalistes. Pétersburg*, **38**, 181–187.

——— (1912), Contributions à l'étude de la digestion intracellulaire chez les protozoaires, *Arch. zool. exptl. et gén.*, **10**, 373–497.

——— (1914), Les Infusoires peuvent-ils apprendre à choisir leur nourriture?, *Arch. Protistenk.*, **34**, 60–78.

——— (1915), Sur la circulation des vacuoles digestives des Infusoires, *Compt. rend. soc. biol. Paris*, **78**, 176–178.

——— (1915 a), Zur Frage nach der intrazellulären Verdauung. Über die Bewegung der Verdauungsvakuolen, *Petrograd Bull. lab. biol.*, **15**, 47–67.

——— (1917), On the question regarding the capability of Infusoria to "learn" to choose their food, *Russ. J. Zool.*, Petersburg, **2**, 397.

——— (1924), Immortalite et rajeunessement dans la biologie moderne, Paris, 283 pp.

——— (1933), Esperimenti sulla moltiplicazione degli infusori sotto l'azione di circuiti oscilanti (Multiplication of infusoria under the action of oscillating circuits), *Atti. accad. nazl. Lincei Classe Sci. fis., Mat. e Nat.*, **18**, 232–234.

——— and M. A. Galadjieff (1916), Le problème de l'immortalité des protozoaires unicellulaires, *Petrograd Bull. Acad. Sci.*, **10**, 1809–1816.

Metcalf, M. M. (1923), The opalinid ciliate infusorians, *Smithsonian Inst. Pubs.*, 120.

Metschnikoff, E. (1864), Über die Gattung Sphaerophrya, *Arch. Anat. u. Physiol.*, **70**, 258.

——— (1882), Zur Lehre über die intrazelluläre Verdauung der Tiere, *Zool. Anz.*, **113**, 310.

——— (1889), Recherches sur la digestion intracellulaire, *Ann. Inst. Pasteur*, **3**, 25–29.

Metz, C. B. (1946), Induction of meiosis and "pseudo-selfing" in *Paramecium aurelia* of one mating type by dead animals of opposite type, *Anat. Record*, **96**, 35–36.

——— (1946 a) Effect of various agents on the mating type substance of *Paramecium aurelia*, Variety 4, *Anat. Record*, **94**, 347.

——— (1947), Induction of "pseudo selfing" and meiosis in *Paramecium aurelia* by formalin killed animals of opposite mating type, *J. Exp. Zoöl.*, **105**, 115–139.

——— (1948), The nature and mode of action of the mating type substances, *Am. Naturalist*, **82**, 85–95.

——— and W. Butterfield (1950), Extraction of a mating reaction inhibiting agent from *Paramecium calkinsi*, *Proc. Nat. Acad. Sci.*, **36**, 268–271.

——— and ——— (1951), Action of various enzymes on the mating type substances of *Paramecium calkinsi*, *Biol. Bull.*, **101**, 99–105.

——— and M. T. Foley (1947), Fertilization studies on a non-conjugating race of *Paramecium aurelia*, *Biol. Bull.*, **93**, 210.

——— and ——— (1949), Fertilization studies on *Paramecium aurelia:* an experimental analysis of a non-conjugating stock, *J. Exp. Zoöl.*, **112**, 505–528.

——— and E. M. Fusco (1948), Inhibition of the mating reaction in *Paramecium aurelia* with antiserum, *Anat. Record*, **101**, 654–655.

Metz, C. B. and E. M. Fusco (1949), Mating reactions between living and lyophilized paramecia of opposite mating type, *Biol. Bull.*, **97**, 245.

Metzner, P. (1921), Zur Kenntnis der photodynamïschen Erscheinungen, *Biochem. Z.*, **113**, 145.

——— (1923), Studien über die Bewegungsphysiologie niederer Organismen, *Naturwissenschaften.*, **11**, 365, 395.

——— (1927 a), Pulsierende Vakuolen, *Tabulae biol.*, **4**, 490–496.

——— (1927 b), Photodynamische Erscheinungen, *Tabulae biol.*, **4**, 496–510.

Meyer, C. and F. Shaffer (1936), Paramecium and its antigenic properties, *Mendel Bull.*, **8**, 97.

Meyers, E. C. (1927), Relation of density of population and certain other factors to survival and reproduction in different biotypes of *Paramecium caudatum*, *J. Exp. Zoöl.*, **49**, 1–43.

Michelson, E. (1928), Existenzbedingungen und Cystenbildung bei *Paramecium caudatum* Ehrenberg, *Arch. Protistenk.*, **61**, 167–184.

Middleton, A. R. (1918), Heritable effects of temperature differences on the fission rate of *Paramecium caudatum*, *Genetics*, **3**, 534–572.

——— (1922), Heritable effects of chemically differing media on the fission rate of *Paramecium caudatum*, *Anat. Record*, **23**, 93.

——— (1928), Heritability of the effects of chemically different media upon the fission rate within the clone of *Paramecium caudatum*, *Anat. Record*, **41**, 50.

——— (1940), Heritability of the effects of active and inactivated liver extract on the fission rate of *Paramecium caudatum*, *J. Exp. Zoöl.*, **84**, 85–112.

——— and G. E. Wakerlin (1939), Effect of parenteral liver extract on the fission rate of *Paramecium caudatum*, *Proc. Soc. Exp. Biol. Med.*, **42**, 442–444.

Minchin, E. A. (1912), Introduction to the study of the Protozoa, London.

Mitchell, W. H. (1929), The division rate of Paramecium in relation to temperature, *J. Exp. Zoöl.*, **54**, 383–410.

Mitrophanow, P. (1903), Nouvelles recherches sur l'appareil nucléaire des Paramécies, *Arch. zool. exp. et. gén.* (*4*), **1**, 411–435.

——— (1904), Sur la structure, l'origine et le mode d'action des trichocystes chez les Paramecium, *Mém. soc. nat. Varsovie, Sect. Biol. Ann.*, **14**, 18.

——— (1904 a), Note sur les corpuscules basaux des formations vibratiles, *Arch. zoöl. exp. et. gén.* (*4*), **2**, 167–169.

——— (1905), Etude sur la structure, le développement et l'explosion des trichocystes des Paramécies, *Arch. Protistenk.*, **5**, 78–91.

Mitrophanowa, J. (1925), The influence of the concentration of the hay infusion on the resistance of Paramecium against poisons, *Bull. inst. recherches biol. Perm*, **3**, 229–240.

Mizuno, F. (1927), Sur la croissance normale du *Paramecium caudatum*, *Sci. Rep. Tohoku. Imp. Univ. Sendai*, **2**, 367–381.

Moment, G. B. (1944), A simple method for quieting paramecium and other small organisms during prolonged observation, *Science*, **99**, 544.

Monod, J. (1933), Mise en évidence du gradient axial chez les infusoires ciliés par photolyse à l'aide des rayons ultraviolets, *Compt. rend. acad. sci.*, **196**, 212–214.

Moore, A. (1903), Some facts concerning geotropic gatherings of Paramaecia, *Am. J. Physiol.*, **9**, 238–244.

Moore, I. (1927), Inheritance of atypical form in *Paramecium aurelia*, *Proc. Soc. Exp. Biol. Med.*, **24**, 770–771.

Morea, L. (1927), Influence de la concentration en ions H sur la culture de quelques infusoires, *Compt. rend. soc. biol.*, **97**, 49–50.

Morgan, T. H. (1914), Heredity and Sex, New York, Columbia University Press.

Morse, M. (1909–10), Shaking experiments with Protozoa, *Proc. Soc. Exp. Biol. Med.*, **7**, 58–60.

Moses, M. J. (1949), Nucleoproteins and the cytological chemistry of Paramecium nuclei, *Proc. Soc. Exp. Biol. Med.,* **71,** 537–539.

———— (1949 a), Nucleic acids and proteins of the nuclei of Paramecium, Thesis, Columbia University.

———— (1950), Nucleic acids and proteins of the nuclei of Paramecium, *J. Morphol.,* **87,** 493–536.

Motolese, F. (1920), Sulle proprietà farmacologiche dell'acido picrico, *Arch. farmacol. sper.,* **9.**

Mottram, J. C. (1940), Benzypyrene, Paramecium and the production of tumors, *Nature,* **145,** 184–185.

———— (1941), Abnormal paramecia produced by blastogenic agents and their bearing on the cancer problem, *Cancer Research,* **1,** 313–323.

———— (1942), The Problem of Tumours, London, H. K. Lewis and Co. Ltd., 91.

———— (1944), *Cancer Research,* **4,** 241.

———— and I. Doniach (1947), Photodynamic action of carcinogenic agents, *Nature,* **140,** 933.

Mouton, H. (1899), Sur le galvanotropisme des Infusoires ciliés, *Compt. rend. soc. biol.,* **128,** 1247–1249.

———— (1902), Recherches sur la digestion chez les amibes. Et sur leur diastase intracellulaire, *Ann. Inst. Pasteur,* **16,** 457–509.

Mukerji, B., N. K. Dutta, and S. C. Ganguly, Studies on some dextro-rotatory hydrocupreidine derivatives. III. Comparative effects on *Paramecium caudatum, Indian J. Med. Research,* **30,** 325–330.

Müller, J. (1856), Beobachtungen an Infusorien, *Monatsber. der Berl. Akad.,* 393.

Müller, O. F. (1773–74), Vermium terrestrium et fluviatilium, *Historia.*

———— (1786), Animalcula Infusoria fluviatilia et marina. Hauniae, Copenhagen.

Müller, P. T. (1912), Über die Rolle der Protozoen bei der Selbstreinigung stehenden Wassers, *Arch. Hyg.,* **75,** 321–352.

Müller, W. (1932), Cytologische und vergleichend-physiologische Untersuchungen über *Paramecium multimicronucleatum* und *Paramecium caudatum,* zugleich ein Versuch zur Kreuzung beider Arten, *Arch. Protistenk.,* **78,** 361–462.

Myers, E. C. (1927), Relation of density of population and certain other factors to survival and reproduction in different biotypes of *Paramecium caudatum, J. Exp. Zoöl.,* **49,** 1–43.

Nadler, J. E. (1929), Notes on the loss and regeneration of the pellicle of *Blepharisma undulans, Biol. Bull.,* **56,** 327–330.

Nagai, H. (1907), Der Einfluss verschiedener Narkotika, Gase und Salze auf die Schwimmgeschwindigkeit von Paramaecien, *Z. allgem. Physiol.,* **6,** 195–212.

Nägeli, C. von (1893), Ueber oligodynamische Erscheinungen in lebenden Zellen, *Denksch. schweiz. naturforsch. Ges.,* **33,** 1.

Nanney, D. (1952), Mating type determination in *Paramecium aurelia:* a model of nucleocytoplasmic interaction, *Proc. Nat. Acad. Sci.,* (In press).

Nardone, R. M. and C. G. Wilber (1950), Nitrogenous excretion in *Colpidium campylum, Proc. Soc. Exp. Biol. Med.,* **75,** 559–561.

Nassonov, D. (1924), Der Exkretionsapparat der Protozoa als Homologon des Golgischen Apparats der Metazoenzellen, *Arch. mikroskop. Anat. Entwicklungsmech.,* **103,** 437–482.

———— (1925), Zur Frage über den Bau und die Bedeutung des lipoiden Exkretionsapparates bei Protozoa, *Z. Zellforsch. u. mikroskop. Anat.,* **2,** 87–97.

Necheles, H. (1924), Section in Winterstein's Handbuch der vergleichenden Physiologie, **2,** 931.

Nelson, E. C. (1933), The feeding reaction of *Balantidium coli* from the chimpanzee and the pig, *Am. J. Hyg.,* **18,** 185–201.

Nelson, E., and K. K. Krueger (1950), Effect of various inhibitors on the respiration of *Paramecium caudatum, Anat. Record,* **108,** 46.

Neresheimer, E. R. (1903), Über die Höhe histologischer Differenzierung bei heterotrichen Ciliaten, *Arch. Protistenk.,* **2,** 305–324.

Neuhaus, H. (1910), Versuche über Gewöhnung an Arsen, Antimon, Quecksilber und Kupfer bei Infusorien, *Arch. intern. pharmaco-dynamie,* **20,** 393–413.

Neuschlosz, S. (1919), Untersuchungen über die Gewöhnung an Gifte, über das Wesen der Chininfestigkeit bei Protozoen, *Pflügers Arch. ges. Physiol.,* **176,** 223–235.

——— (1920), Untersuchungen über die Gewöhnung an Gifte. II, III, *Pflügers Arch. ges. Physiol.,* **178,** 61–79.

Nickson, J. J. (1952), Symposium on Radiobiology, New York, John Wiley and Sons, 1–465.

Nigrelli, R. F. and R. P. Hall (1929), *Paramecium bursaria* as a laboratory demonstration of cyclosis, *Science,* **70,** 311.

Nikitin, S. A. (1930), Untersuchungen über Röntgensensibilisiuung: II. Teil. Über die Chemosensibilisierung der Protozoen (Roentgen sensitization. II. Chemical sensitization of Protozoa), *Strahlentherapie,* **36,** 539–545.

Nikitinsky, J. (1928), Über die wirkung der kohlensäure auf Wasserorganismen, *Centr. Bakt.,* **73,** 481–483.

——— and F. K. Mudrezowa-Wyss (1930), Über die Wirkung des Kohlensäure, des Schwefelwasserstoffs, des Methans und der Abwessenheit des Sauerstoffs auf Wasserorganismen, *Zentr. Bakt.,* **81,** 167–198.

Nirenstein, E. (1905), Beiträge zur Ernährungsphysiologie der Protisten, *Z. allgem. Physiol.,* **5,** 435–510.

——— (1910), Über Fettverdauung und -speicherung bei Infusorien, *Z. allgem. Physiol.,* **10,** 137–149.

——— (1920), Über das Wesen der Vitalfärbung, *Pflügers Arch. ges. Physiol.,* **179,** 233–336.

——— (1925), Über die Natur und Stärke der Säureabscheidung in den Nahrungsvakuolen von *Paramecium caudatum, Z. wiss. Zoöl.,* **125,** 513–518.

——— (1927), Die Nahrungsaufnahme bei Protozoen. Die Verdauungsvorgänge bei Protozoen, *Bethes Handb.,* **3,** 3–20.

Noguchi, H. (1924), Action of certain biological, chemical and physical agents upon cultures of Leishmania, etc., *Proc. Intern. Conf. Health Problems, Kingston, B. W. I.*

Noland, L. E. (1927), Conjugation in the ciliate, *Metopus sigmoides* C. & L., *J. Morphol. Physiol.,* **44,** 341–361.

Nowikoff, M. (1908), Über die Wirkung des Schilddrüsenextraktes und einiger anderer Organstoffe auf Ciliaten, *Arch. Protistenk.,* **11,** 309–325.

——— (1923), Die Bodenprotozoen und ihre Bedeutung für die Bodenkultur, *Heidelb. Akten d. v. Portheim-Stiftg., 3, Biol.,* **1.**

Nussbaum, M. (1886), 1. Über spontane, künstliche Teilung von Infusorien. Verh. naturh. Ver. d. preuss. Rheinlande. Bonn 1884.—2. Über die Teilbarkeit der lebendigen Materie. 1. Die spontane und künstliche Teilung der Infusorien, *Arch. mikroskop. Anat. Entwicklungsmech,* **26,** 485.

Oehler, R. (1916), Amöbenzucht auf reinem Boden, *Arch. Protistenk.,* **37,** 175–190.

——— (1919), Flagellaten- und Ciliatenzucht auf reinem Boden, *Arch. Protistenk.,* **40,** 16–26.

——— (1920 a), Flagellaten- und Ciliatenzucht auf reinem Boden, *Arch. Protistenk.,* **40,** 16–26.

——— (1920 b), Gereinigte Ciliatenzucht, *Arch. Protistenk.,* **41,** 34–49.

——— (1921), Wirkung von Bakteriengiften auf Ziliaten, *Centr. Bakt. Parasitenk.,* **86,** 494–500.

——— (1922 a), Die Zellverbindung von *Paramecium bursaria* mit *Chlorella vulgaris* und anderen Algen, *Arb. Staatsinst. exp. Therap. u. Georg. Speyer-Hause, Frankfurt a. m.,* **15,** 5–18.

—— (1924 a), Weitere Mitteilungen über gereinigte Amöben- und Ciliatenzucht, *Arch. Protistenk.,* **49,** 112–134.

—— (1924 b), Gereinigte Zucht von freilebenden Amöben, Flagellaten und Ciliaten, *Arch. Protistenk.,* **49,** 287–296.

Oettli, M. (1910), Kleine Schulversuche. Die Thermotaxis der Paramaecien, *Monatsh. naturw. Unterricht.,* **3,** 508.

Ogata, M. (1893), Über die Reinkultur gewisser Protozoen, *Centr. Bakt. Parasitenk.,* **14,** 165–169.

Okazaki, Y. (1927 a), Über die Arzneikombination an Paramaecien, *Japan. J. Obstetr. Gynecol.,* **10,** 55.

—— (1927 b), Ein Vergleich der letalen dose der Pharmaka für einzellige lebewesen und für Wirbeltiere, (Lethal doses of drugs in Paramecium and mice), *Folia Pharmacol. Japon.,* **5,** 11–16.

Olifan, W. I. (1936), Variabilität von *Paramecium caudatum* bei langdauernder Einwirkung differenter Temperaturen, *Arch. Protistenk.,* **86,** 427–453.

Oliphant, J. F. (1938), The effect of chemicals and temperature on reversal in ciliary action in Paramecium, *Physiol. Zoöl.,* **11,** 19–30.

—— (1941), Reversal of ciliary action in Paramecium induced by chemicals, *Anat. Record,* **81,** suppl. 67.

—— (1942), Reversal of ciliary action in *Paramecium* induced by chemicals, *Physiol. Zoöl.,* **15,** 443–452.

—— (1943), Effect of some chemicals, which affect smooth muscle contraction, on ciliary activity in Paramecium, *Proc. Soc. Exp. Biol. Med.,* **54,** 62–64.

Opton, E. M. Demonstration of the cytosomic morphology of *Paramecium polycaryum, Anat. Record,* **84,** 485.

Orlova, A. F. (1941), Dlitel'nye modifikatsti u *Paramecium caudatum; Paramecium multimicronucleatum.* (Dauermodifikationen in *Paramecium caudatum* and *Paramecium multimicronucleatum*), *Zool. Zhur.,* **20,** 341–370.

Osthelder, F. (1907), Einige Beobachtungen über die photodynamische Wirkung auf Zellen, Diss. München, 45 pp.

Pace, D. M. (1945), The effect of cyanide on respiration in *Paramecium caudatum* and *Paramecium aurelia, Biol. Bull.,* **89,** 76–83.

—— and K. K. Kimura (1944), The effect of temperature on respiration in *Paramecium aurelia* and *Paramecium caudatum, J. Cellular Comp. Physiol.,* **24,** 173–183.

Pacinotti, G. (1914), Infusorien welche Glykogen in Fett umwandeln, *Boll. Soc. eustachiana,* No. 3.

Packard, C. (1925), The effect of light on the permeability of *Paramecium, J. Gen. Physiol.,* **7,** 363–372.

—— (1926), The effect of sodium on the rate of cell division, *J. Cancer Research,* **10,** 1–14.

Paneth, J. (1890–91), Über das Verhalten von Infusorien gegen Wasserstoffsuperoxyd., *Biol. Zentr.,* **10,** 95–96.

Park, O. (1929), The differential reduction of osmic acid in the cortex of Paramecium and its bearing upon the metabolic gradient conception, *Physiol. Zoöl.,* **2,** 449–458.

Parker, G. H. (1929), The neurofibril hypothesis, *Quart. Rev. Biol.,* **4,** 155–178.

Parker, R. C. (1926), Symbiosis in *Paramecium bursaria, J. Exp. Zoöl.,* **46,** 1–12.

—— (1926), The effect of selection on the rate of reproduction of *Paramecium aurelia, Proc. Soc. Exp. Biol. Med.,* **24,** 227–229.

—— (1927), The effect of selection in pedigree lines of Infusoria, *J. Exp. Zoöl.,* **49,** 401–439.

Parnas, J. K. (1926), Allgameines und Vergleichendes des Wasserhaushalts, *Handb. norm. u. path. Physiol.*, **17**, 137–160.

Parona, C. (1880), Prime ricerche inturno ai Protisti del lago d'Orta con cenno della loro corologia italiana, *Boll. sci.*, **2**, 17–26.

Parpart, A. K. (1928), The bacteriological sterilization of Paramecium, *Biol. Bull.*, **55**, 113–120.

Patrick, R. and H. R. Roberts (1949), Biological survey of the Conestoga Creek Basin and observations on the West Branch Brandywine Creek, Report by Acad. Nat. Sci. of Phila.

Pearl, R. (1900), Studies on electrotaxis: 1. On the reactions of certain Infusoria to the electric current, *Am. J. Physiol.*, **4**, 96–127, *Science*, **13**, 715.

—— (1906), 1. A Biometrical study of conjugation in Paramecium, *Proc. Roy. Soc. (London)*, **77**, 377–383.

—— (1906 a), Variation in Chilomonas under favorable and unfavorable conditions, *Biometrika*, **5**, 533–572.

—— (1907), A biometrical study of conjugation in Paramecium, *Biometrika*, **5**, 213–297.

—— and Dunbar, F. J. (1905), Some results of a study of variation in Paramecium, *Papers Mich. Acad. Sci.*, 77–86.

Pearson, K. (1901), Note on Dr. Simpson's Memoir on *Paramecium caudatum*, *Biometrika*, **1**, 404–408.

Pease, D. C. (1947), The structure of trichocysts revealed by the electron microscope, *J. Cellular Comp. Physiol.*, **29**, 91–94.

Pecker, S. (1915), Die Änderung von Colpoden und deren Cysten unter dem Einfluss von Blutserum, *Pflügers Arch. ges. Physiol.*, **163**, 101–147.

Peebles, F. (1912), Regeneration and regulation in *Paramecium caudatum*, *Biol. Bull.*, **23**, 154–170, *Science*, **35**, 470.

Penard, E. (1922), Etudes sur les infusoires de l'eau douce, Geneva.

Penn, A. B.K. (1937), Reinvestigation into the cytology of conjugation in *Paramecium caudatum*, *Arch. Protistenk.*, **89**, 45–54.

Pensa, A. (1927), Ciliated protozoa and contractility, *Nature*, **119**, 439.

Perty, M. (1852), System der Infusorien, Bern, Mittheil., 57–67.

—— (1852a), Zur Kenntnis der kleinsten Lebensformen, Bern.

Peter, K. (1898-99), Das Zentrum für die Flimmer- und Geisselbewegung, *Anat. Anz.*, **15**, 271–283.

Peters, A. W. (1901), Some methods for use in the study of Infusoria, *Am. Naturalist*, **35**, 553–559.

—— (1904), Metabolism and division in Protozoa, *Proc. Am. Acad. Arts Sci.*, **39**, 441–516.

—— (1905a), The sequence of organisms in Protozoan culture and its irreversibility, *Science*, **21**, 851–852.

—— (1905b), An analysis of physiological conditions in a Protozoan culture, *Science*, **21**, 832.

—— (1907a), Chemical studies on the cell and its medium: Part I. Methods for the study of liquid culture media, *Am. J. Physiol.*, **17**, 443–477.

—— (1907b), Chemical studies on the cell and its medium: Part II. Some chemico-biological relations in liquid culture media, *Am. J. Physiol.*, **18**, 321–346.

—— (1908), Chemical studies on the cell and its medium: Part III. The function of the inorganic salts of the Protozoan cell and its medium, *Am. J. Physiol.*, **21**, 105–125.

—— (1921), Nutrition of the protozoa: The growth in sterile culture medium, *J. Physiol.*, **53**, 108.

—— and B. Opal (1909), Studies on enzymes: II. The diastatic enzyme of Paramecium in relation to the killing concentration of copper sulphate, *J. Biol. Chem.*, **6**, 65–73.

—— and M. H. Rees (1906), Some relations of Protozoa to certain ions in their medium, *Science*, **23**, 527–528.

Peters, F. (1930), Eine einfache Methode zur Betäubung von Rädertieren, Paramaecien usw. in Ausgestrecktem Zustande, *Zool. Anz.*, **87**, 18–20.

Peters, R. A. (1921), The substances needed for the growth of a pure culture of *Colpidium colpoda, J. Physiol.*, **55**, 1–32.

Petersen, W. A. (1927), The axial gradient in Paramecium, *Science*, **66**, 157–158.

—— (1929), The relation of density of population to rate of reproduction in *Paramecium caudatum, Physiol. Zoöl.*, **2**, 221–254.

Peterson, R. E. (1942), Essential factors for the growth of the ciliate protozoan, *Colpidium campylum, J. Biol. Chem.*, **146**, 537–545.

Petschenko, B. (1910), Formations fibro-crystalloïdes dans le corps de la Paramécie, *Trav. lab. zool. Univ. Varsovie*, **38**, 52.

—— (1911), *Drepanospira Mülleri* n. g. n. sp. parasite des Paramaeciums; contribution à l'étude de la structure des bactéries, *Arch. Protistenk.*, **22**, 248.

Pfeffer, W. (1897), Zur Kenntnis der Plasmahaut und der Vakuolen etc., *Abhandl. math.-phys. Klasse sächs. Wiss.*, **16**, 878.

Phelps, A. (1913), Effect of H-ion concentration on the division rate of *Paramecium aurelia, Science*, **74**, 395–396.

—— (1934), Studies on the nutrition of Paramecium, *Arch. Protistenk.*, **82**, 134–163.

—— (1936), Growth of protozoa in pure culture: II. Effect upon the growth curve of different concentrations of nutrient materials, *J. Exp. Zoöl.*, **72**, 479–496.

—— (1946), Growth of Protozoa in pure culture: III. Effect of temperature upon the division rate, *J. Exp. Zoöl.*, **102**, 277–292.

Phillips, R. L. (1922), The growth of Paramecium in infusions of known bacterial content, *J. Exp. Zoöl.*, **36**, 135–183.

Philpott, C. H. (1925), Growth of Paramaecia in pure cultures of *Pseudomonas pyocyaneus, Anat. Record*, **31**, 314.

—— (1926), Growth of Paramecia in the presence of diphtherial toxin, *Anat. Record*, **34**, 156.

—— (1927), Growth of Paramecia in pure culture of pathogenic bacteria, *Anat. Record*, **34**, 156.

—— (1928), Growth of Paramecia in pure cultures of pathogenic bacteria and in the presence of soluble products of such bacteria, *J. Morphol. Physiol.*, **46**, 85–129.

—— (1930), Effect of toxins and venoms on Protozoa, *J. Exp. Zoöl.*, **56**, 167–183.

—— (1931a), Use of protozoa in measuring the potency of cobra antiserum, *Science*, **74**, 157–158.

—— (1931b), Relative resistance of fourteen species of protozoa to the action of *Crotalus atrox* and Cobra venoms, *Biol. Bull.*, **60**, 64–66.

—— (1932), Natural and acquired resistance in Protozoa to the action of antigenic poisons, *J. Exp. Zoöl.*, **63**, 553–571.

Pick, E. and R. Wasicky (1915), Über die Wirkung des Papaverins und Emetins auf Protozoen, *Wien. klin. Wochschr.*, **28**, 6.

Pierson, B. F. (1938), The relation of mortality after endomixis to the prior inter-endomictic interval in *Paramecium aurelia, Biol. Bull.*, **74**, 235–243.

Piffault, C. (1939), Action des rayons X sur l'eau et ses conséquences, *Compt. rend. soc. biol.*, **130**, 43–44.

Platt, J. (1899), On the specific gravity of Spirostomum, Paramecium and the tadpole in relation to the problem of geotaxis, *Am. Naturalist*, **33**, 31–38.

Pollack, H. (1928), Intracellular hydrion concentration studies, *Biol. Bull.*, **55**, 383.

Pollister, A. W. and C. Leuchtenberger (1949), The nature of the specificity of methyl green for chromatin, *Proc. Nat. Acad. Sci.*, **35**, 111–116.

Popoff, M. (1907), Depression der Protozoenzelle und der Geschlechtszellen der Metazoen, *Arch. Protistenk., R.* Hertwig Festband.

—— (1908), Experimentelle Zellstudien. I, *Arch. Zellforsch.,* **1**, 246–379.

—— (1909 a), Experimentelle Zellstudien II, and III, *Arch. Zellforsch.,* **3**, 124–180, **4**, 1–43.

—— (1910), Über den Einfluss chemischer Reagentien auf den Funktionszustand der Zelle, *Sitzber. Ges. morphol. physiol.,* **25**, 55–58.

—— and M. Jeljaskowa (1924), Über die Beschleunigung der Teilungsrate von *Paramecium caudatum* durch zellstimulierende Mittel, *Biol. Zentr.,* **44**, 87–90.

Port, J. (1927), Die Wirkung der Neutralsalze auf die Koagulation des Protoplasmas bei *Paramecium caudatum, Protoplasma,* **2**, 401–419.

—— (1927 a), Beitrag zur Temperaturwirkung auf die Pulsation der Vakuolen bei *Paramecium caudatum, Protoplasma,* **1**, 566–580.

—— (1928), Untersuchungen über die Plasmakoagulation von *Paramecium caudatum, Acta et Commentationes Univ. Tartu.,* **1**, 1–52.

Potts, H. E. (1943a), Effect of propionic acid and of Na propionate on Paramecia, *J. Franklin Inst.,* **236**, 99–100.

—— (1943b), *Paramecium caudatum* as a test animal for organic arsenicals, *J. Franklin Inst.,* **236**, 499–505.

—— (1944), Effects of various chemicals on *Paramecium caudatum, J. Franklin Inst.,* **237**, 227–230.

Powers, E. L. Jr. (1943), The mating types of double animals in *Euplotes patella, Am. Midland Naturalist,* **30**, 175–195.

—— (1948), Death after autogamy in *Paramecium aurelia* following exposure in solution to the radioactive isotopes P^{32} and $Sr^{89,90}$ Y^{90}, *Genetics,* **33**, 120.

—— and C. Raper (1950), Responses induced by a nitrogen mustard compound in *Paramecium aurelia, Genetics,* **35**, 131.

—— and D. Shefner (1950), Effects of high dosages of X-rays in *Paramecium aurelia, Genetics,* **35**, 131.

—— and —— (1951), Tritium-induced effects in *Paramecium aurelia, Proc. Soc. Exptl. Biol. Med.,* **78**, 493–497.

Powers, J. H. and C. Mitchell (1910), A new species of Paramecium (*Paramecium multimicronucleata*) experimentally determined, *Biol. Bull.,* **19**, 324–332.

Poyarkoff, E. (1928), La méthode statistique dans la protophysiologie (The statistical method in protophysiology), *Protoplasma,* **3**, 550–557.

Prandtl, H. (1906), Die Konjugation von *Didinium nasutum* O. F. M., *Arch. Protistenk.,* **7**, 229–258.

Prát, S. (1917), Einige Bemerkungen über *Paramecium bursaria* und über photodynamische Wirkung, Biol. Listy 6, Brünn.

Preer, J. R. (1941), The effect of temperature on the periodicity of uniparental nuclear reorganization in *Paramecium aurelia,* variety 2, *Anat. Record,* **81**, suppl. 131.

—— (1946), Some properties of a genetic cytoplasmic factor in Paramecium, *Proc. Nat. Acad. Sci.,* **32**, 247–253.

—— (1948), A study of some properties of the cytoplasmic factor "Kappa" in *Paramecium aurelia,* Variety 2, *Genetics,* **33**, 349–404.

—— (1948a), The killer cytoplasmic factor kappa: its rate of reproduction, the number of particles per cell, and its size, *Am. Naturalist,* **82**, 35–42.

—— (1950), Microscopically visible bodies in the cytoplasm of the "killer" strains of *Paramecium aurelia, Genetics,* **35**, 344–362.

Prenant, A. (1913), Les appareils ciliés et leurs dérivés, *J. de l'anat. physiol.,* **49**, 150, 545.

Pringsheim, E. G. (1915), Die Kultur von *Paramecium bursaria, Biol. Zentr.,* **35**, 375.

—— (1924), Methodik der Reizversuche von einzelligen Lebewesen, *Abderhaldens Handb.,* **12**.

—— (1925), Über *Paramecium bursaria*. Ein Beitrag zur Symbiosefrage, *Lotos,* Prag. **73,** 185–188.

—— (1928), Physiologische Untersuchungen an *Paramecium bursaria*. Ein Beitrag zur Symbioseforschung, *Arch. Protistenk.,* **64,** 289–418.

—— (1929), Untersuchungen an *Paramecium bursaria* über die Symbiosefrage, *Intern. Congr. Zool.,* **10,** 906–907.

—— (1937), Beiträge zur Physiologie saprotropher Algen und Flagellaten. Die Stellung der Azetatflagellaten in einem physiologischen Ernährungssystem, *Planta,* **27,** 61–92.

—— (1946), Pure Cultures of Algae: Their Preparation and Maintenance, New York, The Macmillan Co., 1–119.

Pritchard, A. (1861), A History of Infusoria, ed. 4, London.

Prosser, C. L., F. A. Brown, et al (1951), Comparative Animal Physiology, Philadelphia, W. B. Saunders.

Prowazek, S. v. (1897), Vitalfärbungen mit Neutralrot an Protozoen, *Z. wiss. Zoöl.,* **63,** 187–194.

—— (1899), Kleine Protozoenbeobachtungen, *Zool. Anz.,* **22,** 339–345.

—— (1900), Protozoenstudien. II, *Wien, A. Hölder,* **2,** 58 pp.

—— (1901 a), Zelltätigkeit und Vitalfärbung, *Zoöl. Anz.,* **24,** 455–460.

—— (1901 b), Beiträge zur Protoplasmaphysiologie, *Biol. Zentr.,* **21,** 88, 144.

—— (1902), Studien zur Biologie der Zelle, *Z. allgem. Physiol.,* **2,** 385–394.

—— (1903), Beitrag zur Kenntnis der Regeneration und Biologie der Protozoen, *Arch. Protistenk.,* **3,** 46–60.

—— (1908), Das Lecithin und seine biologische Bedeutung, *Biol. Zentr.,* **28,** 382–389.

—— (1909), Taschenbuch der mikroskopischen Technik der Protistenuntersuchungen, ed. 2, Leipzig, Barth.

—— (1910 a), Einführung in die Physiologie der Einzelligen, Leipzig-Berlin, B. G. Teubner, 172 pp.

—— (1910b), Giftwirkung und Protozoenplasma, *Arch. Protistenk.,* **18,** 221–244.

—— (1910 c), Studien zur Biologie der Protozoen, V., *Arch. Protistenk.,* **20,** 201.

—— (1913), Fluoreszenz der Zellen. Reicherts Fluoreszenzmikroskop, *Zoöl. Anz.,* **42,** 375.

Pruthi, H. S. (1926–27), On the hydrogen ion concentration of hay infusions with special reference to its influence on the protozoan sequence, *Brit. J. Exp. Biol.,* **4,** 292–300.

Przemycki, A. M. (1894), Über die Zellkörnchen bei den Protozoen, *Biol. Zentr.,* **14,** 620–626.

—— (1897), Über die intravitale Färbung des Kerns und des Protoplasmas, *Biol. Zentr.,* **17,** 321–335, 353–364.

Przibram, K. (1913–17), Über die ingeordnete Bewegung niederer Tiere, I and II, *Pflügers Arch. ges. Physiol.,* **153,** 401–405, *Arch. Entwicklungsmech.,* **43,** 20–27.

Purdy, W. C. and C. T. Butterfield (1918), Effect of plankton animals upon bacterial death rates, *Am. J. Pub. Health,* **8,** 499–505.

Pütter, A. (1900), Studien über die Thigmotaxis bei Protisten, *Arch. Anat. u. Physiol., Physiol. Abt., Suppl.,* 243–302.

—— (1904), Die Flimmerbewegung, *Ergeb. Physiol.*

—— (1904 a), Die Reizbeantwortung der ciliaten Infusorien, *Z. allgem. Physiol.,* **3,** 406–455.

—— (1905), Die Atmung der Protozoen, *Z. allgem. Physiol.,* **5,** 566–612.

—— (1911), Vergleichende Physiologie, Jena, G. Fischer.

—— (1914), Die Anfänge der Sinnestätigkeit bei Protozoen, *Umschau,* 87–94.

Quennerstedt, A. (1865), Bidrag till Sveriges Infusorie-fauna, *Acta. Univ. Lundensis, Math.-naturw. Abt.,* 69.

Raab, O. (1900), Über die Wirkung fluoreszierender Stoffe auf Infusorien, *Z. Biol.,* **21,** 524–546.

Radl, E. (1903), Untersuchungen über die Phototropismus der Thiere, Leipzig, 1–188.

Rafalko, J. S. (1946), A modified feulgen technique for small and diffuse chromatin elements, *Stain Technol.*, **21**, 91–93.

Raffel, D. (1930), The effect of conjugation within a clone of *Paramecium aurelia*, *Biol. Bull.*, **58**, 293–312.

——— (1932), Inherited variation arising during vegetative reproduction in *Paramecium aurelia*, *Biol. Bull.*, **62**, 244–257.

——— (1932a), The occurrence of gene mutations in *Paramecium aurelia*, *J. Exp. Zoöl.*, **63**, 371–412.

——— (1933), A genetic study of the reduction division in *Paramecium aurelia*, *J. Exp. Zoöl.* **66**, 89–123.

Rakieten, M. L. (1928), Effect of serological systems on Paramecium: I. The influence of agglutination upon Paramecium, *J. Immunol.*, **15**, 527–537.

Rammelmeyer, H. (1925), Zur Frage über die Glykogendifferenz bei *Paramecium caudatum*, *Arch. Protistenk.*, **51**, 184–188.

Ramsdel, S. G. (1927), A note on anaphylactic behavior in the Paramecium, *J. Immunol.*, **14**, 197–199.

Rautmann, H. (1909), Der Einfluss der Temperatur auf das Grössenverhältnis des Protoplasmakörpers zum Kern. Experimentelle Untersuchungen an *Paramecium caudatum*, *Arch. Zellforsch.*, **3**, 44–80.

Ray, O. M. (1947), The influence of organic material on the action of a chloroamine on respiration and reproduction in *Paramecium caudatum*, *Anat. Record*, **99**, 47.

Rayl, D. F. and E. H. Shaw, Jr. (1941), A kinetic study of the mechanism of action of surface active toxicants toward *Paramecium caudatum*, *Proc. S. Dakota Acad. Sci.*, **21**, 72–73.

Rees, C. W. (1922 a), The neuromotor apparatus of Paramaecium, *Univ. Calif. Pubs. Zoöl.*, **20**, 333–364.

——— (1922 b), The microinjection of Paramecium, *Univ. Calif. Pubs. Zoöl.*, **20**, 235–242.

——— (1930), Is there a neuromotor apparatus in *Diplodinium ecaudatum?*, *Science*, **71**, 369–370.

Regnard, P. (1891), Recherches expérimentales sur les conditions physiques de la vie dans les eaux, Paris, Ed. Masson.

Rentschler, L. B. (1931), The effects of ultra-violet light on Paramaecium, *Science*, **73**, 480–481.

Reukauf, E. (1930), Zur Biologie von *Didinium nasutum.*, *Z. vergleich. Physiol.*, **11**, 689–701.

Rhumbler, L. (1889), Die verschiedenen Cystenbildungen und die Entwicklungsgeschichte der holotrichen Infusoriengattung Colpoda, *Z. wiss. Zoöl.*, **46**, 549–601.

——— (1898), Physikalische Analyse von Lebenserscheinungen der Zelle, I. (Kontraktile Vakuole.), *Arch. Entwicklungsmech.*, **7**, 103.

——— (1898), Zellleib-, Schalen- und Kern-Verschmelzungen bei den Rhizopoden und deren wahrscheinliche Beziehungen zu phylogenetischen Vorstrufen der Metazoenbefruchtung, *Biol. Centr.*, **18**, 69–86.

——— (1925), Protozoa, *Kükenthals Handb. d. Zoöl.*, **1**, 1–292, (Infusoria, 256–82).

Richards, O. W. (1929), The correlation of the amount of sunlight with the division rate of Ciliates, *Biol. Bull.*, **56**, 298–305.

——— (1936), Killing organisms with chromium as from incompletely washed bichromate-sulfuric-acid cleaned glassware, *Physiol. Zoöl.*, **9**, 246–253.

——— (1941), The growth of the Protozoa, chap. 10 in Calkins, G. N. and F. M. Summers, Protozoa in Biological Research, New York, Columbia University Press, 1–1148.

——— and J. A. Dawson (1927), The analysis of the division rates of ciliates, *J. Gen. Physiol.*, **10**, 853–858.

Riddle, M. C. and H. B. Torrey (1923), The physiological response of Paramecium to thyroxin, *Anat. Record,* **24**, 396.

Robbie, W. A. (1946), The quantitative control of cyanide in manometric experimentation, *J. Cellular Comp. Physiol.,* **27**, 181–209.

Robertson, M. (1934), An in vitro study of the action of immune bodies called forth in the blood of rabbits by the injection of the flagellate protozoon, *Boda caudatus, J. Path. Bact.,* **38**, 363–390.

——— (1939 a), A study of the reactions *in vitro* of certain ciliates belonging to the *Glaucoma-Colpidium* group to antibodies in the sera of rabbits immunised therewith, *J. Path. Bact.,* **48**, 305–322.

——— (1939 b), An analysis of some of the antigenic properties of certain ciliates belonging to the *Glaucoma-Colpidium* group as shown in their response to immune serum, *J. Path. Bact.,* **48**, 323–338.

Robertson, T. B. (1908), On the normal rate of growth of an individual and its biochemical significance, *Arch. Entwicklungsmech.,* **25**, 581–614.

——— (1921), Experimental studies on cellular multiplication: I. The multiple of isolated infusoria, II. The influence of mutual contiguity upon reproductive rate and the part played therein by the "X-substance" in bacterized infusions which stimulates the multiplication in infusoria, *Biochem. J.,* **15**, 595–619.

——— (1922), Reproduction in cell-communities, *J. Physiol.,* **56**, 404–412.

——— (1923), The Chemical Basis of Growth and Senescence, Philadelphia and London, J. B. Lippincott Co., 389 pp.

——— (1924 a), The nature of the factors which determine the duration of the period of lag in cultures of infusoria, *Australian J. Exp. Biol. Med. Sci.,* **1**, 105–120.

——— (1924b), The influence of washing upon the multiplication of isolated infusoria and upon allelocatalytic effect in cultures initially containing two infusoria, *Australian J. Exp. Biol. Med. Sci.,* **1**, 151–173.

——— (1924c), Allelocatalytic effect in cultures of Colpidium in hay infusions and in synthetic medium, *Biochem. J.,* **18**, 1240–1247.

Roesle, E. (1902), Die Reaktion einiger Infusorien auf einzelne Induktionsschläge, *Z. allgem. Physiol.,* **2**, 139–168.

Rohde, K. (1917), Untersuchungen über den Einfluss der freien H-Ionen im Innern lebender Zellen auf den Vorgang, der vitalen Färbung, *Pflügers Arch. ges. Physiol.,* **168**, 411–433.

Rohdenburg, G. L. (1930), The effect of the internal secretions upon the division energy of Paramoecia, *J. Cancer Research,* **14**, 509–515.

Rood, O. (1853), On the *Paramecium aurelia, Am. J. Sci. Arts* (Silliman's J.), **15**, 70–72.

Root, W. S. (1930), The influence of carbon dioxide upon the oxygen consumption of Paramecium and the egg of Arbacia, *Biol. Bull.,* **59**, 48–62.

Rosenberg, L. E. (1932), A culture medium for Paramecium, *Science,* **75**, 364.

——— (1940), Conjugation in *Opisthonecta henneguyi*—A Free Swimming Vorticellid, *Proc. Am. Phil. Soc.,* **82**, 437–449.

Roskin, G. (1929), Die Wirkung du Radiumämanation auf Infusorien. (The effect of radium emanation on Infusoria), *Arch. Protistenk.,* **66**, 340–345.

——— (1945), Distribution of ribonucleic acid in the cytoplasm and nuclei of different cells, *Compt. rend. acad. sci. U.R.S.S.,* **49**, 288–291.

——— (1946), Cytotoxic factor in the blood of cancer patients, *Am. Rev. Soviet Med.,* **4**, 115–117.

——— (1946 a), The toxicity of blood in cancer, *Am. Rev. Soviet Med.,* **4**, 118–127.

——— and E. Dune (1929), Zur Frage über die Wirkung des Chinins auf die Zelle, *Arch. Protistenk.,* **66**, 346–354.

——— and L. Levinsohn (1926), Die Oxydasen und Peroxydasen bei Protozoa, *Arch. Protistenk.,* **56**, 145–166.

Roskin, G. and K. Romanowa (1929), Arzneimittel und ultravioletten Strahlen. I. Mitteilung. Die Chininwirkung auf die Zelle bei gleichzeitiger Bestrahlung Derselben mit ultravioletten Strahlen, *Z. Immunitätsforsch,* **62,** 147–157.

Rossbach, M. J. (1872), Die rhythmischen Bewegungserscheinungen der einfachsten Organismen und ihr Verhalten gegen physikalische Agentien und Arzneimittel, *Verhandl. Würzburger phys.-med. Ges.,* **2.**

Rosser, F. T. (1941), A new method for washing paramecia, *Can. J. Research D,* **19,** 144–149.

Rössle, R. (1905), Spezifische sera gegen Infusorien, *Arch. Hyg.,* **54,** 1–31.

———— (1909), Zur Immunität einzelliger Organismen, *Verhandl. deut. path. Ges.,* **13,** 158–162.

Rothert, W. (1904), Ueber die Wirkung des Aethers und Chloroforms auf die Reizbewegungen der Mikro-organismen, *Jahrb. wiss. Botan.* **39,** 1–70.

Roux, J. (1901), Faune Infusoirienne des Eaux stagnantes des Environs de Geneve, Geneve: H. Kündig.

Rumjantzew, A. and B. Kedrowsky (1927), Untersuchungen über die Vitalfärbung einiger Protisten, *Protoplasma,* **1,** 189–203.

Salomonsen, C. J. (1903), Dr. Georg Dreyers Sensibilisiringsforsög, *Oversigt. Selskab. Virksomhed Kgl. Danske Videnskab. Selskab.,* 393–97.

Sampson, M. M. (1925), Conditions of validity of Macallum's microchemical test for calcium, *Science,* **62,** 400–401.

Sand, R. (1901), Action thérapeutique de l'arsénic, de la quinine, du fer et de l'alcool sur les infusoires ciliés, *Ann. Soc. Roy. Sci. méd et nat. Bruxelles,* **1,** 10.

Sandon, H. (1932), The Food of Protozoa, *Pubs. faculty Sci. Egyptian Univ.,* 1–187.

Sarkar, S. L. (1936), The action of quinine on *Paramecium caudatum, Arch. Protistenk.,* **87,** 268–271.

Sassuchin, D. N. (1928), Zur Frage über die Parasiten der Protozoen. Parasiten von *Nyctotherus ovalis, Arch. Protistenk.,* **62,** 355–407.

———— (1934), Hyperparasitism in Protozoa, *Quart. Rev. Biol.,* **9,** 215–224.

Sato, T. and H. Tamiya (1937), Über die Atmungsfarbstoffe von Paramecium, *Cytologia, Fujii Jubilee Volume,* 1133–1138.

Saunders, J. T. (1925), The trichocysts of Paramoecium, *Proc. Cambridge Phil. Soc.,* **1,** 249–269.

Sawano, E. (1938), Studies on the proteolytic system in the Ciliata, *Paramecium caudatum, Science Repts. Tokyo Bunrika Daigaku, B. Zoöl.,* **3,** 221–241.

Schaefer, G. J. (1922), Studien über den Geotropismus von *Paramecium aurelia, Pflügers Arch. ges. Physiol.,* **195,** 227–244.

Schäfer, E. A. (1904), Theory of ciliary movement, *Anat. Anz.,* **24,** 497–511.

———— (1905), Models to illustrate ciliary action, *Anat. Anz.,* **26,** 517–521.

Scharrer, E. (1933), Über die Wirkung von Karbolsäure auf Paramäcien, *Arch. Protistenk.,* **80,** 349–350.

Schaudinn, F. (1899), Über den Einfluss der Röntgenstrahlen auf Protozoen, *Pflügers Arch. ges. Physiol.,* **77,** 29–44.

Scheer, B. T. (1948), Comparative Physiology, New York, John Wiley and Sons, 1–563.

Schenk, F. (1899), Physiologische Charakteristik der Zelle, *Würzb., Stubers Verl.,* 123 pp.

Schewiakoff, T. (1893), Ueber die geographisiche Verbreitung der süsswasser Protozoen, *Mem. Acad. Imp. Sci. St. Petersburg,* **41,** ser. 7.

———— (1896), Organisation et classification des infusoires *Aspirotricha, Mem. Acad. Imp. Sci. St. Petérsburg,* **4,** ser. 8, 1–395.

———— (1891), Bemerkung zu der Arbeit von Prof. Famintzin über Zoochlorellen, *Biol. Zentr.,* **11,** 475.

———— (1893), Über die Natur der sogenannten Exkretkörner der Infusorien, *Z. wiss. Zoöl.,* **57,** 32–56.

Schlieper, C. (1930), Die Osmoregulation wasserlebender Tiere, *Biol. Rev.*, **5**, 309–356.

Schmalhausen, J. and E. Syngajewskaja (1925), Studien über Wachstum und Differenzierung. 1. Die individuelle Wachstumskurve von *Paramecium caudatum*, *Arch. Entwicklungsmech.*, **105**, 711–717.

Schmidt, O. (1849), Einige neue Beobachtungen über die Infusorien, *Notizen Gebiete Natur-u. Heilkunde (3)*, **9**, 6–7.

——— (1853), Lehrbuch der vergleichenden Anatomie, *Froriep's Notizen Aerzt. Heilk.*, **9**.

Schmidt, W. J. (1939), Über die doppelbrechung der Trichocysten von Paramecium, *Arch. Protistenk.*, **92**, 527–536.

Schneider, A. (1893), The contractile vesicle of Paramecium, *Am. Month. Microscop. J.*, **14**, 80–83.

Schneider, E. (1926), Die biologische Wirkung der Röntgenstrahlen auf einzellige Lebewesen nach Untersuchungen an Paramaecien, *Strahlentherapie*, **22**, 92–106.

Schneider, K. C. (1905), Plasmastruktur und -bewegung bei Protozoen und Pflanzenzellen, *Wien. A. Hölder*, **4**, 118.

Schrank, F. P. (1803), *Fauna boica*, **3**, 65–70.

Schuberg, A. (1905), Über Cilien und Trichocysten einiger Infusorien, *Arch. Protistenk.*, **6**, 61–110.

Schuckmann, J. (1920), Serologische Untersuchung an Kulturamöben, *Berlin klin. Wochschr.*, 545–547.

Schürmayer, C. B. (1890), Über den Einfluss äusserer Agentien auf einzellige wesen, *Jena. Z. Naturw.*, **24**, 402–470.

Schwartz, V. (1934), Versuche über Regeneration und Kerndimorphismus der Ciliaten, *Nachr. Ges. Wiss. Göttingen*, Math-physik. VI, N. F. 1, 143–155.

——— (1939), Konjugation micronucleusloser Paramaecien, *Naturwissenschaften*, **27**, 724.

Seaman, G. R. (1947), Penicillin as an agent for sterilization of Protozoan cultures, *Science*, **106**, 327.

Sellards, A. W. (1911), Immunity reactions with Amoebae, *Philippine J. Sci.*, **6**, 281–298.

Semenoff, W. E. and A. S. Maslowa (1934), Vital'noe okrashivanie infuzorii pri pomoshchi fagotsitoza *B. prodigiosus* (vital staining of Infusoria by means of phagocytosis of *B. prodigiosus*), *Arch. Russes, Anat. Hist. et Embryol.*, **13**, 351–358 Russ., 435–441 Ger.

——— and ——— (1935), Vitale Infusorienfärbung durch Phagozytóse des B. prodigiosus, *Arch. Protistenk.*, **85**, 224–233.

Serrano, F. (1879), Influence des diverses couleurs sur le développement et la respiration des infusoires, *Compt. rend. acad. sci.*, **89**, 959–960.

Seshachar, B. R. (1947), Chromatin elimination and the ciliate macronucleus, *Am. Naturalist*, **81**, 316–320.

Shapiro, N. N. (1927), The cycle of hydrogen-ion concentration in the food vacuoles of Paramecium, Vorticella and Stylonychia, *Trans. Am. Microscop. Soc.*, **46**, 45–53.

Sharp, R. G. (1914), *Diplodinium ecaudatum* with an account of its neuromotor apparatus, *Univ. Calif. Pubs. Zoöl.*, **13**, 42–122.

Sharpe, M. J. (1930), The influence of H_2S on reproduction rate in *Paramecium caudatum*, *Protoplasma*, **10**, 251–252.

Shaw, Jr., E. H. and L. J. Geppert (1937 a), A kinetic study of the death of *Paramecium caudatum* under the influence of various toxic agents, *Biodynamica*, 1–11.

——— and ——— (1937 b), Toxicity of meta-substituted phenols to *Paramecium caudatum*, *Proc. Soc. Exp. Biol. and Med.*, **37**, 320–323.

———, E. N. Ordal and F. J. Wingfield (1949), Preliminary report on the toxic effect of uranium to *Paramecium caudatum*, *Proc. S. Dakota Acad. Sci.*, **28**, 121–127.

Shaw-Mackenzie, J. A. (1916), The action of copper salts on Protozoa, *Med. Press and Circ.,* **102,** 50–52.

Shipley, P. G. and C. F. De Garis (1925), The third stage of digestion in Paramecia, *Science,* **62,** 266–267.

Shishlyaeva, Z. S. and A. P. Muratova (1936), The increase in the parasiticide action of quinine under the action of ultra short waves, *Z. Microbiol., Epidemiol. Immunitätsforsch. (U.S.S.R.),* **16,** 847.

Shoup, C. S. and J. T. Boykin (1931), The insensitivity of Paramecium to cyanide and effects of iron on respiration, *J. Gen. Physiol.,* **15,** 107–118.

Shubnikova, E. (1947), Ribonucleic acid in the life cycle of the protozoon cell, *Compt. rend. acad. sci. U.R.S.S.,* **55,** 517–520.

Shumway, W. (1914), The effect of thyroid on division rate of Paramecium, *J. Exp. Zoöl.,* **17,** 297–314.

———— (1917), The effect of a thyroid diet on Paramecium, *J. Exp. Zoöl.,* **22,** 529–563.

———— (1929), The species of Paramecium and the thyroid question, *Science,* **69,** 622–623.

Sickels, G. M. and M. Shaw (1935), The effect of purified diphtheria toxin on *Planaria maculata* and *Paramecium caudatum, J. Bact.,* **31,** 73.

Siebold, C. T. E. von (1845), Lehrbuch d. vergleich. Anat. d. wirbellosen Thiere.

Simpson, G. G. and A. Roe (1939), Quantitative Zoology, New York, McGraw-Hill Book Co., 1–414.

Simpson, J. Y. (1901), Studies in Protozoa, *Proc. Scott. Microscop. Soc.,* Sec. 3.

———— (1901 a), Observations on binary fission in the life-history of Ciliata, *Proc. Roy. Soc.* Edinburgh, **23,** 401–421.

———— (1901b), The relation of binary fission and conjugation to variation, *Rep. 71, Brit. Asso. Advancement Sci.,* 688–689.

———— (1902), The relation of binary fission to variation, *Biometrika,* **1,** 400–404.

Slifer, E. H., E. C. Herbert, R. Blumenthal, T. P. Sun, and C. C. Wang (1929), The specific conductivity of protozoan cultures, *Proc. Soc. Exp. Biol. Med.,* **26,** 605–606.

Smaragdova, N. P. (1940), Studies on natural selection in Protozoa: III. Natural selection in populations of *Paramecium bursaria, Zool. Zhur.,* **19.**

Smelev, K. (1928), Sensibilisierung chininresistenter Rassen von Protozoen durch Arsen, *J. Exp. Biol. Med.,* 460–466.

Smith, G. A. (1934), Lag in the division time of *Paramecium caudatum, Anat. Record,* **60,** 92.

Smith, J. C. (1904), A preliminary contribution to the protozoan fauna of the Gulf biological station, with notes on some rare species, *Rept. Gulf Biol. St. (New Orleans),* **2,** 43–55.

Smith, J. D. (1944), A technique for mounting free-living protozoa, *Science,* **100,** 62.

Smith, S. (1908), The limits of educability of Paramecium, *J. Comp. Neurol.,* **18,** 499–510.

Soest, H. (1937), Dressurversuche mit Ciliaten und Rhabdocoelen turbellarien, *Z. vergleich. Physiol.,* **24,** 720–748.

Sokolov, B. (1913), Contribution au problème de la régénération des protozoaires, *Compt. rend. soc. biol. Paris,* **75,** 297–298, 299–301.

———— (1924), Das Regenerationsproblem bei Protozoen, *Arch. Protistenk.,* **47,** 143–253.

Sonne, C. (1929), The biological effects of the ultraviolet rays and the investigation as to the part of the spectrum they lie in, *Arch. Phys. Therapy X-Ray, Radium,* **10,** 239–252.

Sonneborn, T. M. (1936), Factors determining conjugation in *Paramecium aurelia:* I. The cyclical factor: The recency of nuclear reorganization, *Genetics,* **21,** 503–514.

———— (1937 a), Induction of endomixis in *Paramecium aurelia, Biol. Bull.,* **72,** 196–202.

———— (1937b), The extent of the interendomictic interval in *Paramecium aurelia* and some factors determining its variability, *J. Exp. Zoöl.,* **75,** 471–502.

———— (1937 c), Sex, sex inheritance and sex determination in *Paramecium aurelia, Proc. Nat. Acad. Sci.,* **23,** 378–385.

——— (1938), The delayed occurrence and total omission of endomixis in selected lines of *Paramecium aurelia, Biol. Bull.,* **74,** 76–82.

——— (1938 a), Mating types, toxic interactions and heredity in *Paramecium aurelia, Science,* **88,** 503.

——— (1938 b), Mating types in *Paramecium aurelia:* Diverse conditions for mating in different stocks; occurrence, number and interrelations of the types, *Proc. Am. Phil. Soc.,* **79,** 411–434.

——— (1939), Genetic evidence of autogamy in *Paramecium aurelia, Anat. Record,* **75,** 85.

——— (1939 a), *Paramecium aurelia:* Mating types and groups; lethal interactions; determination and inheritance, *Am. Naturalist,* **73,** 390–413.

——— (1939 b), Sexuality and related problems in Paramecium, *Collecting Net,* **14,** 77–84.

——— (1940), The relation of macronuclear regeneration in *Paramecium aurelia* to macronuclear structure, amitosis and genetic determination, *Anat. Record,* **78,** 53–54.

——— (1941), The occurrence, frequency and causes of failure to undergo reciprocal cross-fertilization during mating in *Paramecium aurelia,* variety I, *Anat. Record,* **81,** 66–67.

——— (1941 a), The effect of temperature on mating reactivity in *Paramecium aurelia,* Variety 1, *Anat. Record,* **81,** 131.

——— (1941 b), Sexuality in unicellular organisms, Chap. 14 in Protozoa in Biol. Research, New York, Columbia Univ. Press.

——— (1941 c), Relation of macronuclear regeneration in *Paramecium aurelia* to macronuclear structure, amitosis and genetic determination, *Collecting Net,* **16,** 3–4.

——— (1942), A case of the inheritance of environmental effects and its explanation in Paramecium, *Proc. Indiana Acad. Sci.,* **51,** 262–263.

——— (1942 a), Sex hormones in unicellular organisms, *Cold Spring Harbor Symposia Quant. Biol.,* **10,** 111–124.

——— (1942 b), Inheritance of an environmental effect in *Paramecium aurelia,* variety 1, and its significance, *Genetics,* **27,** 169.

——— (1942 c), Double animals and multiple simultaneous mating in variety 4 of *Paramecium aurelia* in relation to mating types, *Anat. Record,* **84,** 479–480.

——— (1942 d), Evidence for two distinct mechanisms in the mating reaction of *Paramecium aurelia, Anat. Record,* **84,** 542–543.

——— (1942 e), More mating types and varieties in *Paramecium aurelia, Anat. Record,* **84,** 542.

——— (1942 f), Inheritance in ciliate Protozoa, *Am. Naturalist,* **76,** 46–62.

——— (1943), A new system of determination and inheritance of characters, *Records Genetics Soc. Am.,* **12,** 53–54.

——— (1943 a), Development and inheritance of serological characters in variety 1 of *Paramecium aurelia, Genetics,* **28,** 90.

——— (1943 b), Acquired immunity to a specific antibody and its inheritance in *Paramecium aurelia, Proc. Indiana Acad. Sci.,* **52,** 190–191.

——— (1943 c), Gene and Cytoplasm: I. The determination and inheritance of the killer character in variety 4 of *Paramecium aurelia.* II. The bearing of the determination and inheritance of characters in *Paramecium aurelia* on the problems of cytoplasmic inheritance, Pneumococcus transformations, mutations and development, *Proc. Nat. Acad. Sci.,* **29,** 329–343.

——— (1944), Exchange of cytoplasm at conjugation in *Paramecium aurelia,* Variety 4, *Anat. Record,* **89,** 49.

——— (1945), Gene action in Paramecium, *Ann. Missouri Botan. Garden,* **32,** 213–221.

——— (1945 a), The dependence of the physiological action of a gene on a primer and the relation of primer to gene, *Am. Naturalist,* **79,** 318–339.

Sonneborn, T. M. (1945 b), Evidence for a bipartite structure of the gene, *Genetics,* **30,** 22–23.

—— (1946), A system of separable genetic determiners in the cytoplasm of *Paramecium aurelia,* variety 4, *Anat. Record,* **94,** 346.

—— (1946a), Inert nuclei: inactivity of micronuclear genes in variety 4 of *Paramecium aurelia, Genetics,* **31,** 231.

—— (1947), Recent advances in the genetics of Paramecium and Euplotes, in Advances in Genetics, New York, Acad. Press, **1,** 263–358.

—— (1947a), Experimental control of the concentration of cytoplasmic genetic factors in Paramecium, *Cold Spring Harbor Symposia Quant. Biol.,* **11,** 236–255.

—— (1947b), Developmental mechanisms in Paramecium, *Growth,* **11,** 291–307.

—— (1948), The determination of hereditary antigenic differences in genically identical paramecium cells, *Proc. Nat. Acad. Sci.,* **34,** 413–418.

—— (1948a), Symposium on plasmagenes, genes and characters in *Paramecium aurelia, Am. Naturalist,* **82,** 26–34.

—— (1948b), Genes, cytoplasm and environment in Paramecium, *Sci. Monthly,* **67,** 154–160.

—— (1949), Beyond the gene, *Am. Scientist,* **37,** 33–59.

—— (1949a), Ciliated Protozoa: cytogenetics, genetics, and evolution, *Ann. Rev. Microbiol.,* 55–80.

—— (1950), Heredity, environment and politics, *Science,* **111,** 529–539.

—— (1950 a), The cytoplasm in heredity, *Heredity,* **4,** 11–36.

—— (1950b), Methods in the general biology and genetics of *Paramecium aurelia, J. Exp. Zoöl.,* **113,** 87–147.

—— (1950c), Paramecium in modern biology, *Bios,* **21,** 31–43.

—— (1950d), Partner of the genes, *Scient. Am.,* **183,** 30–39.

—— (1950 e), Cellular transformations, *Harvey Lect.,* **44,** 145–164.

—— (1950f), The kinetosome in cytoplasmic heredity, *J. Heredity,* **41,** 222–224.

—— (1951), Beyond the gene—two years later, Science in Progress, G. A. Baitsell, New Haven, Conn. Yale Univ. Press, 167–203.

—— (1951 a), Some current problems of genetics in the light of investigations on *Chlamydomonas* and *Paramecium, Cold Spring Harbor Symposia Quant. Biol.,* **16,** 483–503.

—— (1951 b), The role of the genes in cytoplasmic inheritance, Chap. 14, Genetics in the 20th Century, Ed. by L. C. Dunn, New York, Macmillan Co.

—— and B. M. Cohen (1936), Factors determining conjugation in *Paramecium aurelia:* II. Genetic diversities between stocks or races, *Genetics,* **21,** 515–518.

—— and R. V. Dippell (1943), Sexual isolation, mating types and sexual responses to diverse conditions in variety 4, *Paramecium aurelia, Biol. Bull.,* **85,** 36–43.

—— and —— (1946), Mating reactions and conjugation between varieties of *Paramecium aurelia* in relation to conceptions of mating type and variety, *Physiol. Zoöl.,* **19,** 1–18.

—— and —— (1946a), The significance of race 31 as a link between group A and B varieties of *Paramecium aurelia, Anat. Record,* **96,** 19.

——, ——, and W. Jacobson (1946), Some properties of kappa (killer cytoplasmic factor) and of paramecin (killer substance) in *Paramecium aurelia,* variety 4, *Records Genetics Soc. Am.,* **15,** 69.

——, ——, and —— (1946), Paramecin 51, an antibiotic produced by *Paramecium aurelia:* amounts released from killers and taken up by sensitives; conditions protecting sensitives, *Anat. Record,* **96.** 18–19.

—— and A. Lesuer (1948), Antigenic characters in *Paramecium aurelia* (Variety 4): determination, inheritance and induced mutations, *Am. Naturalist,* **82,** 69–78.

—— and R. S. Lynch (1932), Racial differences in the early physiological effects of conjugation in *Paramecium aurelia, Biol. Bull.,* **62,** 258–293.

—— and —— (1934), Hybridization and segregation in *Paramecium aurelia, J. Exp. Zoöl.,* **67,** 1–72.

—— and —— (1937), Factors determining conjugation in *Paramecium aurelia:* III. A genetic factor: The origin at endomixis of genetic diversities, *Genetics,* **22,** 284–296.

Sosnowski, J. (1899), Untersuchungen über die Veränderungen des Geotropismus bei *Paramecium aurelia, Bull. intern. acad. Cracovie,* Mars.

—— (1899a), Beiträge zur Chemie der Zelle, *Centr. Physiol.,* **13,** 267–270.

Spallanzani, L. (1776), Opuscoli de fisica animale e vegetabile, etc., Modena.

Specht, H. (1934), Aerobic respiration in *Spirostomum ambiguum* and the production of ammonia, *J. Cellular Comp. Physiol.,* **5,** 319–333.

Spek, J. (1919), Experimentelle Beiträge zur Physiologie der Zellteilung, *Biol. Zentr.,* **39,** 23–34.

—— (1920), Experimentelle Beiträge zur Kolloidchemie der Zellteilung. Kolloidchem, *Beihefte,* **9,** 259.

—— (1928), Die Struktur der lebenden Substanzen im Lichte der Kolloidforschung (Opalina), *Kolloid-Z.,* **46,** 314–320.

Spencer, H. (1924), Studies of a pedigree culture of *Paramecium calkinsi, J. Morphol.,* **39,** 543–551.

Spencer, R. R. and D. Calnan (1945), Studies in species adaptation: III. Continuous exposure of paramecia to methylcholanthrene and other agents for more than five years, *J. Natl. Cancer Inst.,* **6,** 147–154.

—— and M. B. Melroy (1940), Effect of carcinogens on small free-living organisms: II. Survival value of methylcholanthrene adapted Paramecium, *J. Natl. Cancer Inst.,* **1,** 343–348.

—— and —— (1914), Effect of carcinogens on small organisms: III Cell-division rate and population levels of methylcholanthrene-adapted paramecia, *J. Natl. Cancer Inst.,* **2,** 185–191.

—— and —— (1943), Principles of species adjustment: I Continuous exposure, *J. Natl. Cancer Inst.,* **4,** 249–263.

Staniewicz, W. (1910), Etudes expérimentales sur la digestion de la graisse dans les infusoires ciliés, *Bull. acad. sci. Cracovie,* Sér. B, 199.

Stanley, J. (1945), A method of growing dense cultures of Paramecium, *Science,* **103,** 115–116.

Statkewitsch, P. (1903), Ueber die Wirkung der Induktionsschläge auf einige Ciliata, *Physiol. Russe,* **3,** 1–55.

—— (1903 a), Über die Wirkung von Induktionsschlägen auf einige Ciliata: galvanotropism and galvanotaxis of organisms, *Physiol. Russe,* **3,** 1–55.

—— (1904), Galvanotropismus und Galvanotaxis der Ciliata, *Z. allgem. Physiol.,* **4,** 296–332.

—— (1905), Galvanotropismus und Galvanotaxis der Ciliata, *Z. allgem. Physiol.,* **5,** 511–534.

—— (1905 a), Zur Methodik der biologischen Untersuchungen über die Protisten, *Arch. Protistenk.,* **5,** 17–39.

—— (1907), Galvanotropismus und Galvanotaxis der Ciliata, *Z. allgem. Physiol.,* **6,** 13–43.

Stefanowska, M. (1902), Modifications microscopiques du protoplasma vivant, dans l'anesthésie, *Compt. rend. soc. biol.,* **54,** 545–547.

Stein, F. (1854), Die Infusionstiere auf ihre Entwicklungsgeschichte untersucht, Leipzig.

—— (1859–1878), Der Organismus der Infusionsthiere, Leipzig.

Stempell, W. (1914), Über die Funktion der pulsierenden Vakuole und einen Apparat zur Demonstration derselben, *Zool. Jahrb., Abt. allgem. Zool. Physiol.,* **34,** 437–478.

Stempell, W. (1914 a), Lyotropie und pulsierende Vakuole, *Zool. Anz.*, **58**, 232–233.

——— (1924), Weitere Beiträge zur Physiologie der pulsierenden Vakuole von Paramecium, *Arch. Protistenk.*, **48**, 342–364.

Stockard, C. R. (1921), Developmental rate and structural expression, *Am. J. Anat.*, **28**, 115–227.

Stocking, R. J. (1915), Variation and inheritance in abnormalities occurring after conjugation in *Paramecium caudatum*, *J. Exp. Zoöl*, **19**, 387–449.

Stokes, A. C. (1885), Some new Infusoria, *Am. Naturalist*, **19**, 433–443.

——— (1893), The contractile vesicle, *Am. Monthly Micro. J.* **14**, 182–188.

Stokstad, E. L. R., C. E. Hoffman, M. A. Regan, D. Fordham and T. H. Jukes (1949), Observations on an unknown growth factor essential for *Tetrahymena geleii*, *Arch. Biochem.*, **20**, 75–82.

Stokvis, C. S., and W. H. Swellengrebel (1911), Purification of water by infusories, *J. Hyg.*, **11**, 481–486.

Stone, W. S. and F. H. K. Reynolds (1939), A practical method of obtaining bacteria-free cultures of *Trichomonas hominis*, *Science*, **90**, 91–92.

Stowell, R. E. (1945), Feulgen reaction for thymonucleic acid, *Stain Technol.*, **20**, 45–58.

Stranghöner, E. (1932), Teilungsrate und Kernreorganisationsprozess bei *Paramaecium multimicronucleatum*, *Arch. Protistenk.*, **78**, 302–360.

Strauss, W. L. (1923), Thyroid cultures of Paramecium, *Science*, **58**, 205.

Strelnikow, D. (1924), L'adsorption des colorants basiques par *Paramecium caudatum*, *Compt. rend. soc. biol.*, **100**, 1004.

Strong, R. M. (1916), Culture-media for Paramecium and Euglena, *Science*, **44**, 238.

Subramaniam, M. K. (1947), Is the macronucleus of ciliates endopolyploid?, *Current Sci.*, **16**, 228–229.

Summers, F. M. (1935), The division and reorganization of the macronuclei of *Aspidisca lynceus Muller, Diophrys appendiculata Stein*, and *Stylonychia pustulata Ehrenberg*, *Arch. Protistenk.*, **85**, 173–208.

——— (1941), The protozoa in connection with morphogenetic problems, in chap. 16, Protozoa in Biological Research, New York, Columbia University Press, 772–817.

——— and H. K. Hughes (1940), Experiments with *Colpidium campylum* in high-frequency electric and magnetic fields, *Physiol. Zoöl.*, **13**, 227–242.

Sun, A. (1912), Experimentelle Studien über Infusorien, *Arch. Protistenk.*, **27**, 207–218.

Surber, E. W. and O. L. Meehan (1931), Lethal concentrations of arsenic for certain aquatic organisms, *Trans. Am. Fisheries Soc.*, **61**, 225–239.

Swarzewsky, B. (1914), Zur Chromidienfrage und Kerndualismushypothese, III and IV, *Biol. Zentr.*, **32**, 535–545, 545–564.

Swezey, W. W. and F. O. Atchley (1935), Comparative behavior characteristics of free-living and parasitic protozoa, *Trans. Am. Microscop. Soc.*, **54**, 98–102.

Syngajewskaja, K. (1926), Individual growth of *Paramecium caudatum*, *Mem. Acad. Sci. Ukrain. Socialistic Rep.*, **2**, 437–442.

——— (1929), Growth of *Paramecium caudatum* and the osmotic pressure during its life cycle, *Trav. inst. biol.*, **12**, 323–327.

Szücs, J. and B. Kisch (1912), Über die kombinierte Wirkung von fluoreszierenden Stoffen und Alkohol, *Z. Biol.*, **58**, 558–570.

Takenouchi, M. (1918), Cytolytic action of normal and immune serum on Infusoria, *J. Infectious Diseases*, **23**, 396–414.

Taliaferro, W. H. (1929), The Immunology of Parasitic Infections, New York.

——— (1941), The Immunology of the Parasitic Protozoa, chap. 18 in Protozoa in Biol. Research, New York, Columbia University Press, 830–889.

Tang, P. S. and H. Z. Gaw (1937), Mechanism of death in unicellular organisms: I. Delayed death and change in resistance to ultraviolet radiation in *Paramecium bursaria* with age of culture, *Chinese J. Physiol.*, **11**, 305–314.

Tanzer, C. (1941), Serological studies with free-living Protista, *J. Immunol.*, **42**, 291–312.

Tappeiner, H. v. (1896), Über die Wirkung der Phenylchinoline und Phosphine auf niedere Organismen, *Deut. Arch. klin. Medi.*, **56**.

―――― (1909), Die photodynamische Erscheinung, *Ergeb. Physiol.*, **8**, 698–741.

―――― and A. Jodlbauer (1904), Über die Wirkung der photodynamischen (fluoreszierenden) Stoffe auf Protozoen und Enzyme, *Deut. Arch. klin. Medi.*, **80**, 427–487.

―――― and ―――― (1907), Die sensibilisierende Wirkung fluoreszierender Substanzen, Leipzig.

Tartar, V. (1938), Regeneration in the genus Paramecium, *Anat. Record*, **72**, 52.

―――― (1939), The so-called racial variation in the power of regeneration in Paramecium, *J. Exp. Zoöl.*, **81**, 181–208.

―――― (1940), Nuclear reactions in Paramecium, *Anat. Record*, **78**, 109.

―――― (1941), Intracellular patterns: facts and principles concerning patterns exhibited in the morphogenesis and regeneration of ciliate Protozoa, *Third Growth Symposium*, 21–40.

―――― and T. T. Chen (1940), Preliminary studies on mating reactions of enucleate fragments of *Paramecium bursaria*, *Science*, **91**, 246–247.

―――― and ―――― (1941), Mating reactions of enucleate fragments in *Paramecium bursaria*, *Biol. Bull.*, **80**, 130–138.

Tatum, E. L., L. Garnjobst, and C. V. Taylor (1942), Vitamin requirements of *Colpoda duodenaria*, *J. Cellular Comp. Physiol.*, **20**, 211–224.

Taylor, C. V. (1920), An accurately controllable micropipette, *Science*, **51**, 617–618.

―――― (1923), The effect of the removal of the micronucleus, *Science*, **58**, 308.

―――― (1923 a), The contractile vacuole in Euplotes, an example of sol-gel reversibility of cytoplasm, *J. Exp. Zoöl.*, **37**, 259–290.

―――― (1935), The effects of x-rayed medium on living cells, *Estratto dagli Atti del I, Congresso Internazionale di Elettro-radio-biologia*, **2**.

―――― (1941), Fibrillar systems in ciliates, Protozoa in Biological Research, New York, Columbia University Press.

―――― and W. P. Farber (1924), Fatal effects of the removal of the micronucleus in Euplotes, *Univ. Calif. Pubs. Zoöl.*, **26**, 131–144.

―――― and A. G. R. Strickland (1938), Reactions of *Colpoda duodenaria* to environmental factors: I. Some factors influencing growth and encystment, *Arch. Protistenk.*, **90**, 396–409.

――――, J. O. Thomas, and M. G. Brown (1933), Studies on Protozoa: IV. Lethal effects of the X-radiation of a sterile culture medium for *Colpidium campylum*, *Physiol. Zoöl.*, **6**, 467–492.

Taylor, H. S., W. W. Swingle, H. Eyring, and A. A. Frost (1933), The effect of water containing the isotope of hydrogen upon fresh water organisms, *J. Cellular Comp. Physiol.*, **4**, 1–8.

Ten Kate, C. G. B. (1927), Über das Fibrillensystem der Ciliaten, *Arch. Protistenk.*, **57**, 362–426.

Thapar, G. S. and S. S. Choudhury (1923), The occurrence and significance of a third contractile vacuole in *Paramecium caudatum*, *J. Proc. Asiatic Soc. Bengal*, **18**, 93.

Thompson, D. W. (1945), On Growth and Form, New York, The Macmillan Company.

Tittler, I. A. (1948), An investigation of the effects of carcinogens on *Tetrahymena geleii*, *J. Exp. Zoöl.*, **108**, 309–325.

―――― and M. Kobrin (1941), The reaction of Paramecium to certain carcinogens, *Anat. Record*, **81**, suppl., 130–131.

―――― and ―――― (1942), Effects of carcinogenic agents on *Paramecium caudatum*, *Proc. Soc. Exp. Biol. Med.*, **50**, 95–96.

Tönniges, G. (1914), Die Trichocysten von *Frontonia leucas* und ihr chromidialer Ursprung, *Arch. Protistenk.*, **32**, 298–378.

Torrey, H. B. (1924), The effect of dissolved thyroxin on the division rate of Paramecium, *Anat. Record*, **26**, 367.

——— (1934), Thyroxin and regeneration in the hydroid Pennaria, *Physiol. Zoöl.*, **7**, 586–592.

———, M. C. Riddle and J. L. Brodie (1925), Thyroxin as a depressant of the division rate of Paramecium, *J. Gen. Physiol.*, **7**, 449–460.

Towle, E. W. (1904), A study of the effects of certain stimuli, single and combined, upon Paramecium, *Am. J. Physiol.*, **12**, 220–236.

Tschakhotine, S. (1938), Réactions "conditionées" par microponction ultraviolette dans le comportement d'une cellule isolée (*Paramecium caudatum*), *Arch. Inst. prophylac.*, **10**, 119–131.

Tsukamoto, M. (1895), On the poisonous action of alcohols upon different organisms, *J. Coll. Sci. Imp. Univ. Tokyo*, **7**, 269–281.

Tsvetkov, A. N. (1928), Sur les mouvements spontanés de *Paramecium caudatum*, *Compt. rend. acad. sci. U.R.S.S.*, **7**, 105–106.

Tuan, H. C. (1930), Picric acid as a destaining agent for iron alum hematoxylin, *Stain Technol.*, **5**, 135–138.

Tunnicliff, R. (1928), Use of Paramecium for studying toxins and antitoxins, *Proc. Soc. Exper. Biol. Med.*, **26**, 213–217.

——— (1929), Use of Paramecia for studying toxic substances in urine, *J. Infectious Diseases*, **45**, 244–246.

Turner, J. P. (1930), Division and conjugation in *Euplotes patella Ehrenberg* with special reference to the nuclear phenomena, *Univ. Calif. Pubs. Zoöl.*, **33**, 193–258.

——— (1931), A simple apparatus for washing protozoa, *Science*, **74**, 99–100.

——— (1933), The external fibrillar system of Euplotes with notes on the neuromotor apparatus, *Biol. Bull.*, **64**, 53–66.

Ugata, T. (1926), Biochemical studies on the growth of paramecium: I. The effects of amino acids, beef-extract and nucleic acid upon the division rate in *Paramecium caudatum* and the relation between their action and bacterial flora. II. The effect of splitting products of nucleic acid upon the division rate in *Paramecium caudatum*, *J. Biochem. (Japan)*, **6**, 417–464.

Underhill, F. P. and L. L. Woodruff (1914), Protozoan protoplasm as an indicator of pathological changes: III. In fatigue, *J. Biol. Chem.*, **17**, 9–12.

Unger, W. B. (1925), The relation of contractile and food vacuoles to rhythms in Paramecium, *Proc. Soc. Exp. Biol. Med.*, **22**, 333–334.

——— (1926), The relationship of rhythms to nutrition and excretion in Paramecium, *J. Exp. Zoöl.*, **43**, 373–412.

——— (1929, 1930), The sequence of protozoa in five plant infusions, *Anat. Record*, **44**, 248, **48**, 348.

——— (1931), The protozoan sequence of five plant infusions, *Trans. Am. Microscop. Soc.*, **50**, 143–153.

Urbanowicz, K. (1927), Gurtwitschs mitogenetische Strahlung, an Paramaecienteilungen geprüft, *Arch. Entwicklungsmech. Organ.*, **110**, 417–426.

van Wagtendonk, W. J. (1948), The killing substance Paramecin: chemical nature, *Am. Naturalist*, **82**, 60–68.

——— (1948 a), The action of enzymes on paramecium, *J. Biol. Chem.*, **173**, 691–704.

——— (1949), Antigenic transformation in *Paramecium aurelia* under the influence of proteolytic enzymes, *Federation Proc.*, **8**, 262.

——— and P. L. Hackett (1949), The culture of *Paramecium aurelia* in the absence of other living organisms, *Proc. Nat. Acad. Sci.*, **35**, 155–159.

——— and L. P. Zill (1947), Inactivation of paramecin ("killer" substance of *Paramecium aurelia* 51, variety 4) at different hydrogen ion concentrations and temperatures, *J. Biol. Chem.*, **171**, 595–603.

Vegener, E. E. (1940), A method for total preparations of microscopic organisms, *Arch. Russes anat. histol. et embryol.*, **23**, 185–186.

Versluys, J. (1906), Über dir Konjugation der Infusorien, *Biol. Zentr.*, **26**, 46–62.

Verworn, M. (1889), Psychophysiologische Protistenstudien, Experimentelle Untersuchungen, Jena, 1–219.

——— (1889 a), Die polare Erregung der Protisten durch den galvanischen Strom, *Arch. ges. Physiol.*, **45**, 1–36.

——— (1889 b), Die polare Erregung der Protisten durch den galvanischen Strom, *Arch. ges. Physiol.*, **46**, 281–303.

——— (1892), Die Bewegung der lebendigen Substanz, Jena, 1–103.

——— (1896), Untersuchungen über die polare Erregung der lebendigen Substanz durch den constanten Strom. III. Mittheilung, *Arch. ges. Physiol.*, **62**, 415–450.

——— (1896 a), Die polare Erregung der lebendigen Substanz durch den constanten Strom. IV. Mittheilung, *Arch. ges. Physiol.*, **65**, 47–62.

——— (1915), Allgemeine Physiologie, ed. 6, Jena, G. Fischer.

Vieweger, T. (1912), Recherches sur la sensibilité des infusoires (alcaliooxytaxisme), les reflexes locomoteurs, l'action des sels, *Arch. biol. Liége*, **27**, 723–799.

Visscher, J. P. (1923), Feeding reactions in the cilate *Dileptus gigas*, with special reference to the function of trichocysts, *Biol. Bull.*, **45**, 113–143.

——— (1927), Conjugation in the ciliated protozoon, *Dileptus gigas*, with special reference to the nuclear phenomena, *J. Morphol. and Physiol.*, **44**, 383–415.

——— (1927), A neuromotor apparatus in the ciliate *Dileptus gigas*, *J. Morphol. and Physiol.*, **44**, 373–381.

Volkonsky, M. (1929), Les Phénomènes cytologiques au cours de la digestion intracellulaire de quelques ciliés, *Compt. rend. soc. biol.*, **101**, 133–135.

——— (1934), L'Aspect cytologique de la digestion intracellulaire, *Arch. exp. Zellforsch. Gewebezücht.*, **15**, 355–372.

Volterra, V. (1926), *Mem. reale accad. nazl. Lincei*, **2**, 36.

von Brand, T. (1934), Das Leben ohne Sauerstoff bei wirbellosen Tiere, *Ergeb. Biol.*, **10**, 37.

——— (1946), Anaerobiosis in Invertebrates, Biodynamica, Normandy, Missouri, 1–328.

Vonwiller, P. (1921), Intravitale Färbung von Protozoen, Abderhaldens Handbook, **5**, 87–96.

Wallengren, H. F. S. (1902), Inanitionserscheinungen der Zelle, *Z. allgem. Physiol.*, **1**, 67–128.

——— (1902–3), Zur Kenntnis der Galvanotaxis, I, II, III, *Z. allgem. Physiol.*, **2, 3**.

——— (1903), Die Einwirkung des konstanten Stromes auf die innere Protoplasmabewegung bei den Protozoen, *Z. allgem. Physiol.*, **3**, 22–32.

Wang, C. C. (1928), Ecological studies of the seasonal distribution of protozoa in a freshwater pond, *J. Morphol.*, **46**, 431–478.

Wankell, F. (1922), Über Reduktion basischer Farbstoffe im lebenden Protoplasma, *Ber. Ges. Freiburg*, **23**, 118–144.

Watanabe, H. (1925), Studien über Flimmerbewegung und eine neue Paraffineinbettungsmethode, *Z. Anat. Entwicklungs-geschichte.*, **75**, 433–759.

Weatherby, J. H. (1927), The function of the contractile vacuole in *Paramecium caudatum* with special reference to the excretion of nitrogenous compounds, *Biol. Bull.*, **52**, 208–218.

——— (1929), Excretion of nitrogenous substances in Protozoa, *Physiol. Zoöl.*, **2**, 375–394.

——— (1941), The Contractile Vacuole, in Calkins, G. H. and F. M. Summers, Protozoa in Biological Research, New York, Columbia University Press.

Weber, G. (1912), Die Bewegung der Periostomcilien bei den heterotrichen Infusorien. Sitzber., *Wien. Akad. Math.-naturw. Klasse*, **121**, 46 pp.

Weinstein, I. (1930), Quantitative biological effects of monochromatic ultraviolet light, *J. Optical Soc. Am.*, **20**, 433–456.

Weisbach, W. (1930), Protozoa, *Tabulae biol.*, **6**, 1–37.

Weismann, A. (1884), Zur Frage nach der Unsterblichkeit der Einzelligen, *Biol. Zentr.*, **4**, 651.

—— (1890), Bemerkungen zu einigen Tagesproblemen, *Biol. Zentralbl.*, **10**, 1, 33.

Wenk, P. (1939), Zur Frage der Punkterwärmung im hoch frequenten Wechselfeld, *Strahlentherapie*, **65**, 657–663.

Wenrich, D. H. (1924), The structure and division of *Paramecium trichium*, *Anat. Record.*, **29**, 117.

—— (1926), The structure and division of *Paramecium trichium*, *J. Morphol.* **43**, 81–103.

—— (1928 a), Eight well-defined species of Paramecium, *Trans. Am. Microscop. soc.*, **47**, 274–282.

—— (1928 b), *Paramecium woodruffi nov. spec.* (ciliata), *Trans. Am. Microscop. Soc.*, **47**, 256–261.

—— (1933), A modification of Schaudinn's fixative for Protozoa, *Stain Technol.*, **8**, 158.

—— (1937), Protozoological methods (Chap. 8), in McClung, C. E., Handbook of Microscopical Technique for Workers in Animal and Plant Tissue, New York, P. B. Hoeber, 1–698.

—— (1940), Chromosomes in Protozoa, *Collecting Net*, **15**, 1–6.

—— and C. C. Wang (1928), The occurrence of conjugation in *Paramecium calkinsi*, *Science*, **67**, 270–271.

Wense, T. (1935), Colloidal changes indicated by experiments on *Paramecium caudatum* as the basis of sympathetic nervous processes, *Arch. exptl. Path. Pharmakol.*, **179**, 275.

Wenyon, C. M. (1926), Protozoology, London, Baillière, Tindal and Cox.

Wetzel, A. (1923), Über die cytologischen Differenzierungen am Munde der Ciliaten, *Verhandl. deut. zool. Ges.*, **28**, 85–86.

—— (1925), Vergleichend-cytologische Untersuchungen an Ciliaten, *Arch. Protistenk.*, **51**, 209–304.

Weyland, H. (1914), Versuche über das Verhalten von *Colpidium colpoda* gegenüber reizenden und lähmenden Stoffen, *Z. allgem. Physiol.*, **16**, 123–162.

Wichterman, R. (1937), Division and conjugation in *Nyctotherus cordiformis* (Ehr.) Stein (Protozoa, Ciliata) with special reference to the nuclear phenomena, *J. Morphol.*, **60**, 563–611.

—— (1937 a), Conjugation in *Paramecium trichium Stokes* (Protozoa, Ciliata) with special reference to the nuclear phenomena, *Biol. Bull.*, **73**, 397–398.

—— (1937 b), Studies on living conjugants of *Paramecium caudatum*, *Anat. Record*, **70**, 66.

—— (1937 c), Studies on living conjugants of *Paramecium caudatum*, *Biol. Bull.*, **73**, 396–397.

—— (1938), Does transfer of pronuclei ever occur in conjugation of *Paramecium caudatum?*, *Biol. Bull.*, **75**, 376–377.

—— (1939), Cytogamy: A new sexual process in joined pairs of *Paramecium caudatum*, *Nature*, **144**, 123–124.

—— (1939 a), Division of *Paramecium trichium* in a comparatively constant culture of hay infusion, *Anat. Record*, **75**, 151–152.

—— (1940), Cytogamy: A sexual process occurring in living joined pairs of *Paramecium caudatum* and its relation to other sexual phenomena, *J. Morphol.*, **66**, 423–451. (Note listing under motion picture films of Paramecium).

—— (1940 a), Motion pictures of cytogamy in *Paramecium caudatum*, *Anat. Record*, **78**, 110.

—— (1940 b), Parasitism in *Paramecium caudatum, J. Parasitol.,* **26,** 29.

—— (1941), Studies on zoöchlorellae-free *Paramecium bursaria, Biol. Bull.,* **81,** 304–305.

—— (1943), Conjugation and fate of exconjugants of zoöchlorellae-free *Paramecium bursaria, Anat. Record,* **87,** 50.

—— (1944), Pure-line mass cultures for demonstrating the mating reaction and conjugation in Paramecium, *Turtox News,* **22,** 33–37.

—— (1944 a), Recent discoveries of nuclear processes and sexual phenomena in Paramecium, *Turtox News,* **22,** 105–110. (Note listing under motion picture films of Paramecium)

—— (1945), Schizomycetes parasitic in *Paramecium bursaria, J. Parasitol.,* **31,** 25.

—— (1945 a), A modified Petri dish for continuous temperature observation, *Science,* **101,** 184.

—— (1946 a), Time relationships of the nuclear behavior in the conjugation of green and colorless *Paramecium bursaria, Anat. Record,* **94,** 39–40.

—— (1946 b), The behavior of cytoplasmic structures in living conjugants of *Paramecium bursaria, Anat. Record,* **94,** 93–94.

—— (1946 c), Unstable micronuclear behavior in an unusual race of Paramecium, *Anat. Record,* **94,** 94.

—— (1946 d), A new glass device for staining cover-glass preparations, *Science,* **103,** 23–24.

—— (1946 e), Further evidence of polyploidy in the conjugation of green and colorless *Paramecium bursaria, Biol. Bull.,* **91,** 234.

—— (1946 f), Direct observation of the transfer of pronuclei in living conjugants of *Paramecium bursaria, Science,* **104,** 505–506.

—— (1947 a), *Paramecium nucleatum* versus *Paramecium multimicronucleatum* (Section: Comments by Readers), *Science,* **106,** 491.

—— (1947 b), Action of X-rays on mating types and conjugation of *Paramecium bursaria, Biol. Bull.,* **93,** 201.

—— (1948 a), The time schedule of mating and nuclear events in the conjugation of *Paramecium bursaria, Turtox News,* **26,** 2–10.

—— (1948 b), The biological effects of x-rays on mating types and conjugation of *Paramecium bursaria, Biol. Bull.,* **94,** 113–127.

—— (1948 c), The presence of optically active crystals in *Paramecium bursaria* and their relationship to symbiosis, *Anat. Record,* **101,** 97–98.

—— (1948 d), The hydrogen-ion concentration in the cultivation and growth of the eight species of Paramecium, *Biol. Bull.,* **95,** 272.

—— (1948 e), Mating types and conjugation of four different races of *Paramecium calkinsi* and the effect of x-rays on the mating reaction, *Biol. Bull.,* **95,** 271–272.

—— (1949), The collection, cultivation and sterilization of Paramecium, *Proc. Penna. Acad. Sci.,* **23,** 151–180.

—— (1950), The occurrence of a new variety containing two opposite mating types of *Paramecium calkinsi* as found in sea water of high salinity content, *Biol. Bull.,* **99,** 366.

—— (1950 a), Comparative behavior in the mating and conjugation of opposite mating types in different clones of *Paramecium bursaria, Anat. Record,* **108,** 139.

—— (1950 b), A simple method for obtaining large numbers of fission-stages of Paramecium and certain other negatively geotropic ciliates, *Biol. Bull.,* **99,** 366–367.

—— (1951), The effects of x-rays upon dividing and non-dividing cells: *Paramecium caudatum, Biol. Bull.,* **101,** 232.

—— (1951 a), The effects of x-rays upon conjugant and vegetative stages of *Paramecium calkinsi, Biol. Bull.,* **101,** 232–233.

—— (1951 b), The biology of *Paramecium calkinsi* with special reference to ecology, cultivation, structural characteristics and mating-type phenomena, *Proc. Am. Soc. Protozoöl.*, **2**, 11–12.

Wichterman, R. (1951 c), Effects of high dosage irradiation with x-rays upon dividing and non-dividing cells and conjugant stages: *Paramecium, Anat. Record,* **111**, 111.

———— (1951 d), The ecology, cultivation, structural characteristics and mating types of *Paramecium calkinsi, Proc. Penna. Acad. Sci.,* **25**, 51–65.

———— (1952), A method for obtaining abundant dividing stages of Paramecium. *Trans. Am. Microscop. Soc.,* **71**, 303–305.

———— and F. H. J. Figge (1952), Environmental factors concerned with the lethality of x-rays in *Paramecium, Biol. Bull.,* **103**, 311.

Wieland, H. (1922), Über den Wirkungsmechanismus betäubender Gase, des Stickoxyduls und des Azetylens, *Arch. exp. Path. Pharmakol.,* **92**, 96–152.

Wilczynski, J. (1939), Zur weiteren Analyse der "Lag-Periode" in der Paramecienteilungs-rhythmik, *Arch. f. Protistenk.,* **93**, 87–104.

Willey, A. and C. Lherisson (1930), An interpretation of mass conjugation in paramecium, *Science,* **71**, 367–369.

Wilson. C. N. (1927), A comparative study of several samples of neutral red and their effects on *Paramecium caudatum, Stain Technol.,* **2**, 115–123.

Wilson, E. B. (1928), The Cell in Development and Heredity, New York, The Macmillan Co.

Winterstein, H. (1926), Die Narkose, ed. 2, Berlin, Julius Springer.

Wladimirsky, A. P. (1916), Are the Infusoria capable of "learning" to select their food?, *Russ. J. Zoöl.,* **44**, 4.

Wohlfarth-Bottermann, K. E. (1950), Funktion und Struktur der Paramecium-tricho-cysten, *Naturwiss.,* **37**, 562–563.

Wolfson, C. (1935), Observations on Paramecium during exposure to sub-zero tempera-tures, *Ecology,* **16**, 630–639.

Wolman, M. (1939), A proliferative effect of carcinogenic hydrocarbons upon multiplica-tion of Paramecium, *Growth,* **3**, 386.

Woodruff, L. L. (1905), An experimental study on the life-history of *Hypotrichous in-fusoria, J. Exp. Zoöl.,* **2**, 585–635.

———— (1908), The life-cycle of Paramecium when subjected to a varied environment, *Am. Naturalist,* **42**, 520–526.

———— (1908 a), Effects of alcohol on the division-rate of Infusoria, *Science,* **27**, 412–413.

———— (1908 b), Increased susceptibility of Protozoa to poison due to treatment with alcohol, *Proc. Soc. Exp. Biol.,* **5**, 82–83.

———— (1908 c), Effects of alcohol on the life-cycle of Infusoria, *Biol. Bull.,* **15**, 85–104.

———— (1908 d), Increased susceptibility of Protozoa to poison due to treatment with alcohol, *Proc. Soc. Exp. Biol.,* **5**, 82–83.

———— (1909 a), Studies on the life-cycle of Paramecium, *Proc. Soc. Exp. Biol.,* **6**, 117–118.

———— (1909 b), Duration of the life-cycle of Paramecium, *Science,* **29**, 425.

———— (1909 c), Further studies on the life-cycle of Paramecium, *Biol. Bull.,* **17**, 287–308.

———— (1910), On the power of reproduction without conjugation in Paramecium, *Proc. Soc. Exp. Biol. Med.,* **7**, 144.

———— (1911), *Paramecium aurelia* and *Paramecium caudatum, J. Morphol.,* **22**, 223–237.

———— (1911 a), Two thousand generations of Paramecium, *Arch. Protistenk.,* **21**, 263–266.

———— (1911 b), Evidence of the adaptation of Paramecia to different environments, *Biol. Bull.,* **22**, 60–65.

—— (1911 c), The effect of excretion products of Paramecium on its rate of reproduction, *J. Exp. Zoöl.*, **10**, 557–581.

—— (1911–12), Evidence on the adaptation of Paramecia to different environments, *Biol. Bull.*, **22**, 60–65.

—— (1912), Observations on the origin and sequence of Protozoan fauna of hay infusions, *J. Exp. Zoöl.*, **12**, 205–264.

—— (1912 a), A summary of the results of certain physiological studies on a pedigreed race of Paramecium, *Biochem. Bull.*, **1**, 396–412.

—— (1913 a), The effect of excretion products of Infusoria on the same and on different species, with special reference to the protozoan sequence in infusions, *J. Exp. Zoöl.*, **14**, 575–582.

—— (1913 b), 3300 Generationen von Paramecien ohne Konjugation oder künstliche Reizung, *Biol. Zentr.*, **33**, 34–36.

—— (1914), So-called conjugating and non-conjugating races of Paramecium, *J. Exp. Zoöl.*, **16**, 237–240.

—— (1915), The problem of rejuvenescence in Protozoa, *Biochem. Bull.*, **4**, 371–378.

—— (1917), The influence of general environmental conditions on the periodicity of endomixis in *Paramecium aurelia, Biol. Bull.*, **33**, 437–462.

—— (1917 a), Rhythms and endomixis in various races of *Paramecium aurelia, Biol. Bull.*, **33**, 51–56.

—— (1921), The structure, life history and intrageneric relationship of *Paramecium calkinsi* n. sp., *Biol. Bull.*, **41**, 171–180.

—— (1921 a), Micronucleate and amicronucleate races of Infusoria, *J. Exp. Zoöl.*, **34**, 329–337.

—— (1921 c), The present status of the long-continued pedigree culture of *Paramecium aurelia* at Yale University, *Proc. Nat. Acad. Sci.*, **7**, 41–44.

—— (1925), The physiological significance of conjugation and endomixis in the infusoria, *Am. Naturalist*, **69**, 225–249.

—— (1926), Eleven thousand generations of Paramecium, *Quart. Rev. Biol.*, **1**, 436–438.

—— (1929), Thirteen thousand generations of Paramecium, *Proc. Soc. Exp. Biol. Med.*, **26**, 707–708.

—— (1931), Micronuclear variation in *Paramecium bursaria, Quart. J. Microscop. Sci.*, **74**, 537–545.

—— (1931 a), Variations in the micronuclear apparatus of *Paramecium bursaria, Proc. Soc. Exp. Biol. Med.*, **28**, 818.

—— (1932), *Paramecium aurelia* in pedigree culture for twenty-five years, *Trans. Am. Microscop. Soc.*, **51**, 196–198.

—— (1939), Microscopy before the nineteenth century, *Am. Naturalist*, **73**, 485–516.

—— (1941), Endomixis, Chap. 13 in Calkins, G. N. and F. M. Summers, Protozoa in Biological Research, New York, Columbia University Press.

—— (1943), The pedigreed culture of *Paramecium aurelia* at Yale University, *Proc. Nat. Acad. Sci.*, **29**, 135–136.

—— (1945), The early history of the genus Paramecium with special reference to *Paramecium aurelia* and *Paramecium caudatum, Trans. Conn. Acad. Arts Sci.*, **36**, 517–531.

—— and G. A. Baitsell (1911 a), The reproduction in *Paramecium aurelia* in a "constant" culture medium of beef extract, *J. Exp. Zoöl.*, **11**, 135–142.

—— and —— (1911 b), Rhythms in the reproductive activity of Infusoria, *J. Exp. Zoöl.*, **11**, 339–359.

—— (1911c), The temperature coefficient of the rate of reproduction of *Paramecium aurelia, Am. J. Physiol.*, **19**, 146–155.

—— and H. H. Bunzel (1909), The relative toxicity of various salts and acids toward Paramecium, *Am. J. Physiol.*, **25**, 190–194.

Woodruff, L. L. and R. Erdmann (1914), A normal periodic reorganization process without cell fusion in Paramecium, *J. Exp. Zoöl.*, **17**, 425–518.

―――― and H. Spencer (1922), On the method of macronuclear disintegration during endomixis in *Paramecium aurelia, Proc. Soc. Exp. Biol. Med.*, **19**, 290–291.

―――― and ―――― (1922a), Studies on *Spathidium spathula:* I. The structure and behavior of Spathidium with special reference to the capture and ingestion of its prey, *J. Exp. Zoöl.*, **35**, 189–205.

―――― and ―――― (1923), *Paramecium polycaryum* sp. nov., *Proc. Soc. Exp. Biol. Med.*, **20**, 338–339.

―――― and ―――― (1924), Studies on *Spathidium spathula:* II. The significance of conjugation, *J. Exp. Zoöl.*, **39**, 133–196.

―――― and W. W. Swingle (1922–23), The effect of thyroid products on Paramecium, *Proc. Soc. Exp. Biol. Med.*, **20**, 386.

―――― and ―――― (1924), The effects of thyroid and some other endocrine products on Paramecium, *Am. J. Physiol.*, **69**, 21–34.

Woodward, A. (1928), The effect of Arbacia egg-secretion on the division rate of Paramecia, *Papers Mich. Acad. Sci.*, **10**, 613–616.

Worley, L. G. (1933), The intracellular fiber systems of Paramecium, *Proc. Nat. Acad. Sci.*, **19**, 323–326.

―――― (1934), Ciliary metachronism and reversal in Paramecium, Spirostomum and Stentor, *J. Cellular Comp. Physiol.*, **5**, 53–72.

Wrisberg, H. A. (1765), Observationum de animalculis infusoriis satura, Goettingen.

Wrześniowski, A. (1867), Beitrag zur Naturgeschichte der Infusorien, Kraków, C. K. Uniwers. Druk, **7**, 116 pp.

―――― (1869), Ein Beitrag zur Anatomie der Infusorien, *Arch. mikroskop. Anat. Entwicklungsmech.*, **5**, 25–48.

Wyckoff, R. W. G., A. H. Ebeling, and A. L. TerLouw (1932), A comparison between the ultra violet microscopy and the Feulgen staining of certain cells, *J. Morphol.*, **53**, 189–199.

Yancey, P. H. (1931), Effect of super-centrifuging on fission in Paramecium, *Proc. Soc. Exp. Biol Med.*, **28**, 877–878.

Yasuda, A. (1900), Studien über die Anpassungsfähigkeit einiger Infusorien an konzentrierte Lösungen, *J. Coll. Sci. Imp. Univ. Tokyo*, **13**, 101–140.

Yocom, H. B. (1918), The neuromotor apparatus of *Euplotes patella, Univ. Calif. Pubs. Zoöl.*, **18**, 337–396.

―――― (1928), The effect of the quantity of culture medium on the division rate of Oxytricha, *Biol. Bull.*, **54**, 410.

―――― (1934), Observations on the experimental adaptation of certain fresh-water ciliates to sea water, *Biol. Bull.*, **67**, 273–276.

Young, R. A. (1924), On the excretory apparatus in Paramecium, *Science*, **60**, 244.

Young, R. T. (1917), Experimental induction of endomixis in *Paramecium aurelia, J. Exp. Zoöl.*, **24**, 35–53.

―――― (1918), The relation of rhythms and endomixis, their periodicity and synchronism in *Paramecium aurelia, Biol. Bull.*, **35**, 38–47.

Yung, E. (1913), L'explosion des infusoires, *Arch. sci. phys. et nat.*, **35**, 21.

Zacharias, O. (1888), Zur Kenntnis der Fauna des süssen und salzigen Sees bei Halle a. S., *Z. wiss. Zoöl.*, **46**, 217–232.

Zawoiski, E. J. (1951), A statistical analysis of the size of different clones of *Paramecium calkinsi* and the fate of exconjugants, *Proc. Penn. Acad. Sci.*, **25**, 66–71.

Zenker, K. (1866), Beiträge zur Naturgeschichte der Infusorien, *Arch. mikroskop. Anat. Entwicklungsmech.*, **2**, 332.

Zhalkovskii, B. G. (1938), The differences in biological action of transmitted and reflected light: I. Experiments with *Paramecium caudatum*, *Bull. biol. méd. exptl. U.R.S.S.,* 493–495, *chem. abst.,* **33**, 2544.

Zingher, J. A. (1933), Beobachtungen an fetteinschlüssen bei einigen Protozoen, *Arch. Protistenk.,* **81**, 57–87.

—— (1935), Biometrische Untersuchungen an Infusorien: III. Ueber die Wirkung der Temperatur und des Lichts auf Grosse und die Teilung von *Paramecium caudatum Ehrenberg* and *Stylonychia pustulata Ehrenberg, Arch. Protistenk.,* **85**, 341–349.

——, K. J. Narbutt, and W. A. Zingher (1932), Biometrische Untersuchungen an Infusorien: II. Ueber die Mittelgrösse und Variabilität von *Paramecium caudatum Ehrenberg* und *Stylonychia pustulata Ehrenberg, Arch. Protistenk.,* **77**, 73–91.

Zirkle, R. E. (1936), Modification of radiosensitivity by means of readily penetrating acids and bases, *Am. J. Roentgenol. Radium Therapy,* **35**, 230–237.

Züelzer, M. (1905), Über die Einwirkung der Radiumstrahlen auf Protozoen, *Arch. Protistenk.,* **5**, 358–369.

—— (1910), Der Einfluss des Meerwassers auf die pulsierende Vakuole, *Arch. Entwicklungsmech. Organ.,* **29**, 632–640.

Zweibaum, J. (1912), Les conditions nécessaires et suffisantes pour la conjugaison du *Paramecium caudatum, Arch. Protistenk.,* **26**, 275–393.

—— (1921), Ricerche sperimentali sulla conjugazione degli Infusori: I. Influenza della conjugazione sull'assorbimento dell'O_2 nel *Paramecium caudatum, Arch Protistenk.,* **44**, 99–114.

—— (1922), Ricerche sperimentale sulla conjugazione degli Infusori: II. Influenza della conjugazione sulla produzione dei materiale di riversa nel *Paramecium caudatum, Arch. Protistenk.,* **44**, 375–396.

Motion picture films of Paramecium

Harrison, J. A., E. H. Fowler, O. W. Richards and J. A. Maurer (1946), Serologic and mating reactions in *Paramecium bursaria,* Phila., Temple University (Color, silent, titled, 16 mm., 375 feet).

Lloyd, F. E., Simple animal forms: Paramecium and the ciliates. Cambridge, Mass., Harvard Film Service, Graduate School of Education, Harvard University, (Black and white, silent, titled, 16 mm., 275 feet).

Schaeffer, A. A. and R. Wichterman (1938), *Chaos chaos* Linnaeus, 1767: the giant rare ameba discovered by Roesel in 1755 (showing ingestion of paramecia by amebas), Philadelphia, Temple University and the Biological Institute, (Black and white, silent, titled, 16 mm., 475 feet).

Wichterman, R. (1940), *CYTOGAMY:* a new sexual process occurring in living joined pairs of *Paramecium caudatum.* Philadelphia, Temple University and the Biological Institute, (Black and white, silent, titled, 16 mm., 400 feet).

Wichterman, R. (1944), The mating reaction and conjugation in *Paramecium bursaria,* Philadelphia, Temple University and the Biological Institute, (Black and white, silent, titled, 16 mm., 150 feet).

Author Index

Page numbers set in **boldface** indicate illustrations in the text. Those followed by t refer to tables while n refers to footnotes.

Subject Index

Page numbers set in **boldface** indicate illustrations in the text. Those followed by t refer to tables while n refers to footnotes.

N

Narcotics, effects of, 147–148
Neufeld quellung reaction, 387n
Neuromotor center (neuromotorium, neuro-
  motor apparatus), **49**–50, 53, 53n,
  54–55, 249–250
Neutral red granules, 82, 169–170
Nile blue sulfate, 81n, 413
Nitrogen mustard, 324
Novocain, 59
Nuclear abnormalities, 295
Nuclear processes, 253–312
Nuclei, 86–88 (*see also* Macronucleus;
  Micronucleus)
  diploid, 321, 323, 332
  haploid, 321–323, 346
  stains for, 415, 421–425
  variation in, 88–92
Nutrition, 40, 155–174
  types of, 156
Nyctotherus, 20, 91, 377

O

Oikomonas, 187–188
Oligodynamic action, 150–151
Opalina, 156, 377
Optical properties, 117
Oral groove, 15, 41, **42**, 158, 234
Organic extracts, effects of, 152–154, 429–
  430
Osmic impregnation method, 426
Osmium-toluidin staining, 425
Osmotic concentration, 206–211, 242
Oxidation, 217
Oxygen
  carbon dioxide relationship, 196–197
  reaction to, 241
  requirements, 94–95, 217–226
Oxygen consumption, 219t, 220–226
  physiological condition, 221
  respiration studies, 218–224
  respiratory enzymes, 224–226
  temperature, effect of, 221

P

Papilla pulsatoria, 72
Parameciidae, 14
Paramecin, 340–341, 345
  action of on *P. aurelia,* **341**, 344
  action of on *P. bursaria,* 341–**342**

Paramecium,
  amicronucleate races of, 90–91, 279, 282,
    314, 351, 375
  as food source 394
  classification of, 13–14
  discovery of, 1–2
  early nomenclature of, 1–12
  fossil of, 94n
  origin, name, 2
  potential immortality of, 360, 364, 373,
    428–429
  "slipper animalcule," first use of, 2
  spelling of, 6–7
  usefulness in research, 126, 428–432
Paramecium ambiguum, 11
Paramecium aurelia, **18**, 22
  autogamy in, 297–**299, 300**
  conjugation in, 271, **272, 273,** 274
  contractile vacuole pores in, 76
  cytogamy in, 311
  endomixis in, 260, **261, 262,** 293
  fission in, 258
  genetics of, 316–324
  hemixis in, 262, **263,** 264
  immunologic studies of, 379
  macronuclear regeneration in, **264**–265
  mating types and varieties in, 316–319
  name origin and discovery of, 6
  spiraling in, **23**
Paramecium bursaria, **19, 24**–25
  colorless races of, 26, 314, 371, 387, 404,
    407–409
  conjugation in, 277–**280, 281, 282, 283,**
    284
  cytogamy in, 311
  discovery of, 9
  genetics of, 329, 332
  immunologic studies of, 387–390
  spiraling in, **27**
Paramecium calkinsi, **19,** 28, **29**–30
  contractile vacuoles of, **29**
  cultivation of, 99
  mating reaction, dead animals, 348–350
  mating types of, 336t–337
  spiraling in, 27
Paramecium caudatum, **18,** 21–22
  conjugation in, 274, **275, 276,** 277, **278**
  cytogamy in, 301, **302**–304, 307–**308, 309**
  discovery of, 7
  fission in, **255**–**258**
  mating reaction and types, 332–334t, 335
  spiraling in, **23, 42**
Paramecium chilodonides, 36–37